STRESS PROTEINS IN MEDICINE

edited by

Willem van Eden

University of Utrecht
Utrecht, The Netherlands

Douglas B. Young

St. Mary's Hospital Medical School
Imperial College of Science, Technology and Medicine
London, England

Marcel Dekker, Inc. New York • Basel • Hong Kong

ISBN: 0-8247-9623-3

The publisher offers discounts on this book when ordered in bulk quantities. For more information, write to Special Sales/Professional Marketing at the address below.

This book is printed on acid-free paper.

Marcel Dekker, Inc.
270 Madison Avenue, New York, New York 10016

Current printing (last digit):
10 9 8 7 6 5 4 3 2 1

PRINTED IN THE UNITED STATES OF AMERICA

Preface

All living cells, from the simplest prokaryote to the most complex multicellular organism, contain stress proteins—molecular chaperones that are responsible for management of unfolded polypeptides within the cell. Unfolded polypeptides are generated during protein synthesis, and also as a result of breakdown associated with protein turnover. Interaction of polypeptides with the chaperone proteins plays an essential role in their folding and assembly into functionally mature oligomers and regulates trafficking between intracellular compartments. Particularly high levels of molecular chaperones are required to maintain protein homeostasis in cells subjected to stress conditions—such as heat shock, nutrient deprivation, malignant transformation, and anoxia—when intracellular proteins are destabilized, or when a major alteration is required in overall cell protein composition. The major molecular chaperones were, in fact, initially identified by the dramatic induction of their expression level in stressed cells and are consequently often referred to as "heat shock proteins" or "stress proteins." It is important to realize that stress proteins are critical to the maintenance of cell integrity during normal physiological growth as well as in response to the stressful event of imminent loss of normal physiology. The stress proteins discussed in this book belong for the most part to two major protein classes, commonly referred to as hsp 70 and hsp 60, in reference to their approximate molecular weight and their initial identification as heat shock proteins. Generally, bacteria express one or at most two homologues from each of these families, while eukaryotic cells have multiple homologues localized in different intracellular compartments and regulated in response to differing signals. Homologues within each family share a high degree of sequence homology, with almost 50% amino acid identity between corresponding bacterial and mammalian proteins.

In view of the fundamental role of stress proteins in maintenance of protein homeostasis, it seems likely that malfunctions associated with members of stress protein families would have pathological effects. Such effects might be minimal under normal physiological conditions, but could be exacerbated at times when other disease stimuli trigger the requirement for local alterations in stress protein function, in particular afflicted cells or tissues. During infection, it can be anticipated that the requirement of stress proteins for cell viability will be equally essential for the pathogen and for the infected host. Just as stress proteins are essential in "normal" as well as stressed cells, it is clear that changes in stress protein expression will be associated with physiologically normal events accompanying infection as well as with any subsequent pathological events. In studying the role of stress proteins in medicine it is important to understand the normal physiology of stress proteins in order to detect and characterize aberrations associated with disease.

In addition to the direct role of stress proteins in cell physiology, their potential medical influence is compounded by their ability to act as potent immunogens. Responses to microbial stress proteins are a prominent feature of the immune repertoire in patients and in experimental animals, and there has been wide discussion of the possibility that recognition of conserved, self-like, epitopes on such antigens could influence infectious and other diseases. Three broad hypotheses have been proposed concerning the relevance of autoreactivity to stress proteins. In some instances, this can be seen as an example of "mimicry" in which an initial response against a pathogen component cross-reacts with a self-protein, triggering autoimmune pathology. Alternatively, the detection of autoreactive responses in individuals without disease suggests they may be integral to normal immune function—possibly with a role in "immune surveillance," acting as a generalized recognition system for cells stressed by infection, malignancy, or other factors. Finally, responses to stress proteins can be viewed as a secondary event, reflecting tissue breakdown and release of intracellular proteins following any pathological disturbance. This book provides a state-of-the-art overview from experts in a wide range of disciplines, setting out currently available evidence on the role of stress proteins in medicine. The reader is invited to consider whether the involvement of stress proteins can be described by a single set of general principles or is best viewed within the specific context of each individual disease, and to evaluate the potential contribution of such research to the development of novel diagnostic, prophylactic, and therapeutic approaches to control of human disease.

The book opens with introductory chapters providing an overview of the cell biology and immunology of stress proteins. This is followed by 15 chapters focusing predominantly on immunological responses to stress proteins in a range of immune-mediated diseases, such as arthritis, lupus, Crohn's disease, and multiple sclerosis, and in infectious diseases including tuberculosis, leishmaniasis, and fungal infection. Three chapters deal with infertility, experimental studies of tumor immunology, and organ transplantation. The next section contains 12 chapters that focus on local expression of stress proteins in diseased tissues. In addition to diseases covered in the immunological section, this section includes sarcoidosis, liver diseases, gastritis, celiac disease, artherosclerosis, allergic asthma, and muscular diseases. One chapter is devoted to myocardial protection in heart disease. A final section focuses on some exciting new aspects of stress proteins, including their use as carrier molecules for vaccines and their relationship with the gamma/delta subset of T cells.

The book is the first comprehensive compilation of the wide variety of observations made on the topic of stress proteins in disease. It is intended to provide a useful resource for readers as diverse as medical doctors, pathologists, cell biologists, immunologists, microbiologists, and all who wish to stay informed in this rapidly evolving field.

Willem van Eden
Douglas B. Young

Contents

1

Stress Proteins as Molecular Chaperones

R. John Ellis
University of Warwick, Coventry, England

I. INTRODUCTION

All cells contain groups of highly conserved proteins that increase rapidly in concentration when the cells are exposed to environmental stresses. The most studied stress is a temperature 5–10°C higher than that optimal for the growth of the cell being studied, and thus these proteins are often called heat shock proteins (or hsps). However, the same set, or in some cases a subset, of these proteins increases in amount when cells are subjected to a whole range of other insults, including oxidants, transition series metals, ethanol, amino acid analogues, metabolic poisons such as arsenite and azide, glucose starvation, calcium ionophores, and pathogenic microorganisms, as well as in disease states such as cancer, fever, ischemia, cardiac hypertrophy, and oxidant injury. There is much indirect evidence that these proteins have a protective function, allowing cells to recover from the inducing stress, and to survive subsequent stronger stresses that would otherwise be lethal. This common cellular defense mechanism is now referred to as the stress response, and the proteins that accumulate are called stress proteins.

The study of stress responses at the molecular level has until recently concentrated on the mechanisms that cause these proteins to accumulate, because they are such splendid examples of inducible gene expression. A significant component of these mechanisms is the rapid increase in transcription of genes encoding stress proteins, whereas that of genes encoding many other proteins and ribosomal RNA is often reduced. Much detailed information is available about the regulatory DNA sequences and DNA-binding proteins involved in stress protein accumulation in both prokaryotes and eukaryotes (1), although the mechanisms by which stresses are sensed to trigger transcriptional activation are still under debate.

Even more unclear until recently are the cellular functions of the diverse groups of stress proteins, especially since so many different environmental treatments cause induction of these proteins. A common result of some of these treatments is the production of misfolded proteins, and this realization led to the abnormal protein hypothesis that proposes that stress proteins accumulate in response to the production of proteins with abnormal structures (2,3). Supporting evidence came from studies showing that an increase in the concentrations of denatured proteins in both *Escherichia coli* and eukaryotic cells triggers the activation of stress protein genes. However, the high degree of conservation of aminoacyl sequences among stress proteins in all organisms, together with the observation that many of these proteins are present when the cells are not subject to stress, suggests that at least some of these proteins share functions that are essential to normal cellular operations but are required to a higher degree under stress conditions. Proteins related in sequence to stress proteins, but which are expressed constitutively, are sometimes called heat shock cognate (hsc) proteins. Thus, some stress proteins are members of multigene superfamilies of which a subset is expressed in the absence of stress to serve basic cellular functions. What are these functions?

A considerable stimulus was given to the study of the functions of stress proteins by the suggestion (4) that some of them assist protein assembly/disassembly processes fundamental to all types of cell by acting as molecular chaperones (5). This view proposes that protein assembly processes that can occur spontaneously in vitro do not occur spontaneously in vivo but are assisted by preexisting proteins whose role can be accurately described in certain respects as analogous to that of human chaperones in regulating interactions between people (6,7). Molecular chaperones were originally postulated to function by ensuring that the folding of other proteins and their association into higher order structures occur correctly (5). They achieve this not by contributing steric information necessary for these processes but by preventing incorrect interactions that would otherwise produce misassembled structures. It was in addition proposed, following the suggestion of Pelham (4), that molecular chaperones are involved in disassembly of oligomeric structures; disassembly is an essential step in the function of some oligomers, such as those involved in DNA replication, the recycling of clathrin coats, and the activation of transcription factors.

This molecular chaperone concept can accommodate at least some aspects of the stress response by supposing that the need for these chaperones increases when proteins are damaged by stress; if the cell is to survive stress, such damaged proteins need to be refolded correctly or removed by proteolysis, whereas undamaged proteins need to be protected against subsequent stresses that might create incorrect assemblies. Thus, the stress response can be viewed in many respects as an amplification of a basic chaperone function that all cells require under normal conditions rather than as a unique function required only under stress conditions. In this light, the hsp terminology places an inappropriate emphasis on just one aspect of these proteins, but it is now too embedded in the literature and too convenient to use for alternatives to be popular. It also remains possible that some stress proteins have functions which are uniquely required in the stress situation.

The view that some stress proteins act as molecular chaperones has gained increasing experimental support in recent years, especially with respect to the folding of newly-synthesized or transported polypeptide chains (7–14), but it is important to note that not all stress proteins act as molecular chaperones and conversely, not all molecular chaperones accumulate after stress. For example, only 5 of the 20 or so stress proteins of *E. coli* whose expression is controlled by the sigma 32 transcription factor are known to act as molecular chaperones (15), whereas nucleoplasmin, the first protein to be described as a molecular chaperone (16), is not a stress protein.

In this introductory chapter, the main types of stress proteins that act as molecular chaperones are briefly described and their functions in assisting protein folding are discussed. There is a vast literature on stress proteins, and other references should be consulted for discussion of their diverse aspects (17–21); an excellent book discussing both biological and medical aspects of stress proteins is available (22).

II. MAIN TYPES OF STRESS PROTEIN

Since the functions of stress proteins are only now becoming defined, a variety of names exist in the literature. A convenient terminology is based on their migration during electrophoresis on sodium dodecyl sulphate polyacrylamide gels, and on this basis, four main groups can be recognized; namely hsp 90, hsp 70, hsp 60, and the small hsps (or hsps 15–30). These groups all contain conserved protein families, and the numbers represent rounded-off values for the apparent relative molecular masses of their subunits; for example, the hsp 70 group contains highly related proteins of subunit size 66–78 kDa. The small hsps are especially prominent in plants which contain four multigene families (23), whereas only one small hsp of 28 kDa has been identified in mammals (24). The fifth group is used to encompass a mixture of less well-studied stress proteins that are unrelated in sequence to either one another or to the other groups. The hsp 60 proteins are often called the chaperonins (cpns), since they were the first group identified as functioning as molecular chaperones (25). Table 1 lists some of the properties of these groups. A more comprehensive list of stress proteins found in mammalian cells is available (24). Some key features

Table 1 Properties of Some Types of Stress Protein

Group	Other names: Prokaryotes	Other names: Eukaryotes	Location	Native form	Comments	Key reference
Hsp 90	HptG	hsp 83,87	Cytosol	Dimer	Binds to other proteins and regulates their activity; prevents aggregation of refolding polypeptides in vitro	13,33
		grp 94	ER			
Hsp 70	DnaK	Ssa 1-4	Cytosol	Monors and dimers	Binds to extended polypeptides; dissociates some oligomers; complexes; has ATPase activity	12,13
		Ssb 1-2	Cytosol			
		Ssc 1p	Mitochondrion			
		Scd 1p a Kar2	ER			
		BiP or grp 78	ER			
		hsp 72	Cytosol/nucleus			
		hsp 73	Cytosol/nucleus			
		clathrin-uncoating ATPase	Cytosol			
		hsp 70	Chloroplast			

Hsp 60 or chaperonins (cpn)	Cpn60 (GroEL)	—	Cytosol	14—mer	Binds to partially folded polypeptides and assists their correct folding; has weak ATPase activity	11,14,15
	Cpn10 (GroEL)	—	Cytosol	7—mer		
		TCP-1	Cytosol	16 or 18—mer		
		hsp 60	Mitochondrion	14—mer		
		Rubisco subunit binding protein (Cpn 60)	Chloroplast	14—mer		
		cpn 21	Chloroplast	?		
Small hsps	None (?)	hsp 27 & 28	Cytosol	Large oligomers	Function obscure but some may be molecular chaperones	17,77
		In yeast and mammals; hsp 15–30 in plants	Cytosol/ chloroplasts/ ER	Large oligomers		
Others	Dna J	Homologues	Cytosol	Dimer		13,78,81
	Grp E	Homologues	Cytosol/ mitochondrion	Dimer		

of each group will now be discussed briefly; it will become apparent that in many cases stress proteins from different families cooperate with one another in chaperoning a variety of cellular processes.

A. Hsp 90

Hsp 90 is the most abundant constitutively expressed stress protein in the eukaryotic cytosol (13); mammalian cells have in addition the related glucose-regulated protein (grp) 94 in the lumen of the endoplasmic reticulum (ER). This group of stress proteins is highly conserved in bacteria, yeast, and mammals. Yeast contains two related types of cytosolic hsp 90, one constitutive and the other inducible by heat shock; deletion of both genes is lethal, whereas deletion of either alone is not but prevents growth above 37.5°C (26). Postulated functions of the hsp 90 proteins include binding to partly folded proteins, thereby preventing their aggregation, and forming specific complexes with other proteins, thereby regulating their function.

The vertebrate cytosolic hsp 90 is bound to a wide range of proteins such as viral tyrosine kinases and steroid hormone receptors (27). These complexes appear to function as transient control devices which restrict the functional properties of the protein bound to the hsp 90. Thus, newly synthesized viral oncogene protein pp60 binds to hsp 90 and another protein of 50 kDa, which has been identified as a member of the peptidyl proline isomerase family; in this complex, the src protein does not exhibit its normal tyrosine kinase activity until the complex dissociates when the kinase binds to the inner side of the plasma membrane (28). In analogous fashion, hsp 90 and hsp 70 are found in murine cells as a complex with the glucocorticosteroid receptor; formation of this complex prevents the receptor binding to DNA and is, in addition, a prerequisite for the receptor to develop the ability to bind steroid hormone (29). Hormone binding triggers release of hsp 90 and allows the receptor to bind to its DNA target sequence (30).

The abundance of hsp 90 suggests it is likely to have a widespread role in regulating the functions of a range of proteins by transiently interacting with them; immunoprecipitation of Hepa cell extracts with monoclonal antibodies to hsp 90 reveals heteromeric complexes with hsp 70 and three other proteins (31). The observation that hsp 90 can prevent the aggregation of refolding proteins in vitro suggests that it may play a role in assisting the folding of newly synthesized polypeptides in vivo (32). The related grp 94 protein of the ER lumen may in similar fashion assist the folding of polypeptides as they emerge in the lumen after transport across the ER membrane. Newly synthesized light and heavy chains of immunoglobulin bind first to hsp 70 and then to grp 94 after transport into the ER lumen (93). The grp 94 protein contains phosphate and carbohydrate groups and occurs in the Golgi body as well as in the ER (33), so it is likely to have additional functions.

Mechanistic details about hsp 90 functions are scarce and binding specificities are undefined, but like the hsp 70 and hsp 60 proteins, hsp 90 proteins possess ATPase activity (34). Multiple isoforms of hsp 90 bearing different numbers of phosphorylated side chains have been found in animal cells (13), perhaps produced by autophosphorylation (34). A most interesting recent finding is that hsp 90 proteins bind transcriptional factors that regulate the expression of stress protein genes; this observation supports the operation of a feedback control system that matches the concentration of hsp 90 proteins with that of its target proteins (34).

B. Hsp 70

Hsp 70 homologues occur in all the known groups of organism, that is, the archebacteria, the eubacteria, and the eukaryotes, and are the most highly conserved proteins found so far. For example, the eukaryotic hsp 70 homologues are all between 50 and 98% identical in aminoacyl sequence. This high conservation makes these proteins ideal for studies of evolutionary relationships; a recent comparison (35) concludes that the archebacteria are more closely related to the gram-positive bacteria than to the eukaryotes, a view different from that based on studies of ribosomal RNA sequences.

The aminoterminal two thirds of the hsp 70s contains an ATPase site, and is more highly conserved than the carboxyterminal portion that is thought to contain a peptide binding site (36). The crystal structure of a 44-kDa proteolytic fragment of a mammalian hsc 70 protein (the clathrin-uncoating ATPase) reveals a two-lobed domain enclosing a deep cleft in which ATP binds; the structure is very similar to the ATP-binding domain of G actin (37,38). The structure of the carboxyterminal region is not known, but from sequence comparisons is suggested to resemble the major histocompatibility complex class I protein which binds peptides in an extended conformation (39,40). Nmr measurements show that the hsp 70 protein of *E. coli* (called DnaK) also binds peptides in an extended conformation (41).

All eukaryotes contain multiple hsp 70s, unlike *E. coli*, in which only one has been reported so far. The best-studied organism in this regard is a yeast (*Saccharomyces cerevisiae)*, which contains eight hsp 70s, six in the cytosol called Ssa1-4 and Ssb1-2, one in the mitochondrial matrix called Ssc1p, and one in the ER lumen called Ssd1p or Kar2p (12). In animal cells, the ER hsp 70 homologue is referred to as BiP, since it binds to a wide range of newly transported proteins entering the ER lumen from the cytosol. This binding is normally transient and reversed by ATP, but it is prolonged if the protein is unable to fold or oligomerize correctly owing to mutation (9). Synthetic peptide binding studies show that BiP prefers to bind heptameric peptides enriched in aliphatic amino acid residues (42). A more detailed study employed the affinity panning of bacteriophage libraries displaying random octapeptides to characterize the sequences and binding constants of peptides that bind to BiP (43). Sequence analyses show that BiP prefers to bind peptides containing a subset of aromatic and hydrophobic residues in alternating positions, with the side chains of alternating residues pointing into four pockets in a cleft in the BiP molecule (43). This preference for hydrophobic residues confirms the original speculation of Pelham (4). Since BiP is known to bind to a wide variety of unrelated polypeptides in the ER lumen, the strong but redundant hydrophobicity of the peptide motif identified in this study is consistent with the idea that, in the cell, BiP binds transiently to these sequences before they become buried inside fully folded proteins. Studies using a mutant gene for the mitochondrial hsp 70 of yeast similarly suggests that this protein binds transiently in a ATP-reversible fashion to proteins being translocated from the cytosol; this binding is essential for both the transport of extended polypeptides into the mitochondrial matrix and their subsequent correct folding (44–46).

The six cytosolic hsp 70 homologues in yeast fall into two distinct subfamilies, called Ssa1-4 and Ssb1-2. These subfamilies are not functionally equivalent, and their regulation is different; expression of three of the four *SSA* genes is increased by heat shock, whereas that of the *SSB* genes is inhibited. There is some functional overlap between the Ssa proteins because only one of the four members of this subfamily is essential for cell viability, provided it is expressed at a high level (47). A mutant yeast in which both *SSB* genes are deleted grows slowly at all temperatures and is cold sensitive. Yeast cells depleted of Ssa proteins accumulate both mitochondrial and ER precursor proteins in the cytosol, whereas addition of Ssa proteins in vitro stimulates protein translocation into isolated mitochondria and mammalian microsomal vesicles (48). Proteins hsc 72 and hsp 73 of the mammalian cytosol bind to newly synthesized nascent polypeptide chains labeled in vivo (49,50), as do the Ssb proteins of yeast (51). The interactions of hsp 70 proteins with other stress proteins in assisting the correct folding of newly synthesized polypeptide chains is discussed in Section IV.

All these observations are consistent with a model in which the normal role of constitutive hsp 70 proteins is to bind transiently to hydrophobic regions of extended polypeptides as these emerge in several processes; either during synthesis from a tunnel in the ribosome, during transport from a lipid bilayer, or on the surface of folded proteins as a result of conformational changes that occur during their normal functioning (as in the dissociation of clathrin coats). Under normal growth conditions, this binding serves a basic chaperone function by reducing incorrect interactions between such transiently exposed interactive surfaces, but under stress conditions, this binding is required to a greater extent, because such surfaces increase in concentration as a result of the denaturing effects of stress. An additional factor with specific respect to the stress of heat shock is that hydrophobic interactions increase in strength as the temperature rises, so the need to restrict these interactions becomes more urgent.

C. Hsp 60 or Chaperonins

These proteins were originally identified in several laboratories in the 1970s as host-encoded proteins required for the morphogenesis of several bacteriophages in *E. coli*. An operon called *groE* was identified encoding two protein subunits, one around 60 kDa called GroEL and the other around 10 kDa called GroES (reviewed in Ref. 52). Both proteins are essential for the growth of *E. coli* at all temperatures, and it was known that GroEL is immunologically similar to a dominant immunogen of unknown function called the bacterial common antigen found in many human bacterial infections. The precise function of GroEL did not become clearer until a subunit of a plastid protein implicated in the assembly of the chloroplast enzyme ribulose bisphosphate carboxylase-oxygenase (rubisco) was found to be 46% identical in aminoacyl sequence to GroEL (25). Contemporaneous work with the protozoan *Tetrahymena thermophila* revealed a 60-kDa heat shock protein in the mitochondrial matrix that shows high sequence similarity to both GroEL and the chloroplast homologue (53). Because the chloroplast protein was regarded as a molecular chaperone (54), and the action of GroEL in phage morphogenesis could be regarded in the same fashion, it was proposed (25) to call the whole group of highly conserved proteins the chaperonins (cpns), and the two subunits are now known as cpn 60 and cpn 10.

Subsequent work (55,56) revealed faint sequence similarity between cpn 60 and TCP-1, a protein found in the eukaryotic cytosol, with a homologue in archebacteria. These similarities led to the proposal (57) that the chaperonins be defined as a family of related molecular chaperones containing two distinct subfamilies; the GroE subfamily found in the eubacteria, plastids, and mitochondria, and the TCP-1 subfamily found in the eukaryotic cytosol and the archebacteria. The sequence similarities within each subfamily are high compared with those between subfamilies, consistent with an endosymbiont type of origin which groups the eubacteria with plastids and mitochondria, and the archebacteria with the eukaryotic cytosol. The GroE members in the eubacteria and mitochondria and the TCP-1 homologue in archebacteria are all strongly heat inducible, but the chloroplast cpn 60 and the TCP-1 in eukaryotes are not; this is another example supporting the conclusion that stress proteins and molecular chaperones are not synonymous terms.

There is now evidence from biochemical and genetic studies that both chaperonin subfamilies assist the correct folding of newly synthesized polypeptides in the cytosol of prokaryotes and eukaryotes but at a later stage in the folding process than the hsp 70 family (reviewed in refs. 11-15); in addition, the GroE subfamily assists the folding of polypeptides imported from the cytosol into plastids and mitochondria or synthesized within these organelles (58). It was the identification of the chaperonins as a ubiquitous highly conserved group of molecular chaperones (25) that sparked the current resurgence of both biochemical and genetic experimentation on how proteins fold inside cells (59).

GroE cpn 60 molecules purified from eubacteria, mitochondria, and chloroplasts are large oligomeric structures made from two stacked heptameric rings (the "double donut") of 60-kDa subunits, each ring enclosing a central cavity about 6 nm in diameter as revealed by negative staining. The cavity length is either 7.5 or 15.0 nm, depending on whether the two cavities are continuous or separate, as one structural study suggests (60). Each 60-kDa subunit contains an ATPase site. The mitochondrial and bacterial GroE cpn 60 is homo-oligomeric, but the chloroplast homologue contains equal amounts of two different but related sequences whose distribution in the oligomer is unknown (63).

Cpn 10 molecules occur as single heptameric rings of 10-kDa subunits, except for the chloroplast homologue which occurs as a double-headed molecule called cpn 21 that contains two cpn 10 sequences in tandem (61). The cpn 10 from *E. coli* binds ATP but does not hydrolyze it (62). Cpn 10 is sometimes referred to as co-chaperonin, but this is strictly incorrect, since cpn 10 shares some sequence similarity to cpn 60 and is thus a true member of the chaperonin family by definition (63). In the presence of either MgATP or MgADP, and under physiological conditions of pH and ionic strength, cpn 10 forms an asymmetrical binary complex by binding to one end of the cpn 60 cylinder (64,65). The role of the central cavities and ATP hydrolysis by this binary complex in assisting the folding of a wide variety of polypeptide chains is discussed in Section IV.

The TCP-1 chaperonin (for tailless complex polypeptide) was originally identified in mice as a cytosolic protein of 57-kDa subunits found in all cell types but especially abundant in testes (66). TCP-1 resembles GroE cpn 60 in occurring as a large oligomeric complex of two stacked rings containing a central cavity, but it differs in that each ring contains eight or nine subunits rather than seven and is heteromeric (67,68); the cavity is also larger than that found in the GroE cpn 60. Nine related but different subunits of TCP-1–containing chaperonin complexes have been identified from mammalian testis, and seven of the genes have been cloned. Because at least three of the genes for these subunits are not part of the mouse tailless gene complex, the TCP-1 complex has been renamed as the chaperonin-containing TCP-1 (CCT), and the TCP-1 subunit itself has been designated CCT alpha (68). All these subunits contain a motif resembling the ATP-binding domain of the cAMP-dependent protein kinases. The CCT subunits are highly diverged from one another in sequence (around 30% identity), but are each conserved from mammals to yeast, so it is possible that each subunit long ago evolved an independent function that has been maintained in many eukaryotes. There may also be several kinds of CCT complexes containing different combinations of different subunits, and it has been suggested that this diversity arose early in eukaryotic evolution to allow this complex to assist the folding of the wide variety of proteins found in the eukaryotic cytosol compared with that of prokaryotes (68). This suggestion supports the proposal that protein evolution involves constraints dependent not just on the suitability of altered sequences for functional purposes but also on their compatibility with the chaperonin protein folding machinery (59).

Much less information is available about the ability of these CCT complexes to assist the folding of polypeptides compared with the GroE chaperonins, and they are more difficult to study, but there is evidence that they assist the folding of actin and tubulin both in vitro and in vivo (69–71). There appears to be no separate oligomer equivalent to cpn 10 required for CCT to function, but hsp 70 specifically copurifies with the CCT (67).

Chaperonins appear to be absent from four intracellular compartments where proteins are known to fold; that is, the periplasmic space of some bacteria, the endoplasmic reticulum, the intramitochondrial membrane space, and the chloroplast thylakoid lumen. All secreted proteins pass through the endoplasmic reticulum in animal cells, and many of these are of medical interest. This compartment contains both a machinery for creating the disulfide bonds common in secreted proteins and stress proteins of the hsp 70 and 90 families. There is a need for more studies on how proteins fold within these compartments in the apparent absence of the chaperonins, which genetic studies show are essential for the correct folding of many proteins in the cytosol of *E. coli* (89).

D. Small Hsps

There are a diverse group of stress proteins, and different organisms contain very different numbers of them, with the extremes being one protein of about 27–28 kDa in mammals and yeast to over 30 proteins of 15–30 kDa in higher plants; *E. coli* appears to lack any member of this group. The small hsps are also the least understood in terms of specific functions, to the extent that it has been proposed that they may be products of "selfish" or ancient viral DNA and serve no function (72). This may be the case for yeast, where deletion of the single gene produces no discernible phenotype (72), but seems unlikely to be true for plants, where they occur in the ER and chloroplasts as well as the cytosol, and are the most highly induced of all the stress proteins (17). In many organisms, the small hsps occur as large oligomers in the cytosol which aggregate into granules on heat shock; for example, mammalian 28-kDa protein occurs in Hela cells with a size of 200–400 kDa at 37°C, but with a size over 2000 kDa after heat shock (73). Possible enzymatic activities of any of these forms have not been studied, but recent work shows that mouse and human hsp 28 will act like hsp 90 in preventing the aggregation of heat- or urea-denaturated enzymes in vitro in an ATP-independent manner (74). These small hsps show sequence similarity to the carboxyterminal region of the major eye lens protein alpha-B-crystallin, which also protects proteins against aggregation in vitro (74–76). These protective effects suggest that at least some of the small hsps act as molecular chaperones and raise the intriguing possibility that the evolution of an eye lens protein from such a chaperone reflects not just its suitability for optical purposes, but that its chaperone activity is important in maintaining the integrity of lens proteins against aggregation that could impair their optical properties (77). The lens carries out little or no protein synthesis or turnover, so the long-term properties of its proteins must be safeguarded by some other means, and the idea that the structural proteins have chaperone properties is appealing. These ideas require testing by genetic approaches for their relevance to the in vivo situation.

E. Other Types of Stress Protein

This group is a ragbag of unrelated stress proteins, and only two known to act as molecular chaperones in polypeptide folding (DnaJ and GrpE) are considered here. For information about others such as hsp 75, hsp 104, hsp 110, lon, lysU, rpoD, Clp, and ubiquitin, reference 21 is a good starting point.

The genes encoding the DnaJ and GrpE proteins were discovered, together with that for the DnaK protein, because mutations in any of them block replication of phage lambda DNA in *E. coli* (52). Subsequent in vitro studies showed that these three unrelated proteins act together in the presence of ATP in dissociating DnaB helicase from a tight inactive complex with other proteins; the released active helicase then initiates the unwinding of the DNA double helix at the origin of replication (78). This ability of these three proteins to disassemble protein complexes is aptly described by the term "molecular crowbar" (13).

A family of proteins sharing some similarity to the 70 aminoterminal residues of DnaJ has been found in eukaryotes (12,79), including one that is anchored in the ER membrane of yeast and is essential for protein secretion (80). The GrpE protein of *E. coli* is essential for growth at all temperatures, and recently the first eukaryotic homologue of GrpE was discovered, copurifying with the hsp 70 of yeast mitochondria; this protein is also essential for viability (81).

Besides acting as a molecular crowbar to disassemble protein complexes, the DnaK/DnaJ/GrpE trio has a separate function. These three proteins interact with one another in the presence of ATP to protect denatured polypeptides refolding in vitro, and by inference nascent polypeptides elongating in vivo, from premature misfolding and aggregation before subsequent correct folding is assisted by the chaperonin family (82,83). The cooperation of these molecular chaperones in assisting polypeptide folding is discussed in Section IV after a brief outline of the molecular chaperone concept.

III. MOLECULAR CHAPERONE CONCEPT

A. Origins

The term *molecular chaperone* was published first in 1978 to describe the properties of the nuclear protein nucleoplasmin in assisting the in vitro assembly of nucleosomes from isolated histones and DNA (16). Nucleosomes can be dissociated into histones and DNA by exposure to high salt concentrations that disrupt the electrostatic interactions holding the nucleosome components together. At the time, the prevailing paradigm about the assembly of such complex structures was based on the principle of self-assembly; this states that both the folding of a polypeptide chain and any subsequent association with other molecules is a spontaneous process requiring neither energy expenditure nor molecules other than the components of the assembled structure itself. This principle derives from experiments where certain viruses were reconstituted in infectious form by simply mixing together the component protein and nucleic acid molecules (84), and from the classic observations of Anfinsen that some denatured pure proteins will spontaneously refold into their active original conformations if the denaturing agent is removed (85). The self-assembly principle is an important corollary of the central dogma of molecular biology, according to which the primary structure of proteins is encoded in the base sequence of nucleic acids; the in vitro self-assembly experiments thus confirm the expectation that all the information for protein folding and association is contained within the primary structures of the proteins themselves and is not derived from an external source. It was thus anticipated that nucleosomes should self-assemble when histones and DNA are mixed together.

Attempts to assemble nucleosomes by mixing isolated histones and DNA under physiological conditions of ionic strength fail spectacularly; large nonspecific aggregates form instead of nucleosomes because of the strong electrostatic interactions between the positively charged histones and the negatively charged DNA that occur once the salt concentration is lowered to physiological levels. Laskey and his colleagues found that prior addition of the acidic nuclear protein nucleoplasmin to the histones reduces their strong positive charge density to the point where productive interactions to form nucleosomes predominate when DNA is subsequently added (16). This reduction of charge density to allow correct interactions to predominate over incorrect ones appears to be the only function of nucleoplasmin; the possibility that nucleoplasmin also carries steric information essential for nucleosome assembly was ruled out by experiments whereby nucleosomes could be assembled in the absence of nucleoplasmin but only under nonphysiological conditions; for example, during slow reduction of the high salt concentration by dialysis. Nor is nucleoplasmin a component of the assembled nucleosome—it is required only during the assembly process.

Thus, the principle of self-assembly remains intact in that the steric information for nucleosome structure resides in the histones themselves, but the principle is incomplete as a description of the in vivo situation, because assistance from a preexisiting protein is required in order for self-assembly to prevail over incorrect interactions; that is, those that produce nonfunctional structures. Laskey and his colleagues coined the term *molecular chaperone* to describe this function of nucleoplasmin because of the analogy with the human chaperone, whose traditional role is to prevent incorrect interactions between pairs of human beings, without either providing steric information necessary for their correct interaction or the necessity to be present during their married life.

The author came across the term *molecular chaperone* in 1985 while searching for a precedent for the observation that the assembly of the photosynthetic enzyme rubisco (ribulose bisphosphate carboxylase-oxygenase) in chloroplasts isolated from higher plants also seems to involve the transient assistance of another protein that is not a component of the assembled enzyme. The essential observation was that rubisco large subunits, newly synthesized in vitro inside isolated chloroplasts, are bound noncovalently to another protein before transfer into the assembled holoenzyme. It was proposed that binding to this so-called rubisco large subunit binding protein might be an obligatory step in the assembly of rubisco (86). This proposal did not meet with much approval at the time because of the general acceptance of the principle of self-assembly, so the discovery of a precedent was encouraging and led to the suggestion that the rubisco large subunit binding protein could be a second example of a molecular chaperone (54).

Up to this point, it was thought that nucleosomes and rubisco are special cases because it was obvious that the tendency of their subunits to form unspecific aggregates in vitro is unusually great compared with that of other protein complexes. This restricted view changed as the result of a seminal paper by Pelham (4) in which, although he did not use the term *molecular chaperone*, he proposed that the need for what is, in essence, a molecular chaperone function might be widespread. Pelham proposed that members of the hsp 70 and 90 families in animal and microbial cells are involved in the assembly and disassembly of proteins in the cytosol, nucleus, and ER compartments under nonstress growth conditions, but are required in increased amounts during stress, both to unscramble protein aggregates that then have to refold correctly and to prevent subsequent aggregation by binding to hydrophobic surfaces exposed as a result of the stress. This speculative paper emboldened the author to propose that all cells contain a variety of proteins that act as molecular chaperones, in that they prevent incorrect interactions during the operation of several basic cellular processes under normal growth conditions (5). Strong support was given to this proposal by the discovery that the rubisco large subunit binding protein is highly related in sequence to the GroEL protein of *E. coli*, which led to the introduction of the term *chaperonin* for this particular family of molecular chaperones (2).

B. Current View of Molecular Chaperones

Molecular chaperones are currently defined as proteins that assist the correct noncovalent assembly of other protein-containing structures in vivo but are not permanent components of these structures when they are performing their normal biological functions (10). In this definition "assembly" is used in a broad sense to encompass not only the folding of newly synthesized polypeptide chains and any association into oligomers that they undergo but also any changes in the degree of either folding or association that may occur when proteins cross membranes, perform their normal functions, or are repaired or removed after damage by stress. An alternative definition that expresses the same idea has been proposed more recently (11): "a molecular chaperone is a protein that binds to and stabilizes an otherwise unstable conformer of another protein, and by controlled binding and release of the substrate protein facilitates its correct fate in vivo, be it folding, oligomeric assembly, transport to another subcellular compartment, or controlled switching between active/inactive conformations."

The suggestion that the function of at least some stress proteins is to act as molecular chaperones derives logically from both these definitions, since the requirement for such stabilization increases when proteins are denatured by stress.

An interesting question to consider is why molecular chaperones exist at all; the principle of self-assembly is well supported by many in vitro experiments, so why is it apparently insufficient to account for the in vivo situation? A possible answer stems from the observation that many fundamental cellular processes involve the transient exposure of interactive surfaces to the intracellular medium, and so run the risk that these surfaces may interact incorrectly to generate nonfunctional structures. Interactive surfaces are here defined as any regions of intra- or intermolecular contact that are important in stabilizing the structure. It has long been known that the success of an in vitro refolding experiment of the type pioneered by Anfinsen (85) increases as the temperature and protein concentration decreases, because these conditions reduce the probability of incorrect interactions between hydrophobic surfaces transiently exposed during the refolding process (87). Such refolding experiments are commonly carried out at protein concentrations in the 1 μg/ml to 1 mg/ml range, whereas in vivo protein concentrations are much higher—in the range 200–300 mg/ml. Thus, although the self-assembly principle implies that all the interactions that occur in vivo are both necessary and sufficient to produce the correct structures, the molecular chaperone concept suggests that on the contrary there is an intrinsic risk of incorrect interactions producing abnormal structures because of the sheer concentration of interacting surfaces. All cells thus require a molecular chaperone function to reduce this risk. On this basis, we can summarize the new view by the statement that in vivo, protein self-assembly is not spontaneous and energy independent but is assisted by preexisting proteins acting as molecular chaperones, some of which hydrolyze ATP (8,10).

The most detailed information currently available about the mechanism of action of some molecular chaperones concerns their role in the folding of newly synthesized polypeptide chains; this information is too extensive to present in full, but is summarized in Section IV.

IV. STRESS PROTEIN–ASSISTED PROTEIN FOLDING

Since the identification of the chaperonins (25), several laboratories have repeated the basic in vitro protein refolding experiments of the type pioneered by Anfinsen, with the modification that molecular chaperones of various types have been added to the refolding buffer (8–11,62,82,83); the author calls these "neo-Anfinsen" experiments. These experiments show that the yield of correctly folded proteins increases as a result of these additions but that the rate of folding does not. The vast majority of these experiments use the GroE chaperonins and the DnaK/DnaJ/GrpE trio from *E. coli*.

The results of these neoAnfinsen experiments are commonly extrapolated to the more complex in vivo situation, and there is strong genetic evidence (88,89) in the case of *E. coli* to support the validity of this extrapolation but only in a general sense. This genetic evidence shows that these chaperones are required in active form to prevent protein misfolding in vivo, but the genetic approach does not have the resolution to confirm that the precise details of chaperone action deduced from neoAnfinsen experiments apply in vivo. With this caveat, two basic models of chaperone action appear to this reviewer to be the best supported by most of the published information. These models are illustrated in Figures 1 and 2.

First, monomeric or dimeric proteins of the DnaK/DnaJ/GrpE type bind transiently to extended nascent polypeptide chains; this binding prevents premature folding before synthesis of the polypeptide chain is complete (see Fig. 1). This model is based on refolding studies of chemically-denatured rhodanese using purified chaperones (82,83), but consistent with this model is the binding of eukaryotic hsp 70 proteins to nascent polypeptide chains in vivo (49–51),

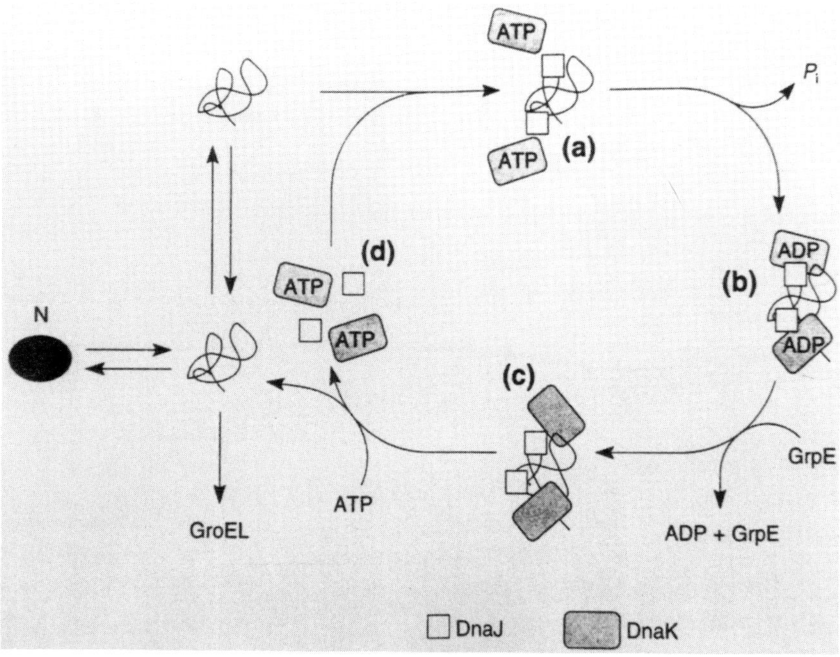

Figure 1 Model for the reaction cycle of DnaK, DnaJ, and GrpE in protein folding based on studies on the in vitro refolding of rhodanese after denaturation by 6 M guanidinium chloride (82). (a) Unfolded protein binds to DnaJ. (b) DnaK binds to DnaJ and hydrolyses bound ATP to ADP. (c) Addition of GrpE causes ADP to dissociate and weakens the interaction between DnaK and the protein. (d) Binding of ATP to the ternary complex of unfolded protein/DnaK/DnaJ causes it to dissociate. The protein can then either rebind to DnaJ, or be transferred to the GroE chaperonin system for final folding, or possibly fold spontaneously. N = native protein. (From Ref. 82.)

and the report that DnaJ can be cross-linked in cell-free extracts to nascent chains of both cytosolic proteins and those that are transported into mitochondria and the ER lumen (90). Second, oligomeric chaperones of the chaperonin family bind partially folded intermediates of whole polypeptide chains (often termed compact intermediates or molten globules) in a cavity inside the end of each cpn 60 oligomer. Each cavity binds one polypeptide chain. This cavity provides a protected environment where the chain can be released after ATP binding and start to fold correctly according to the still unknown rules that govern protein folding, thus reducing the risk of aggregation with similar partially folded intermediates. It has in addition been proposed that the binding of the compact intermediate in the cpn 60 cavity breaks some noncovalent interactions that would

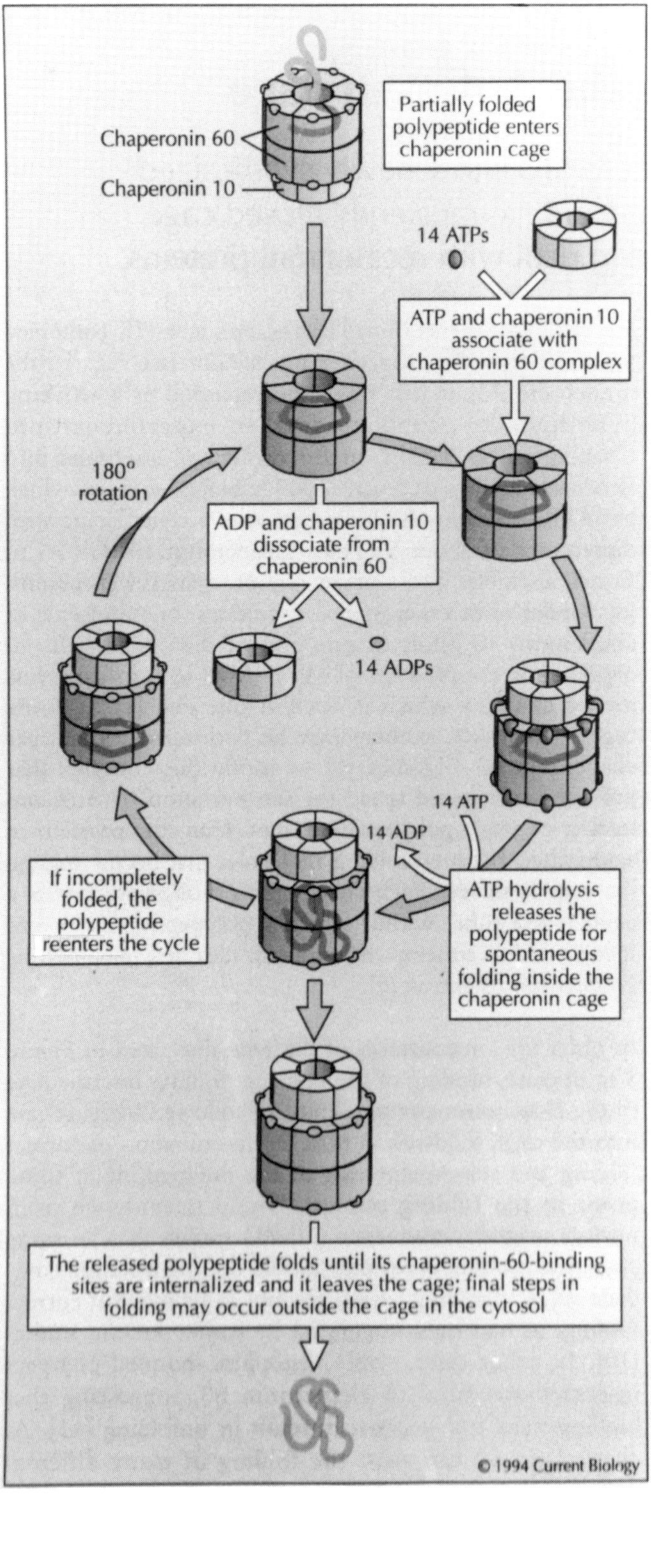

Chaperonin 60
Chaperonin 10
Partially folded polypeptide enters chaperonin cage
14 ATPs
ATP and chaperonin 10 associate with chaperonin 60 complex
180° rotation
ADP and chaperonin 10 dissociate from chaperonin 60
14 ADPs
14 ATP
14 ADP
If incompletely folded, the polypeptide reenters the cycle
ATP hydrolysis releases the polypeptide for spontaneous folding inside the chaperonin cage
The released polypeptide folds until its chaperonin-60-binding sites are internalized and it leaves the cage; final steps in folding may occur outside the cage in the cytosol
© 1994 Current Biology

otherwise prevent misfolded intermediates from reaching the correct tertiary conformation (91). The protein is eventually released because the groups that are recognized by the inside walls of the cpn 60 cavity are now buried within the correctly folded protein (see Fig. 2).

This view of chaperonin action has been termed the Anfinsen cage model to emphasize the point that the chaperonins do not violate the principle of protein self-assembly but permit it to operate with reduced risk of aggregation (60, 92). It is important to note that in this model cpn 60 does not catalyze protein folding in the sense of enhancing its rate, nor is ATP binding and hydrolysis used to drive the folding process. Instead the cpn 60 provides a dynamic cage inside which the chain can be allowed to start to fold spontaneously in the absence of competing aggregation reactions with other folding chains.

The mode of action of the chaperonins is currently a very active research area, and alternative models have been proposed. Thus, it has been noted that cpn 60 oligomers with cpn 10 bound to both ends simultaneously can be observed under some in vitro conditions, and it has been proposed that such intermediates may be obligatory components in chaperonin action (94,95); no compelling evidence for this view has yet been published. Studies with mutant forms of cpn 60 from *E. coli* that bind partially folded protein but do not release it have prompted the suggestion that the protein is released and rebound by different molecules of cpn 60 during chaperonin action, and that therefore most of the folding takes place outside the cage in free solution (96). This latter view may describe what is happening under the particular in vitro conditions used, but it seems unlikely on theoretical grounds to describe the in vivo situation. Folding in free solution runs the risk of running into the aggregation problem at the high protein concentrations obtaining in vivo, whereas these same conditions can be calculated to so reduce the diffusion of the released protein that is very likely to rebind to the same cpn 60 molecule. Even in the dilute solutions used in vitro, the rate of diffusion of released protein is so much faster than the rate of folding that the protein will spend most of the time bound to cpn 60. Thus, until stronger evidence appears as to where the majority of the folding takes place, this reviewer favors the type of model presented in Figure 2.

Both the DnaK/DnaJ/GrpE trio and the chaperonins have little specificity for the protein substrates so far tried, and so may act as general protein folding machines. However, there is no information about the precise nature of the interactions between the chaperonins and their substrates other than that hydro-

---

Figure 2 Model for the action of the chaperonins in protein folding based on studies of the in vitro refolding of citrate synthase and rhodanese after chemical denaturation, taking into account several previous studies (62). (From Ref. 92.)

phobic sidechains in the cpn 60 cage are essential (97). There are some indications that specialized chaperonins may exist in chloroplasts and some bacteria to deal with proteins that present particular folding difficulties (59). Both types of chaperones use ATP binding and hydrolysis as part of their reaction cycles, and they are presumed to act sequentially. There is no information how the polypeptide chain is transferred from the small to the large chaperones, but this sequential action may be directed by the successive exposure of aminoacyl sequences with different binding specificities in the folding chain. The same binding specificities could operate in the in vivo refolding of proteins damaged by stress.

The models illustrated in Figures 1 and 2 represent current working hypotheses, and they will doubtless be altered as more data are obtained, especially since the crystal structure of the cpn 60 from *E. coli* has now been published (98). Most uncertainty concerns precisely how and where the polypeptide is bound to chaperonin 60 and where folding takes place. For example, it is possible that some initial folding occurs while the polypeptide is still partially bound rather than free inside the cage, whereas final stages in folding may occur safely in free solution outside the cage once the risk of aggregation is small. Some polypeptides are too large to fit entirely inside the cage, so they may be handled one domain at a time. There is also an urgent need to test the details of these models in cell-free protein-synthesizing systems and to extend them to the TCP-1 system in the eukaryotic cytosol, where many proteins of medical interest achieve their active conformations.

V. CONCLUSIONS

The idea that proteins in living cells do not fold spontaneously, as once thought, but require assistance from preexisting proteins acting as molecular chaperones has a potential medical implication. Perhaps chaperone diseases exist, in which there is a failure of protein assembly and/or import due to either mutation or a change in the amount of one of the growing number of molecular chaperones that are being identified. There is already a report of a human patient who died 2 days after birth; cultured skin fibroblasts from this patient exhibit abnormal mitochondria containing only about 20% of the normal concentration of cpn 60, so this may be an example of such a chaperone disease (99).

REFERENCES

1. R.I. Morimoto, Cells in stress: transcriptional activation of heat shock genes, *Science 259*:1409 (1993).
2. P.M. Kelley and M.J. Schlesinger, The effect of amino acid analogs and heat shock on gene expression in chicken embryo fibroblasts, *Cell 15*:1277 (1978).

3. L.E. Hightower, Cultured cells exposed to amino acid analogs or puromycin rapidly synthesize several polypeptide, *J. Cell Physiol. 102*:407 (1980).
4. H.R.B. Pelham, Speculations on the functions of the major heat shock and glucose regulated stress proteins, *Cell 46*:959 (1986).
5. R.J. Ellis, Proteins as molecular chaperones, *Nature 328*:378 (1987).
6. R.J. Ellis and S.M. Hemmingsen, Molecular chaperones; proteins essential for the biogenesis of some macromolecular structures, *Trends Biochem. Sci. 14*:339 (1989).
7. R.J. Ellis and S.M. van der Vies, Molecular chaperones, *Annu. Rev. Biochem. 60*:321 (1991).
8. R.J. Ellis (ed.), Molecular chaperones. *Semin. Cell Biol. 1*:1–73 (1990).
9. M-J. Gething and J. Sambrook, Protein folding in the cell, *Nature 355*:33 (1992).
10. *Molecular Chaperones* (R.J. Ellis, R.A. Laskey, and G.H. Lorimer, eds.), Chapman and Hall, London, 1992, p. 121.
11. J.P. Hendrick and F-U. Hartl, Molecular chaperone functions of heat shock proteins, *Annu. Rev. Biochem. 62*:349 (1993).
12. E.A. Craig, D.B. Gambill, and R.J. Nelson, Heat shock proteins: molecular chaperones of protein biogenesis, *Microbiol. Rev. 57*:402 (1993).
13. C. Georgopoulos and W.J. Welch, Role of the major heat shock proteins as molecular chaperones, *Ann. Rev. Cell Biol. 9*:601 (1993).
14. A.A. Gatenby and P.V. Viitanen, Structural and functional aspects of chaperonin-mediated protein folding, *Ann. Rev. Plant Physiol. Plant Mol. Biol. 45*:469 (1994).
15. C. Georgopoulos, D. Ang, K. Liberek, and M. Zylicz, Properties of the *Escherichia coli* heat shock proteins and their role in bacteriophage lambda growth, *Stress Proteins in Biology and Medicine* (R. I. Morimoto, A. Tissieres, and C. Georgopoulos eds.), Cold Spring Harbor Laboratory Press, New York 1990, p. 191.
16. R.A. Laskey, B.M. Honda, A.D. Mills, and J.T. Finch, Nucleosomes are assembled by an acidic protein which binds histones and transfers them to DNA, *Nature 275*:416 (1978).
17. S. Lindquist and E.A. Craig, The heat shock proteins, *Annu. Rev. Genet. 22*:631 (1988).
18. M.J. Schlesinger, Heat shock proteins, *J. Biol. Chem. 265*:12111 (1990).
19. E.A. Craig and C.A. Gross, Is hsp 70 the cellular thermometer? *Trends Biochem. Sci. 16*:135 (1991).
20. W.J. Welch, Mammalian stress response: cell physiology structure/function of stress proteins and implications for medicine and disease, *Physiol. Rev. 72*:1063 (1992).
21. S. Lindquist, Heat shock proteins and stress tolerance in microorganisms, *Curr. Opin. Gene Dev. 2*:748 (1992).
22. *The Biology of Heat Shock Proteins and Molecular Chaperones* (R.I. Morimoto, A. Tissieres, and G. Georgopoulos eds). Cold Spring Harbor Laboratory Press, New York, 1994,
23. E. Vierling, The roles of heat shock proteins in plants, *Annu. Rev. Plant Physiol. Plant Mol. Biol. 42*:579 (1991).
24. W.J. Welch, The mammalian stress response: cell physiology and biochemistry of stress proteins, *Stress Proteins in Biology and Medicine* (R.I. Morimoto, A.

Tissieres, and C. Georgopoulos eds.) Cold Spring harbor laboratory Press, New York 1990, p. 223.

25. S.M. Hemmingsen, C. Woolford, S.M. van der Vies, K. Tilly, D.T. Dennis, C.P. Georgopoulos, R.W. Hendrix, and R.J. Ellis, Homologous plant and bacterial proteins chaperone oligomeric protein assembly, *Nature 333*:330 (1988).
26. K. Borkovich, F. Farrelly, D. Finkelstein, J. Taulein, and S. Lindquist, Hsp 82 is an essential protein that is required in higher concentrations for growth of cells at higher temperatures, *Mol. Cell Biol. 9*:3919 (1989).
27. W.B. Pratt, L.C. Scherrer, K.A. Hutchinson , and F.C. Dalman, A model of glucocorticoid receptor unfolding and stabilization by a heat shock protein complex, *J. Steroid Biochem. 41*:223 (1992).
28. J. Brugge, E. Erikson, and R.L. Erikson, The specific interaction of the Rous sarcoma virus transforming protein pp60 with cellular proteins, *Cell 25*:363 (1981).
29. D. Picard, M. Khurseed, M. Garabedian, M. Fortin, S. Lindquist, and K. Yamamoto, Reduced levels of hsp 90 compromise steroid receptor action in vivo, *Nature 348*:166 (1990).
30. J.S. Brugge, Interaction of the Rous sarcoma virus protein, $pp60^{src}$, with the cellular proteins pp50 and pp90, *Curr. Topics Microbiol. Immunol. 123*:1 (1986).
31. G.H. Perdew and M.L. Whitelaw, Evidence that the 90 kDa heat shock protein (hsp 90) exists in cytosol in heteromeric complexes containing hsp 70 and three other proteins with Mr of 63,000, 56,000 and 50,000, *J. Biol. Chem. 266*:6708 (1991).
32. H. Wiech, J. Buchner, R. Zimmerman, and U. Jakob, Hsp 90 chaperones protein folding in vitro, *Nature 358*:169 (1992).
33. R.A. Mazarella and M. Green, ERp99, an abundant conserved glycoprotein of the endoplasmic reticulum is homologous to the 90 kDa heat shock protein (hsp 90) and the glucose regulated protein (grp 94), *J. Biol. Chem. 262*:8875 (1987).
34. K. Nadeau, A. Das, and C.T. Walsh, Hsp 90 chaperonins possess ATPase activity and bind heat shock transcription factors and peptidyl prolyl isomerases, *J. Biol. Chem. 268*:1479 (1993).
35. R.S. Gupta and B. Singh, Cloning of the hsp 70 gene from *Halobacterium marismortui*: relatedness of archaebacterial HSP70 to its eubacterial homologs and a model for the evolution of the HSP70 gene, *J. Bacteriol. 174*:4594 (1992).
36. T.G. Chappell, B.B. Konforti, S.L. Schmid, and J.E. Rothman, The ATPase core of a clathrin uncoating ATPase, *J. Biol. Chem. 262*:746 (1987).
37. K.M. Flaherty, C. DeLuca-Flaherty, and D.B. McKay, Three dimensional structure of the ATPase fragment of a 70K heat shock cognate protein, *Nature 346*:623 (1990).
38. K.M. Flaherty, D.B. McKay, W. Kabash and K. Holmes, Similarity of the three-dimensional structure of actin and the ATPase fragment of a 70 kDa heat shock cognate protein, *Proc. Natl. Acad. Sci. USA 88*:5041 (1991).
39. M. Flajnik, C. Canel, J. Kramer, and M. Kasahara, Hypothesis: which came first, MHC class I or class II, *Immunogenetics 33*:295 (1991).
40. F. Rippman, W. Taylor, J. Rothbard, and N.M. Green, A hypothetical model for the peptide binding domain of hsp 70 based on the peptide binding domain of HLA, *EMBO J. 10*:1053 (1991).

41. S.J. Landry, R. Jordan, R. MacKaken, and L. Gierasch, Different conformations of the same polypeptide bound to chaperones DnaK and GroEL, *Nature 355*:455 (1992).
42. G.C. Flynn. K, Pohl, M.T. Flocco, and J.E. Rothman, Peptide-binding specificity of the molecular chaperone BiP, *Nature 353*:726 (1992).
43. S. Blond-Elguindi, S.E. Cwirla, W.J. Dower, R.J. Lipshutz, S.R. Sprang, J.F. Sambrook, and M-J. Gething, Affinity panning of a library of peptides displayed on bacteriophages reveals the binding specificity of BiP, *Cell 75*:717 (1993).
44. P.J. Kang, J. Ostermann, J. Shilling, W. Neupert, E.A. Craig, and N. Pfanner, Hsp 70 in the mitochondrial matrix is required for translocation and folding of precursor proteins, *Nature 348*:137 (1990).
45. P. Scherer, U. Krieg, S. Hwang, D. Vestweber, and G. Schatz, A precursor protein partially translocated into yeast mitochondria is bound to as 70 kd mitochondrial stress protein, *EMBO J. 9*:4315 (1990).
46. B.D. Gambill, W. Voos, P.J. Kang, B. Miao, T. Langer, E.A. Craig, and N. Pfanner, A dual role for mitochondrial heat shock protein 70 in membrane translocation of preproteins, *J. Cell Biol. 123*:109 (1993).
47. M. Werner-Washburne, D.E. Stone, and E.A. Craig, Complex interactions among members of an essential subfamily of hsp 70 genes in *Saccharomyces cerevisiae*, *Mol. Cell Biol. 7*:2568 (1987).
48. T. Dierks, P. Klappa, H. Wiech, and R. Zimmerman, The role of molecular chaperones in protein transport into the endoplasmic reticulum, *Molecular Chaperones* (R.J. Ellis, R.A. Laskey, and G.C. Lorimer, eds.), Chapman and Hall, London, 1993, p. 79.
49. R.P. Beckmann, L. Mizzen, and W.J. Welch, Interaction of Hsp 70 with newly-synthesized proteins: implications for protein folding and assembly, *Science 248*:850 (1990).
50. C.R. Brown, R.L. Martin, W.J. Hansen, R.P. Beckmann, and W.J. Welch, The constitutive and stress inducible forms of hsp 70 exhibit functional similarities and interact with one another in an ATP-dependent fashion, *J. Cell Biol. 120*:1101 (1993).
51. R.J. Nelson, T. Ziegelhoffer, C. Nicolet, M. Werner-Washburne, and E.A. Craig, The translation machinery and seventy kilodalton heat shock protein cooperate in protein synthesis, *Cell 71*:97 (1992).
52. D.I. Friedman, E.R. Olson, K. Tilly, C. Georgopoulos, I. Herskowitz, and F. Banuett, Interactions of bacteriophage and host macromolecules in the growth of bacteriophage lambda, *Microbiol. Rev. 48*:299 (1984).
53. T.W. McMullin and R.L. Hallberg, A highly evolutionary conserved mitochondrial protein is structurally related to the protein encoded by the *Escherichia coli* groEL gene, *Mol. Cell. Biol. 8*:371 (1988).
54. J.E. Musgrove and R.J. Ellis, The rubisco large subunit binding protein, *Philos. Trans. R. Soc. Lond. (Biol.) 313*:419 (1986).
55. R.J. Ellis, Molecular chaperones: the plant connection, *Science 250*:954 (1990).
56. R. Gupta, Sequence and structural homology between a mouse t-complex protein TCP-1 and the chaperonin family of bacteria (groEL, 60-65 kDa heat shock antigen) and eukaryotic proteins, *Biochem. Int. 20*:833 (1990).

57. R.J. Ellis, Cytosolic chaperonin confirmed, *Nature 358*:191 (1992).
58. R.A. Stuart, D.M. Cyr, E.A-. Craig and W. Neupert, Mitochondrial molecular chaperones: their role in protein translocation, *Trends Biochem. Sci. 19*:87 (1994).
59. R.J. Ellis, Roles of molecular chaperones in protein folding, *Curr. Opin. Struct. Biol. 4*:117 (1994).
60. H.R. Saibil, D. Zheng, A.M. Roseman, A.S. Hunter, G.M.F. Watson, S. Chen,S. auf der Mauer, B.P. O'Hara, S.P. Wood, N.H. Mann, L.K. Barnett, and R.J. Ellis, ATP induces large quaternary rearrangements in a cage-like chaperonin structure, *Curr. Biol. 3*:265 (1993).
61. U. Bertsch, J. Soll, R. Seetharam, and P.V. Viitanen, Identification, characterisation and DNA sequence of a functional 'double' GroES-like chaperonin from chloroplasts of higher plants, *Proc. Natl. Acad. Sci. USA 89*:8696 (1992).
62. J. Martin, M. Mayhew, T. Langer, and F-U. Hartl, The reaction cycle of GroEL and GroES in chaperonin-assisted protein folding, *Nature 366*:228 (1993).
63. R. Martel, L.P. Cloney, L.I. Pelcher, and S.M. Hemmingsen, Unique composition of plastid chaperonin 60; alpha and beta polypeptide-encoding genes are highly divergent, *Gene 94*:181 (1990).
64. H.R. Saibil and S.P. Wood, Chaperonins, *Curr. Opin. Struct. Biol. 3*:207 (1993).
65. T. Langer, G. Pfeifer, J. Martin, W. Baumeister, and F-U. Hartl, Chaperonin-mediated protein folding: GroES binds to one end of the GroEL cyclinder, which accommodates the protein substrate within its central cavity, *EMBO J. 11*:4757 (1992).
66. L.M. Silver, K. Artzt, and D. Bennet, A major testicular cell protein specified by a mouse T/t complex gene *Cell 17*:275 (1979).
67. V.A. Lewis, G.M. Hynes, D. Zheng, H.R. Saibil, and K. Willison, T-complex polypeptide-1 is a subunit of a heteromeric particle in the eukaryotic cytosol, *Nature 358*:249 (1992).
68. H. Kubota, G. Hynes, A. Carne, A. Ashworth, and K. Willison, Identification of six TCP-1-related genes encoding divergent subunits of the TCP-1-containing chaperonin, *Curr. Biol. 4*:89 (1994).
69. Y. Gao, J.O. Thomas, R.L. Chow, G-H. Lee, and N.J. Cowan, A cytoplasmic chaperonin that catalyzes beta-actin folding, *Cell 69*:1043 (1992).
70. M.B. Yaffe, G.W. Farr, D. Miklos, A.L. Horwich, M.L. Sternlicht, and H. Sternlicht, TCP-1 complex is a molecular chaperone in tubulin biogenesis, *Nature 358*:245 (1992).
71. H. Sternlicht, G.W. Farr, M.L. Sternlicht, J.K. Driscoll, K. Willison, and M.B. Yaffe, The t-complex polypeptide 1 is a chaperonin for tubulin and actin in vivo, *Proc. Natl. Acad. Sci. USA 90*:9422 (1993).
72. R.E. Susek and S.L. Lindquist, Hsp26 of *Saccharomyces cerevisiae* is related to the superfamily of small heat proteins but is without a demonstrable function, *Mol. Cell. Biol. 9*:5265 (1989).
73. A.P. Arrigo and W.J. Welch, Purification and characterisation of the small 28 kd mammalian stress protein, *J. Biol. Chem. 262*:15359 (1987).
74. J. Jakob, M. Gaestel, K. Engel, and J. Buchner, Small heat shock proteins are molecular chaperones, *J. Biol. Chem. 268*:1517 (1993).

75. J. Horwitz, Alpha-crystallin can function as a molecular chaperone, *Proc. Natl. Acad. Sci. USA 89*:10449 (1992).
76. K.B. Merck, P.J.T.A. Groenen, C.E.M. Voorter, W.A. Haard-Hoekman, J. Horwitz, H. Bloemendal, and W.W. De Jong, Structural and functional similarities of bovine alpha-crystallin and mouse small heat shock proteins. A family of chaperones, *J. Biol. Chem. 268*:1046 (1993).
77. R. Jaenicke and T.E. Creighton, Junior chaperones, *Curr. Biol. 3*:234 (1993).
78. M. Zylicz, *The Escherichia coli* chaperones involved in DNA replication, *Molecular Chaperones* (R.J. Ellis, R.A. Laskey and G.C. Lorimer, eds.), Chapman and Hall, London, 1993, p. 15.
79. P. Bork, C. Sander and A. Valencia, A module of the DnaJ heat shock proteins found in malaria parasites, *Trends Biochem. Sci. 17*:129 (1992).
80. I. Sadler, A. Chiang, T. Kurihara, J.J. Rothblatt, J.P. Way, and P. Silver, A yeast gene important for protein assembly into the endoplasmic reticulum and the nucleus has homology to DnaJ, an *Escherichia coli* heat shock protein, *J. Cell Biol. 109*: 2665 (1989).
81. L. Bolliger, O. Deloche, B.S. Glick, C. Georgopoulos, P. Jeno, N. Kronidou, M. Horst, N. Morishima, and G. Schatz, A mitochondrial homolog of bacterial GrpE interacts with mitochondrial hsp 70 and is essential for viability, *EMBO J.* in press.
82. T. Langer, C. Lu, H. Echols, J. Flanagan, M.K. Hayer, and F-U. Hartl, Successive action of DnaK. DnaJ and GroEL along the pathway of chaperone-mediated protein folding, *Nature 356*:683 (1992).
83. F-U. Hartl, R. Hlodan, and T. Langer, Molecular chaperones in protein folding: the art of avoiding sticky situations, *Trends Biochem. Sci. 19*:20 (1994).
84. D.L.D. Caspar and A. Klug, Physical principles in the construction of regular viruses, Cold Spring Harbor *Quant. Biol. 27*:1 (1962).
85. C.B. Anfinsen, Principles that govern the folding of protein chains, *Science 181*: 223 (1973).
86. R. Barraclough and R.J. Ellis, Protein synthesis in chloroplasts XI. Assembly of newly-synthesized large subunits of ribulose bisphosphate carboxylase in intact isolated pea chloroplasts, *Biochim. Biophys. Acta. 608*:19 (1980).
87. R. Seckler and R. Jaenicke, Protein folding and protein refolding, *FASEB J. 6*:2545 (1992).
88. A. Gragerov, E. Nudler, N. Komissarova, G.A. Gaitanaris, M.E. Gottesman, and V. Nikiforov, Cooperation of GroEL/GroES and DnaK/DnaJ heat shock proteins in preventing protein misfolding in *Escherichia coli, Proc. Natl. Acad. Sci. USA 89*:10341 (1992).
89. A.L. Horwich, K.B. Low, W.A. Fenton, I.N. Hirshfield, and K. Furtak, Folding in vivo of bacterial cytoplasmic proteins: role of GroEL, *Cell 74*:909 (1993).
90. J.P. Henrick, T. Langer, T.A. Davis, F-U. Hartl, and M. Wiedmann, Control of folding and membrane translocation by binding of chaperone DnaJ to nascent polypeptides, *Proc. Natl. Acad. Sci. USA 190*:10216 (1993).
91. G.S. Jackson, R.A. Staniforth, D.J. Halsall, A. Atkinson, J.J. Holbrook, A.R. Clarke, and S.G. Burston, Binding and hydrolysis of nucleotides in the chaperonin catalytic cycle: implications for the mechanism of assisted protein folding, *Biochemistry 32*:2554 (1993).

92. R.J. Ellis, Opening and closing the Anfinsen cage, *Curr. Bio. 4*:633 (1994).
93. J. Melnick, J.L. Dul, and Y. Argon, Sequential interaction of the chaperones BiP and GRP 94 with immunoglobulin chains in the endoplasmic reticulum, *Nature 370*:373 (1994).
94. A. Azem, M. Kessel and P. Goloubinoff, Characterisation of a functional $GroEL_{14}$ $(GroES_7)_2$ chaperonin hetero-oligomer, *Science 265*:653 (1994).
95. M. Schmidt, K. Rutkat, R. Rachel, G. Pfeifer, R. Jaenicke, P. Viitanen, G. Lorimer, and J. Buchner, Symmetric complexes of GroE chaperonins as part of the functional cycle, *Science 265*:656 (1994).
96. J.S. Weismann, Y. Kashi, W.A. Fenton and A.L. Horwich, GroEL-mediated protein folding proceeds by multiple rounds of binding and release of nonnative forms, *Cell 78*:693 (1994).
97. W.A. Fenton, Y. Kashi, K. Furtak, and A.L. Horwich, Residues in chaperonin GroEL required for polypeptide binding and release, *Nature 371*:614 (1994).
98. K. Braig, Z. Otwinowski, R. Hegde, D.C. Boisvert, A. Joachimiak, A.L. Horwich, and P.B. Sigler, The crystal structure of the bacterial chaperonin GroEL at 2.8 A, *Nature 371*:578 (1994).
99. E. Agsteribbe, A. Huckriede, M. Veenhula, M.H.J. Ruiters, et al, A fatal, systemic mitochondrial disease with decreased enzyme activities, abnormal ultrastructure of the mitochondria and deficiency of heat shock protein 60, *Biochem. Biophys. Res. Commun. 193*:146 (1993).

2

The Unique Role of Heat Shock Proteins in Infections

Bernd Schoel and Stefan H. E. Kaufmann
University of Ulm, Ulm, and Max Planck-Institute for Infection Biology, Berlin, Germany

I. INTRODUCTION

Parasitic microorganisms have played a tremendous role in the evolution of humans (1). Parasites are microbes which live at the expense of their mammalian hosts. Although some of these parasites interact with their hosts loosely, others achieve intimate contact, thus causing stable infection. As a direct or indirect consequence, the host is harmed and clinical disease evolves. Whether infection progresses in a stable or an abortive form and, as a corollary, whether disease develops or not is markedly influenced by the host immune system. This immune system has the capacity to identify small molecular entities which are characteristic of foreign invaders. Thus, the immune system is first able to distinguish its own molecular entities (termed "self") from those of foreign intruders (termed "nonself"), and second, it is able to differentiate specifically among the enormous diversity of microbial pathogens. The whole molecule to which the immune system responds is an antigen, and the small entity within the antigen which is actually "seen" by the immune system is an epitope. For protein antigens, these epitopes are short regions encompassing 6–12 amino acids (AAs).

Evolutionary thinking holds that the length of the epitope has not evolved arbitrarily but rather represents the outcome of a balanced process between avoidance of antiself responses on the one hand and broad covering of as many foreign entities as possible (2). Obviously, longer peptides would improve distinction of self from nonself, whereas shorter ones would facilitate broad responsiveness even in the face of ongoing microbial mutations. Although the focus of the immune system on epitopes ranging from 6 to 12 AAs seems to work fine in most cases, problems may arise in certain instances. Some antigens are highly abundant, because they are markedly conserved across species barriers. On the one hand, these antigens are frequently presented to the immune system, so that they cannot be ignored. On the other hand, such antigens appear similar in host and invader and hence may be confused with each other. The heat shock proteins (hsps) are paradigmatic members of this group.

First, hsps exist in all living cells. Second, high sequence homology not only exists among hsps of various microbial pathogens but even between hsps of the invader and host. Third, many hsps are produced at elevated levels in response to various insults and, therefore, represent characteristic markers of stress situations which constantly arise during host pathogen interactions. Fourth, by acting as "chaperones," hsps interact with denatured proteins. Because protein denaturation represents a first step of antigen processing, hsps may encounter the immune system more frequently than the average self protein.

These features of hsps may be to the benefit of the host. Thus, focus on proteins which are shared by and abundant in all infectious agents can improve anti-infectious immune responses. Second, focus on a marker of mutilated cells which cannot be rescued will contribute to host surveillance. Such features, however, may also turn against the host as soon as they become exaggerated, thus causing destruction of cells which are essential to host integrity. The situation may be complicated further in that hsps are not restricted to parasite and host but also exist abundantly in a third group of biomaterial with which humans are in constant contact; that is, our everyday nutrition.

In this chapter, we focus on some features of hsps as they are of potential relevance to the host-pathogen relationship, maintenance of host integrity, and eventual emergence of autoaggressive reactions.

II. IMMUNE SYSTEM AS AN ADOPTIVE RESPONSE TO THE DIVERSITY OF MICROBIAL PATHOGENS

The host immune system encounters the world of pathogenic microbes at a nonspecific and a specific level. The specific arm focuses on and improves the response initiated by the nonspecific effector system. Subsequently, it activates nonspecific effector mechanisms so that they fulfill their tasks more appropriately. This cooperation provides to the host an adequate defense against its invaders. The specific immune response encompasses a regulatory and an effector system. Specificity relies on the recognition of unique molecular entities by receptors of homologous structures. These receptors are the surface immunoglobulins (Ig) on B lymphocytes and the T-cell receptors (TCR) on T lymphocytes.

For both lymphocyte subsets specificity is generated by somatic genetic rearrangements, and antigen stimulation leads to the activation of a particular clone expressing a unique receptor. In principle, the receptors have to recognize epitopes derived from nonself antigens, and to ignore self epitopes. To achieve this goal, lymphocytes undergo selection processes. Most T lymphocytes recognizing self are deleted in the thymus before they mature, and self reactive clones which evade deletion are inactivated in the periphery (3). It is this inactivation of physically existing T cells expressing self-reactive TCR which bears the potential risk of autoimmune disease. B cells are deleted, but not positively selected, in the bone marrow when pre-B cells encounter membrane-bound antigens on adjacent cells. Evading self-reactive B cells are rendered anergic in the periphery and have a remarkably short life span. After antigenic contact, mature B cells undergo another phase of specificity variation by somatic mutation. This process of affinity maturation may result in the generation of potentially self-reactive B cells which require further control. At this stage, positive selection of mature B cells by foreign antigen and specific T-cell help are necessary for their survival. Still, such mechanisms allow for the existence of self-reactive B cells, the activation of which becomes possible through appropriately presented antigens, such as highly cross-reactive foreign antigens (4,5).

Memory is another common feature shared by both lymphocyte subsets. Phenotypically immune memory appears as an improved response to repeated contact with the same or a similar antigenic structure. If this structure is expressed by a microbial pathogen, a booster provides the basis for preventive vaccination. On the other hand, if a cross-reactive structure becomes expressed by self, autoimmune responses may arise (6).

The two lymphocyte subsets differ in the kind of antigen recognition and in the type of ensuing response. B lymphocytes, after initial activation through antigenic contact, produce antibodies of the same specificity as that of their surface receptor. The antibodies and surface Ig directly react with their antigens in the extracellular space. T lymphocytes only respond to antigenic peptides associated with cells in the context of proteins encoded by the major histocompatibility complex (MHC) (7). These polymorphic genes encode two different kinds of proteins: the MHC class I gene product, which is a single transmembrane polypeptide chain noncovalently linked with β_2-microglobulin, and the MHC class II encoded proteins encompassing two noncovalently linked transmembrane polypeptide chains. Besides the classic polymorphic class I genes—termed class Ia—another set of nonpolymorphic class Ib genes is encoded within the MHC (8). Each antigen is processed by the presenting cell and ends up as a peptide located in MHC-encoded "peptide receptors." The class, locus, and allele inherited by an individual convey a certain degree of specificity for the peptides bound within the cleft which determines length and anchoring amino acids of the peptide (9–12). After interaction with the MHC peptide complex, T cells activate effectors of the immune system by secreting cytokines which activate nearby cells. Cytokine secretion is primarily a function of $CD4^+$, MHC class II restricted T helper cells (13). Another subset of T cells, the $CD8^+$, MHC class I restricted T lymphocytes, function mostly as cytolytic cells (7).

Because of their MHC restriction, activation of the two responding types of T cells is ultimately controlled by the class of peptide-presenting MHC molecules (14). Generally, MHC class II is confined to mononuclear phagocytes and B cells, and it is charged through the endosomal compartment of the cell. This can ideally defend against microbes living in extracellular spaces, because phagocytosis and antibodies are the major effector mechanisms for this kind of pathogens. Antibodies neutralize toxic compounds secreted by certain extracellular pathogens. Furthermore, they facilitate phagocytosis by coating the invader (opsonization). These mechanisms lead to the eradication of many pathogens responsible for acute infections (15). MHC class I products are expressed by virtually any host cell and they are loaded with peptides derived from cytosolic proteins (14). Therefore, this presentation pathway can indicate alterations in a particular cell; for example, caused by infection with an intracellular pathogen. Living within host cells, these pathogens are shielded from antibodies. Therefore, destruction of their habitat and/or replication machinery represents an effective means for their combat (15–17).

In addition to this segregation of CD4 and CD8 T lymphocytes determined by MHC restriction, a second division exists which is based on the nature of the TCR. The majority of peripheral T cells expresses an α/β TCR and a minor pool of T lymphocytes expresses a γ/δ TCR (7). The former cells are almost exclusively $CD4^+$ or $CD8^+$, and the latter cells are mostly $CD4^-CD8^-$. The MHC restriction of these double-negative γ/δ T cells remains a matter of controversy. Restriction by products of the nonpolymorphic MHC class Ib gene has been suggested (18). Alternatively, it has been proposed that the TCR recognizes antigen directly similarly as surface immunoglobulins do (19). Finally, conventional MHC restriction has been observed. The selection of γ/δ T cells during maturation is less well understood than that for α/β T cells. Evidence for positive and negative thymic selection of γ/δ T cells has been presented. Apart from this, extrathymic selection was observed in some experiments (18). Major stimulatory ligands for γ/δ T cells are common structural motifs such as phosphate on certain small-sized carriers of microbial origin, or hsps (20–26). These and other findings have led to the proposal that regulatory and scavenger functions are a major task of γ/δ T cells. Because of the less specific nature of these ligands, γ/δ T cells bear a higher risk of causing detrimental autoimmune reactions (6,18,27).

III. HSPS ARE EXCEPTIONAL IN DEFENSE AGAINST AND INVASION OF PATHOGENS

Within the interplay between invading pathogen and defending host, hsps perform an unique role as compared with the average protein. This exceptional impact occurs at two levels: (1) in the area of protein turnover due to the functioning of certain hsps as chaperones, and (2) in the area of antigen-specific immune responses due to the extraordinary homology of hsps across species barriers. The following paragraphs briefly describe general principles, and specific examples are given in Section V.

A. Functional Aspects of Hsps in Defense and Invasion

Hsps of the 60- and 70-kDa families participate in the folding/unfolding of other proteins and hence are termed chaperones. This chaperone function aids in the synthesis of proteins, the formation of protein complexes, and in protein translocation from one cellular compartment to another by means of transporter systems in the separating membrane (28–30). During the immune response, BiP and grp (glucose-regulated protein) 94 (hsp 70 and hsp 90 in the endoplasmic reticulum [ER]) participate in the assembly of Ig (31,32). Furthermore, hsps are involved in antigen processing and presentation. Hsps promote refolding of proteins which became partly denatured by heat or other "stress stimuli." This capability of renaturing other proteins is beneficial for the infected host cell, because the mobilization of defense mechanisms causes a stressful environment not only for the invader but also for the host. The pathogen uses its own hsps to maintain its proteins in a functional state. Therefore, hsps are related to microbial virulence to the disadvantage of the host. One step further, cellular hsps may be abused by certain viruses to promote their own assembly.

B. Immunological Aspects of Hsps in Defense and Invasion

Hsps are highly conserved among species. Generally, this is advantageous for the immune system: Repeated contact with different microbes may expose the host to conserved regions of microbial hsps so that the immune system is already alerted when it encounters a pathogen. Because pathogens express elevated hsp levels during infection, hsps may serve as early targets of the immune response (33–35). The possibility to protect against microbes by immunization with hsp has been utilized for several pathogens. Hsps were not only used as target antigens but also as carrier molecules for unrelated antigens (36,37). Similarities between prokaryotic and eukaryotic hsps, however, may activate a cross-reactive immune response to self epitopes derived from cellular hsps. This cross-reactivity may promote immune surveillance and elimination of abnormally altered cells. But to the detriment of the host, this cross-reactivity may result in autoimmune reactions as a sequelae of infection (6).

IV. HSPS ARE UNIQUE TARGETS OF AUTOIMMUNE RESPONSES

Autoimmunity, generally considered unfavorable for the host, at its root means recognition of self and hence has to be viewed in an unbiased form (3,6,38). The outcome of self-recognition depends on the context: Elimination of aberrant cells via recognition of self-epitopes indicating an abnormally altered state may be to the benefit of the host. When such an autoimmune reaction, however, becomes chronic or exacerbated owing to permanent stimulation by the ongoing inflammation, autoimmune disease may arise. The γ/δ T cells have been proposed to play a unique role in recognizing certain antigens indicative for stress such as hsps (20–24,39). Hsp epitopes which are expressed after "stress" (e.g., after viral infection) become detectable for the specific immune system and thus can serve as an indicator for abnormality. The structures visible to the immune system are MHC/peptide complexes and perhaps also hsps on the cell surface (6). Either kind of recognition would promote immune surveillance.

The possibility that hsps proper are expressed on the cell surface remains a matter of controversy. Several experimental systems have provided circumstantial evidence to support this notion principally based on the identification of hsps by specific monoclonal antibodies (MAbs) (40–43). Immune precipitation and SDS-PAGE analyses revealed that proteins, which were specifically recognized by such MAbs, have the expected size of 60 kDa (22,24). Final proof, however, for the identity of these proteins requires amino acid sequencing. Passive cotranslocation of hsps together with typical membrane-bound proteins is possible, since coprecipitation of hsp 60 with a 70-kDa molecule and of hsp-like proteins with MHC class I molecules has been described (24,44). The immune system neglects, whether this protein is an hsp proper, or alternatively has a similar image to allow for cross recognition by hsp-specific antibodies (45,46).

Not only antibodies but possibly also γ/δ T lymphocytes may recognize such surface-expressed proteins directly. Several studies have provided evidence for recognition of hsps by γ/δ T cells (22,24,39). Particularly, the intraepithelial γ/δ lymphocytes in the intestine, which are entrapped between epithelial cells and hence are unable to recirculate, fail to contact a great variety of antigens (16, 39,47,48). Therefore, they must be biased to a restricted number of markers of alterations, such as hsps. It appears an interesting possibility for γ/δ T cells to focus on only few antigens instead of expressing a broad spectrum of specificity to any kind of antigen (6,49).

α/β T cells will recognize conventional MHC/peptide complexes, and T cells with specificity for cross-reactive regions of hsps shared between microbe and host have been identified (50–52). CD4 T cells recognizing such epitopes have been frequently observed, indicating the existence of self-reactive lymphocytes in the periphery. Although hsp-reactive CD8 T cells appear to be less frequent, such T-cell lines have been established and shown to recognize stressed host cells (52). Similarly, conventional recognition of antigenic peptides including those derived from hsps in the context of MHC gene products by some γ/δ T cells has been described (21,23,53).

Recent progress in peptide elution and sequencing have made it possible to isolate and identify peptides from MHC-encoded proteins (11,54–57). Frequently, epitopes derived from hsps have been identified. When compared with a protein sequence database, several hsp peptides were found identical with or highly similar to sequences from hsps of parasites, and furthermore to hsps of some common nutrients. Such peptides may concern T cells cross-reactive to self-hsp. Since self-reactive T cells can evade thymic deletion, peripheral silencing mechanisms must exist. One such mechanism is tolerization by oral immunization. This could mean that the route by which hsps contact the immune system influences the kind of immune response that develops: either activation of immunity with the potential risk of autoimmune disease by highly similar epitopes from microbial pathogens or, alternatively, by induction of oral tolerance of cross-reactive food-derived epitopes through the intestine (6,58).

V. REPRESENTATIVE EXAMPLES FOR HSP INVOLVEMENT IN HOST/PATHOGEN RELATIONSHIP

For numerous microbial pathogens the expression of hsps and eventual development of a specific immune response have been described (for extensive reviews see refs. 6,33,and 59–61). Although viruses do not encode their own hsps, they frequently induce hsp synthesis in host cells or at least maintain normal cellular levels of particular hsps. Table 1 summarizes representative examples of various types of pathogens together with the major hsp families. The physiological effect of an hsp cognate can be beneficial or detrimental for the host depending upon its specific function during infection.

A. Beneficial Effects of Hsps

Beneficial effects of hsps based on their chaperone functions have been reported. Evidence for autoprotection of the infected cell by elevated hsp synthesis during infection has been obtained with human cells (62–67). Increased expression and supply of the inducible hsp 72 by astrocytes during scrapie may contribute to neuronal homeostasis by preventing the deposition of neurotoxic amyloid proteins. This could retard damage of the central nervous system, at least in the initial phase of disease (68).

Table 1 Hsps and Infection

Type of pathogen	Hsp cognate	Disease	Reference
Viruses			
HTLV-I	Self-hsp 70	T-cell leukemia	91
HIV	Self-hsp 70	AIDS	45
Bacteria			
Mycobacterium tuberculosis	Hsp 70	Tuberculosis	123
	Hsp 60	Tuberculosis	123, 124
	Small hsps	Tuberculosis	125
	GroES	Tuberculosis	126
Mycobacterium leprae	Hsp 70	Leprosy	127
	Hsp 60	Leprosy	123, 124
	Small hsps	Leprosy	128
	GroES	Leprosy	129
Chlamydia trachomatis	Hsp 70	Trachoma	130
	Hsp 60	Trachoma	115
Borrelia burgdorferi	Hsp 70	Lyme disease	131
	Hsp 60	Lyme disease	132
Coxiella burnetii	Hsp 60	Q-fever	133
Treponema pallidum	Hsp 60	Syphilis	134
Legionella pneumophila	Hsp 60	Legionnaire's disease	78, 135
Brucella abortus	Hsp 60	Brucellosis	136
Bordetella pertussis	Hsp 60	Pertussis	77
Helicobacter pylori	Hsp 60	Gastritis, ulcer	117

Neisseria meningitidis	Hsp 60	Meningitis	137
Yersinia enterocolitica	Hsp 60	Yersiniosis	79
Fungi			
Candida albicans	Hsp 90	Candidosis	84, 86
Aspergillus fumigatus	Hsp 90	Aspergillosis	138
Histoplasma capsulatum	Hsp 70	Histoplasmosis	139, 140, 141
Protozoa			
Plasmodium falciparum	Hsp 90	Malaria	142
	Hsp 70	Malaria	143, 144
Trypanosoma cruzi	Hsp 90	Chagas' disease	146
	Hsp 70	Chagas' disease	147,148
Trypanosoma brucei brucei	Hsp 90	Trypanosomiasis of cattle	149
	Hsp 70	Trypanosomiasis of cattle	150
Leishmania amazonensis	Hsp 90	Leishmaniasis	151
Leishmania brasiliensis	Hsp 70	Leishmaniasis	148
Leishmania donovani	Hsp 70	Leishmaniasis	152, 153
Leishmania major	Hsp 70	Leishmaniasis	154
Helminths			
Schistosoma mansoni	Hsp 90	Schistosomiasis	155
	Hsp 70	Schistosomiasis	156, 157
	Small hsps	Schistosomiasis	158
Brugia malayi	Hsp 70	Filariasis	159
Onchocerca volvulus	Hsp 70	Onchocercosis	160

In these examples, hsps were helpful by preventing protein denaturation. During antigen processing, proteins become denatured and are cleaved into peptides. Consistent with a possible participation of hsps in this process, several reports have been published verifying this concept. A peptide-binding protein (PBP 72/74), which is involved in the intracellular assembly of the antigen/MHC class II complex, has been characterized as a member of the hsp 70 family (69). Cytosolic hsp 70 and, similarly, gp 96 in the ER (homologous to p84/86hsp) are associated with tumor-specific peptides (70–72). Antigenic peptides may be transferred to MHC class I directly or through internalization of the hsp/peptide complexes by antigen-presenting cells (72,73). Finally, MHC class I presentation of certain antigenic peptides independent from peptide transporters and dependent on constitutively expressed hsp 73 has been described (74). These latter examples provide first hints as to how chaperones participate in basic immune functions.

Beneficial effects to the host of hsps functioning as immunogen have also been described. Because any kind of microbial pathogen synthesizes hsps (see Table 1), and because hsps are highly conserved among different species, frequent contact with microorganisms of even low virulence could cause a strong immune response to these proteins. Antibody and T-cell responses to hsps have been described using peripheral blood and cord blood cells from healthy individuals (33,50,75,76). High anti-hsp antibody titers were observed in children vaccinated with the trivalent vaccine against tetanus, diphtheria, and pertussis, suggesting immunodominance of hsp antigens (77). Therefore, a more rapid immune response to virulent pathogens displaying hsp cognates may occur in healthy individuals.

Consistent with this idea a purified major cytoplasmic membrane protein (a member of the hsp 60 family) of *Legionella pneumophila* has been shown to be protective in guinea pigs (78). Adoptive transfer of CD4 α/β T-cell clones with specificity for hsp 60 of *Yersinia enterocolitica* protects mice from lethal infection with this pathogen (79). A high degree of protection was obtained with a monocytic tumor cell line transfected with the mycobacterial hsp 60 gene as vaccine. Such immunized mice were protected against challenge with *Mycobacterium tuberculosis*. The responsible T cells were $CD8^+$ and $CD4^+$ (80). Infection with *Candida albicans* may result in superficial candidosis which is prevented by cell-mediated immunity or systemic candidosis, where humoral immunity against a breakdown product of hsp 90 prevents dissemination beyond the mucosa. In systemic candidosis, low titers of anti–hsp 90 antibodies are fatal to the host (81–85). Studies in a mouse model of systemic candidosis showed that prophylactic administration of anti–hsp 90 antibodies with reactivity to a defined and conserved epitope reduced mortality of mice by about 50% (86). The fungal hsp 90 itself seems to contribute to pathogenesis, and anti–hsp 90 autoantibodies neutralize this unusual effect (87). This reamphasizes the possibility that hsp-specific immunity mediates protection. In another model, hsp was used as carrier of unrelated antigens for vaccination against malaria. This hsp-malaria antigen vaccine showed improved efficacy against malaria plasmodia in mice preimmunized with BCG (36,37).

Self-epitopes of hsps become expressed on the cell surface subsequent to increased hsp production, and thereby the immune system becomes allerted. Stressed cells are detected by $CD8^+$ T lymphocytes cross reactive for bacterial and self hsp 60 in a mouse model, whereby not only heat but also cytokines and even virus induced MHC I presentation of self epitopes which were then recognized by T cells (52,63,88–90). Furthermore, cell surface hsp 60 induced by human immunodeficiency virus (HIV) promoted lysis by antibody-dependent cellular cytotoxicity (ADCC) (45), and HTLV-1–infected cell lines stimulated formation of anti-hsp autoantibodies in rabbits (91). The expression of self-hsp could promote immunosurveillance of virus-infected cells (6,92).

B. Detrimental Effects of Hsps

Detrimental corollae of hsps to the host are also associated with their function as chaperones or immunogens. Microorganisms face drastic environmental changes such as temperature alterations when they invade the host (34,35). Already such changes necessitate adaptive mechanisms in the pathogen. In many parasites, a sudden temperature shift induces stage differentiation. In promastigotes of *Leishmania* an hsp response occurs in vitro when the temperature is increased from 26 to 35°C, which is then followed by stage differentiation (93). In the fungal pathogen *Histoplasma capsulatum*, a shift in temperature from 25 to 37°C leads to a morphological change from the mycelial form of the fungus as environmental saprophyte to the yeast form as parasite in the host (94,95). Strains of *H. capsulatum* differ in their virulence, and this difference correlates with temperature sensitivity and the level of hsps synthesized at the time of infection (96–99). The ratio of unsaturated to saturated fatty acids in the membrane of the organism controls membrane fluidity at different temperatures. Because strains differing in their virulence express varying fatty acid ratios and varying levels of hsps, it has been suggested that the lipid profile not only modifies the perception of temperature changes but also hsp responses and adaption to the new environment (95).

After successful infection, the invader is exposed to nonspecific host resistance mechanisms, most importantly phagocytosis by professional phagocytes (17). Pathogen survival in these hostile cells is supported by several mechanisms, including hsp induction. In *Salmonella typhimurium*, levels of hsps increase after phagocytosis or after experimental treatment with reactive oxygen intermediates, one of the defense mechanisms of phagocytes (100,101). Hsp deletion mutants of *S. typhimurium* do not survive such defense mechanisms and lose their virulence in vivo (102,103).

Also during late stages of infection, chaperone functions of hsps may be harmful to the host. Self-hsp can promote virus assembly in the infected cell. Numerous viruses induce a stress response (60) either directly (such as adenovirus where the viral gene product E1A activates hsp 70 transcribtion [104]) or indirectly (such as vaccinia virus by maintaining expression of the major cellular hsp while retarding synthesis of most cellular proteins [105]). Association of hsp 70 with viral components has been shown in several instances (105,106). A striking homology between the aminoterminal residues of polyoma virus T antigens and the conserved "J domain" of DnaJ cognates has been found (107). This domain is essential for association of DnaJ with hsp 70 (DnaK). Therefore, interaction of viral T antigen with hsp 70 cognates mediated by the "J domain" motif appears possible (107). Similarly, bacteriophage assembly is promoted by the hsp 60 cognate, GroEL, in *Escherichia coli*, which is the first example for contribution of host hsps to assembly of viral invaders (108–111). Therefore, hsps could promote virus assembly and replication and support viral infections in mammals (92).

Another example of harmful chaperone effect is given by the fungal pathogen *C. albicans*. Here circulating fungal hsp 90 contributes to pathogenesis. It has been suggested that hsp 90 binds to important serum proteins, thereby interfering with the function of the latter. Because lysis of yeast cells by antimycotic drugs such as amphotericin B increases the levels of fungal hsp 90, toxic effects are possible which could be circumvented by concomittant application of anti–hsp 90 antibodies (87).

So far, we have discussed the detrimental role for the host of hsps functioning as chaperones. The detrimental role of hsps as immunogens may preponderate in autoimmune reactions, when hsp-specific immune responses turn toward self-hsp. Some infections are accompanied by damage of specific tissue, such as destruction of the peripheral nerves in tuberculoid leprosy. Hsp-specific T cells can be primed during infection with *M. leprae* and then cause irreversible nerve cell damage because of cross recognition of self-antigens (88). In an in vitro model, Schwann cells expressed MHC class I on the cell surface and were lysed by CD8 T lymphocytes with specificity for mycobacterial hsp 60 after interferon gamma (IFN-γ) activation (88). In another study Schwann cells presenting *M. leprae* antigens were lysed by *M. leprae*–specific CD8 T cells, and although the antigens recognized were not studied, a contribution of hsps in this destructive process is likely (112). Schwann cells exposed to heat or infected with *M. leprae* synthesize hsp 70, and they are able to present mycobacterial hsp 70 to specific T cells (113,114). These and other examples provide further evidence for hsp involvement in pathology (115–117).

C. Hsps and Cross Reactivity Between Nonself and Self

Numerous studies have reported autoimmune recognition of hsp (for summary, see ref. 6), but the relation between priming of hsp-specific lymphocytes by nonself-hsp and subsequent response to cross-reactive self-hsp epitopes is poorly understood. T cells recognizing epitopes from hsp 60 shared among microbes and humans have been identified in healthy individuals (50,51). By using an in vitro stimulation system with a tryptic peptide digest of mycobacterial hsp 60, CD8 T lymphocytes could be generated from naive mice which lysed stressed macrophages (52). In order to address the question, whether such cross-reactive T lymphocytes can be activated in vivo, immunization with mycobacterial hsp 60 in an appropriate adjuvant was employed to activate CD8 T cells, as is the case in infections with intracellular bacteria (17,90,117,118). Two types of CD8 T cells were identified: those recognizing mycobacterial hsp 60 only and those cross reactive to self-epitopes expressed by stressed macrophages.

For the cross-reactive CD8 T cells, an unusual peptide specificity was found (89). The T-cell clone reacts with a cluster of peptides which are closely related to, although not identical with, the predicted nonamer peptide of hsp 60, best fulfilling the requirements for H-2D^b binding. The characteristic features of this peptide cluster includes carboxyterminal AA elongation from the anchoring AA Asn in position 5 of the optimal H-2D^b binding nonamer peptide (89). The homologous part of the mouse hsp 60 does not bind to H-2D^b, because it contains an Asp instead of the anchoring Asn. Because this AA is negatively charged, it is probably repulsed from the negatively charged binding pocket of H-2D^b (90,120,121). The T-cell clone, however, responded to a self–hsp 60 peptide embodying the H-2D^b motif but distinct from the homologous part of mycobacterial hsp 60. Consistent with these findings are results from experiments in which target cells were treated with hsp 60–specific antisense oligonucleotides. Hsp 60 expression was reduced, and concomitantly recognition by the hsp 60–reactive CD8 T-cell clone was diminished (122). This strongly suggests that self–hsp 60 is the source of the self-epitope presented by stressed host cells and cross recognized by CD8 T cells primed with mycobacterial hsp 60.

Although a single hsp 60 peptide was recognized best, the clone responded to several peptides from murine hsp 60. Hence, it may sense a mosaic of related peptides presented by stressed cells. Because of this, the cell surface density of epitopes derived from a single protein can be amplified, thus promoting TCR aggregation. T-cell activation can therefore occur even if the density of a single MHC/peptide complex is lower than normally required for activation of a highly specific T cell. Hence, reactivity to a cluster of peptides increases the chance of cross-reactive responses with the potential risk of autoaggression. It remains to be established whether such mosaic-type recognition of related peptides to-

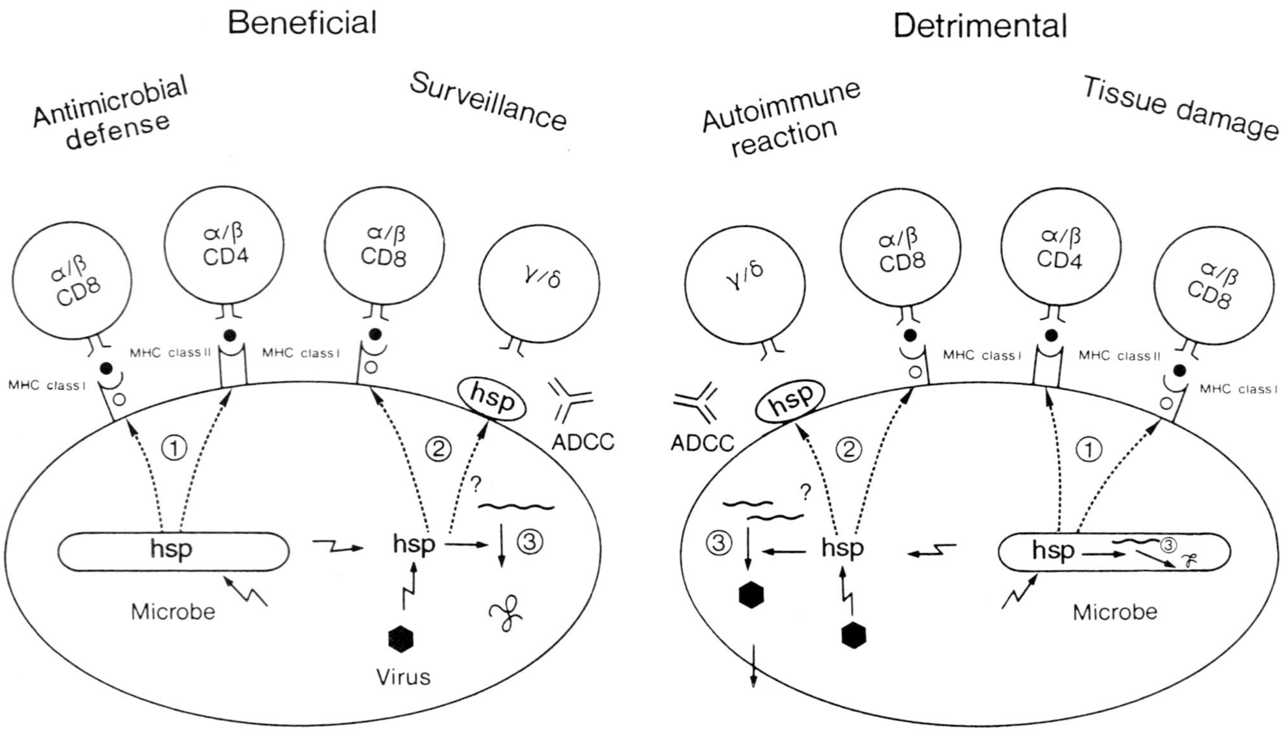

Figure 1 Left: microbial and cellular hsp can participate in host defense as antigens (1 and 2) or as chaperones (3). Right: immune recognition of microbial hsps can cause irreversible tissue damage (1), and immune recognition of self-hsp can lead to autoimmune reactions (2). Chaperones can be exploited by microbial pathogens to stabilize their own proteins within a hostile environment, or by viral pathogens to promote their own assembly (3). ADCC, antibody-dependent cellular cytotoxicity; ↗ stress; (1) presentation of microbial hsps by MHC class I and class II encoded proteins. For simplicity, cellular compartments and pathways of MHC loading are not shown; (2) presentation of self-hsp by MHC class I; (3) chaperone function of hsp.

gether with the high conservation of hsps provide general rules for hsp-related autoimmune responses.

VI. CONCLUSIONS

We have tried to outline the unparalleled features of hsps, as compared with ordinary proteins, in relation to host defense against microbial pathogens on two levels: (1) their chaperone functions in the pathogen and in the host, (2) their high conservation causing cross-reactive immune responses. The effects of hsps as chaperones and antigens following infections with microbial pathogens are summarized in Figure 1. Because hsps contribute to the survival of pathogens, hsp-gene deletions could be useful for attenuating pathogens to be used as vaccines. This assumption is based on the chaperone function of hsps, where microbial hsps refold proteins during host defense and thereby promote survival of the invader. Owing to boosting of anti-hsp immune responses by frequent encounter with different microbes expressing hsps of high sequence similarity, hsps can rapidly mobilize immunity against pathogens. This is being exploited for protein-vaccines, where hsps serve as immunogenic carriers to improve the antigenicity of an unrelated antigen. Hsps themselves can be protective antigens, as was illustrated in this chapter by several examples. An obstacle for the general use of hsps as vaccine antigens or carriers are their similarities with their mammalian hsp cognates. The development of autoimmune reactions has been documented in several instances. Such autoimmune reactions may contribute to surveillance of abnormally altered cells, but segregation from autoimmune disease remains an essential task difficult to achieve.

ACKNOWLEDGMENTS

This work received financial support by German Research Foundation SFB 322 (B7). We thank A. Gritzan for secretarial help.

REFERENCES

1. P.W. Ewald, *Evolution of Infectious Disease*, Oxford University Press, New York, 1994.
2. A. Mitchison, Will we survive, *Sci. Am. 269*:136 (1993).
3. P. Marrack and J.W. Kappler, How the immune system recognizes the body, *Sci. Am. 269*:80 (1993).
4. G.J.V. Nossal, Negative selection of lymphocytes, *Cell 76*:229 (1994).
5. H. von Boehmer, Positive selection of lymphocytes, *Cell 76*:219 (1994).
6. S.H.E. Kaufmann and B. Schoel, Heat shock proteins as antigens in immunity against infection and self, *The Biology of Heat Shock Proteins and Molecular*

Chaperones (R.I. Morimoto, A. Tissiéres and C. Georgopoulos, eds.), Cold Spring Harbor Laboratory Press, Cold Spring Harbor, NY, 1994, p. 495.
7. C.A. Janeway, How the immune system recognizes invaders, *Sci. Am. 269*:72 (1993).
8. E.G. Pamer, M.J. Bevan, and K. Fischer Lindahl, Do nonclassical, class I β MHC molecules present bacterial antigens to T cells, *Trends Microbiol. 1*:35 (1993).
9. P.J. Bjorkman, M.A. Saper, B. Samraoui, W.S. Bennett, J.L. Strominger, and D. C. Wiley, Structure of the human class I histocompatibility antigen, HLA-A2, *Nature 329*:506 (1987).
10. J.H. Brown, T.S. Jardetzky, J.C. Gorga, L.J. Stern, R.G. Urban, J.L. Strominger, and D. C. Wiley, Three-dimensional structure of the human class II histocompatibility antigen HLA-DR1, *Nature 364*:33 (1993).
11. H.G. Rammensee, K. Falk, and O. Rötzschke, Peptides naturally presented by MHC class I molecules, *Annu. Rev. Immunol. 11*:213 (1993).
12. V.H. Engelhard, Structure of peptides associated with class I and class II MHC molecules, *Annu. Rev. Immunol. 12*:181 (1994).
13. W.E. Paul, and R.A. Seder, Lymphocyte responses and cytokines, *Cell 76*:241 (1994).
14. R.N. Germain, MHC-dependent antigen processing and peptide presentation: providing ligands for T lymphocyte activation, *Cell 76*:287 (1994).
15. W.E. Paul, Infectious diseases and the immune system, *Sci. Am. 269*:90 (1993).
16. S.H.E. Kaufmann, Immunity to intracellular bacteria, *Fundamental Immunology* (W. E. Paul, ed.), Raven Press, New York, 1993, p. 1251.
17. S.H.E. Kaufmann, Immunity to intracellular bacteria, *Annu. Rev. Immunol. 11*:129 (1993).
18. W. Haas, P. Pereira, and S. Tonegawa, Gamma/delta cells, *Annu. Rev. Immunol. 11*:637 (1993).
19. H.J. Schild, N. Mavaddat, C. Litzenberger, E.W. Ehrich, M.M. Davis, J.A. Bluestone, L. Matis, R.K. Draper, and Y.-H. Chien, The nature of major histocompatibility complex recognition by γδ T cells, *Cell 76*:29 (1994).
20. A. Haregewoin, G. Soman, R.C. Hom, and R.W. Finberg, Human γ/δ^+ T cells respond to mycobacterial heat-shock protein, *Nature 340*:309 (1989).
21. W. Born, L. Hall, A. Dallas, J. Boymel, T. Shinnick, D. Young, P. Brennan, and R. O'Brien, Recognition of a peptide antigen by heat shock-reactive γ/δ T lymphocytes, *Science 249*:67 (1990).
22. P. Fisch, M. Malkovsky, S. Kovats, E. Sturm, E. Braakman, B.S. Klein, S. D. Voss, L.W. Morissey, R.DeMars, W.J. Welch, R.L.H. Bolhuis, and P.M. Sondel, Recognition by human Vγ9/Vδ2 T cells of a GroEL homolog on Daudi Burkitt's lymphoma cells, *Science 250*:1269 (1990).
23. Y.-X. Fu, R. Cranfill, M. Vollmer, R. Van der Zee, R. L. O'Brien, and W. Born, In vivo responses of murine γ/δ T cells to heat shock protein-derived peptide, *Proc. Natl. Acad. Sci. USA 90*:322 (1993).
24. I. Kaur, S.D. Voss, R.S. Gupta, K. Schell, P. Fisch, and P.M. Sondel, Human peripheral gamma/delta T cells recognize hsp60 molecules on Daudi Burkitt's lymphoma cells, *J. Immunol. 150*:2046 (1993).

25. P. Constant, F. Davodeau, M.-A. Peyrat, Y. Poquet, G. Puzo, M. Bonneville, and J.-J. Fournie, Stimulation of human γδ T cells by nonpeptidic mycobacterial ligands, *Science 264*:267 (1994).
26. B. Schoel, S. Sprenger, and S.H.E. Kaufmann, Phosphate is essential for stimulation of Vγ9Vδ2 T lymphocytes by mycobacterial low molecular weight ligand, *Eur. J. Immunol. 24*:1886 (1994).
27. S.H.E. Kaufmann, C. Blum, and S. Yamamoto, Crosstalk between α/β T cells and γ/δ T cells in vivo: Activation of α/β T cell responses after γ/δ T cell modulation with the monoclonal antibody GL3, *Proc. Natl. Acad. Sci. USA 90*:9620 (1993).
28. F.U. Hartl, J. Martin, and W. Neupert, Protein folding in the cell: the role of molecular chaperones hsp70 and hsp60, *Annu. Rev. Biophys. Biomol. Struct. 21*:293 (1992).
29. E.A. Craig, B.D. Cambill, and R.J. Nelson, Heat shock proteins: Molecular chaperones of protein biogenesis, *Microbiol. Rev. 57*:402 (1993).
30. E.A. Craig, J.S. Weissman, and A.L. Horwich, Heat shock proteins and molecular chaperones: mediators of protein conformation and turnover in the cell, *Cell 78*:365 (1994).
31. I.G. Haas, and M. Wabl, Immunoglobulin heavy chain binding protein, *Nature 306*:387 (1983).
32. J. Melnick, J.L. Dul, and Y. Argon, Sequential interaction of the chaperones BiP and GRP94 with immunoglobulin chains in the endoplasmic reticulum, *Nature 370*:373 (1994).
33. S.H.E. Kaufmann, Heat shock proteins and the immune response. *Immunol. Today 11*:129 (1990).
34. S.H.E. Kaufmann, B. Schoel, A. Wand-Württenberger, U. Steinhoff, M.E. Munk, and T. Koga, T cells, stress proteins and pathogenesis of mycobacterial infections, *Curr. Topics Microbiol. Immunol. 155*:125 (1990).
35. S.H.E. Kaufmann, Heat shock proteins and pathogenesis of bacterial infections, *Springer Semin. Immunopathol. 13*:25 (1991).
36. A.R. Lussow, C. Barrios, J.D.A. Van Embden, R. Van der Zee, A.S. Verdini, A. Pessi, J.A. Louis, P.-H. Lambert, and G. Del Giudice, Mycobacterial heat-shock proteins as carrier molecules, *Eur. J. Immunol. 21*:2297 (1991).
37. C. Barrios, A.R. Lussow, J.D.A. Van Embden, R. Van der Zee, R. Rappuoli, P. Costantino, J.A. Louis, P.-H. Lambert, and G. Del Giudice, Mycobacterial heat-shock proteins as carrier molecules. II. The use of the 70-kDa mycobacterial heat-shock protein as carrier for conjugated vaccines can circumvent the need for adjuvants and Bacillus Calmette Guérin priming, *Eur. J. Immunol. 22*:1365 (1992).
38. L. Steinman, Autoimmune disease, *Sci. Am. 269*:106 (1993).
39. R. Rajasekar, G.-K. Sim, and A. Augustin, Self heat shock and γ/δ T-cell reactivity, *Proc. Natl. Acad. Sci. USA 87*:1767 (1990).
40. A. VanBuskirk, B.L. Crump, E. Margoliash, and S.K. Pierce, A peptide binding protein having a role in antigen presentation is a member of the hsp 70 heat shock familiy, *J. Exp. Med. 170*:1799 (1989).
41. W. Jarjour, L.A. Mizzen, W.J. Welch, S. Denning, M. Shaw, T. Mimura, B.F. Haynes, and J.B. Winfield, Constitutive expression of a groEL-related protein on the surface of human cells, *J. Exp. Med. 172*:1857 (1990).

42. A. Wand-Württenberger, B. Schoel, J. Ivanyi, and S.H.E. Kaufmann, Surface expression by mononuclear phagocytes of an epitope shared with mycobacterial heat shock protein 60, *Eur. J. Immunol. 21*:1089 (1991).
43. F.M. Erkeller-Yueksel, D.A. Isenberg, V.B. Dhillon, D.S. Latchman, and P.M. Lydyard, Surface expression of heat shock protein 90 by blood mononuclear cells from patients with systemic lupus erythematosus. *J. Autoimmun. 5*:803 (1992).
44. M. Ferrarini, S. Heltai, M.R. Zocchi, and C. Rugarli, Unusual expression and localization of heat-shock proteins in human tumor cells, *Int. J. Cancer 51*:613 (1992).
45. S. Di Cesare, F. Poccia, A. Mastino, and V. Colizzi, Surface expressed heat-shock proteins by stressed or human immunodeficiency virus (HIV)–infected lymphoid cells represent the target for antibody-dependent cellular cytotoxicity, *Immunology 76*:341 (1992).
46. F. Poccia, P. Piselli, S. Di Cesare, S. Bach, V. Colizzi, M. Mattei, A. Bolognesi, and F. Stirpe, Recognition and killing of tumor cells expressing heat shock protein 65 kD with immunotoxins containing saporin, *Br. J. Cancer 66*:427 (1992).
47. P. Brandtzaeg, L.M. Sollid, P.S. Thrane, D. Kvale, K. Bjerke, H. Scott, K. Kett, and T.O. Rognum, Lymphoepithelial interactions in the mucosal immune system, *Gut 29*:1116 (1988).
48. J.P. Allison and W.L. Havran, The immunobiology of T cells with invariant γ/δ antigen receptors, *Annu. Rev. Immunol. 9*:679 (1991).
49. N. Kobayashi, G. Matsuzaki, Y. Yoshikai, R. Seki, J. Ivanyi, and K. Nomoto, $V\delta5^+$ T cells of BALB/c mice recognize the murine heat shock protein 60 target cell specificity, *Immunology 81*:240 (1994).
50. M.E. Munk, B. Schoel, S. Modrow, R.W. Karr, R.A. Young, and S.H.E. Kaufmann, Cytolytic T lymphocytes from healthy individuals with specificity to self epitopes shared by the mycobacterial and human 65 kDa heat shock protein, *J. Immunol. 143*:2844 (1989).
51. J.R. Lamb, V. Bal, P. Mendez-Samperio, A. Mehlert, J. Rothbard, S. Jindal, R. A. Young, and D.B. Young, Stress proteins may provide a link between the immune response to infection and autoimmunity, *Int. Immunol. 1*:191 (1989).
52. T. Koga, A. Wand-Württenberger, J. DeBruyn, M.E. Munk, B. Schoel, and S.H.E. Kaufmann, T cells against a bacterial heat shock protein recognize stressed macrophages, *Science 245*:1112 (1989).
53. R.L. O'Brien, Y.-X. Fu, R. Cranfill, A. Dallas, C. Ellis, C. Reardon, J. Lang, S.R. Carding, R. Kubo, and W. Born, Heat shock protein hsp60-reactive γ/δ cells: a large, diversified T-lymphocyte subset with highly focused specificity, *Proc. Natl. Acad. Sci. USA 89*:4348 (1992).
54. T.S. Jardetzky, W.S. Lane, R.A. Robinson, D.R. Madden, and D.C. Wiley, Identification of self peptides bound to purified HLA-B27, *Nature 353*:326 (1991).
55. C.A. Nelson, R.W. Roof, D.W. McCourt, and E.R. Unanue, Identification of the naturally processed form of hen egg white lysozome bound to the murine major histocompatibility complex class II molecule I-A^k, *Proc. Natl. Acad. Sci. USA 89*:7380 (1992).
56. R.M. Chicz, R.G. Urban, J.C. Gorga, D.A.A. Vignali, W.S. Lane, and J.L.

Strominger, Specificity and promiscuity among naturally processed peptides bound to HLA-DR alleles. *J. Exp. Med. 178*:27 (1993).

57. J.R. Newcomb, and P. Cresswell, Characterization of endogenous peptides bound to purified HLA-DR molecules and their absence from invariant chain-associated α/β dimers, *J. Immunol. 150*:499 (1993).
58. H.L. Weiner, A. Friedman, A. Miller, S. J. Khoury, A. Al-Sabbagh, L. Santos, M. Sayegh, R.B. Nussenblatt, D.E. Trentham, and D.A.Hafler, Oral tolerance: immunologic mechanisms and treatment of animal and human organ-specific autoimmune diseases by oral administration of autoantigens, *Annu. Rev. Immunol. 12*:809 (1994).
59. R.A. Young and T.J. Elliott, Stress proteins, infection, and immune surveillance, *Cell 59*:5 (1989).
60. R.A. Young, Stress proteins and immunology, *Annu. Rev. Immunol. 8*:401 (1990).
61. D.B. Young and A. Mehlert, Stress proteins and infectious diseases, *Stress Proteins in Biology and Medicine* (R. Morimoto et al., ed.), Cold Spring Harbor Laboratory Press, Cold Spring Harbor, NY, 1990, p. 131.
62. M. Jäättela, Effects of heat shock on cytolysis mediated by NK cells, LAK cells, activated monocytes and TNFs-α and β, *Scand. J. Immunol. 31*:175 (1990).
63. S.H.E. Kaufmann, B. Schoel, T. Koga, A. Wand-Württenberger, M.E. Munk, and U. Steinhoff, Heat shock protein 60: implications for pathogenesis of and protection against bacterial infections, *Immunol. Rev. 121*:67 (1991).
64. U. Steinhoff, A. Wand-Württenberger, A. Bremerich, and S.H.E. Kaufmann, *Mycobacterium leprae* renders Schwann cells and mononuclear phagocytes susceptible or resistant against killer cells, *Infect. Immun. 59*:684 (1991).
65. S. Kantengwa, Y.R.A. Donati, M. Clerget, I. Parini Maridonneau, F. Sinclair, E. Marethoz, A.D.M. Rees, D.O. Slosman, and B.S. Polla, Heat-shock proteins: An autoprotective mechanism for inflammatory cells? *Semin. Immunol. 3*:49 (1991).
66. M. Jäättela and D. Wissing, Heat-shock proteins protect cells from monocyte cytotoxicity: possible mechanism of self-protection, *J. Exp. Med. 177*:231 (1993).
67. W.R. Schwan, and W. Goebel, Host cell responses to *Listeria monocytogenes* infection include differential transcription of host stress genes involved in signal transduction, *Proc. Natl. Acad. Sci. USA 91*:6428 (1994).
68. J.F. Diedrich, R.I. Carp, and A.T. Haase, Increased expression of heat shock protein, transferrin, and β20-microglobin in astrocytes during scrapie, *Microbiol. Pathogen. 15*:1 (1993).
69. S.K. Pierce, D.C. De Nagel, and A.M. van Buskirk, A role for heat shock proteins in antigen processing and presentation, *Curr. Topics Microbiol. Immunol. 167*:83 (1991).
70. H. Udono, and P.K. Srivastava, Heat shock protein 70-associated peptides elicit specific cancer immunity, *J. Exp. Med. 178*:1391 (1993).
71. Z. Li, and P.K. Srivastava, Tumor rejection antigen gp96/grp94 is an ATPase: implications for protein folding and antigen presentation, *EMBO J. 12*:3143 (1993).
72. H. Udono and P.K. Srivastava, Comparison of tumor-specific immunogenicities of stress-induced proteins gp96, hsp90, and hsp70, *J. Immunol. 152*:5398 (1994).

73. Z. Li and P.K. Srivastava, A critical contemplation on the role of heat shock proteins in transfer of antigenic peptides during antigen presentation, *Behring Inst. Mitt. 94*:37 (1994).
74. R. Schirmbeck and J. Reimann, Peptide transporter-independent, stress protein-mediated endosomal processing of endogenous protein antigens for major histocompatibility complex class I presentation, *Eur. J. Immunol. 24*:1478 (1994).
75. M.E. Munk, B. Schoel, and S.H.E. Kaufmann, T cell responses of normal individuals towards recombinant protein antigens of *Mycobacterium tuberculosis*, *Eur. J. Immunol. 18*:1835 (1988).
76. H.P. Fischer, C.E.M. Sharrock, and G.S. Panayi, High frequency of cord blood lymphocytes against mycobacterial 65-kDa heat shock protein, *Eur. J. Immunol. 22*:1667 (1992).
77. G. Del Giudice, A. Gervaix, P. Costantino, C.-A. Wyler, C. Tougne, E.R. De Graeff-Meeder, J. van Embden, R. Van der Zee, L. Nencioni, R. Rappuoli, S. Suter, and P.-H. Lambert, Priming to heat shock proteins in infants vaccinated against pertussis. *J. Immunol. 150*:2025 (1993).
78. S.J. Blander and M.A. Horwitz, Major cytoplasmic membrane protein of *Legionella pneumophila*, a genus common antigen and member of the hsp 60 family of heat shock proteins, induces protective immunity in a guinea pig model of Legionnaire's disease, *J. Clin. Invest. 91*:717 (1993).
79. A. Noll, A. Rogenkamp, J. Heesemann, and I.B. Autenrieth, Protective role for heat shock protein-reactive αβ T cells in murine yersiniosis, *Infect. Immun. 62*: 2784 (1994).
80. C.L. Silva and D.B. Lowrie, A single mycobacterial protein (hsp 65) expressed by a transgenic antigen-presenting cell vaccinates mice against tuberculosis, *Immunology 82*:244 (1994).
81. R.C. Matthews, J.P. Burnie, and S. Tabaqchali, Immunoblot analysis of the serological response in systemic candidosis, *Lancet 2*:1415 (1984).
82. R. Matthews, J.P. Burnie, and W. Lee, The application of epitope mapping in the development of a new serological test for systemic candidosis, *J. Immunol. Methods 143*:73 (1991).
83. R. Matthews, D. Smith, J. Midgley, J. Burni, I. Clark, M. Conolly, and B. Gazzard, Candida and AIDS: evidence for protective antibody, *Lancet 2*:263 (1988).
84. R.C. Matthews, and J.P. Burnie, Cloning of a DNA sequence encoding a major fragment of the 47 kilodalton stress protein homologue of *Candida albicans*, *FEMS Microbiol. Lett. 60*:25 (1989).
85. R.C. Matthews, J.P. Burnie, and S. Tabaqchali, Isolation of immunodominant antigens from sera of patients with systemic candidiasis and characterization of serological response to *Candida albicans, J. Clin. Microbiol. 25*:230 (1987).
86. R.C. Matthews, J.P. Burnie, D. Howat, T. Rowland, and F. Walton, Autoantibody to heat shock protein 90 can mediate protection against systemic candidosis, *Immunology 74*:20 (1991).
87. R.C. Matthews and J. Burnie, The role of hsp90 in fungal infection, *Immunol. Today 13*:345 (1992).
88. U. Steinhoff, B. Schoel, and S.H.E. Kaufmann, Lysis of interferon-γ activated

Schwann cells by crossreactive CD8α/β T cells with specificity to the mycobacterial 65 kDa heat shock protein, *Int. Immunol. 2*:279 (1990).
89. B. Schoel, U. Zügel, T. Ruppert, and S.H.E. Kaufmann, Elongated peptides, not the predicted nonapeptide stimulate a major histocompatibility complex class I-restricted cytotoxic T lymphocyte clone with specificity for a bacterial heat shock protein, *Eur. J. Immunol. 24*:3161 (1994).
90. U. Zügel, B. Schoel, S. Yamamoto, H. Hengel, B. Morein, and S.H.E. Kaufmann, Crossrecognition by CD8 TcR α/β CTL of peptides in the self and the mycobacterial hsp60 which share intermediate sequence homology, *Eur. J. Immunol., 25*:451 (1995).
91. L. Chouchane, F.S. Bowers, S. Sawasdikosol, R.M. Simpson, and T.J. Kindt, Heat-shock proteins expressed on the surface of human T cell leukemia virus type I-infected cell lines induce autoantibodies in rabbits, *J. Infect. Dis. 169*:253 (1994).
92. S. Jindal and M. Malkovsky, Stress responses to viral infection, *Trends Microbiol. 2*:89 (1994).
93. M. Shapira, J.G. McEwen, and C.L. Jaffe, Temperature effects on molecular processes which lead to stage differentiation in *Leishmania, EMBO J. 7*:2895 (1988).
94. B. Maresca and G.S. Kobayashi, Dimorphism in *Histoplasma capsulatum*: a model for the study of cell differentiation in pathogenic fungi, *Microbiol. Rev. 53*:186 (1989).
95. B. Maresca, L. Carratù, and G.S. Kobayashi, Morphological transition in the human fungal pathogen *Histoplasma capsulatum, Trends Microbiol. 2*:110 (1994).
96. A.M. Lambowitz, G.S. Kobayashi, A. Painter, and G. Medoff, Possible relationship of morphogenesis in pathogenic fungus, *Histoplasma capsulatum*, to heat shock response, *Nature 303*:806 (1983).
97. G. Medoff, B. Maresca, A.M. Lambowitz, G. Kobayashi, A. Painter, M. Sacco, and L. Carratu, Correlation between pathogenicity and temperature sensitivity in different strains of *Histoplasma capsulatum, J. Clin. Invest. 78*:1638 (1986).
98. G. Medoff, G.S. Kobayashi, A. Painter, and S. Travis, Morphogenesis and pathogenicity of *Histoplasma capsulatum, Infect. Immun. 55*:1355 (1987).
99. E.J. Patriarca, G.S. Kobayashi, and B. Maresca, Mitochondrial activity and heat-shock response during morphogenesis in the pathogenic fungus *Histoplasma capsulatum, Biochem. Cell Biol. 70*:207 (1992).
100. M.F. Christman, R.W. Morgan, F.S. Jacobson, and B.N. Ames, Positive control of a regulon for defenses against oxidative stress and some heat shock proteins in *Salmonella typhimurium, Cell 41*:753 (1985).
101. N.A. Buchmeier and F. Heffron, Induction of *Salmonella* stress proteins upon infection of macrophages, *Science 248*:730 (1990).
102. P.I. Fields, R.V. Swanson, C.G. Haidaris, and F. Heffron, Mutants of *Salmonella typhimurium* that cannot survive within the macrophage are avirulent. *Proc. Natl. Acad. Sci. USA 83*:5189 (1986).
103. K. Johnson, I. Charles, G. Dougan, D. Pickard, P. O'Gaora, G. Costa, T. Ali, I. Miller, and C. Hormaeche, The role of a stress-response protein in *Salmonella typhimurium* virulence, *Mol. Microbiol. 5*:401 (1991).

104. J.R. Nevins, Induction of the synthesis of a 70,000 Dalton mammalian heat shock protein by the adenovirus E1A gene product, *Cell 29*:913 (1982).
105. S. Jindal and R.A. Young, Vaccinia virus infection induces a stress response that leads to association with hsp70 with viral proteins, *J. Virol. 66*:5357 (1992).
106. D.G. Macejak and P. Sarnow, Association of heat shock protein 70 with enterovirus capsid precursor P1 in infected human cells, *J. Virol. 66*:1520 (1992).
107. W.L. Kelley and S.J. Landry, Chaperone power in a virus? *TIBS 19*:277 (1994).
108. C.P. Georgopoulos, R.W. Hendrix, S.R. Casjens, and A.D. Kaiser, Host participation in bacteriophage lambda heat assembly, *J. Mol. Biol. 76*:45 (1973).
109. N. Sternberg, Properties of a mutant of *Escherichia coli* defective in bacteriophage λ head formation (*groE*), *J. Mol. Biol. 76*:25 (1973).
110. C.P. Georgopoulos, and B. Horn, Identification of a host protein necessary for bacteriophage morphogenesis (the *groE* gene product), *Proc. Natl. Acad. Sci. USA 75*:131 (1978).
111. R.W. Hendrix and L. Tsui, Role of the host in virus assembly: Cloning of the *Escherichia coli groE* gene and identification of its protein product, *Proc. Natl. Acad. Sci. USA 75*:136 (1978).
112. U. Steinhoff and S.H.E. Kaufmann, Specific lysis by $CD8^+$ T cells of Schwann cells expressing *Mycobacterium leprae* antigens, *Eur. J. Immunol. 18*:969 (1988).
113. Y. Mistry, D.B. Young, and R. Mukherjee, Hsp70 synthesis in Schwann cells in response to heat shock and infection with *Mycobacterium leprae, Infect. Immun. 60*:3105 (1992).
114. A.L. Ford, W.J. Britton, and P.J. Armati, Schwann cells are able to present exogenous mycobacterial hsp70 to antigen-specific T lymphocytes, *J. Neuroimmunol. 43*:151 (1993).
115. R.P. Morrison, R.J. Belland, K. Lyng, and H.D. Caldwell, Chlamydial disease pathogenesis. The 57-kD chlamydial hypersensitivity antigen is a stress response protein, *J. Exp. Med. 170*:1271 (1989).
116. H.R. Taylor, I.W. Maclean, R.C. Brunham, S. Pal, and J. Wittum-Hudson, Chlamydial heat shock proteins and trachoma, *Infect. Immun. 58*:3061 (1990).
117. B.E. Dunn, R.M. Roop II, C.C. Sung, S.A. Sharma, G.I. Perez-Perez, and M.J. Balser, Identification and purification of a cpn60 heat shock protein homolog from *Helicobacter pylori, Infect. Immun. 60*:1946 (1992).
118. H. Takahashi, T. Takeshita, B. Morein, S. Putney, R.N. Germain, and J.A. Berzofsky, Induction of $CD8^+$ cytotoxic T cells by immunization with purified HIV-1 envelope protein in ISCOMs, *Nature 344*:873 (1990).
119. K. Lovgren, and Bror Morein, The ISCOM: an antigen delivery system with built-in adjuvant, *Mol. Immunol. 28*:285 (1991).
120. M. Matsumura, D.H. Fremont, P.A. Peterson, and I.A. Wilson, Emerging principles for the recognition of peptide antigens by MHC class I molecules, *Science 257*:927 (1992).
121. A.C.M. Young, W. Zhang, J.C. Sacchettini, and S.G. Nathenson, The three-dimensional structure of H-$2D^b$ at 2.4 A resolution: implications for antigen-determinant selection, *Cell 76*:39 (1994).
122. U. Steinhoff, U. Zügel, A. Wand-Württenberger, H. Hengel, R. Rösch, M.E. Munk, and S. H. E. Kaufmann, Prevention of autoimmune lysis by T cells with

specificity for a heat shock protein by antisense oligonucleotide treatment, *Proc. Natl. Acad. Sci. USA 91*:5085 (1994).

123. D.B. Young, R.B. Lathigra, R.W. Hendrix, D. Sweetser, and R.A. Young, Stress proteins are immune targets in leprosy and tuberculosis, *Proc. Natl. Acad. Sci. USA 85*:4267 (1988).
124. T.M. Shinnick, M.H. Vodkin, and J.C. Williams, The *Mycobacterium tuberculosis* 65-kilodalton antigen is a heat shock protein which corresponds to common antigen and to the *Escherichia coli* GroEL protein, *Infect. Immun. 56*:446 (1988).
125. A. Verbon, R.A. Hartskeerl, A. Schuitema, A.H.J. Kolk, D.B. Young, and R. Lathigra, The 14,000-molecular-weight antigen of *Mycobacterium tuberculosis* is related to the alpha-crystallin family of low-molecular-weight heat shock proteins, *J. Bacteriol. 174*:1352 (1992).
126. P.N. Baird, L.M.C. Hall, and A.R.M. Coates, A major antigen from *Mycobacterium tuberculosis* which is homologous to the heat shock proteins groES from *E. coli* and the htpA gene product of *Coxiella burneti, Nucleic Acids Res. 16*:9047 (1988).
127. R.J. Garsia, L. Hellqvist, R.J. Booth, A.J. Radford, W.J. Britton, L. Astbury, R.J. Trent, and A. Basten, Homology of the 70-kilodalton antigens from *Mycobacterium leprae* and *Mycobacterium tuberculosis* 71-kilodalton antigen and with the conserved heat shock protein 70 of eucaryotes. *Infect. Immun. 57*:204 (1989).
128. A.N. Nerland, A.S. Mustafa, D. Sweetser, T. Godal, and R.A. Young, A protein antigen of *Mycobacterium leprae* is related to a family of small heat shock proteins. *J. Bacteriol. 170*:5919 (1988).
129. V. Mehra, B.R. Bloom, A.C. Bajardi, C.L. Grisso, P.A. Sieling, D. Alland, J. Convit, X. Fan, S.W. Hunter, P.J. Brennan, T.H. Rea, and R.L. Modlin, A major T cell antigen of *Mycobacterium leprae* is a 10-kD heat-shock cognate protein, *J. Exp. Med. 175*:275 (1992).
130. S.L. Danilition, I.W. Maclean, R. Peeling, S. Winston, and R.C. Brunham, The 75-kilodalton protein of *Chlamydia trachomatis*: a member of the heat shock protein 70 family, *Infect. Immun. 58*:189 (1990).
131. J. Anzola, B.J. Luft, G. Gorgone, R.J. Battwyler, C. Soderberg, R. Lahesmaa, and G. Peltz, *Borrelia burgdorferi* HSP70 homolog: characterization of an immunoreactive stress protein, *Infect. Immun. 60*:3704 (1992).
132. K. Hansen, J.M. Bangsborg, H. Fjordvang, N.S. Pedersen, and P. Hindersson, Immunochemical characterization of, and isolation of the gene for a *Borrelia burgdorferi* immunodominant 60-kilodalton antigen common to a wide range of bacteria, *Infect. Immun. 56*:2047 (1988).
133. M.H. Vodkin and J.C. Williams, A heat shock operon in *Coxiella burnetii* produces a major antigen homologous to a protein in both *Mycobacteia* and *Escherichia coli, J. Bacteriol. 170*:1227 (1988).
134. P. Hindersson, J.D. Knudsen, and N.H. Axelsen, Cloning and expression of *Treponema pallidum* common antigen (Tp-4) in *E. coli* K-12, *J. Gen. Microbiol. 133*:587 (1987).
135. P.S. Hoffman, C.A. Butler, and F.D. Quinn, Cloning and temperature-dependent expression in *Escherichia coli* of a *Legionella pneumophila* gene coding for a genus-common 60-kilodalton antigen, *Infect. Immun. 57*:1731 (1989).

136. R.M. Roop III, M.L. Price, B.E. Dunn, S.M. Boyle, N. Sriranganathan, and G.G. Schurig, Molecular cloning and nucleotide sequence analysis of the gene encoding the immunoreactive *Brucella abortus* hsp60 protein, BA60K, *Microbiol. Pathogen. 12*:47 (1992).
137. Y. Pannekoek, I.G. Schuurman, J. Dankert, and J.P. van Putten, Immunogenicity of the meningococcal stress protein MSP63 during natural infection, *Clin. Exp. Immunol. 93*:377 (1993).
138. J.P. Burnie and R.C. Matthews, Heat shock protein 88 and Aspergillus infection, *J. Clin. Microbiol. 29*:2099 (1991).
139. G. Minchiotti, S. Gargano, and B. Maresca, Molecular cloning and expression of hsp82 gene of the dimorphic pathogenic fungus *Histoplasma capsulatum, Biochim. Biophys. Acta 1131*:103 (1992).
140. F.J. Gomez, A.M. Gomez, and G.S.J. Deepe, An 80-kilodalton antigen from *Histoplasma capsulatum* that has homology to heat shock protein 70 induces cell-mediated immune responses and protection in mice, *Infect. Immun. 60*:2565 (1992).
141. G. Shearer, Jr., C.H. Birge, P.D. Yuckenberg, G.S. Kobayashi, and G. Medoff, Heat-shock proteins induced during the mycelial-to-yeast transitions of strains of *Histoplasma capsulatum, J. Gen. Microbiol. 133*:3375 (1987).
142. M. Jendoubi and S. Bonnefoy, Identification of a heat shock-like antigen in *P. falciparum*, related to the heat shock protein 90 family, *Nucleic Acids Res. 16*: 10928 (1988).
143. A.E. Bianco, J.M. Favaloro, T.R. Burkot, J.G. Culvenor, P.E. Crewther, G.V. Brown, R.F. Anders, R.L. Coppel, and D.J. Kemp, A repetitive antigen of *Plasmodium falciparum* that is homolgous to heat shock protein 70 of *Drosophila melanogaster, Proc. Natl. Acad. Sci. USA 83*:8713 (1986).
144. F. Ardeshir, J.E. Flint, S.J. Richman, and R.T. Reese, A 75kd merozoite surface protein of *Plasmodium falciparum* which is related to the 70kd heat-shock proteins, *EMBO J. 6*:493 (1987).
145. D. Mattei, L.S. Ozaki, and L. Pereira da Silva, A *Plasmodium falciparum* gene encoding a heat shock-like antigen related to the rat 78 kD glucose-regulated protein, *Nucleic Acids Res. 16*:5204 (1988).
146. E.A. Dragon, S.R. Sias, E.A. Kato, and J.D. Gabe, The genome of *Trypanosoma cruzi* contains a constitutively expressed tandemly arranged multicopy gene homologous to a major heat shock protein. *Mol. Cell. Biol. 7*:1271 (1987).
147. D.M. Engman, L.V. Kirchhoff, and J.E. Donelson, Molecular cloning of mtp70, a mitochondrial member of the hsp70 family. *Mol. Cell. Biol. 9*:5163 (1989).
148. P.L. Yeyati, S. Bonnefoy, G. Mirkin, A. Debrabant, S. Lafon, A. Panebra, E. Gonzalez-Cappa, J.P. Dedet, M. Hontebeyrie-Joskowicz, and M.J. Levin, The 70-kDa heat-shock protein is a major antigenic determinant in human Trypanosoma cruzi/Leishmania braziliensis braziliensis mixed infection, *Immunol. Lett. 31*: 27(1992).
149. J. Mottram, W. Murphy, and N. Agabian, A transcriptional analysis of the *Trypanosoma brucei* hsp83 gene cluster, *Mol. Biochem. Parasitol. 37*:115 (1989).
150. D.J. Glass, R.I. Polvene, and L.H.T. van der Ploeg, Conserved sequences and

transcription of the hsp70 gene family in *Trypanosoma brucei, Mol. Cell. Biol. 6*:4657 (1986).

151. M. Shapira and G. Pedraza, Sequence analysis and transcriptional activation of heat shock protein 83 of *Leishmania mexicana amazonensis, Mol. Biochem. Parasitol. 42*:247 (1990).
152. J. MacFarlane, M.L. Blaxter, R.P. Bishop, M.A. Miles, and J.M. Kelly, Characterization of a *Leishmania donovani* antigen similar to heat shock protein 70, *Biochem. Soc. Trans. 17*:168 (1989).
153. J. MacFarlane, M.L. Blaxter, R.P. Bishop, M.A. Miles, and J.M. Kelly, Identification and characterization of a *Leishmania donovani* antigen belonging to the 70-kDa heat-shock protein family, *Eur. J. Biochem. 190*:377 (1990).
154. M.G. Lee, B.L. Atkinson, S.H. Giannini, and L.H.T. van der Ploeg, Structure and expression of the hsp 70 gene family of *Leishmania major, Nucleic Acids Res. 16*:9567 (1988).
155. K.S. Johnson, K. Wells, J.V. Bock, V. Nene, D.W. Taylor, and J.S. Cordingley, The 86-kilodalton antigen from *Schistosoma mansoni* is a heat-shock protein homologous to yeast hsp60, *Mol. Biochem. Parasitol. 36*:19 (1989).
156. R. Hedstrom, J. Culpepper, R.A. Harrison, N. Agabian, and G. Newport, A major immunogen in *S. mansoni* infections is homologous to the heat-shock protein hsp70, *J. Exp. Med. 165*:1430 (1987).
157. R. Hedstrom, J. Culpepper, V. Schinski, N. Agabian, and G. Newport, Schistosome heat-shock proteins are immunologically distinct host-like antigens, *Mol. Biochem. Parasitol. 29*:275 (1988).
158. V. Nene, D.W. Dunne, K.S. Johnson, D.W. Taylor, and J.S. Cordingley, Sequence and expression of a major egg antigen from *Schistosoma mansoni*: homologies to heat shock proteins and alpha-crystallins, *Mol. Biochem. Parasitol. 21*:179 (1986).
159. M.E. Selkirk, D.A. Denham, F. Partono, and R.M. Maizels, Heat shock cognate 70 is a prominent immunogen in Brugian filariasis, *J. Immunol. 143*:299 (1989).
160. N.M. Rothstein, G. Higashi, J. Yates, and T.V. Rajan, *Onchocerca volvulus* heat shock protein 70 is a major immunogen in amicrofilaremic individuals from a filariasis-endemic area, *Mol. Biochem. Parasitol. 33*:229 (1989).

3

Clinical Implications of the Stress Response

George Minowada and William J. Welch
University of California, San Francisco, San Francisco, California

I. INTRODUCTION

The heat shock response, first observed in fruit fly *Drosophila melanogaster* over 30 years ago, provided investigators a relatively simple way to study rapid changes in gene expression. Simply raising the temperature of *Drosophila* above its physiological norm resulted in the decreased expression of those genes which were active prior to the temperature shock and the increased expression of genes encoding a group of proteins referred to as the heat shock proteins (hsps). Over the last 30 years, similar changes in gene expression following relevant temperature shocks have been observed in cells from all organisms, be they derived from bacteria, plants, yeasts, or mammals. Moreover, the heat shock proteins from various organisms appear highly conserved with respect to their primary structure, mode of regulation, and biochemical function. In addition to heat shock treatment, many other types of metabolic insults, including exposure to heavy metals, amino acid analogues, different metabolic poisons as well as a variety of relevant insults in vivo (e.g., ischemia/reperfusion), also elicit increased expression of the hsps. Accordingly, many investigators now refer to the response more generally as the stress response and the proteins whose expression increases as the stress proteins.

As one might predict, the stress response represents a universally conserved cellular defense program. Perhaps the best example of how the stress response

This chapter is reproduced from *The Journal of Clinical Investigation*, 1995, 95:3–12, by copyright permission of the Society for Clinical Investigation.

provides for increased cellular protection is illustrated by the phenomenon of "acquired thermotolerance." Cells subjected to a sublethal heat shock treatment, if provided a subsequent recovery period at their normal growth temperature, now are able to survive a second and what would otherwise be a lethal heat shock challenge. Acquired thermotolerance is usually transient, lasting about 24 hr in cells grown in culture, and appears dependent on a number of changes induced by the initial or "priming" heat shock treatment, including the increased expression and accumulation of the stress proteins. We now know that most of the hsps also are expressed constitutively in normal or "unstressed" cells where they play a role in a number of fundamental biological processes. For example, many of the constitutively expressed hsps function as so-called "molecular chaperones," facilitating various aspects of protein maturation throughout the cell. The stress response/proteins also play an important role in a number of clinically relevant phenomenon, including tissue and organ trauma and the immune response. In this chapter, we discuss the clinical implications of the heat shock response/proteins beginning with a brief summary of the structure/function of the hsps and then turn toward more recent developments with potential clinical ramifications. (Because of space limitation, we cannot cite all of the appropriate references and, therefore, refer the reader to other recent reviews whenever possible.)

II. INDUCTION OF THE HEAT SHOCK RESPONSE.

As mentioned above, a diverse array of metabolic insults, including the exposure of cells to heavy metals, various ionophores, amino acid analogues, and metabolic poisons which target ATP production, all result in similar changes in gene expression leading to the accumulation of the stress proteins. Many of these agents/treatments which induce a stress response share the common property of being "protein chaotropes"; they adversely affect the proper conformation and therefore the function of proteins. Consequently, it was proposed (1) and later demonstrated that under conditions where abnormally folded proteins might begin to accumulate in the cell, a stress response would be initiated (2). Presumably the resultant increases in the levels of the stress proteins would somehow facilitate the identification, removal, and/or restoration of those proteins adversely affected by the particular stress event; again a prediction which now appears correct. The intracellular accumulation of abnormally folded proteins initiates the stress response by activating a specific transcription factor, referred to as the heat shock factor (HSF-1) (reviewed in ref. 3). HSF-1, present in the normal, unstressed cell as an inactive monomer, rapidly trimerizes in response to metabolic stress. Trimerization enables HSF-1 to bind to a consensus nucleotide sequence, referred to as the heat shock element (HSE), located within the promoter element of those genes encoding the stress proteins, thereby resulting in the high-level transcription of the heat shock genes. Recently, another related transcription factor, HSF-2, has been identified. Shown to be important in regulating the expression of hsp transcription during hemin-induced differentiation of K562 cells (4), HSF 2 also may turn out to be important for controlling the activities of hsp gene expression in the normal or unstressed cell.

III. OVERVIEW OF THE STRUCTURE AND FUNCTION OF THE STRESS PROTEINS

A. Types of Stress Proteins

The mammalian stress proteins often are divided into two groups based on their classic mode of induction: the heat shock proteins (hsps) and the glucose-regulated proteins (grps) (reviewed in refs. 5 and 6). In general, both the hsps and grps are identified by their apparent molecular mass as determined by SDS-PAGE (Table 1). Members of the grp family were first observed to exhibit increased expression in cells starved of glucose. Subsequently, the grps were shown to also undergo high-level expression in cells treated with calcium ionophores, when subjected to anoxic-like conditions, or in response to added reducing agents such as B-mercaptoethanol. Interestingly, many of the agents/treatments which induce the grps, most of which reside within the secretory pathway of cells, adversely affect protein secretion. We now know that the various grps are in fact related to members of the hsp family, with the two families of stress proteins exhibiting considerable similarities with respect to both their structure and function.

Our understanding of the structure and function of a number of the stress proteins has advanced rapidly in recent years, sometimes from unexpected sources. For example, several proteins whose functions had already been defined subsequently were shown to be synthesized at higher levels in the cell under stress (reviewed in ref. 6). Ubiquitin, a small polypeptide of approximately 8 kDa known to be involved in targeting proteins for degradation, is synthesized at relatively high rates after heat shock. Hsp 56, characterized initially as part of a larger protein complex that binds steroid hormone receptors (reviewed in ref. 7), and now known to bind to the immunosuppressants FK506 (8) and rapamycin, shows increased expression after stress (9). Like other identified FK506 binding proteins, hsp 56 appears to function as a peptidylprolyl *cis-trans* isomerase (rotamase). Other examples of proteins whose function was already known and subsequently shown to represent members of the stress protein family include heme oxygenase, involved in catalyzing the breakdown of heme into bilverdin; the multiple drug-resistant gene product or P-glycoprotein, a plasma membrane transporter involved in multidrug resistance to many cancer chemotherapeutic agents; alpha B-crystallins, integral structural components of the lens

Table 1

Name	Size (kDa)[a]	Bacterial homologue	Locale	Remarks
Ubiquitin	8	–	Cytosol/nucleus	Involved in nonlysosomal protein degradation pathway
Hsp 10	10	GroES	Mitochondria/chloroplast	Cofactor for Hsp 60
Low molecular weight hsps	20–30	Possible homologues	Cytosol/nucleus recently identified	Proposed regulator of actin cyto-skeleton; proposed molecular chaperone
Hsp 47	47	–	Endoplasmic reticulum	Collagen chaperone
Hsp 56	56	–	Cytosol	Part of steroid hormone receptor complex; binds FK506
Hsp 60	60	GroEL	Mitochondria/chloroplast	Molecular chaperone ("chaperonin")
TCP-1	60	GroEL	Cytosol/nucleus	Molecular chaperone related to Hsp 60
Hsp 72	70	DnaK	Cytosol/nucleus	Highly stress inducible
Hsp 73	70	DnaK	Cytosol/nucleus	Constitutively expressed molecular chaperone
Grp 75	70	DnaK	Mitochondria/chloroplast	Constitutively expressed molecular chaperone
Grp 78 (BiP)	70	DnaK	Endoplasmic reticulum	Constitutively expressed molecular chaperone
Hsp 90	90	htpG	Cytosol/nucleus	Part of steroid hormone receptor complex; chaperone (?) for retrovirus-encoded tyrosine protein kinases
Hsp 104/110	104/110	Clp family	Cytosol/nucleus	Required to survive severe stress; molecular chaperone (?)

[a]Approximate size by SDS-PAGE, native molecular weight is often very different.

and now known to be present in other cell types; and at least two glycolytic enzymes enolase and glyceraldehyde 3-phosphate dehydrogenase. For at least some of these proteins, we have some insights as to why they are upregulated after stress. For example, the glycolytic pathway provides an essential energy source in the event that aerobic respiration becomes uncoupled, which is often a consequence of metabolic stress. Consequently, increases in the synthesis of the aforementioned glycolytic enzymes may facilitate the increased demand on the glycolytic pathway after heat shock. Similarly, higher levels of ubiquitin may provide for an increased capacity to recognize and degrade irreparably damaged proteins after heat shock and/or other metabolic insults. We suspect that other proteins whose function have already been established also will be found to be stress inducible, perhaps in a cell type–specific manner.

1. *Hsp 60 and Hsp 70 Families*

As is presented in Table 1, the number of proteins induced in eukaryotic cells after stress is rather extensive and continues to grow. Instead of providing an overview of all of thesc proteins, we will focus our discussion on those stress proteins which have been best characterized and/or which have generated the most surprises as it relates to their structure/function. Perhaps the most exciting development regarding the function of the stress proteins involves the role of some of these proteins in protein biogenesis (reviewed in refs. 10 and 11). Here two families of stress proteins, the hsp 60 and hsp 70 families, each consisting of multiple and related members expressed constitutively in all cells, have been shown to participate directly in various aspects of protein maturation. Members of the hsp 70 family, distributed throughout various subcellular compartments and expressed in cells grown under normal conditions, include hsp 73 (also known as constitutive hsp 70, hsc 70, or hsp 70 cognate), present within the cytoplasm and nucleus; grp75, a component of the mitochondria; and grp 78 (also known as BiP) a resident of the endoplasmic reticulum (ER). In addition, under conditions of metabolic stress, another form of hsp 70, referred to as the highly stress-inducible hsp 70 (or hsp 72), is synthesized at very high levels, exhibits considerable homology to hsp 73, and like hsp 73 resides within the nucleus and cytoplasm. Although too extensive to discuss here, other proteins related to the hsp 70 family also have been identified, but their functions remain somewhat obscure. All members of the hsp 70 family have been shown to bind ATP through a highly conserved amino-terminal nucleotide binding domain,whose overall structure appears very similar to that of two other ATP binding proteins, actin and hexokinase. In addition, presumably via their carboxyterminal domain, hsp 70 family members appear to bind to both unfolded proteins and short polypeptides in vitro. Results from many laboratories have shown that members of the hsp 70 family, within their own distinct subcellular compartment, interact with other cellular proteins undergoing synthesis on the ribosome or translocation into organelles. These observations have led to the suggestion that the hsp 70 family members function in the early stages of protein maturation by binding to and stabilizing the unfolded state of a newly synthesized protein. Once synthesis or translocation of the target protein has been completed, the particular hsp 70 family member is released in a process requiring ATP and likely other proteinacious cofactors, thereby allowing the target protein to commence folding and/or assembly.

In a similar manner, members of the hsp 60 family also bind ATP and interact transiently with unfolded polypeptides (reviewed in refs. 10 and 11). In plant and animal cells related forms of hsp 60 have been observed in both mitochondria and chloroplasts. Like their bacterial counterpart, the GroEL protein, members of the hsp 60 family are characterized by their distinctive seven-membered ring-like structure, often being found in vivo as two rings stacked one on top of the other. In addition to binding to unfolded proteins, members of the hsp 60 family appear to "catalyze" protein folding and/or protein assembly. In conjunction with a cofactor of approximately 10 kDa (referred to as GroES in bacteria), which like hsp 60 assembles into a seven-membered ring particle, hsp 60 binds to an unfolded protein and facilitates the subsequent folding of the target protein. Recently, electron microscopy studies have revealed that the target protein undergoing folding is actually present within the central cavity of the hsp 60 "double-donut." It has been suggested that the target protein, via multiple rounds of release and rebinding, likely fueled by ATP hydrolysis, eventually acquires its final folded structure and then is released from the central cavity (12). As one might expect, other related members of the hsp 60 family are rapidly being identified. For example, recent studies indicate that there may be as many as eight or more hsp60-like homologues (referred to as the TCP-1 family) within the cytoplasm of yeast and animal cells (13).

These observations regarding the structure and apparent function of both hsp 60 and hsp 70 have led to new proposals regarding the mechanisms of protein folding/assembly. Earlier work by Anfinsen and others demonstrated the principle of protein self-assembly: All of the information necessary for the proper folding of a protein is provided by its primary amino acid sequence (14). This conclusion was based on the simple but elegant observations that a protein, when denatured by a protein chaotrope (e.g., urea), could spontaneously refold on removal of the chaotrope. Results from studies of the hsp 60 and hsp 70 families have led to a modification of these earlier concepts. In vivo protein folding and/or assembly may in fact occur by assisted self-assembly, with such assistance provided by so-called molecular chaperones such as members of the hsp 60 and hsp 70 families (reviewed in refs. 11 and 15). Molecular chaperones function in ways that do not contradict the principle of self-assembly. For example, although not conveying any specific information for folding, molecular chaperones participate in the process by preventing improper or nonproductive intra- or intermolecular interactions that could lead to protein misfolding and/or aggregation. Molecular chaperones do not become a part of the final, properly folded protein. By facilitating productive folding and assembly pathways, molecular chaperones ensure high fidelity and efficiency in the protein folding/assembly process. In fact, many investigators have suggested that molecular chaperones like hsp 70 and hsp 60 may work in tandem to facilitate the folding process. For example a newly synthesized protein, stabilized in an unfolded state via interaction with a member of the hsp 70 family, is transferred to a member of the hsp 60 family where protein folding commences. In a similar but somewhat distinct fashion, both hsp 70 and hsp 60 also may play critical roles in facilitating the orderly assembly of oligomeric protein complexes.

2. *Hsp 90 Family*

In perhaps a variation of the chaperone concept, another extremely abundant constitutively expressed stress protein, hsp 90, also appears to interact transiently with other proteins. For example, newly synthesized $pp60^{src}$, the virally encoded tyrosine protein kinase of Rous sarcoma virus, has been observed to interact with hsp 90 and a 50-kDa protein. Although present in such a complex in the cytoplasm, $pp60^{src}$ appears inactive and unable to function as a tyrosine protein kinase. Once the complex reaches the inner side of the plasma membrane, $pp60^{src}$ is released and becomes integrated into the plasma membrane where it appears to be biologically active. Although less developed, a similar maturation pathway appears to characterize several other retrovirus-encoded oncogenic tyrosine protein kinases (reviewed in ref. 16). To date, the exact biological role served by hsp 90 in the proper maturation and/or regulation of these various protein kinases remains unclear. However, hsp 90, in conjunction with several other members of the stress protein family (e.g., hsp 70, hsp 56), also has been shown to interact with steroid hormone receptors, intracellular proteins which when activated by their appropriate steroid ligand become active as a transcription factor (reviewed in ref. 17). Similar to the situation with the virally encoded protein kinases, steroid receptors appear to be maintained in an inactive state when bound to the oligomeric protein complex that includes hsp 90. Binding of the steroid hormone appears to initiate a series of events wherein hsp 90 is released from the complex, thereby enabling the receptor to oligomerize and acquire a DNA binding conformation. These observations in sum, point toward a role for hsp 90 in regulating the biological activities of target proteins with whom it transiently interacts. In particular, hsp 90 may bind to or mask domains of target proteins which are critical for their biological activation.

3. *Low Molecular Weight Hsps*

Recently, members of the low molecular weight hsps (molecular masses of 20–30 kDa) were reported to exhibit molecular chaperone-like properties and to facilitate, at least in vitro, accelerated folding of a target polypeptide (18). However, unlike the other more well-defined chaperones, the low molecular weight hsps do not appear to bind nucleotides like ATP. A number of recent studies have concluded that the low molecular weight hsps may be important regulatory components of the actin-based cytoskeleton. In vitro, the single mammalian member, hsp 28, has been reported to act as both an inhibitor of actin polymerization as well as a promoter of disassembly of already formed actin filaments (19). Overexpression of mammalian hsp 28 in vitro resulted in an apparent stabilization of and/or increase in the actin containing stress fibers (20). Moreover, a subpopulation of hsp 28 was shown to colocalize with actin within the "ruffling" membrane present at the leading edge of the cell (21). Because so many of the agents (including heat shock treatment, a wide variety of cytokines, mitogens, and tumor promoters) that induce the rapid phosphorylation of hsp 28 also induce a rapid rearrangement of the actin cytoskeleton, particularly the cortical actin network, the hypothesis that hsp 28 somehow is involved in regulating the actin cytoskeleton remains attractive and is under further study by a number of laboratories.

Our progress in elucidating the functions of the stress proteins in the normal cell also have been enlightening in terms of our understanding why many of these proteins are upregulated in cells undergoing metabolic stress. Specifically, under conditions in which protein folding is perturbed or preexisting proteins begin to unfold and denature (e.g., in response to heat shock treatment), the cell responds by increasing the synthesis of stress proteins, many of which function as molecular chaperones. Presumably the resultant increased levels of the stress proteins affords the cell a means by which to (1) identify and perhaps facilitate refolding of those proteins adversely affected by the metabolic insult; (2) identify and bind to abnormally folded proteins for their eventual targeting to an appropriate proteolytic system; and (3) facilitate the synthesis and maturation of new proteins needed to replace those which were adversely affected during the particular metabolic insult. We suspect that all three of these possible scenarios are likely correct. At the cellular level, however, the exact role of the stress proteins as it pertains to the acquisition of cellular thermotolerance remains less well defined. Controversy exists as to which members of the stress protein family are the most important "contributors" to the phenomenon of thermotolerance. Moreover, whether other physiological changes which accompany the induction of the stress response are also important for the acquisition of the thermotolerant state continues to be debated. Our own prejudice is that the increase in the levels of the stress proteins are in fact important but by no means the sole basis by which to explain the complicated phenomenon of cellular thermotolerance.

IV. CLINICAL IMPLICATIONS

A. Thermotolerance

Are the biochemical and functional characteristics of the heat shock response/ proteins as defined by cell culture experiments relevant to organs and tissues in the whole organism? Examination of tissues and organs subjected to various metabolic insults such as ischemia or fever revealed that the stress response does occur in different tissues/organs in vivo (reviewed in ref. 6). In addition, it has been observed that seizures and excitatory amino acids such as glutamate can induce a stress response in the brain (22,23). Increased synthesis of stress proteins was also observed in the rat heart subjected to hemodynamic overload (24). These findings that physiologically relevant insults can induce the stress response in vivo have led to studies examining whether the phenomenon of cellular thermotolerance is operative in the animal. Preliminary work from a number of laboratories appears promising in this regard. For example, rodents subjected to whole body hyperthermia (which was demonstrated to result in increased levels of the stress proteins in the heart) suffered less myocardial damage in response to a subsequent ischemia/reperfusion episode (25). Similar results were found in a rabbit model (26). In contrast, Yellon et al. did not observe such a protective effect, but their study used a longer ischemia/reperfusion insult (27). Taken together, these studies on the heart indicate that hyperthermic treatment can render the myocardium more resistant to ischemia/reperfusion–induced damage up to a certain point. Similarly, rat brains first rendered thermotolerant appeared less vulnerable to ischemia reperfusion–induced damage as well as to the deleterious effects of glutamate stimulation (23). In rodents first subjected to hyperthermia, the extent of retinal damage due to a subsequent intense light exposure was greatly diminished. Moreover, these investigators showed that maximal protection correlated with the overall levels of the highly stress-inducible hsp 72 in the retina (28). Last, in a rat model of the human adult respiratory distress syndrome, rats first made thermotolerant (via whole body hyperthermia) suffered no mortality as compared with 27% mortality observed for the nonheated control group (29).

Results such as these have stimulated efforts at possibly harnessing the protective effects of the thermotolerant phenotype in clinically relevant situations. One important issue is how to develop the thermotolerant phenotype in a timely and clinically relevant fashion. In vitro, the phenotype requires anywhere from 8 to 18 hr from the time of induction of the stress response to develop fully, which is likely the time period necessary to synthesize and accumulate maximal levels of the stress proteins. In the case of the heart, it is not possible to predict the occurrence of an acute myocardial infarction. Those individuals at high risk for an imminent infarction are typically already experiencing severe ischemia. The extent to which these ischemic episodes in humans induce a stress response remains an open question. Even if a pharmacological means of rapidly inducing the thermotolerant phenotype (on the order of minutes to perhaps a few hours) is developed, the applicability to acute clinical situations is unclear. Application of the stress response to clinical medicine is more promising in nonemergent situations such as scheduled surgery. For example, reconstructive surgeons have found that skin flap survival improves if the flap is first made thermotolerant (30). Other possible applications include making donor transplant organs thermotolerant and thereby possibly increasing the window of time that the organ can be transplanted and/or possibly even improving survivability once transplanted. The enthusiasm for exploiting the stress response in clinical medicine must be tempered, however, by the paucity of information regarding the price the cell pays for becoming thermotolerant. In particular, we still do not know all of the other cellular consequences associated with the acquisition of the thermotolerant phenotype.

B. Role of Stress Proteins in Cancer Therapy

The ability to modulate the stress response also has therapeutic implications as it relates to cancer. The increased expression of hsp 28 and hsp 72 has been associated with enhanced survival of tumor cells subjected to some cancer chemotherapeutic agents (reviewed in ref. 31). The increased expression of the multidrug resistance protein (MDR), whose corresponding gene contains an appropriate heat shock element, has been clearly shown to underlie the development of resistance to many cancer chemotherapeutic agents (reviewed in ref. 32). Thus, finding a way by which to downregulate or even prevent the expression of stress proteins in malignant cells may enhance the efficacy of many chemotherapeutic agents. Such an advance might allow the use of lower doses of chemotherapy, and thus perhaps minimize the toxic side effects of these agents on normal cells. In preliminary experiments, a flavonoid compound, querecetin, reported to be an inhibitor of protein kinases, was shown specifically to inhibit the expression of the stress proteins, although the mechanism by which such inhibition is manifested is not known (33).

C. Neuroendocrine Mechanisms as Stress Protein Modulators

The recent evidence that the expression of the stress proteins also may be regulated, at least in part, by neuroendocrine mechanisms represents an exciting new development. Activation of the hypothalamic-pituitary axis was shown to induce specific expression of the most highly stress induced protein, hsp 72, in the rat adrenal cortex. Hypophysectomy ablated the response, and the addition of adrenocorticotropic hormone restored specific expression in the hypophysectomized rats. In contrast, the sympathetic nervous system appeared to be important in the regulation of both hsp 72 and hsp 27 expression in the rat aorta. Adrenergic antagonists were found to block such expression, whereas adrenergic agonists induced their expression (34–36). Last, a dopamine agonist induced expression of hsp 72 in both the adrenal cortex and aorta of the rat (37). Clearly, these findings suggest that adrenergic and dopaminergic agents, as well as drugs that affect the hypothalamic-pituitary axis, have potential as modulators of the heat shock response in humans.

D. Role of Stress Proteins in Immunity

Investigation of possible clinical implications of the stress response/proteins gained added impetus as evidence began to accumulate that one or more of the stress proteins play a role in various aspects of the immune system (reviewed in ref. 6 and 38). First, genes encoding two members of the hsp 70 family were found to reside within the major histocompatibility complex (MHC). Moreover, computer modeling of the carboxy-terminal domain of hsp 70 family members, thought to be involved in the binding of both small peptides as well as unfolded polypeptides, revealed a possible binding motif very similar to the peptide binding cleft of the MHC class I proteins. Second, a peptide binding protein, termed PBP 74, was identified and shown to be related to the other members of the hsp 70 family. PBP 74 has been proposed be involved in peptide loading of MHC class II molecules. Third, deoxyspergulain, an immunosuppressant agent whose mechanism of action is unknown but appears to be distinct from that of both cyclosporin A and FK506 was found to specifically bind to hsp 73 (39). As mentioned earlier, FK506 appears to bind to hsp 56 (a protein with rotamase activity) and together the complex appears to have immunosuppressive activity (8). Finally, two of the 11 self-peptides isolated from purified class I HLA-B27 were shown to be peptides derived from hsp 90. Although these types of observations clearly are intriguing, they remain primarily phenomenological in nature, and they therefore will require further study to ascertain their biological relevance.

More compelling and scientifically developed is the observation that stress proteins from a variety of pathogens act as immunodominant antigens in animals (reviewed in refs. 40 and 41). For many years, it had been known that bacteria produced an approximately 60-kDa genus-specific protein; an antibody raised against this protein from one species tended to recognize the protein in all other species of the genus but not in any species from another genus. This protein was shown to be GroEL, the bacterial homologue of hsp 60, and it is a major target of the mammalian humoral response to bacterial infections. Interestingly, in many parasitic infections, it is the parasitic form of hsp 70, and in some cases hsp 90, which represents a major target for the humoral arm of the immune response. Recent evidence indicates that at least some parasitic and bacterial stress proteins can also induce a relatively strong T-cell response. Moreover, roughly 10–20% of γ/δ T cells, a poorly understood population of T cells, have been shown to be specific for stress proteins of various pathogens (42). A particularly attractive hypothesis is that this class of T cells, which by lining the airway, gut, and epidermal epithelium, is well positioned to provide an early line of immune defense at major body-environment interfaces where pathogen entry into the host is likely to occur.

Does the immune response to bacterial and parasitic stress proteins protect the host from infection? A number of studies suggest that the answer is yes. Protection against chlamydial diseases was associated with an immune response to chlamydial DnaK, the bacterial homologue of hsp 70 (43). In a guinea pig model of legionnaire's disease, immunization with GroEL purified from *Legionella pneumophila* was effective in preventing disease (44). Immunization of mice with an 80-kDa protein from *Histoplasma capsulatum*, a protein related to hsp 70, resulted in improved resistance to infection (45). In a monkey model of malaria, immunization with *Plasmodium falciparum* hsp 70 prevented infection with blood stages of *P. falciparum* (46). In addition, infected hepatocytes appeared to express a cell surface epitope from this same hsp 70. An anti–hsp 70 antibody recognizing this epitope appeared to be effective in an antibody-dependent cytoloysis of the infected hepatocyte (47). Thus, these examples indicate that the immune response to stress proteins from various infectious pathogens is associated with protection for the host.

Why does the immune system appear preferentially to target the stress proteins of infectious pathogens? Several possible reasons that are not mutually exclusive have been suggested (38,40,41). The stress proteins represent a relatively abundant set of proteins within the invading pathogen, and, therefore, on a purely statistical basis, the immune system may be more likely to recognize these "foreign" proteins rather than a less abundant one. In this regard, it also has been reported that infection not only induces a stress response in the host but also in the invading pathogen (thereby increasing the levels of "pathogenic stress proteins") (48). Alternatively, faced with a vast number of potential pathogens, the immune system may have simplified the problem of detection by taking advantage of the fact that stress proteins are essential components of any organism (i.e., most of the genes encoding the stress proteins are essential for growth) and they exhibit a high degree of homology. For example, recognizing genus-specific GroEL may have enabled the immune system to strike a balance between the need for sensitivity (i.e., the ability quickly to recognize the presence of a pathogen) and the need for specificity (i.e., responding only to pathogenic bacteria as opposed to normal commensal flora such as exists within the intestines). Similarly, by preferentially recognizing parasitic hsp 70, the immune system can distinguish between a bacterial versus a fungal infection.

The association of a protective humoral response against pathogen stress proteins lies at the crux of an apparent conundrum. Stress proteins have not been characterized as being localized to the cell surface or as being secreted. How then does the humoral response confer protection? One mechanism could be the earlier described antibody-dependent cell cytolysis of hepatocytes infected with *P. falciparum* which present the pathogen's hps 70 on the surface of the infected cell. Yet, many bacterial and parasitic pathogens produce their deleterious effects without ever entering host cells. One intriguing study suggested that GroEL from *Salmonella typhimurium* mediated binding of the bacterium to intestinal mucus (49). Although GroEL appeared to be secreted, it apparently was also present on the cell surface. Interestingly, antibodies against GroEL blocked *Salmonella* aggregation on the intestinal mucus. No biochemical evidence was presented demonstrating a direct interaction between GroEL and the previously identified 15-kDa glycoprotein component of intestinal mucus that mediated binding of *S. typhimurium.* Nevertheless, this report suggests that GroEL, by a change in its locale to the cell surface and/or its secretion, may act as a "virulence factor" in *S. typhimurium* infection. Other pathogens may have evolved similar mechanisms of infection which may account for the observation that the humoral immune response to stress proteins has been associated with protection. Indeed, chlamydial DnaK has been observed on the surface of elementary bodies, although its function at this site is unclear (43). As mentioned earlier, antibodies against chlamydial DnaK were associated with protection from the disease.

What about cell-mediated immunity against pathogenic stress proteins? Most bacterial and parasitic agents appear to induce such an immune response, but in some cases, rather than protecting the host, the response tends to exacerbate or contribute to the disease. For example, T-cell response to chlamydial GroEL has been associated with infection of the female reproductive tract as well as the respiratory system (50). Similarly, T cells that recognize GroEL from *Borrelia burgdorferi*, the etiological agent of Lyme disease, line the synovium of joints affected by the arthritic component of the disease (51). These types of observations have raised the question of whether these T cells were part of a general inflammatory response in these illnesses, or instead directly caused damage, perhaps by cross reacting with self-stress proteins. If the latter were true, it would imply that the well-conserved stress proteins contribute in some way to autoimmune diseases.

The idea that stress proteins play some role in different autoimmune diseases remains highly controversial (52). There are a few animal studies whose results are suggestive but by no means support a direct role for stress proteins in autoimmune disease. For example, T-cell reactivity to mycobacterial GroEL appeared to be important in the development of disease in the nonobese diabetic

(NOD) mouse model of insulin-dependent diabetes (reviewed in ref. 53). More recent work, however, suggested that these GroEL-reactive T cells somehow modulated the immune response rather than directly causing diabetes in the NOD mouse (54,55). Those T cells that recognized mycobacterial GroEL in the adjuvant-induced arthritis rat model probably play a similar modulatory role (reviewed in ref. 53). With regard to human autoimmune diseases and stress proteins, other than juvenile chronic arthritis (JCA), most studies have demonstrated guilt only by association. To date, only the synovial T cells from patients with JCA have been shown to respond strongly to human hsp 60 (56). Otherwise, the literature is filled with studies that are mainly of a phenomenological nature; either they refute (several) the presence of an association between stress proteins and most human autoimmune diseases, or they confirm an association (many) but do not answer the fundamental question of cause and effect.

The discovery that T cells reactive to self stress proteins are present normally in otherwise healthy individual led to the proposal that these T cells may recognize other cells undergoing a stress response (e.g., due to some type of an infection or transformation) and thereby help to eliminate these cells (38). Whether a particular stress event, be it infection, transformation, or some other type of metabolic insult, results in the processing and presentation of peptides derived from self stress proteins to the immune system remains an extremely interesting but unanswered question. Studies with established cell lines have demonstrated that the intracellular locale of many stress proteins changes in response to stress, but localization to the cell surface, either as an intact protein or as a peptide associated with the histocompatibility complex, has not been clearly demonstrated. However, those observations mentioned earlier in which 2 of 11 self-peptides present within the MHC class I molecule were derived from hsp 90 implies that cells may normally present self peptides derived from stress proteins on a routine basis. That different forms of stress in vivo may lead to an increase and/or alteration in the presentation of self stress peptides, thereby providing for some type of activating signal to the immune system is an interesting question which deserves additional study.

Further fueling the idea that stress proteins are important in immune cell recognition are studies reporting that both grp 94 and hsp 73 behaved as tumor-rejection antigens. Both of these stress proteins, when purified from a particular tumor and subsequently used as an immunogen, conferred protection to challenge from that same tumor but not to an antigenically distinct tumor. Grp 94 and hsp 73 from normal cells provided no such tumor immunity, and sequence analysis of the purified stress proteins isolated from the tumors revealed no differences when compared with the proteins isolated from nontumorigenic cells. Subsequent work, however, reported that both hsp 73 and grp 94 bound to a heterogeneous population of peptides (57,58). Presumably, the associated peptides found with grp 94 and hsp 73 from the tumor cells were responsible for the successful tumor immunity, but the direct demonstration that the bound peptides, when used as the immunogen, could confer tumor resistance remains to be shown. Thus, these data suggest that stress proteins are involved in processing proteins for antigenic presentation rather than being immunogenic themselves.

E. Stress Proteins as Vaccines and Adjuvants

There is increasing excitement and enthusiasm for the use of mycobacterial and parasitic forms of the stress proteins as novel acellular vaccines and/or carrier-free adjuvants. For example, mice primed with BCG and then immunized with a hapten conjugated to tuberculin PPD produced long-lasting and high titers of anithapten antibodies without the use of adjuvants (59). The same effect was observed if the hapten was conjugated directly to mycobacterial GroEL or DnaK. Importantly, this effect occurred even in the animal showing high titers of antibody against the mycobacterial GroEL or DnaK stress proteins used as the adjuvant. In fact, high doses of GroEL were shown to be as effective a primer as BCG. However, the most provocative and exciting finding was that an effective T cell–mediated response could be induced by hapten conjugated to DnaK, without the need for either an adjuvant or for previous priming (60). Clearly, these results have broad implications for vaccine development against infectious diseases and tumors.

F. Role of Stress Proteins in Toxicology

Although too extensive in scope to adequately present here, we should mention briefly that changes in stress protein expression may prove useful as it relates to toxicology. In the hope of developing rapid assays as well as reducing the use of animals, toxicologists are exploring the use of changes in the expression of one or more of the stress proteins, in cells grown in vitro, as a sensitive and reliable indicator of the possible toxic effects of different compounds. As a further extension of this type of technology, investigators are developing transgenic stress reporter organisms. Using well-defined heat shock promoter elements to drive the expression of a reporter gene (luciferase, B-galactosidase, chloramphenicol acetyl transferase), the reporter organism carrying such a construct would be employed, for example, in the monitoring of environmental pollutants (61). Although such approaches are still at an exploratory stage, requiring substantial validation efforts, exploiting changes in the expression of the stress proteins as well as other gene products associated with cellular injury (metallothionines, cytochrome 450 system) may revolutionize the field of toxicology.

V. CONCLUSIONS

A field of research that began with a curious observation in *Drosophila* has resulted in a new understanding of how cells respond to sudden and adverse changes in their environment. In addition, through the study of the structure/function of the stress proteins, especially those which function as molecular chaperones, new insights into the details by which proteins are synthesized and acquire their final biologically active conformation have been realized. Equally exciting is the progress being made as it relates the potential diagnostic and therapeutic applications of the stress response/proteins. The use of stress proteins as the next generation of vaccines and/or their use as potentially powerful adjuvants capable of stimulating both T- and B-cell responses to a particular antigen of interest appear close to becoming a reality. One wonders how many more surprises are in store for us as we continue to explore this evolutionarily conserved cellular response.

ACKNOWLEDGMENTS

We thank members of the lab for reviewing the manuscript at various stages of preparation; Joel Ernst and David Erle for critical reading of the final draft; and Christine Mok for secretarial assistance. GM has received past fellowship support from the Cardiovascular Research Institute (NIH HL07185) and presently is supported by GM15526. WJW is supported by grants from NIH (GM3-3551), NSF (MCB9018320) and ACS (CB-91A).

REFERENCES

1. L.E. Hightower, Cultured animal cells exposed to amino acid analogues or puromycin rapidly synthesize several polypeptides, *J. Cell Physiol. 102*:407 (1980).
2. J. Anathan, A.L. Goldberg, and R. Voellmy, Abnormal proteins serve as eukaryotic stress signals and trigger the activation of heat shock genes, *Science 232*:252 (1986).
3. J. Lis, and C. Wu, Protein traffic on the heat shock promoter: parking, stalling, and trucking along, *Cell. 74*:1 (1993).
4. L. Sistonen, K.D. Sarge, B. Phillips, K. Abravaya, and R.I. Morimoto, Activation of heat shock factor 2 during hemin-induced differentiation of human erythroleukemia cells, *Mol. Cell. Biol. 12*:4104 (1992).
5. S.C. Lindquist, The heat shock response, *Ann. Rev. Biochem. 55*:1151 (1986).
6. W.J. Welch, Mammalian stress response: cell physiology, structure/function of stress proteins, and implications for medicine and disease, *Physiol. Rev. 72*:1063 (1992).
7. W.B. Pratt, The role of heat shock proteins in regulating the function, folding, and trafficking of the glucocorticoid recepto, *J. Biol. Chem. 268*:21455 (1993).

8. A.W. Yem, A.G. Tomasselli, R. L. Heinrikson, H. Zurcher-Neely, V.A. Ruff, R.A. Johnson, and M.R. Deibel, Jr., The hsp56 component of steroid receptor complexes binds to immobilized FK506 and shows homology to FKBP-12 and FKBP-13, *J. Biol. Chem. 267*:2868 (1992).
9. E.R. Sanchez, Hsp56: A novel heat shock protein associated with untransformed steroid receptor complexes, *J. Biol. Chem. 265*:22067 (1990).
10. F.U. Hartl, J. Martin, and W. Neupert, Protein folding in the cell: the role of chaperones hsp70 and hsp 60. *Ann. Rev. Biophys. Biomol. Struct. 293* (1992).
11. C. Georgopoulos, and W.J. Welch, Role of the major heat shock proteins as molecular chaperones, *Ann. Rev. Cell. Biol.* 601 (1993).
12. T. Langer, G. Pfeifer, J. Martin, W. Baumeistere, and F.U. Hartl, Chaperonin-mediated protein folding: GroES binds to one end of the GroEL cylinder, which accomodates the protein within its central cavity, *EMBO J. 11*:4757 (1992).
13. H. Kubota, G. Hynes, A. Canne, A. Ashworth, and K. Willison, Identification of six TCP-1 related genes encoding divergent subunits of the TCP-1 containing chaperonin, *Curr. Biol. 4*:89 (1994).
14. C.B. Anfinsen, Principles that govern the folding of protein chains, *Science 8*:223 (1973).
15. R.J. Ellis, and S.M. van der Vies, Molecular chaperones, *Ann. Rev. Biochem.* 321 (1991).
16. J.S. Brugge, Interaction of the Rous sarcoma Virus protein, $pp60^{src}$, with the cellular proteins pp50 and pp90, *Curr. Topics Microbiol. Immunol. 123*:1 (1986).
17. D.F. Smith, and D.O. Toft, Steroid receptors and their associated proteins, *Mol. Endocrinol. 7*:4 (1993).
18. U. Jakob, M. Gaestel, K. Engel, and J. Buchner, Small heat shock proteins are molecular chaperones, *J. Biol. Chem. 268*:1517 (1993).
19. T. Miron, K. Vancompernnolle, J. Vandkerckhove, M. Wilchek, and B.A. Geiger, 25-kD inhibitor of actin polymerization is a low molecular mass heat shock protein, *J. Cell Biol. 114*:255 (1991).
20. J.N. Lavoie, G. Gingras-Breton, R.M. Tanguay, and J. Landry, Induction of chinese hamster hsp27 gene expression in mouse cells confers resistance to heat shock; hsp 27 stabilization of the microfilament organization, *J. Biol. Chem. 268*:3420 (1993).
21. J.N. Lavoie, E. Hickey, L.A. Weber, and J. Landry, Modulation of actin microfilament dynamics and fluid phase pinocytosis by phosphorylation of heat shock protein 27, *J. Biol. Chem. 268*:24210 (1993).
22. K. Vass, M.L. Berger, T.S.J. Nowak, W.J. Welch, and H. Lassmann, Induction of stress protein Hsp 70 in nerve cells after status epilepticus in the rat, *Neurosci. Lett. 100*:259 (1989).
23. R.S. Sloviter, and D.H. Lowenstein, Heat shock protein expression in vulnerable cells of the rat hippocampus as an indicator of excitation-induced neuronal stress, *J. Neurosci. 12*:3004 (1992).
24. C.J. Delcayre, L. Samuel, F. Marotte, F. Best-Belpomme, J.J. Mercadier, and L. Rappaport, Synthesis of stress proteins in rat cardiac mycocytes 2–4 days after the imposition of hemodynamic overload, *J. Clin. Invest. 83*:460 (1988).

25. T.J. Donnelly, R.E. Sievers, F.L.J. Vissern, W.J. Welch, and C.L. Wolfe, Heat shock protein induction in rat hearts: a role for improved myocardial salvage after ischemia and reperfusion?, *Circulation 85*:769 (1992).
26. R.W. Currie, R.M. Tanguay, and J.G. Kingma, Heat shock response and limitation of tissue necrosis during occlusion/reperfusion in rabbit hearts, *Circulation 87*:963 (1993).
27. D. Yellon, M.E. Iliodromitis, D.S. Latchman, D.M. Van Winkle, J.M. Downey, F.M. Williams, and T.J. Williams, Whole body heat stress fails to limit infarct size in the reperfused rabbit heart, *Cardiovasc. Res. 26*:342 (1992).
28. M.F. Barbe, M. Tytell, D.J. Gower, and W.J. Welch, Hyperthermia protects against light damage in the rat retina, *Science 241*:1817 (1988).
29. J. Villar, J.D. Edelson, M. Post, B. Mullen, and A. Slutsky. Induction of heat stress proteins is associated with decreased mortality in an animal model of acute long injury, *Am. Rev. Respir. Dis. 147*:177 (1993).
30. W.J. Koenig, R.A. Lohner, G.A. Perdrizet, M.E. Lohner, R.T. Schweitzer, and V.L. Lewis, Jr, Improving acute skin-flap survival through stress conditioning using heat shock and recovery, *Plast. Reconstuct. Surg. 90*:659 (1992).
31. G.M. Hahn, and G.C. Li, Thermotolerance, thermoresistance, and thermosensitization, *Stress Proteins in Biology and Medicine* (R.I. Morimoto, A. Tissieres, and C. Georgopoulos, eds.), Cold Spring Harbor Laboratory Press, Cold Spring Harbor, NY, pp. 79–100.
32. M.M. Gottesman, and I. Pastan, The multidrug transporter, a double-edged sword, *J. Biol. Chem. 263*:12163 (1988).
33. N. Hosokawa, K. Hirayoshi, A. Nakai, Y. Hosokawa, N. Marui, M. Yoshida, T. Sakai, H. Nishino, A. Aoike, K. Kawai, and K. Nagata, Flavonoids inhibit the expression of heat shock proteins, *Cell Struct. Funct. 15*:393 (1990).
34. M.J. Blake, R. Udelsman, G.J. Feulner, D.D. Norton, and N.J. Holbrook, Stress induced heat shock protein 70 expression in adrenal cortex: an adrenocorticotropic hormone-sensitive, age-dependent response, *Proc. Natl. Acad. Sci. USA. 88*:9873 (1991).
35. R. Udelsman, M.J. Blake, and N.J. Holbrook, Molecular response to surgical stress: Specific and simultaneous heat shock protein induction in the adrenal cortex, aorta, and vena cava, *Surgery 110*:1125 (1991).
36. R. Udelsman, M.J. Blake, C.A. Stagg, D-G. Li, D.J. Putney, and N.J. Holbrook, Vascular heat shock protein expression in response to stress. Endocrine and autonomic regulation of this age-dependent response, *J. Clin. Invest. 91*:465 (1993).
37. M.J. Blake, D.J. Buckley, and A.R. Buckley, Dopaminergic regulation of heat shock protein-70 expression in adrenal gland and aorta, *Endocrinology 132*:1063 (1993).
38. D.C. DeNagel, and S.K. Pierce, Heat shock proteins in immune responses, *Crit. Rev. Immunol. 13*:71 (1993).
39. S.G. Nadler, M.A. Tepper, B. Schacter, and C.E. Mazzucco, Interaction of the immunosuppressant Deoxyspergualin with a member of the hsp70 family of heat shock proteins, *Science 258*:484 (1992).
40. R.A. Young, and T.J. Elliot, Stress proteins, infection, and immune surveillance, *Cell 59*:5 (1989).

41. D.B. Young, Heat-shock proteins: immunity and autoimmunity, *Curr. Opin. Immunol. 4*:396 (1992).
42. W.M. Born, P. Happ, A. Dallas, C. Reardon, R. Kubo, T. Shinnick, P. Brennan, and R. O'Brien, Recognition of heat shock proteins and gamma/delta function, *Immunol. Today 11* (1990).
43. G. Zhong, and R.C. Brunham, Antigenic analysis of the chlamydial 75-kilodalton protein, *Infect. Immun. 60*:323 (1992).
44. S.J. Blander, and M.A. Horwitz, Major cytoplasmic membrane protein of Legionella pneumophila, a genus common antigen and member of the hsp60 family of heat shock proteins, induces protective immunity in a guinea pig model of Legionnaire's disease, *J. Clin. Invest. 91*:717 (1993).
45. F.J. Gomez, A.M. Gomez, and G.S. Deepe, Jr., An 80-kilodalton antigen from Histoplasmia capsulatum that has homology to heat shock prtoein 70 induces cell-mediated immune responses and protection in mice, *Infect. Immun. 60*:2565 (1992).
46. P. Dubois, J.P. Dedet, T. Fandeur, C. Roussilhon, M. Jendoubi, S. Pauillac, O. Mercereau-Puijalon, and L. Pereira da Silva, Protective immunization of the squirrel monkey against asexual blood-stages of Plasmodium falciparum by use of parasite protein fractions, *Proc. Natl. Acad. Sci. USA 81*:229 (1984).
47. L. Renia, D. Mattei, J. Goma, S. Pied, P. Dubois, F. Miltgen, A. Nussler, H. Matile, F. Menegaux, M. Gentilini, and D. Mazier, A malaria heat-shock-like determinant expressed on the infected hepatocyte surface is the target of antibody dependent cell-mediated cytotoxic mechanisms by non-parenchymal liver cells, *Eur. J. Immunol. 20*:1445 (1990).
48. N.A. Buchmeier, and F. Heffron, Induction of Salmonella stress proteins upon infection of macrophages, *Science 248*:730 (1990).
49. M. Ensgraber, and M. Loos, A 66-kilodalton heat shock protein of Salmonella typhimurium is responsible for binding of the bacterium to intestinal mucus, *Infect. Immun. 60*:3072 (1992).
50. R.P. Morrison, R.J. Belland, K. Lyng, and H.D. Caldwell, Chlamydial disease pathogenesis: the 57-kD chlamydial hypersensitivity antigen is a stress response protein, *J. Exp. Med. 170*:1271 (1989).
51. M-C. Shanafelt, P. Hindersson, C. Soderberg, N. Mensi, C.W. Turck, D. Webb, H. Yssel, and G. Peltz, T cell and antibody reactivity with the Borrelia burgdorferi 60 kDa heat shock protein in Lyme arthritis, *J. Immunol. 146*:3985 (1991).
52. S.H.E. Kaufmann, Heat shock proteins and autoimmunity: fact or fiction?, *Curr. Biol. 1*:359 (1991).
53. I.R. Cohen, Autoimmunity to chaperonins in the pathogenesis of arthritis and diabetes, *Ann. Rev. Immunol. 567* (1991).
54. D.L. Kaufman, M. Clare-Salzer, J. Tian, T. Forsthuber, G.S.P. Ting, P. Robinson, M.A. Atkinson, E.E. Sercarz, A.J. Tobin, and P.V. Lehmann, Spontaneous loss of T-cell tolerance to glutamic acid decarboxylase in murine insulin-dependent diabetes, *Nature 365*:69 (1993).
55. R. Tisch, X-D. Yang, S.M. Singer, R.S. Liblau, L. Fugger, and H.O. McDevitt, Immune response to glutamic acid decarboxylase correlates with insulitis in non-obese diabetic mice, *Nature 365*:72 (1993).
56. E.R. De Graeff-Meeder, R. Van Der Zee, G.T. Rijkes, H-J. Schuurman, W. Kuis,

J.M.J. Bijlsma, B.J. M. Zegers, and W. Van Eden, Recognition of human 60 kD heat shock protein by mononuclear cells from patients with juvenile chronic arthritis, *Lancet 337*:1368 (1991).

57. Z. Li, and P.K. Srivastava, Tumor rejection antigen gp96/grp94 is an ATPase: implications for protein folding and antigen presentation. *EMBO J. 12*:3143 (1993).
58. H. Udono, and P.K. Srivastava, Heat shock protein 70-associated peptides elicit specific cancer immunity, *J. Exp. Med. 178*:1391 (1993).
59. A.R. Lussow, C. Barrios, J. van Embden, R. Van der Zee, A.S. Verdini, A. Pessi, J.A. Louis, P-H. Lambert, and G. Del Guidice, Mycobacterial heat-shock proteins as carrier molecules, *Eur. J. Immunol. 21*:2297 (1991).
60. C.A. Barrios, R. Lussow, J. Van Embden, R. Van der Zee, R. Rappuoli, P. Constantino, J.A. Louis, P-H. Lambert, and G. Del Guidice, Mycobacterial heat shock proteins as carrier molecules. II: The use of the 70 kDa mycobacterial heat shock proteins as carrier for conjugated vaccines can circumvent the need for adjuvants and Bacillus Calmette Guerin priming, *Eur. J. Immunol. 22*:1365 (1992).
61. W.J. Welch, How cells respond to stress, *Sci. Am. 268*:56 (1993).

4

T-Lymphocyte Recognition of Hsp 60 in Experimental Arthritis

Stephen M. Anderton and Willem van Eden
Institute of Infectious Diseases and Immunology, University of Utrecht, Utrecht, The Netherlands

I. INTRODUCTION

The pathogenic mechanisms underlying human rheumatoid arthritis (RA) and other arthritic conditions remain unknown. In the search for a better understanding of these mechanisms and the development of potential therapeutic agents several rodent models of arthritis have been exploited. These models can be divided into four categories based on the method of induction of disease:

1. Spontaneous onset; for example, the MRL-lpr/lpr mouse (1).
2. Immunization with joint antigens; for example, native type II collagen in rats (2) and mice (3) and cartilage proteoglycan in mice (4).
3. Immunization with bacterial components; for example, adjuvant arthritis (AA, induced by injection of the heat-killed *Mycobacterium tuberculosis*) (5), and streptococcal cell wall (SCW) arthritis induced using cell walls isolated from group A,B, or C streptococci (6). A similar model uses cell walls from *Lactobacillus casei* (7).
4. Administration of mineral oil or synthetic adjuvants in the absence of bacterial or protein antigens. This group of models can be induced by injecting rats with incomplete Freund's adjuvant (IFA) or paraffin oil alone (8), injecting rats with the lipoidal amine CP20961 (avridine) in mineral oil (9), or injecting mice with the paraffin oil pristane (10).

Although the pathogenic mechanisms of these different models clearly vary, in general the diseases appear to have some T-lymphocyte involvement. This has been demonstrated by the passive transfer of disease to naive animals using purified T cells isolated from the spleen or lymph nodes of arthritic animals. In several models, arthritis development can be prevented by treatment with cyclosporine, or administration of anti-αβ T-cell receptor (TCR) or anti-CD4 antibodies. Thus, T cells play a pivotal role in the immunological basis of arthritis models.

The discovery that T-cell reactivity to mycobacterial heat shock protein (hsp) 65 was of critical importance in the pathogenesis (both induction and protection) of AA resulted in many investigations of T-cell responses to hsp 60 molecules in arthritis and other autoimmune conditions. In this chapter, advances in understanding of the role(s) of T cells specific for hsp 60 in experimental arthritis are reviewed.

II. THE CELLULAR BASIS OF ADJUVANT ARTHRITIS: FIRST EVIDENCE FOR HSPS AS ANTIGENS IN ARTHRITIS

Adjuvant arthritis (AA), the most extensively studied model of arthritis, is induced in several susceptible strains of rat, usually the Lewis rat. The arthritogenic agent, heat-killed *M. tuberculosis* is suspended in incomplete Freund's adjuvant (IFA) and injected intradermally at the base of the tail (0.5–1.0 mg *M. tuberculosis* in 100 μl IFA). Ten to 14 days later, a severe polyarthritis develops in the distal joints of the limbs which resembles RA histologically. There is inflammation of the synovium with a predominantly mononuclear cell infiltrate, pannus formation, cartilage destruction, and bone erosion. The inflammation follows a remitting course and has cleared by 35–40 days after *M. tuberculosis* injection, but the bone deformities remain (5).

The cellular basis of AA was demonstrated by the transfer of the disease to naive recipients using thoracic duct lymphocytes (11), spleen cells, or lymph node cells (12) from arthritic rats. Removal of $CD4^+$ T cells from the transferred populations using the W3/25 antibody abrogated transfer of disease (13). A major advance in understanding AA was made by the generation of the *M. tuberculosis*–specific T-cell line A2 from *M. tuberculosis*–immunized rats (14). This line was shown to induce an arthritis indistinguishable from AA when administered to sublethally irradiated naive syngeneic rats. Paradoxically, when A2 was administered to nonirradiated rats, it did not induce arthritis. Instead the recipient rats were rendered resistant to subsequent attempts to induce AA with *M. tuberculosis*. This divergence in the activity of *M. tuberculosis*–reactive T cells in AA was underlined when individual T-cell clones were generated from the A2 line (15). One clone, A2b, was capable of inducing arthritis in the same way as the parent A2 line. In contrast a second clone, A2c, was nonarthritogenic but conferred protection against AA.

The next important step was to identify precisely the epitope recognized by the arthritogenic T cells. The cloning of *Mycobacterium bovis* BCG DNA and production of recombinant proteins in *Escherichia coli* (16) provided the opportunity to test for responses of the *M. tuberculosis*–T-cell clones against individual mycobacterial proteins. These recombinant mycobacterial proteins included a 65-kDa protein which proved to contain the epitope recognized by clone A2b (17). The use of deletion mutants of the recombinant 65-kDa protein and synthetic peptides identified the epitope as residues 180–188. Comparison of the sequence of the mycobacterial 65-kDa protein with that of *E. coli* chaperone GroEL showed a high degree of homology. Thus, the 65-kDa protein was designated as a heat shock protein: hsp 65 (18).

Using Western blot analysis polyclonal and monoclonal antibodies raised against mycobacterial hsp 65 were found to cross react with similar sized proteins in many other bacterial species (19), indicating the conserved nature of this group of molecules (the hsp 60 family). Hsp 60 cognates are also found in eukaryotic species and show extensive homology with bacterial hsp 60 molecules. The cloning and sequencing of mammalian hsp 60—most notably human (20), mouse (21), and rat (22) hsp 60, also known as P1—revealed approximately 50% amino acid identity with mycobacterial hsp 65 (20).

Thus, the arthritogenic T-cell clone A2b was specific for an epitope of mycobacterial hsp 65. This finding caused great excitement among workers investigating T-cell involvement in human arthritic diseases, and the availability of recombinant hsp 65 prompted many studies of human T-cell reactivity to hsp 65 (reviewed in other chapters). There was now formal evidence that T cells recognizing mycobacterial antigens were capable of inducing arthritis. The mechanism by which these T cells induce arthritis is believed to be via "molecular mimicry" (23): T cells activated by an epitope in mycobacteria also recognize a cross-reactive epitope expressed by joint tissue, thereby initiating inflammation and joint destruction. In support of this, both clones A2b and A2c were shown to respond to preparations of cartilage proteoglycans (24). It may well be the case that once tissue damage is initiated, other T cells specific for antigens unique to the joint are activated and are responsible for perpetuating disease. The activation of the cross-reactive T-cells being the critical first step in disease development. The exact nature of the potential cross-reactive epitope remains unknown. Sequence comparison of the 180–188 epitope with characterized cartilage antigens revealed some homology (four of nine residues identical) with rat proteoglycan link protein. However, A2b did not respond to a synthetic peptide based on this sequence. An alternative possibility was that A2b may recognize the highly conserved rat hsp 60. However, the sequence 180–

188 is not highly conserved between hsp 65 and rat hsp 60 (three of nine residues identical) and again A2b did not respond to a synthetic peptide of rat hsp 60 180–188.

Although A2b was arthritogenic, its sister clone A2c conferred protection against subsequent AA induction using *M. tuberculosis*. These different effects were not due to recognition of different epitopes in *M. tuberculosis* as A2c responded to exactly the same epitope as A2b (17). The two clones have also been shown to utilize identical TCR α and β chains (25). The demonstration that $CD4^+$ T-cell clones could be subdivided on the basis of cytokines they secreted into the T_H1 and T_H2 subgroups provided a possible explanation for why two T-cell clones with identical TCRs and epitope specificities should have opposite effects in AA (26). T_H1 cells secrete the proinflammatory cytokines interferon gamma (IFN-γ) and lymphotoxin, whereas T_H2 cells secrete interleukin-4 (Il-4) and Il-10, which have suppressive effects on inflammation through their inhibitory influence on T_H1 cells and macrophages, respectively. Thus, if A2b was a T_H1-like clone and A2c a T_H2-like clone, they would be expected to have proinflammatory and anti-inflammatory activities, respectively.

So far, the two clones have only been compared in terms of IFN-γ production (27). These tests revealed that A2c produced around 10-fold higher amounts of IFN-γ than A2b. Also, addition of recombinant IFN-γ enhanced antigen-specific proliferation of A2c but inhibited A2b. These results suggested that A2b and A2c were T_H2 and T_H1-like clones, respectively; the opposite of what might be expected. However, the T_H1/T_H2 differentiation is less well defined in the rat than in murine systems, and it is possible that A2c might inhibit AA by the release of other suppressive cytokines such as transforming growth factor (TGF-β) (28). The potential of A2b to induce tissue damage was illustrated by coculture of human articular cartilage explants with A2b and rat thymocytes as APC. Antigen-specific activation of A2b by addition of hsp 65 or the 180–188 peptide resulted in inhibition of cartilage proteoglycan synthesis (29).

An alternative explanation for the protective effect of administration of the A2c clone prior to AA has been proposed involving an idiotypic network of regulatory T cells. A2c administration would result in the generation of T cells specific for the A2b/A2c T-cell receptor (TCR), and following subsequent *M. tuberculosis* immunization these TCR-specific T cells would suppress the generation of A2b-like responses to the 180–188 epitope (30). The finding that A2b could similarly confer protection after hydrostatic shock provided support for this hypothesis (31). Recently, it has been shown that A2b can present endogenously derived peptides of its TCR to $RT1.D^1$–restricted T-cell lines generated from rats immunized with synthetic peptides based on sequences of the A2b TCR α and β chains (32).

III. RESPONSES TO HSP 65 IN ADJUVANT ARTHRITIS AND OTHER ARTHRITIS MODELS

As T-cell responses to hsp 65 180–188 were associated with arthritis, it was of interest to examine T-cell reactivity to hsp 65 during AA and other arthritis models (Table 1). Experiments in AA rats showed clear proliferative responses to hsp 65 in lymph node and splenocyte populations (33). Responses became significant around 10 days after *M. tuberculosis*/IFA immunization and could be detected throughout the course of AA and into remission (i.e., up to day 50). There was no correlation between magnitude of anti–hsp 65 proliferative responses and arthritis score, with responses being similar in all rats. This clear responsiveness to hsp 65 after *M. tuberculosis* immunization is consistent with a previous report suggesting that hsp 65 was an immunodominant T-cell antigen after *M. tuberculosis* immunization in mice. Frequency analysis by limiting dilution revealed that approximately 20% of *M. tuberculosis*-reactive T cells responded to hsp 65 (34). In contrast to T-cell responsiveness to intact hsp 65, T-cell reactivity to a synthetic peptide of the 180–188 epitope proved more elusive and could only be measured at the polyclonal level when inguinal lymph node cells were tested 14 days after immunization (33). However, a more recent study using a longer synthetic peptide 176–190 containing the 180–188 sequence has found significant responses to this epitope using popliteal and inguinal lymph nodes up to 42 days after *M. tuberculosis* immunization (S.M. Anderton, in preparation).

T-cell responsiveness to hsp 65 has also been investigated in other models of arthritis and most notably has been found to be present in streptococcal cell wall arthritis and pristane arthritis. Splenocytes from Lewis rats with SCW arthritis 30 days after SCW injection showed clear proliferative responses to hsp 65 (35). This might be explained by the presence of the streptococcal hsp 60

Table 1 Roles for Hsp 65 in Experimental Arthritis

Model	Anti–hsp 65 T-cell responses	Protection with hsp 65	References
Adjuvant arthritis	+	+	17,33,48,61
SCW	+	+	35
Collagen type II	?	+	48
CP20961	?	+	42,48
Pristane	+	+	36,52,53
Proteoglycan	–	?	38
MRL-lpr/lpr mice	?	?	

homologue in the SCW. T cells activated by the presence of streptococcal hsp 60 might then recognize cross-reactive epitopes in the mycobacterial hsp 65. The findings that primed lymph node cells from rats immunized with SCW/IFA responded to hsp 65 and vice versa support this idea, but formal proof of cross reactivity between the two antigens by use of T-cell clones remains to be provided. Also, this investigation did not examine responses to the AA-associated 180–188 epitope in SCW arthritic rats.

The presence of shared antigenic epitopes in the arthritogenic preparation might account for hsp 65 recognition in SCW arthritis. A model in which this cannot be the case is pristane arthritis in the mouse. This model is induced in susceptible strains (BALB/c and CBA/Ig^{b}) by intraperitoneal injection of the mineral oil pristane (2,6,10,14-tetramethylpentadecane) (10). In this model, several studies by Thompson et al. have revealed clear T-cell responsiveness to hsp 65 using splenocytes from arthritic mice. The magnitude of responses to hsp 65 were clearly higher in mice with gross arthritic disease compared with mice that received pristane but did not develop disease (36). As this model uses no antigenic component capable of inducing hsp 65 cross-reactive T cells, the priming for responses to hsp 65 must be the result of activation by endogenous antigens. These antigens may be the elusive cross-reactive joint-specific antigens which appear to be involved in the AA model or cross-reactive epitopes in mouse hsp 60. However, an alternative might be that the intraperitoneal injection pristane results in leakage of normal gut flora into the peritoneal cavity. It has been proposed that this eventually leads to arthritis as a result of immune activation by the enteric bacteria after either migration of bacteria to the joint or T cell activation in the peritoneal cavity followed by recognition of cross-reactive joint antigens. This hypothesis is supported by the finding that mice maintained in germ-free conditions (and therefore devoid of enteric bacteria) do not develop pristane arthritis, but they do if rehoused in conventional "dirty" environments (37). Clearly, as part of the T-cell activation by enteric bacteria, responses to the immunodominant bacterial hsp 60 homologues will be prevalent and could involve recognition of T-cell epitopes shared with mycobacterial hsp 65. Thus, it is possible that the anti–hsp 65 T-cell responses demonstrated in pristane arthritis are not directly involved in the pathogenesis of the disease but act more as a "marker" for the initiation of the arthritogenic mechanism.

Therefore, in addition to the original finding that T-cell responses to hsp 65 are involved in the pathogenesis of AA, responses have been demonstrated in two other models: SCW and pristane-induced arthritis. Information on the possible role of hsp 65 responses in other arthritis models is limited. Repeated immunization of BALB/c mice with human fetal cartilage proteoglycan induces a polyarthritis and spondylitis associated with humoral and cellular immune responses to human and mouse proteoglycan (4). Analysis of a panel of proteoglycan-reactive T-cell lines and hybridomas revealed responses to proteoglycans but a total lack of responses to hsp 65 (38). To date, no information is available on T-cell reactivity to hsp 65 in collagen type II arthritis, oil-induced arthritis, or spontaneous onset arthritis.

IV. MODULATION OF ADJUVANT ARTHRITIS USING THE ARTHRITIS-ASSOCIATED 180–188 EPITOPE AND PEPTIDE ANALOGUES

Once the epitope recognized by the arthritogenic T-cell clone A2b had been identified, attempts were made to induce AA with synthetic peptides of the 180–188 sequence alone. These efforts failed. Instead, repeated intraperitoneal immunization of rats with the 180–188 peptide (100 μg, emulsified in IFA, 35, 10, and 5 days prior to *M. tuberculosis* immunization) was reported to protect against AA induction (39). T-cell reactivity to 180–188 and *M. tuberculosis* was evident and protection was transferred to naive rats using splenic T cells (40). In experiments using the Lewis rat model of type II collagen–induced arthritis, no protection was found after 180–188 preimmunization. Thus, the protective effects of 180–188 preimmunization corresponded with epitope-specific T-cell activation and were dependent on the use of mycobacteria. Why immunization with 180–188 protects against AA is unclear. Although the protocol adopted has also been used to induce antigen-specific T-cell tolerance (41), this appears not to be the case as T-cell responses to 180–188 are present and protection can be transferred with T cells. An alternative is that immunization with peptide might induce an "A2c-like" protective response (see above), perhaps mediated through cytokine production, which then dominates the response to 180–188 on subsequent *M. tuberculosis* immunization.

Attempts to reproduce these findings have failed because it has been found that preimmunization with IFA alone without antigen is sufficient to protect against AA (W. van Eden et al., unpublished observations, and Ref. 42). Indeed IFA-induced protection has also been found in SCW and CP20961-induced arthritis models using Lewis rats. The mechanism of this protective activity remains unknown, but it is intriguing to note that IFA-immunization without antigen has been reported to induce T-cell reactivity against hsp 65 both in mice (43) and Lewis rats (42).

An alternative approach to the induction of epitope-specific therapy of AA has been identified by a series of experiments by Wauben et al. A panel of nine peptide analogues of the 180–188 epitope were synthesized with individual alanine substitutions at each residue (44). One particular peptide, A183 with a 183L- > A substitution, was not recognized by clone A2b, but effectively inhibited 180–188–induced proliferation, indicating that A183 competed for major histocompatibility complex (MHC) class II (RT1.B^l) binding but did not stimulate A2b. The development of a direct peptide binding assay using isolated RT1.B^l molecules and biotinylated competitor peptides allowed semiquantitative analysis of relative MHC binding affinities. This showed that A183 had an RT1.B^l-binding affinity significantly higher than the native 180–188 peptide (44).

Coimmunization of rats with *M. tuberculosis* and 250 μg of A183 strongly inhibited AA. This might have been due to MHC blockade by A183 preventing AA onset. However, this could be excluded as coimmunization with a peptide analogue of the encephalitogenic MBP(72-85) epitope, which had a significantly higher RT1.B^l binding affinity than A183, had a minimal effect on AA onset and severity but could protect against experimental autoimmune encephalomyelitis. Also, preimmunization of rats with 100 μg A183 7 days prior to *M. tuberculosis* immunization also induced strong inhibition of AA. Thus, the protection found using A183 was not simply a function of MHC binding but also appeared to have an antigen-specific component.

Immunization with A183 induced T-cell responses to both A183 and 180–188, indicating that the residues involved in T-cell recognition were in the native hsp 65 sequence (i.e., T cells were not recognizing the substituted alanine at position 183) (45). However, in contrast to A2b, a T-cell line generated against A183 (line ATL11) failed to respond to intact hsp 65 and *M. tuberculosis.* A series of analogue peptides of 180–186 containing every possible substitution at individual residues was synthesized and tested for stimulatory activity using A2b and ATL11. This analysis revealed that A2b would not allow substitutions at positions 182, 183, 184, and 185, whereas no or limited substitutions were allowed by ATL11 positions 180, 182, and 186. Thus, immunization with A183 induced T-cell responses against a cryptic epitope within 180–188 which is not recognized by A2b. It remains to be determined whether T cells recognizing this epitope are involved in the A183-induced protection against AA.

An alternative explanation for the protective effect lies in recent reports that T-cell clones show differential patterns of responsiveness when stimulated with peptide analogues of their specific epitopes. The same clone can proliferate, secrete cytokines, or revert to an anergic state when stimulated with different peptides containing substitutions in defined TCR-contact residues (46). This might account for the activity of A183 if the "A2b-like" (AA-inducing) T-cell repertoire of the rat can be made nonresponsive. This type of mechanism has been suggested to be involved in similar protective effects observed in murine EAE using analogue peptides of the encephalitogenic MBP(1-11) epitope (47).

V. MODULATION OF ARTHRITIS USING RECOMBINANT MYCOBACTERIAL HSP 65

The localization of the arthritis-associated T-cell epitope to hsp 65 180–188 prompted attempts to induce AA by immunization with hsp 65 alone; all of which failed. Instead, immunization with hsp 65 protected rats against subsequent attempts to induce AA with whole *M. tuberculosis*/IFA (17,48). Subsequently, vaccination with hsp 65 was found to protect against other forms of arthritis (see Table 1). The protective activity is believed to be a function of the T-cell compartment. This is based on findings using T cells derived from inguinal lymph nodes of arthritic rats 14 days after *M. tuberculosis* immunization. After restimulation in vitro with hsp 65, administration of these T cells to *M. tuberculosis*-immunized syngeneic rats (14, 17, and 21 days after *M. tuberculosis*) produced a less severe form of AA compared with control rats (33). This protective effect could not be mimicked by administration of hsp 65. In contrast, intravenous injection of 250 μg hsp 65 7, 9, and 11 days after *M. tuberculosis* immunization induced a more severe form of AA.

A further interesting approach to vaccination against AA was facilitated by the construction of a recombinant vaccinia virus expressing *M. bovis* hsp 65 gene. Injection of rats with the recombinant virus 7 days after *M. tuberculosis* immunization (i.e., during AA development) produced a marked decrease in AA severity compared with control rats receiving wild-type virus (49). Again the protective effect was linked with increased T-cell reactivity to hsp 65.

This correlation of T-cell responsiveness to hsp 65 and protection against AA might account for results of experiments using Fisher rats. Although Fisher rats and Lewis rats share the same MHC class II alleles ($RT1.B^1$ and D^1), Lewis rats are susceptible to AA, whereas Fisher rats are resistant. Thus, there appears to be a non-MHC genetic control over susceptibility. Interestingly, Fisher rats maintained in germ-free conditions are susceptible to AA (50). This was reversed by oral administration of *E. coli* 4 weeks prior *M. tuberculosis* immunization (51). It is reasonable to assume that this *E. coli* treatment might increase immune reactivity to GroEL (the *E. coli* hsp 60 homologue), and this might account for the AA-protective activity. One report suggested that the discrepancy between Lewis and Fisher rats regarding AA susceptibility was due to a lack of T-cell reactivity to the 180–188 epitope in Fisher rats. However, we have demonstrated recently that following immunization with hsp 65 Lewis and Fisher rats both show T-cell reactivity to an identical set of hsp 65 epitopes (S.M. Anderton, unpublished observations). Thus, the resistant Fisher rats do possess T cells which recognize the 180–188 epitope.

The other model in which protective effects of immunization with mycobacterial hsp 65 have been studied is pristane artritis in mice. Immunization with hsp 65 had clear protective effects (36) that were transferable with spleen cells

(52). In experiments to investigate this effect, responses of splenic T cells to overlapping synthetic peptides covering the hsp 65 sequence were tested. Arthritic mice (i.e., not preimmunized with hsp 65) showed proliferative responses to several groups of peptides representing potential epitopes throughout the molecule. In contrast, hsp 65–protected mice showed a more restricted pattern of responsiveness. Furthermore, arthritic mice showed responses to recombinant human hsp 60, whereas hsp 65–protected mice did not. This led to the suggestion that the pathogenesis of pristane arthritis involved activation of T cells recognizing epitopes in bacterial hsp 60. Some of these T cells would then cross react with endogenous self–hsp 60 in the joint to induce arthritis through molecular mimicry. The lack of response to human hsp 60 in hsp 65–protected mice provoked the hypothesis that the protective effect is due to prevention of "epitope spreading" of T-cell reactivity to include self cross-reactive epitopes after subsequent pristane injection while allowing responses to bacterial unique epitopes to develop. This mechanism has been termed "repertoire limitation" (54).

VI. T–CELL REACTIVITY TO SELF–HSP 60 AS A PROTECTIVE MECHANISM IN ARTHRITIS

Vaccination with hsp 65 protects rats against AA (17,33,48) and arthritis models induced using streptococcal cell walls (35), collagen type II (48), and CP20-961 (48). Hsp 65 shares 48% amino acid identity with the homologous rat hsp 60, and there have been numerous reports of cross-reactive immunorecognition of mycobacterial hsp 65 and endogenous self–hsp 60 at the T-cell level (43, 55–57). Expression of mammalian hsp 60 is known to be upregulated as a physiological response to various stressful stimuli, and it has been shown to be elevated in inflamed synovia (58–60). This led to the hypothesis that T cells specific for self–hsp 60 might in some way play a role in the regulation of the inflammatory response (43). Further, this might account for the protective effect of preimmunization with hsp 65 via activation of T cells capable of recognizing self–hsp 60 and leading to improved regulation of future inflammatory episodes induced by administration of arthritogenic substances. To test this hypothesis, we analyzed T-cell epitopes in mycobacterial hsp 65 and rat hsp 60 recognized by AA-susceptible Lewis rats. To determine which epitopes are involved in protection against AA, we immunized rats with synthetic peptides containing individual epitopes and administered epitope-specific T-cell lines and assessed their protective effect.

To identify T-cell epitopes in hsp 65, we analyzed responses of primed lymph node cells and short-term hsp 65–specific T-cell lines derived from rats immunized with hsp 65 to synthetic peptides covering the entire hsp 65 sequence (61). Proliferation assays identified nine epitopes (Table 2), with responses to epitopes 176–190, 211–225, and 226–240 being dominant. The pattern of epitope recognition differed following immunization with *M. tuberculosis* compared with hsp 65 (Table 2). After one in vitro stimulation with hsp 65, T cells from *M. tuberculosis*/IFA-immunized rats showed dominant responses to the 176–190 epitope, whereas response to 211–225 and 226–240 were minor, and responses to 86–100 and 396–410 were absent. Therefore, the AA-inducing protocol (*M. tuberculosis*/IFA immunization) led to a shift in responsiveness toward the AA-associated 180–188 epitope (contained within the 176–190 peptide). Responses to the 180–188 peptide were consistently found to be lower than when the longer 176–190 peptide was used. This stronger stimulation with 176–190 was presumably due to higher binding affinity for MHC or longer half-life in culture.

We generated hsp 65 epitope-specific T-cell lines from rats immunized with hsp 65 by repeated in vitro restimulation with peptides containing individual epitopes as defined above. This approach produced eight $\alpha\beta$ TCR$^+$, CD4$^+$, RT1.B^l-restricted T-cell lines. All responded to their specific peptide and hsp 65. Lines were tested for cross-reactive recognition of corresponding peptides of rat hsp 60. This showed that only one line, H.52 recognizing 256–270, cross reacted with rat hsp 60. This line recognized the core epitope 256–265, a sequence highly conserved between hsp 65 and rat hsp 60, with 7 of 10 residues identical and the other three representing conservative substitutions (Fig. 1).

We tested the ability of individual epitope-specific T-cell lines and peptide vaccination to influence AA development (62). Intravenous administration of the cross-reactive T-cell line H.52 at the time of *M. tuberculosis* immunization reduced the severity of AA, whereas other lines had no effect (Fig. 2). Also, we

Table 2 Mycobacterial Hsp 65 Epitopes Recognized by Lewis Rat T Cells

Immunization		Hsp 65	Mt	T-cell line
Epitope sequence[a]				
86–100	**DGTTTATVLAQALVR**	+	–	H.18
176–190	EESNTFGLQLELTEG	++	++	H.36
211–225	AVLEDPYILLVSSKV	++	+	H.43
226–240	STVKDLLPLLEKVIG	++	+	H.46
256–270	ALSTLVVNKIRGTFK	+	+	H.52
386–400	ELKERKHRIEDAVRN	+	+	–
396–410	DAVRNAKAAVEEGIV	+	–	H.80
446–460	APLKQIAFNSGLEPG	+	+	H.90
511–525	FLTTEAVVADKPEKE	+	+	H.103

[a]The sequences of hsp 65 peptides used to generate each line are shown. Core epitopes, as defined by responses to overlapping peptides, are denoted by underlined residues. Residues sharing identity with the corresponding sequence of rat hsp 60 are in **boldface**.

Differential recognition of epitopes following differing immunization and restimulation protocols are summarized: – no response; + minor response; ++ dominant response. Modified from Ref. 61.

Mt 256-270 ALSTLVVNKI RGTFK
Rat 256-270 ALSTLVLNRL KVGLQ

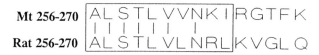

Figure 1 Comparison of the 256–270 sequences of mycobacterial hsp 65 and rat hsp 60. Amino acid sequences of mycobacterial hsp 65 and rat hsp 60 in the 256–270 region recognized by the cross-reactive T-cell line H.52 are shown. The core epitope of 256–265 is boxed. (From Refs. 61 and 62.)

immunized rats with 100 μg of synthetic peptides containing individual epitopes described above. Rats were immunized 7 days prior to AA induction using *M. tuberculosis*. This approach showed that immunization with 256–270 containing the cross-reactive epitope had a strong protective effect against AA (Fig. 3). In five independent experiments using 24 rats, immunization with 256–270

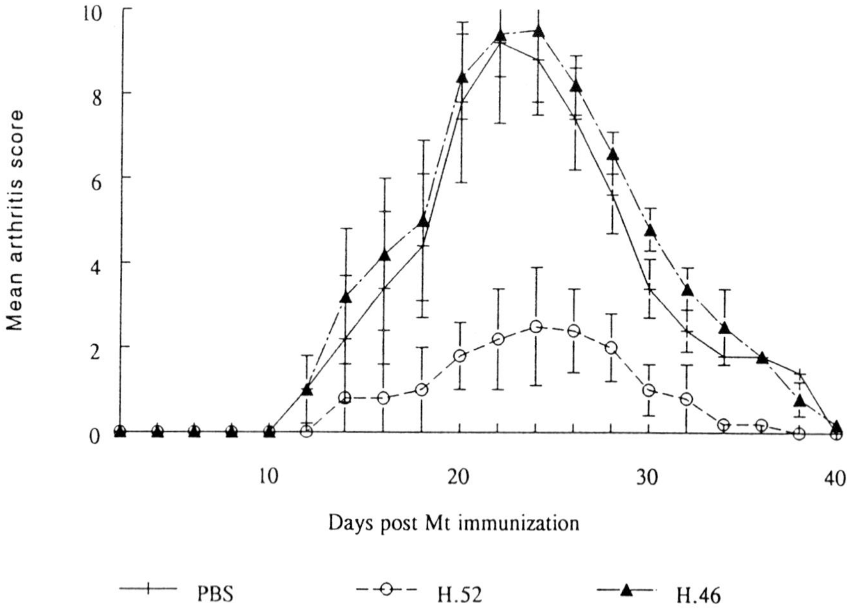

Figure 2 T cells recognizing self–hsp 60 can reduce AA severity. Lewis rats were administered with hsp 65 epitope–specific T-cell lines at the same time as AA induction with *M. tuberculosis* and effects on AA measured. Line H.52, recognizing the cross-reactive 256–270 epitope in rat hsp 60, clearly reduced severity of AA. All other T-cell lines tested specific for epitopes unique to mycobacterial hsp 65 (line H.46, specific for epitope 226–240 shown here) had no effect on AA. (From Ref. 62.)

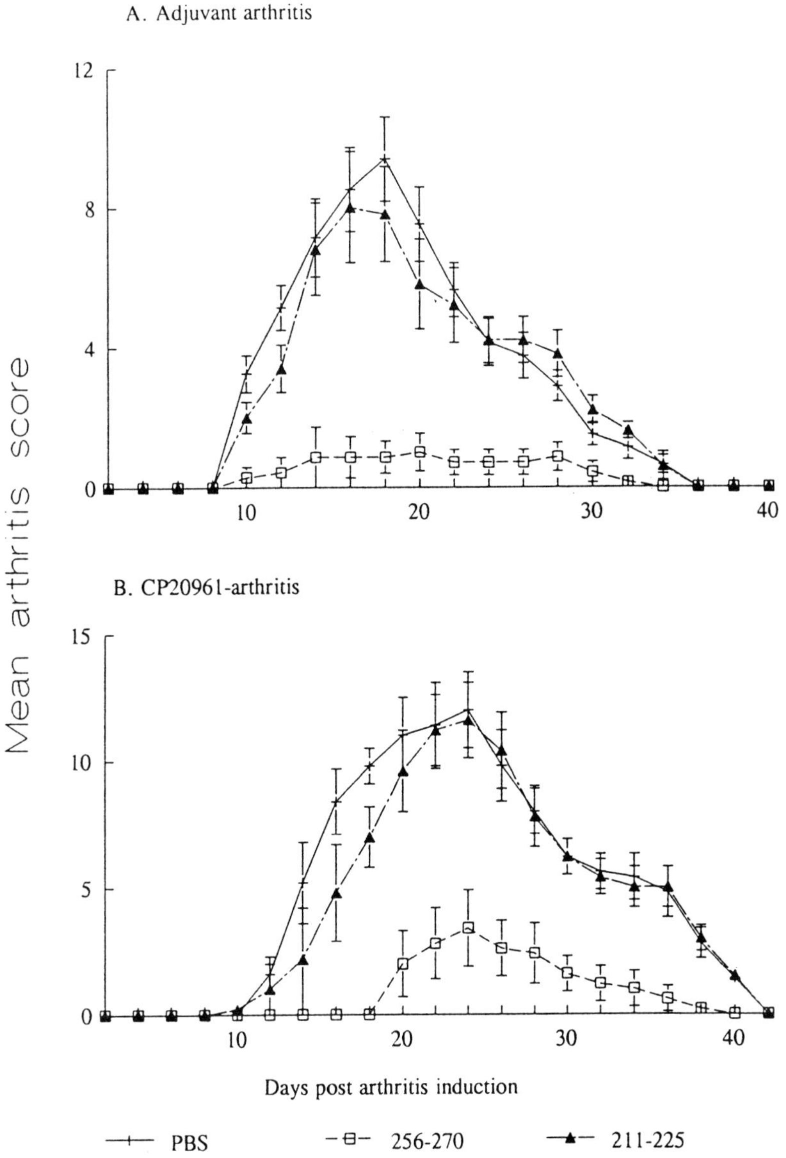

Figure 3 Vaccination with the cross-reactive 256–270 epitope protects against arthritis. Lewis rats were immunized with 100 μg of synthetic peptides corresponding to individual mycobacterial hsp 65 epitopes (as defined in Table 2). Seven days later arthritis was induced by injection of either 0.5 mg *M. tuberculosis* (A) or 2 mg CP20961 (B). Only peptide 256–270 could protect against AA. Peptide 256–270 (but not control peptide 211–225) could protect against CP20961 arthritis. (Modified from Ref. 62.)

resulted in a mean maximum arthritis score of 2.4 compared with 11.5 for control rats receiving PBS. No other hsp 65 epitope had any significant effect on AA. A T-cell line generated from protected rats responded to rat 256–270 and heat shocked antigen-presenting cells, indicating that the cross-reactive epitope was presented in association with MHC class II following processing of endogenous self–hsp 60.

The hypothesis that vaccination against arthritis with hsp 65 involves activation of T cells recognizing a shared epitope in rat hsp 60 does not require the "protective" T cell to recognize a mycobacterial component, thereby accounting for protection in models not using bacterial-derived arthritogens. The most notable of these models in the Lewis rat uses CP20961, a synthetic lipoidal amine with no antigenic cross reactivity with hsp 65. Immunization with 256–270 (but not the mycobacterial unique 211–225 epitope) protected rats against CP20961-induced arthritis (see Fig. 3). Therefore, activation of T cells that recognized the cross-reactive 256–265 epitope in rat hsp 60 protected rats against AA and CP20961 arthritis.

These results provide clear evidence that T cells recognizing a self–hsp 60 epitope can have beneficial effects on arthritis development via a mechanism independent of mycobacteria as the arthritogenic agent. These findings suggest strongly that the mechanism underlying the protective effect involves T-cell recognition of the self–hsp 60 epitope expressed by MHC class II–bearing cells in the inflamed joint. Raised expression due to cell stress during inflammation would make hsp 60 a reliable target antigen for regulatory T-cell recognition controlling inflammatory responses. T cells with low affinity for self–hsp 60, maintained in the periphery by exposure to shared epitopes in bacterial hsp 60 cognates, could be utilized as an antigen-specific mechanism of regulating inflammation. This is in accordance with observations that exposure to endogenous gut flora and BCG can protect against arthritis (51) and diabetes (63). Cross-reactive T-cell responses to an epitope in the highly conserved 243–265 regions of mycobacterial and human hsp 60 have been reported in rheumatoid arthritis (57). Also patients with juvenile chronic arthritis with mild oligoarticular disease of a remitting course show T-cell responses to recombinant human hsp 60 (64), whereas patients with severe disease do not (A.B.J. Prakken, personal communication). If these findings can be extended to show that patients with severe disease lack appropriate responses to self hsp 60, an exciting opportunity for specific immunotherapy by vaccination for T-cell responsiveness to self–hsp 60 would be available.

VII. CONCLUSIONS

The identification of residues 180–188 of mycobacterial hsp 65 as the epitope recognized by arthritogenic T cells in the AA model proved a major advance in the dissection of the molecular mechanisms underlying arthritis pathogenesis. T-cell reactivities to hsp 60 molecules in several (although as yet not all) other arthritis models have been found. The ubiquitous and highly conserved nature of hsp 60 cognates makes them candidate "autoantigens" in human arthritis, particularly reactive arthritis as a consequence of bacterial infection. The AA model has been exploited to demonstrate that immunization with modified analogues of the 180–188 epitope can be used to subvert the development of overt arthritis, presumably via an influence on potentially arthritogenic T cells. Vaccination with mycobacterial hsp 65 can protect against several arthritis models. The mechanisms underlying protection may vary; however, recent data indicate that activation of T cells recognizing self–hsp 60 epitopes in inflamed synovia can mediate protection. These findings suggest that modulation of T-cell recognition to self–hsp 60 can be used to improve regulation of inflammatory processes. These advances in understanding of the roles of T-cell recognition of hsp 60 provide exciting opportunities for antigen-specific therapy in human arthritis.

ACKNOWLEDGMENT

S. M. Anderton is supported by a Wellcome Trust Travelling Research Fellowship to Western Europe.

REFERENCES

1. L. Hang, A.N. Theofilopoulus, and F.J. Dixon, A spontaneous rheumatoid arthritis-like disease in MRL/l mice, *J. Exp. Med. 155*:1690 (1982).
2. D.E. Trentham, A.S. Townes, and A.H. Kang, Autoimmunity to collagen: an experimental model of arthritis, *J. Exp. Med. 146*:857 (1977).
3. P.H. Wooley, Collagen-induced arthritis in the mouse, *Methods Enzymol. 162*:361 (1988).
4. K. Mikecz, T.T. Glant, and A.R. Poole, Immunity to cartilage proteoglycans in BALB/c mice with progressive polyarthritis and ankylosing spondylitis induced by injection of human cartilage proteoglycan, *Arthritis Rheum. 30*:306 (1987).
5. C.M. Pearson, Development of arthritis, periarthritis and periostitis in rats given adjuvant, *Proc. Soc. Exp. Biol. Med. 91*:101 (1956).
6. W.J. Cromartie, J.G. Craddock, J.H. Schwab, S.K. Anderle, and C. Yang, Arthritis in rats after systemic injection of streptococcal cells or cell walls, *J. Exp. Med. 146*:1585 (1977).
7. T.J.A. Lehman, J.B. Allen, P.H. Plotz, and R.L. Wilder, *Lactobacillus casei* cell wall induced arthritis in rats: cell wall fragment distribution and persistence in chronic arthritis-susceptible LEW/N and resistant F344/N rats, *Arthritis Rheum. 27*:939 (1984).
8. S. Kleinau, H. Erlandsson, R. Holmdahl, and L. Klareskog, Adjuvant oils induce

arthritis in the DA rat. I. Characterization of the disease and evidence for immunological involvement, *J. Autoimmun. 4*:871 (1991).
9. Y-H. Chang, C.M. Pearson, and C. Abe, Adjuvant Polyarthritis IV. Induction by a synthetic adjuvant: Immunologic, histopathologic, and other studies. *Arthritis Rheum. 23*:62 (1980).
10. P.H. Wooley, J.R. Seibold, J.D. Whalen, and J.M. Chapdelaine, Pristane induced arthritis. The immunologic and genetic features of an experimental murine model of autoimmune disease, *Arthritis Rheum. 32*:1022 (1989).
11. D.J. Whitehouse, M.W. Whitehouse, and C.M. Pearson, Passive transfer of adjuvant induced arthritis and allergic encephalomyelitis in rats using thoracic duct lymphocytes, *Nature 224*:1332 (1969).
12. C.M. Pearson and F.D. Wood, Passive transfer of adjuvant arthritis by lymph node or spleen cells, *J. Exp. Med. 120*:547 (1964).
13. J.D. Taurog, G.P. Sanberg, and M.L. Mahowald, The cellular basis of adjuvant arthritis. II. Characterization of the cells mediating passive transfer, *Cell. Immunol. 80*:198 (1983).
14. J. Holoshitz, Y. Naparstek, A. Ben-Nun, and I.R. Cohen, Lines of T lymphocytes induce or vaccinate against autoimmune arthritis, *Science 219*:56 (1983).
15. J. Holoshitz, A. Matitiau, and I.R. Cohen, Arthritis induced in rats by clones of T lymphocytes responsive to mycobacteria but not to collagen type II, *J. Clin. Invest. 73*:211 (1984).
16. J.E.R. Thole, H.G. Dauwerse, P.K. Das, D.G. Groothuis, L.M. Schouls, and J.D.A. van Embden, Cloning of the *Mycobacterium bovis* BCG DNA and expression of antigens in *Escherichia coli, Infect. Immun. 50*:800 (1985).
17. W. Van Eden, J.E.R. Thole, R. van der Zee, A. Noordzij, J.D.A. van Embden, E. J. Hensen, and I.R. Cohen, Cloning the mycobacterial epitope recognized by T lymphocytes in adjuvant arthritis. *Nature 331*:171 (1988).
18. T.M. Shinnick, M.H. Vodkin, and J.C. Williams, The *Mycobacterium tuberculosis* 65-kilodalton antigen is a heat shock protein which corresponds to common antigen and to the *Escherichia coli* GroEL protein, *Infect. Immun. 56*:446 (1988).
19. J.E.R. Thole, P. Hindersson, J.De Bruyn, F. Cremers, R. van der Zee, H. De Cock, J. Tommassen, W. van Eden, and J.D.A. van Embden, Antigenic relatedness of a strongly immunogenic mycobacterial protein with a similarly sized ubiquitous bacterial common antigen, *Microb. Pathol. 4*:71 (1988).
20. S. Jindal, A.K. Dubani, B. Singh C.B. Harley, and R.S. Gupta, Primary structure of a human mitochondrial protein homologous to the bacterial and plant chaperonins and to the 65-kilodalton mycobacterial antigen, *Mol. Cell. Biol. 9*:2279 (1989).
21. T.J. Venner and R.S. Gupta, Nucleotide sequence of mouse HSP60 (chaperonin, GroEL homolog) cDNA, *Biochim. Biophys. Acta 1087*:336 (1990).
22. T.J. Venner and R.S. Gupta, Nucleotide sequence of rat hsp60 (chaperonin, GroEL homolog) cDNA, *Nucleic Acids Res. 18*:5309 (1990).
23. M.B.A. Oldstone, Molecular mimicry and autoimmune disease, *Cell 50*:819 (1987).
24. W. Van Eden, J. Holoshitz, Z. Nevo, A. Frenkel, A. Klajman, and I.R. Cohen, Arthritis induced by a T-lymphocyte clone that responds to *Mycobacterium tuberculosis* and to cartilage proteoglycans, *Proc. Natl. Acad. Sci. USA 82*:5117 (1985).

25. C.P.M. Broeren, G.M. Verjans, W. van Eden, J.G. Kusters, J.A. Lenstra, and T. Logtenberg, Conserved nucleotide sequences at the 5′ end of T cell receptor variable genes facilitate polymerase chain reaction amplification, *Eur. J. Immunol. 21*:569 (1991).
26. W.E. Paul and R.A. Sieber, Lymphocyte responses and cytokines, *Cell 76*:241 (1994).
27. C.O. Jacob, J. Holoshitz, P. van der Meide, S. Strober, and H.O. McDevitt, Heterogeneous effects of IFN-gamma in adjuvant arthritis, *J. Immunol. 142*:1500 (1989).
28. A. Miller, A. Al-Sabbagh, L.M.B. Santos, M. Prabhu Das, and H.L. Weiner, Epitopes of myelin basic protein that trigger TGF-β release after oral tolerization are distinct from encephalitogenic epitopes and mediate epitope-driven bystander suppression, *J. Immunol. 151*:7307 (1993).
29. B. Wilbrink, M. Holewijn, J.W.J. Bijsma, C.J.P. Boog, W. den Otter, and W. van Eden, Antigen-specific T cell activation induces inhibition of cartilage proteoglycan synthesis independent of T cell proliferation, *Scand. J. Immunol. 36*:733 (1992).
30. I.R. Cohen, Regulation of autoimmune disease: physiological and therapeutic, *Immunol. Rev. 94*:5 (1986).
31. O. Lider, N. Karin, M. Shinitzky, and I.R. Cohen, Therapeutic vaccination against adjuvant arthritis using autoimmune T cells treated with hydrostatic pressure, *Proc. Natl. Acad. Sci. USA 84*:4577 (1987).
32. C.P.M. Broeren, M.A. Lucassen, M.J.B. van Stipdonk, R. van der Zee, C.J.P. Boog, J.G. Kusters, and W. van Eden, CDR1 T cell receptor β-chain peptide induces MHC class II restricted T-T cell interactions, *Proc. Natl. Acad. Sci. USA 91*:5997 (1994).
33. E.J.M. Hogervorst, J.P.A. Wagenaar, C.J.P. Boog, R. van der Zee, J.D.A. van Embden, and W. van Eden, Adjuvant arthritis and immunity to the mycobacterial 65kD heat shock protein, *Intern. Immunol. 4*:719 (1992).
34. S.H.E. Kaufmann, U. Vath, J.E.R. Tole, J.D.A. van Embden, and F. Emmerich, Enumeration of T cells reactive with *Mycobacterium tuberculosis* organisms and specific for the recombinant mycobacterial 64-kDa, *Eur. J. Immunol. 17*:351 (1987).
35. M.F. Van den Broek, E.J.M. Hogervorst, M.C.J. van Bruggen, W. van Eden, R. van der Zee, and W.B. van der Berg, Protection against streptococcal cell wall-induced arthritis by pretreatment with the 65-kD mycobacterial heat shock protein, *J. Exp. Med. 170*:449 (1989).
36. S.J. Thompson, G.A.W. Rook, R.J. Brealey, R. van der Zee, and C.J. Elson, Autoimmune reactions to heat-shock proteins in pristane-induced arthritis, *Eur. J. Immunol. 20*:2479 (1990).
37. S.J. Thompson and C.J. Elson, Susceptibility to pristane-induced arthritis is altered with changes in bowel flora, *Immunol. Lett. 36*:227 (1993).
38. J-Y. Leroux, A.R. Poole, C. Webber, V. Vipparti, H.U. Choi, L.C. Rosenberg, and S. Banerjee, Characterization of proteoglycan-reactive T cell lines and hybridomas from mice with proteoglycan-induced arthritis, *J. Immunol. 148*:2090 (1992).
39. X-D. Yang, J. Gasser, and U. Feige, Prevention of adjuvant arthritis in rats by a

nonpeptide from the 65-kD mycobacterial heat-shock protein, *Clin. Exp. Immunol. 81*:189 (1990).
40. X-D. Yang, J. Gasser, and U. Feige, Prevention of adjuvant arthritis in rats by a nonpeptide from the 65-kD mycobacterial heat shock protein: specificity and mechanism, *Clin. Exp. Immunol. 87*:99 (1992).
41. A. Gaur, B. Wiers, A. Liu, J. Rothbard, and C.G. Fathman, Amelioration of autoimmune encephalomyelitis by myelin basic protein synthetic peptide-induced anergy, *Science 258*:1491 (1992).
42. C.A. Hicks, A.E. Kingston, M.J. Colston, and M.E.J. Billingham, Oil vehicles have a similar protective effect against adjuvant arthritis as heat shock proteins, *Br. J. Rheumatol. 32(Supl. 1)*:33 (1993).
43. S.M. Anderton, R. van der Zee, and J.A. Goodacre, Inflammation activates self hsp60-specific T cells, *Eur. J. Immunol. 23*:33 (1993).
44. M.H.M. Wauben, C.J.P. Boog, R. van der Zee, I. Joosten, A. Schlief, and W. van Eden, Disease inhibition of major histocompatibility complex binding peptide analogues of diseases-associated epitopes: More than blocking alone, *J. Exp. Med. 176*:667 (1992).
45. M.H.M. Wauben, R. van der Zee, I. Joosten, C.J.P. Boog, A.M.C. van Dijk, M.C. Holewijn, R.H. Meloen, and W. van Eden, A peptide variant of an arthritis-related T cell epitope induces T cells that recognize this epitope as a synthetic peptide but not in its naturally processed form, *J. Immunol. 150*:5722 (1993).
46. B.D. Evavold, J. Sloan-Lancaster, and P.M. Allen, Tickling the TCR: selective T cell functions stimulated by altered peptide ligands, *Immunol. Today 14*:602 (1993).
47. D.E. Smilek, D.C. Wraith, S. Hodgkinson, S. Dwivedy, L. Steinman, and H.O. McDevitt, A single amino acid change in a myelin basic protein peptides confers the capacity to prevent rather than induce experimental autoimmune encephalomyelitis, *Proc. Natl. Acad. Sci. USA 88*:9633 (1991).
48. M.E.J. Billingham, S. Carney, R. Butler, and M.J. Colston, 1990, A mycobacterial heat shock protein induces antigen-specific suppression of adjuvant arthritis, but is not itself arthritogenic, *J. Exp. Med. 171*:339 (1990).
49. E.J.M. Hogervorst, L. Schouls, J.P.A. Wagenaar, C.J.P. Boog, W.J.M. Spaan, J.D.A. van Embden, and W. van Eden, Modulation of experimental autoimmunity: treatment of adjuvant arthritis by immunization with a recombinant vaccinia virus, *Infect. Immun. 59*:2029 (1992).
50. O. Kohashi, J. Kutawa, K. Umehara, F. Uemara, T. Takahashi, and A. Ozawa, Susceptibility to adjuvant-induced arthritis among germfree, specific pathogen free and conventional rats. *Infect. Immun. 26*:791 (1979).
51. Kohashi, O., Y. Kohashi, T. Takahashi, A. Ozawa, and N. Shigematsu, Suppressive effect of *Escherichia coli* on adjuvant-induced arthritis in germ-free rats, *Arthritis Rheum. 29*:547 (1986).
52. M. Ghoraishian, C.J. Elston, and S.J. Thompson, Comparison between the protective effects of mycobacterial 65-kD heat shock protein and ovomucoid in pristane-induced arthritis: relationship with agalactosyl IgG, *Clin. Exp. Immunol. 94*:247 (1993).
53. Deleted in proof.

54. C.J. Elston, R.N. Barker, S.J. Thompson, and N.A. Williams, Immunologically autoreactive T cells, epitope spreading and repertoire limitation, *Immunol. Today 16*:71 (1995).
55. M.E. Munk, B. Schoel, S. Modrow, R.W. Karr, R.A. Young, and S.H.E. Kaufmann, T lymphocytes from healthy individuals with specificity to self-epitopes shared by the mycobacterial and human 65-kilodalton heat shock protein, *J. Immunol. 143*:2844 (1989).
56. J.R. Lamb, V. Bal, P. Mendez-Samperio, A. Mehlert, A. So, J. Rothbard, S. Jindal, R.A. Young, and D.B. Young, Stress proteins may provide a link between the immune response to infection and autoimmunity, *Intern. Immunol. 1*:191 (1989).
57. A.J. Quayle, K.B. Wilson, S.G. Li, J. Kjeldsen-Kragh, F. Oftung, T. Shinnick, M. Sioud, O. Forre, J.D. Capra, and J.B. Natvig, Peptide recognition, T cell receptor usage and HLA restriction elements of human heat-shock protein (hsp) 60 and mycobacterial 65-kDa hsp-reactive T cell clones from rheumatoid synovial fluid, *Eur. J. Immunol. 22*:1315 (1992).
58. A. Karlsson-Parra, K Soderstrom, M. Ferm, J. Ivanyi, R. Kiessling, and L. Klareskog, Presence of human heat shock protein (hsp) in inflamed joints and subcutaneous nodules of RA patients, *Scand. J. Immunol. 31*:283 (1990).
59. S. Kleinau, K. Soderstrom, R. Kiessling, and L.A. Klareskog, A monoclonal antibody to the mycobacterial 65 kD heat shock protein (ML 30) binds to cells in normal and arthritic joints of rats, *Scand. J. Immunol. 33*:195 (1991).
60. C.J.P. Boog, E.R. de Graeff-Meeder, M.A. Lucassen, R. van der Zee, M.M. Voorhorst-Ogink, P.J.S. van Kooten, H.J. Geuze, and W. van Eden, Two monoclonal antibodies generated against human hsp60 show reactivity with synovial membranes of patients with juvenile chronic arthritis, *J. Exp. Med. 175*:1805 (1992).
61. S.M. Anderton, R. van der Zee, A. Noordzij, and W. van Eden, Differential mycobacterial 65-kDa heat shock protein T cell epitope recognition after adjuvant arthritis inducing or protective immunization protocols, *J. Immunol. 152*:3656 (1994).
62. S.M. Anderton, R. van der Zee, A.B.J. Prakken, A. Noordzij, and W. van Eden, Activation of T cells recognizing self 60 kD heat-shock protein can protect against experimental arthritis, *J. Exp. Med. 181*:943 (1995).
63. N. Shehadeh, F. Calcinaro, B.J. Bradley, I. Bruchim, P. Vardi, and K.J. Lafferty, Effect of adjuvant therapy on development of diabetes in mouse and man, *Lancet 343*:706 (1994).
64. E.R. De Graeff-Meeder, R. van der Zee, G.T. Rijkers, H-J. Schuurman, W. Kuis, J.W.J. Bijlsma, B.J.M. Zegers, and W. van Eden, Recognition of human 60 kD heat shock protein by mononuclear cells from patients with juvenile chronic arthritis, *Lancet 337*:1368 (1991).

5

Heat Shock Protein 60 and the Regulation of Autoimmunity

Irun R. Cohen
The Weizmann Institute of Science, Rehovot, Israel

I. INTRODUCTION

The consolidation of an established position on the front line is a prerequisite for a military or political advance into new territory. The point is obvious; in geographical space, old holdings need to be strong to win new holdings. Ideas, however, differ radically from real estate; the conquest of a new idea requires not the consolidation of an established position but rather its abandonment. As Popper has taught, good scientific theories are falsifiable theories; we advance by taking them apart (1). The subject of this book is stress proteins in various fields of medicine: inflammation, degenerative diseases, cancer, infectious diseases, and autoimmune diseases. The details are fascinating and informative, but in the area of immunology, stress proteins are revolutionary: The way the immune system relates to them appears to contradict the expectations implied in the established paradigm of the science. My aim here is to recount explicitly some of the paradoxes of stress protein immunology. I will use the example of heat shock protein (hsp 60) and its role in autoimmune diabetes and other conditions. However, the lessons of hsp 60 immunology are likely to hold for other stress proteins and for other highly conserved molecules that are not stress proteins. By dismantling an established position, we may hope to make some progress toward a new idea.

II. THE CLONAL SELECTION PARADIGM

According to the classic formulation of the clonal selection theory (2), the essence of the immune system and its distinctive feature is its repertoire of antigen receptors. These receptors, which include the T- and B-cell antigen receptors and the antibodies, are produced by somatic genetic recombinations and mutations (3) and so are created anew in each individual. The selective force that determines the frequencies of the various receptors in the repertoire is exerted by the antigens that happen to activate clones of receptor-bearing cells in the individual. The dominant specificities of the receptor repertoire are the products of cell selection by antigens, a selection that takes place in the individual and not in the germ-line of the species where Darwinian selection usually proceeds. Thus, the elaboration of the receptor repertoire, according to classic clonal selection, reflects the immune history of the individual rather than that of the species.

The classic clonal selection theory proposed that tolerance to self-antigens required that the repertoire be purged of all cells bearing receptors capable of recognizing self-molecules (4). The recent demonstration of negative selection in the thymus is seen as the modern vindication of the classic notion of clonal deletion of anti–self reactivity as the cornerstone of self-tolerance (5). Thus, the potentially open-ended receptor repertoire of the immune system is hedged by a developmental bias against autoimmunity. But the bias against self-recognition does not weaken the classic view that the structure of the immune system is in the hands of the arrant antigens that select those clones of lymphocytes bearing complementary receptors. The fundamental view of the clonal selection paradigm is that the immune response is regulated by two factors: the concentration of availablc antigen and the concentration of lymphocytes with specific receptors.

A. The Clonal Selection View of Autoimmunity

The concept of positive selection of particular lymphocyte clones by the foreign antigens and the negative selection by the self-antigens has fostered a number of derivative ideas about autoimmunity and autoimmune diseases:

1. The healthy immune system, by definition, can have no autoimmune lymphocytes because autoimmune lymphocytes, by their nature, must cause autoimmune disease.
2. An autoimmune disease expressed against a particular organ should be attributed to a self-antigen specific for that organ; a target antigen expressed in many organs should lead to a disease in each of them.
3. An autoimmune disease can best be terminated by getting rid of the autoimmune lymphocyte clone or by paralyzing or blocking its activity;

the cure of a disease is contingent on the removal of its cause. Therapy should be based on inactivating the autoimmune lymphocytes.

III. THE CONTRADICTIONS OF HSP 60 AUTOIMMUNITY

The conservation of stress proteins throughout evolution may be explained by their performance of essential functions and by the dependence of this performance on particular structures encoded in unique amino acid sequences; to maintain the functions of stress proteins, evolution, during its meandering course, has had to preserve their key sequences. Other chapters in this book discuss the functions and structures of stress proteins. Whatever these functions may be, the conservation of the amino acid sequences of stress proteins and their universal expression in apparently all cells makes these proteins immunologically interesting; any parasitic cell, prokaryote or eukaryote, will express stress proteins that are sure to have epitopes similar or identical to those expressed by the cells of the host. Therefore, stress proteins challenge the host's immune system to distinguish between the invader and the self (6). An example is hsp 60, in which the mammalian host and its prokariotic invaders express proteins that are about 50% identical in amino acid sequence (7). How does the immune system relate to such a selflike molecule as is hsp 60? The answers to this question seem to contradict all the three principles of autoimmunity derivable from the clonal selection paradigm outlined above.

1. Autoimmunity to hsp 60 is compatible with health. Humans were found to manifest T-cell proliferative responses to peptides with amino acid sequences shared by mycobacterial hsp 60 and human hsp 60 (8). Such anti–hsp 60 reactivity appears to be present at birth in cord blood (9). We have extended the scope of hsp 60 autoimmunity and find that naive, healthy BALB/c mice bear helper T cells primed to a peptide epitope of mouse hsp 60 that is specific for the mouse and is not cross reactive with analogous bacterial hsp 60 sequences present in *Mycobacterium* or in *Escherichia coli* (10). Thus, healthy immune systems do manifest hsp 60 autoimmunity that appears to be specific for the self.

2. Autoimmunity to hsp 60 is also present regularly in diseases restricted to certain organs. We found, for example, that T-cell autoimmunity to a particular peptide epitope of hsp 60 is commonly present in mice of the NOD strain spontaneously developing insulin-dependent diabetes mellitus (IDDM) (11,12), in humans newly diagnosed as suffering from the same disease (13), and in mice of the C57BL/ksj strain induced to develop IDDM by exposure to a low dose of the β-cell toxin streptozotocine (14). This peptide, designated p277 by its order number in the peptide synthesizer, is composed of the string of amino acids in positions 437–460 of the human or mouse hsp 60 sequences. A functional role of p277 immunity in IDDM also was supported by the observation that hyperglycemia and insulitis could be induced in several nondiabetic strains of

mice by active immunization to p277 covalently conjugated to a carrier protein such as bovine serum albumin or ovalbumin (15). Indeed T-cell lines specific for peptide p277 can adoptively transfer diabetes into prediabetic NOD mice (12) or into naive C57BL/6 mice (15). Thus, autoimmune T cells reactive to an epitope of the self–hsp 60 molecule appear to be a cause of diabetes. The fact that hsp 60, like other stress proteins, is expressible in every cell raises the question of how T-cell immunity to the p277 peptide can cause IDDM. Moreover, hsp 60 is not the only self antigen associated with IDDM; glutamic acid decarboxylase (GAD) (16) and other self antigens are targeted along with hsp 60 by the immune systems of humans and mice with IDDM (17). Thus, a specific collective of self-antigens is characteristic of IDDM across species.

But the paradox of autoimmune organ specificity is even more complicated. Autoimmunity to hsp 60 is associated not only with IDDM but also with inflammatory or autoimmune diseases of the joints (18), the eyes (19), the blood vessels (20), the nervous system (21), the skin (22), and probably other organs. Other chapters in this book review the details of the various conditions. The discussion here relates to the observations that autoimmunity to one self-antigen, hsp 60, may be associated with different autoimmune diseases; that autoimmunity to hsp 60 within a collective of other self-antigens, all of which are not organ specific, may be associated with organ-specific diseases; and that autoimmunity to hsp 60 may mark both health and disease.

3. Equally intriguing was the finding that a single administration of 50–100 μg of hsp 60 in saline or of p277, now without conjugation to a carrier protein, could induce a cure of IDDM. Treatment with peptide p277 prevented the development of IDDM in prediabetic NOD mice (12) and even reversed the insulitis in overtly diabetic NOD mice (13). Hence, an autoimmune disease can be controlled without abusing or blocking the autoimmune lymphocytes but by activating the immune system in a particular way (23). The context of autoimmune activation, whether or not peptide p277 is conjugated to a "helper" molecule and not self-recognition alone, seems to determine the biological consequences of the autoimmune response toward disease or toward health. How might we explain all these unexpected observations about hsp 60 autoimmunity?

IV. THE IMMUNOLOGICAL HOMUNCULUS

The paradoxes of hsp 60 autoimmunity can be ascribed to a principal misconception of the classic clonal selection theory: the behavior of the immune system is not controlled solely by chance contacts with antigens; it is no less constrained by the internal organization of the immune system (24). This organization reflects the evolutionary experience of the species. For example, the capacity of the receptor repertoire to interact with particular antigens is dictated by various elements encoded in the germ-line, such as the machinery of antigen processing and presentation, and the major histocompatibility complex

(MHC). Only those peptides that are processible and presentable in MHC clefts are visible to T cells. These elements direct the receptor-bearing cells to *focus* on certain motifs of certain antigens. In this way, the immune system avoids the potential chaos inherent in the ocean of antigens that could activate too many lymphocytes and so swamp the repertoire. Clonal selection, the embodiment of the individual's immune experience, is restricted by the germ-line wisdom of the species. This wisdom extends to the *context* of antigen recognition which is supplied by the cytokine milieu and the molecules controlling cell migration and adherence. Lymphocytes responsive to a particular antigen will produce different biological effects performing, for example, as Th1, Th2, or suppresser cells, depending on the cytokines present at the time of contact with the antigen (25). The organs in which the contact occurs can also influence the nature of the immune response (24). Since the context in which an antigen is seen determines the nature of the immune response, it is the context that endows antigen molecules with their biological meaning (26).

This cognitive theory of the immune system proposes that the system, through its internal organization, is programmed to respond to certain molecules in certain contexts. The details of this view of the immune system have been outlined elsewhere (24,27,28). The point I wish to stress here relates to natural autoimmunity; hsp 60 and the other members of the stress protein families belong to the set of conserved self-antigens to which there exists in healthy individuals a kind of benign autoimmunity. These natural self-antigens define the immunological homunculus; the immune system's internal picture of the individual's body encoded in the set of natural autoimmune lymphocytes that recognize these antigens (24,27,28). The immunological homunculus focuses autoimmunity on a limited and standard set of self-molecules. I believe that the paradoxes of hsp 60 autoimmunity (and of the autoimmunity to the other conserved stress proteins) may be explained by their membership within the set of self-antigens comprising the immunological homunculus. This membership is defined by two conditions: Individuals are born with a high frequency of T and B cells with receptors specific for hsp 60 epitopes and these autoimmune T and B cells are subject to control by anti-idiotypic and other regulatory cells activated in the system. This combination of natural priming and natural regulation sets the stage for hsp 60 autoimmunity as amplifier of inflammation, hsp 60 autoimmunity as restrainer of inflammation, and hsp 60 autoimmunity as a cause of specific disease.

V. HSP 60 AUTOIMMUNITY CAN AMPLIFY OR RESTRAIN INFLAMMATION

My colleagues and I are beginning to investigate the immunology of natural autoimmunity to hsp 60 by constructing transgenic mice with hsp 60 hyperexpressed under the regulation of different promoters. We are also studying the effects of natural autoimmunity to hsp 60 on the immune response to bacterial antigens. These studies are not yet completed, but it appears that natural autoimmunity to hsp 60 can indeed modify significantly both autoimmune diseases and immune responses to foreign antigens.

Interestingly, autoimmunity to hsp 60 can manifest seemingly contradictory effects: Some immune responses can be enhanced and others inhibited. Since expression of hsp 60 is upregulated by the stress of inflammation, it is not surprising that natural autoimmunity to hsp 60 can "help" the initiation of immune responses to other antigens that happen to be at the site of any hyperexpression of hsp 60. The high frequency of anti–hsp 60 T cells present from birth (9) guarantees that some of these anti–hsp 60 will be activated at the site of heightened expression of hsp 60. Thus hsp 60 autoimmunity will be detectable at some stage at the site of every inflammatory response. The natural autoimmune T cells that respond to hsp 60 can add their cytokines to the inflammatory cascade and so can amplify the inflammatory response to other antigens. The automatic activation of anti–hsp 60 cells in inflammatory sites probably accounts for the bewildering associations of hsp 60 autoimmunity with so many disease conditions. In these conditions, the hsp 60 autoimmunity acts as the immune system's internal "adjuvant" to amplify the immune response "nonspecifically."

In contrast to amplification, natural hsp 60 autoimmunity can also aid in downregulating or restraining inflammatory reactions. This is because the natural anti–hsp 60 autoimmune T cells can also activate the specific anti-idiotypic regulatory T cells that seem to exist for each of the autoimmune reactivities present in the immunological homunculus set (27,28). Through a mechanism of bystander suppression (29), the T cells that naturally regulate hsp 60 autoimmunity can shut off responses by other T cells reacting to adjacent antigens. The mechanisms responsible for these effects will become clearer as we and others pursue investigation of the hsp 60 transgenic and other new models. In any case, the central role of self-hsp 60 in stress is an obvious reason for having the immune system born with a bias toward hsp 60 reactivity; hsp 60 autoimmunity can provide both positive and negative feedback to the immune system in the stress that signifies the need for immune intervention.

VI. HSP 60 AUTOIMMUNE DISEASES

In addition to the "nonspecific" adjuvant and regulatory effects of natural hsp 60 autoimmunity, it is clear that certain autoimmune diseases can be caused by "specific" hsp 60 autoimmunity. The development of type I diabetes in the NOD mouse seems to fulfill the criteria for assigning such a causal relationship: anti–hsp 60 T cells can transfer diabetes (12), active immunization to the p277 epitope of hsp 60 can induce diabetes, even in mice that are not prone to diabetes (15), and the p277 peptide can be used to downregulate hsp 60 autoimmunity and cure diabetes (13). Another example of the causal relationship between hsp 60 autoimmunity and a specific disease seems to be a form of uveitis (19). The relationship between hsp 60 autoimmunity and arthritis is less clear (30); mimicry between an epitope of a bacterial specific hsp 60 sequence and a cartilage proteoglycan may be involved (31). Thus, adjuvant arthritis may combine bacteria-hsp 60 mimicry of a joint-specific molecule, along with a "nonspecific adjuvant" effect of true hsp 60 autoimmunity.

In any case, the role of hsp 60 autoimmunity in initiating an autoimmune disease would seem to require fulfillment of at least two conditions. The target organ would have to express hsp 60 in a way different from that of other tissues to allow the anti–hsp 60 T cells (or antibodies) to recognize hsp 60 target epitopes in the absence of stress. This seems to be true for insulin-producing β cells that express hsp 60 in their secretory granules unrelated to stress (32). A second condition for a specific disease would be a lapse or failure of the natural regulatory mechanisms that turn off the inflammatory process (27,28). A prolonged failure in regulation should lead to a chronic disease, such as is the insulitis in untreated NOD mice. Reinstatement of regulation would terminate the inflammation, causing the disease to remit "spontaneously." The diabetes actively induced by hsp 60 autoimmunity is indeed self-limited (15), like experimental autoimmune encephalomyelitis (EAE) and other induced autoimmune diseases (33).

A form of chronic autoimmune disease could conceivably be caused by unregulated hsp 60 autoimmunity that follows the induction of damage to the target tissue by some other primary cause. For example, we have reported that hsp 60 autoimmunity and diabetes can be induced in mice of the C57BL/ksj strain by a low dose of the β-cell toxin streptozotocine (14). Here the toxic insult initiates an inflammatory response that seems to be perpetuated by hsp 60 autoimmunity. Indeed, downregulating the hsp 60 autoimmunity *after* the toxic insult can cure the diabetes (34). Thus, the existence of natural autoimmunity to hsp 60, its membership in the homunculus set of self-antigens, can account for the many of the paradoxical associations of hsp 60 autoimmunity with various types of inflammation. The fundamental questions, however, have yet to be answered.

VII. OPEN QUESTIONS

The first question is directed to the development of the high frequency of "naturally" autoimmune lymphocytes specific for hsp 60 (9). How do these cells develop? The same question must be asked of the cells responsive to all of the self-antigens in the homunculus set. The classic clonal selection paradigm, because of its abhorrence of autoimmunity, drew attention to the process of *negative* selection of T cells during their development in the thymus (5). The immunological homunculus, which is the conceptualization of the empirical existence of natural autoimmunity, leads to the consideration of *positive* thymic selection of some autoreactive T cells. Certain exons of the gene for myelin basic protein have been found to be expressed in the thymus during particular developmental periods (35). Therefore, it is conceivable that hsp 60, along with the other homunculus self-antigens, is programmed to be expressed in the thymus to positively select for reactive T cells.

Another question relates to the development and activity of the regulatory circuits that control hsp 60 autoimmunity. What is the nature of these regulatory cells, where do they originate, and how do they affect the behavior of the hsp 60 autoimmune cells?

A third question relates to the epitopes of hsp 60 and their "specific" roles in particular situations such as diabetes, uveitis, arthritis, atherosclerosis, and in inflammation in general. Why and how are different peptides associated with different inflammatory conditions?

Obviously, questions cannot be answered until they are asked. At this stage, we have just become aware of the existence of hsp 60 autoimmunity within the context of the immunological homunculus. The value of this awareness is the development of an agenda of experimentation that makes our questions both legitimate and compelling.

ACKNOWLEDGMENTS

I thank Doris Ohayon for preparing this manuscript. I am the incumbent of the Mauerberger Chair in Immunology and the director of the Robert Koch-Minerva Center for Research in Autoimmune Disease, The Weizmann Institute of Science.

REFERENCES

1. K.R. Popper, *Conjectures and Refutations*, Basic Books, New York, 1962.
2. F.M. Burnet, *The Clonal Selection Theory of Acquired Immunity*, Vanderbilt University Press, Nashville, 1959.
3. J.L. Jorgensen, P.A. Recay, E.W. Ehrich, and M.M. Davis, Molecular components of T-cell recognition, *Annu. Rev. Immunol. 10*:835 (1992).
4. M. Burnet, *Self and Not-Self*, Cambridge University Press, Cambridge, England, 1969.
5. J.W. Kappler, N. Roehm, and P. Marrack, T cell tolerance by clonal elimination in the thymus, *Cell 49*:273 (1987).
6. I.R. Cohen, Autoimmunity to hsp65 and the immunologic paradigm, *Advances in Internal Medicine* (G.H. Stollerman, ed.), Mosby-Year Book, Chicago, 1992, Vol. 37, p. 295.

7. S. Jindal, A.K. Dudani, C.B. Harley, B. Singh, and R.S. Gupta, Primary structure of a human mitochondriloa protein homologous to bacterial and plant chaperonins and to hsp 65 K D mycobacterial antigen, *Mol. Cell. Biol. 9*:2279 (1989).
8. M.E. Munk, B. Schoel, S. Modrow, R.W. Karr, R.A. Young, and S.H.E. Kaufmann, *J. Immunol. 143*:2844 (1989).
9. H.P. Fisher, C.E. Sharrock, and G.S. Panayi, High frequency of cord blood lymphocytes against mycobacterial 65-kDa heat-shock protein, *Eur. J. Immunol. 22*:1667 (1992).
10. S. Könen-Waisman, M. Fridkin, and I.R. Cohen, Self and foreign hsp60 T-cell epitope peptides serve as immunogenic carriers for a T-cell independent sugar antigen, *J. Immunol.* In press.
11. D. Elias, D. Markovits, T. Reshef, R. van der Zee, and I.R. Cohen, Induction and therapy of autoimmune diabetes in the non-obese diabetic (NOD/LT) mouse by a 65-kDa heat shock protein, *Proc. Natl. Acad. Sci. USA 87*:1576 (1990).
12. D. Elias, T. Reshef, O.S. Birk, R. van der Zee, M.D. Walker, and I.R. Cohen, Vaccination against autoimmune mouse diabetes with a T-cell epitope of the human 65 kDa heat shock protein, *Proc. Natl. Acad. Sci. USA 88*:3088 (1991).
13. D. Elias, and I.R. Cohen, Peptide therapy for diabetes in NOD mice, *Lancet 343*: 704 (1994).
14. D. Elias, H. Prigozin, N. Polak, M. Rapoport, A.W. Lohse, and I.R. Cohen, Autoimmune diabetes induced by the β-cell toxin STZ. Immunity to the 60-kDa heat shock protein and to insulin, *Diabetes 43*:992 (1994).
15. D. Elias, H. Marcus, T. Reshef, V. Ablumunits, and I.R.Cohen, Induction of diabetes in standard mice by immunization to the p277 peptide of a 60kDa heat shock protein (submitted).
16. R. Tisch, X.D. Yang, S.M. Singer, R.S. Liblav, L. Fuggar, and H.O. McDevitt, Immune response to glutamic acid decarboxylase correlates with insulitis in non-obese diabetic mice, *Nature 366*:72–75 (1993).
17. L.C. Harrison, Islet cell antigens in insulin-dependent diabetes in NOD mice: implications for therapeutic intervention in human disease, *Immunol. Today 15*: 115–120 (1994).
18. W. van Eden, J. Thole, R. van der Zee, A. Noordzij, J.D.A. van Embden, E.J. Hensen, and I.R. Cohen, Cloning of the mycobacterial epitope recognized by T lymphocytes in adjuvant arthritis, *Nature 331*:171 (1988).
19. M.R. Stanford, E. Kasp, R. Whiston, A. Hasan, S. Todryk, T. Shinnick, Y. Miozushima, D.C. Dumonde, R. van der Zee, and T. Lehner, Heat shock protein peptides reactive in patients with Behçet's disease are uveitogenic in Lewis rats, *Clin. Exp. Immunol. 97*:226 (1994).
20. G. Wick, G. Schett, A. Amberger, R. Kleindienst, and Q. Xu, Is atherosclerosis an immunologically mediated disease?, *Immunol. Today 16*:27 (1995).
21. F. Mor and I.R. Cohen, T cells in the lesion of experimental autoimmune encephalomyelitis. Enrichment for reactivities to myelin basic protein and to heat shock proteins., *J. Clin. Invest. 90*:2447 (1992).
22. M.G. Danieli, M. Candela, A. M. Ricciatti, R. Reginelli, G. Danieli, I.R. Cohen,

and A. Gabrielli, Antibodies to mycobacterial 65 kDa heat shock protein in systemic sclerosis (Scleroderma), *J. Autoimmun.* *5*:443 (1992).
23. I.R. Cohen, Treatment of autoimmune disease: to activate or to deactivate? *Chem. Immunol.* *60*:150 (1995).
24. I.R. Cohen, The cognitive principal challenges clonal selection, *Immunol. Today* *13*:441 (1992).
25. T.R. Mosmann, and R.L. Coffman, TH1 and TH2 cells: Different patterns of lymphokine secretion lead to different functional properties, *Annu. Rev. Immunol.* *7*:145 (1989).
26. I.R. Cohen, Language, meaning and the immune system, *Isr. J. Med. Sci.* *31*:36 (1995).
27. I.R. Cohen, The cognitive paradigm and the immunological homunculus, *Immunol. Today* *13*:490 (1992).
28. I.R. Cohen and D.B. Young, Autoimmunity, microbial immunity, and the immunological homunculus, *Immunol. Today* *12*:105 (1991).
29. H.L. Weiner, A. Friedman, A. Miller, S.J. Khoury, Al-Sabbagh, L. Santos, M. Sayegh, R.B. Nussenblatt, D.E. Trentham, and D.A. Hafler, Oral tolerance: immunologic mechanisms and treatment of murine and human organ specific autoimmune disease by oral administration of autoantigens, *Ann. Rev. Immunol.* *12*:809 (1994),.
30. I.R. Cohen, Autoimmunity to chaperonins in the pathogenesis of arthritis and diabetes, *Annu. Rev. Immunol.* *9*:567 (1991).
31. W. van Eden, J. Holoshitz, Z. Nevo, A. Frenkel, A. Klajman, and I.R. Cohen, Arthritis induced by a T lymphocyte clone that responds to mycobacterium tuberculosis and to cartilage proteoglycans, *Proc. Natl. Acad. Sci. USA* *82*:5117 (1985).
32. K. Bruzynski, V. Martinez, R.S. Gupta, Secretory granule autoantigen in insulin-dependent diabetes mellitus is related to 62 kDa-heat shock protein (hsp60), *J. Autoimmun.* *5*:453 (1992).
33. A. Ben-Nun, and I.R. Cohen, Spontaneous remission and acquired resistance to autoimmune encephalomyelitis (EAE) are associated with suppression of T cell reactivity: Suppressed EAE effector T cells recovered as T cell lines, *J. Immunol.* *128*:1450 (1982).
34. D. Elias and I. R. Cohen, The hsp60 peptide p277 arrests the autoimmune diabetes induced by the toxin streptozotocin (submitted).
35. P. M. Mathisen, S. Pease, J. Garvey, L. Hood, and C. Readhead, Identification of an embryonic isoform of myelin basic protein that is expressed widely in the mouse embryo, *Proc. Natl. Acad. Sci. USA* *90:*10125 (1993).

6

Immune Responses to Heat Shock Proteins in Reactive Arthritis

J. S. H. Gaston and J. H. Pearce
University of Birmingham, Birmingham, England

I. DEFINITION OF REACTIVE ARTHRITIS

Reactive arthritis (ReA) is defined as an inflammatory arthritis which occurs in relation to specific infection at a site other than the joint (1,2). The occurrence of arthritis is strongly associated with the tissue antigen HLA-B27 (3), and the classic triggering infections which have been considered are those of the gastrointestinal and genitourinary tracts. In principle, infection at any other site could give rise to a reactive process in the joints, and there are reports of ReA in association with bronchopulmonary infection (4,5). Likewise, the reactive arthritis which is seen in association with mycobacterial infection (Poncet's disease) has received increasing attention (6,7), with new cases being described in relation to the use of BCG as immunotherapy in bladder cancer (8–10).

By definition, the organism causing infection at a site other than the joint cannot be isolated from the joint, since culture of an organism from the joint defines septic rather than reactive arthritis. This definition serves well as long as it is conceded that failure to culture the organism from the joint by conventional means does not necessarily imply that no organisms are ever present in joint tissue. Molecular and immunological techniques have been used in recent years to demonstrate that organisms or antigenic material from organisms may well be present within the ReA joint (11–18). Even when intact organisms such as chlamydiae are demonstrated, they would appear to be present in very small numbers and possibly in a noncultivable state. Since the pathological process in the joint of these patients has none of the hallmarks of sepsis as conventionally understood, it seems justifiable to retain the term *reactive arthritis* when organisms have not been isolated by conventional culture. This would allow inclusion of Lyme disease, since *Borrelia burgdorferi* has been demonstrated in synovium but not cultured from it (19) (see Dr. Sigal's article for further discussion on Lyme disease).

II. CURRENT VIEWS ON THE PATHOGENESIS OF REA

Synovial membrane and fluid show extensive infiltration by immunocompetent cells; T lymphocytes are present and activated, there are abundant antigen-presenting cells (macrophages and dendritic cells) which have been shown to have enhanced functional capacity as compared with their peripheral blood counterparts (20), and B cells/plasma cells are also to be seen. It seems safe to assume that the joint is the site of a specific immune response, and the main question concerns the nature of the antigens driving the response. There are two broad possibilities. The first possibility is that the process is entirely driven by antigens of the organism responsible for the previous infection. The second possibility is that the immune response triggered outside the joint has, in susceptible individuals (i.e., those that are $B27^{+}$), some "autoimmune" properties which result in the joint becoming a target for the immune response. Heat shock proteins (hsps) could feature in both scenarios: bacterial hsps could act as targets of the immune response to triggering infection, whereas human hsps (self) could be target autoantigens within the joint, recognized by T cells with a cross-reactive specificity for microbial and human hsps. Both these possibilities will be considered further below.

In distinguishing between these two hypotheses, the question of whether bacterial antigen is present in the joint is crucial. Accumulating evidence strongly suggests that this is so; *Chlamydia* organisms have been demonstrated by immunofluorescence (11), which has been recently backed up by detection of chlamydial nucleic acid using polymerase chain reaction (PCR) (13). For gram-negative organisms such as *Salmonella, Yersinia, and Shigella*, specific LPS, protein antigens (15–17), and synovial immune complexes containing bacterial antigen (21) have all been shown. Although structures with bacterial morphology have been seen in synovial membrane of patients with yersinia-induced ReA, PCR does not detect nucleic acid (22), suggesting that the organism is not intact in the joint even though antigenic material, usually phagocyte-associated, clearly is.

Given the presence of bacterial antigenic material in the joint, there is no pressing need to invoke cross reactivity with self-antigens as a hypothesis to explain the occurrence of arthritis, since there will be a conventional immune response to the intra-articular antigen. However, we are still left with the difficulty of explaining why reactive arthritis is so strongly associated with HLA-

B27. If bacterial antigen is disseminated to the joint in all infected individuals, why is arthritis not the rule rather than the exception? It is possible that B27 could have an effect on the dissemination process, possibly by affecting the immune response at the original mucosal site of infection, or by affecting the invasiveness of the organism. There are published data showing that B27$^+$ transfected L cells are *less* readily infected by salmonellae and shigellae compared with untransfected or HLA-A3–transfected cells (23), so the possibility that B27 exerts its effect in a way which is quite unrelated to the immune response to infection cannot be excluded. However, it is also possible that arthritis is related to the nature of the response mounted by B27$^+$ patients, including the generation of an autoimmune response. In this case, it can be argued that local bacterial antigen is not sufficient to maintain joint inflammation unless accompanied by an autoimmune response to a joint constituent.

Thus, in relation to heat shock proteins, we can ask:

1. Are hsp targets of the immune response to ReA-associated bacteria?
2. If so, are there any unusual properties of this response in patients who develop arthritis as compared with those with uncomplicated infection?

However, before addressing these questions directly, it may be worthwhile examining the stages in the immune response to ReA-associated bacteria in which recognition of hsp might play an important role.

III. IMMUNE RESPONSES TO REA–ASSOCIATED BACTERIA

There are at least two important properties of the principal ReA-associated infections which have implications for the immune response directed against them. First, these organisms are first encountered at mucosal surfaces. Second, the organisms are either obligate or facultative intracellular parasites, so that the organism may be compelled to alter its antigenic structure in order to survive intracellularly.

A. Mucosal Immune Responses to ReA-Associated Bacteria

Several immune mechanisms operate at mucosal surfaces, particularly the production of specific immunoglobulin (IgA). Indeed patients with yersinia-induced ReA have notably prolonged and high titers of yersinia-specific IgA (24,25), suggesting persistence of the organism at a mucosal site and continued stimulus for IgA production. Among the T cells active at mucosal sites are those bearing the $\gamma\delta$ T-cell receptor (TCR) and also CD8$^+$ TCR-$\alpha\beta^+$ cells which differ from their peripheral blood counterparts in their expression of CD8 (they express the CD8 $\alpha\alpha$ homodimer rather than the $\alpha\beta$ heterodimer), and in their differentiation, which may not require thymic selection (26,27). Nothing is known about the responses of either of these T-cell subsets to ReA-associated organisms. However, several known properties of $\gamma\delta$ cells are relevant.

First, $\gamma\delta$ T cells are able to detect cells infected by intracellular organisms. This has been shown most clearly with respect to mycobacteria (28) but also for yersiniae (29). These findings are in line with the proposal that at epithelial surfaces $\gamma\delta$ cells have an immunosurveillance role. The mechanisms whereby infected cells are detected by $\gamma\delta$ T cells remain obscure, but hsps have been proposed as target antigens. Hsp-specific $\gamma\delta$ T cells have been described in both mouse and humans, again particularly with respect to mycobacterial hsp 60 (30,31). Indeed, some of the earliest characterized clones with this specificity were isolated from the joint, albeit from a case of rheumatoid arthritis (32). The way in which hsps are recognized by $\gamma\delta$ T cells remains obscure, since conventional major histocompatibility complex (MHC) restriction is not evident. The class I MHC-like antigen CD1 may play a role in some instances (33,34), but whether this molecule also presents peptides, and their processing and transport requirements (if any), are also unknown. Recently, we have isolated from synovial fluid of a patient with ReA $\gamma\delta$ T cell clones which are specifically cytolytic toward yersinia-infected cells; these clones were still able to kill yersinia-infected cells deficient in class I MHC antigens or peptide transporters (J. Young and J.S.H. Gaston, manuscript in preparation).

In view of our lack of knowledge of the role of mucosal T cells in the normal immune response to ReA-associated pathogens, it is idle to speculate how this might be abnormal in $B27^+$ individuals. In the HLA-B27 transgenic rat, B27 need only be present on bone marrow–derived cells, immunocytes, and not on the enterocytes (35). Therefore, in this model, it is less likely that B27 plays its role by affecting antigen presentation to intraepithelial lymphocytes. Likewise, since neither $\gamma\delta$ T cells nor intraepithelial $CD8^+$ cells require thymic education, B27 is unlikely to act by influencing the TCR repertoire of these cells.

B. Immune Responses to Intracellular Organisms

Immune responses to intracellular infection have mainly been studied in relation to viral infections where synthesis of viral proteins is clearly cytoplasmic. Under these circumstances, viral peptides are efficiently loaded onto class I MHC molecules during their assembly and then displayed to the immune system on the cell surface (36). Intracellular bacteria differ, since many are confined to membrane-enclosed vesicles following their ingestion, so that synthesis of bacterial proteins is not occurring in the cytoplasm. In other cases, bacteria have mechanisms, such as the *Listeria* hemolysin, for the bacteria to enter the cytoplasm (37). Nevertheless, even where bacteria remain within vesicles, it is becoming clear that there are mechanisms for ensuring access of bacteria-derived peptides to the class I MHC presentation pathway (38). The existence of such a pathway accounts for bacteria-specific responses by the CD8$^+$ T-cell subset (39). Mice which are unable to generate CD8 T-cell responses, because of knockout of the β_2-microglobin gene required to assemble class I MHC molecules, are indeed more sensitive to intracellular organisms such as mycobacteria (40). CD4$^+$ T cells are also active in the response to intracellular organisms, since their antigens after ingestion will intersect with endosomal class II MHC molecules en route to the cell surface.

What are the principal target antigens of the immune response to intracellular bacteria, and specifically those organisms associated with ReA? This question has been approached by cloning organism-specific T cells from joint or peripheral blood of patients with ReA and investigating the nature of the antigens they recognize. Although this work is developing rapidly, organism-specific clones have been isolated from relatively few patients, and the number of antigens which have been precisely defined is even smaller.

IV. BACTERIAL ANTIGENS RECOGNIZED BY CD4$^+$ T CELLS FROM PATIENTS WITH REA

A. ReA-Associated with Enteric Organisms

T-cell clones specific for *Yersinia* and *Salmonella* (41–44) have been reported, and they can be divided into two broad groups; those showing specificity for the infecting organism and those recognizing an antigen common to the infecting organism and other enteric bacteria. Those specific for hsps would be expected to be in the latter category, and indeed hsp 60–specific clones derived from *Yersinia*- or *Salmonella*-infected patients are able to respond to other gram-negative organisms including nonpathogenic *Escherichia coli* (42,43). Thus far epitopes within hsp 60 have not been mapped for any of these clones. However, clones reported by Hermann and colleagues showed cross reactivity with both mycobacterial and human hsp 60 (43). Although for recombinant proteins prepared from *E. coli* there is always the possibility of contamination with *E. coli* hsp 60, there appeared to be true cross reactivity with mycobacterial hsp 60, since *Mycobacterium tuberculosis* organisms were stimulatory. Additional evidence that the human hsp was also being recognized came from experiments in which autologous cells were heat shocked and were then able to stimulate the clone in the absence of added yersinial antigen. Interestingly, synovial fluid mononuclear cells were able to stimulate the clone partially without being subject to heat shock. This could indicate that human hsp 60 expression is increased in the synovial population, perhaps due to stresses experienced within the joint. Alternatively, SFMC might contain some yersinial hsp 60 if they have taken up yersinial organisms. The epitope recognized by the cross-reactive clone was mapped to the final 170 amino acids of the protein. More precise mapping of this region using synthetic peptides will be required to demonstrate the cross reactivity unequivocally.

The frequency of clones specific for hsp 60 among the organism-specific population has not been addressed systematically. Probst and colleagues tested a panel of yersinia-specific clones from two patients for their response to yersinial antigens from which hsp 60 had been removed (by using a specific monoclonal antibody) (45). A minority (4/32) failed to respond to the hsp 60–depleted antigens and showed responses to a preparation enriched for hsp 60. This result was interpreted to mean that hsp 60 was not the major target antigen, although it can equally be argued that for 13% of the yersinia-specific clones to respond to a single antigen is quite striking. However, the clones were derived from a T-cell line after several rounds in vitro stimulation which could have allowed dominance of the population by a few clones (either hsp 60 specific or specific for other antigens); thus, the estimate of 13% carries considerable uncertainty.

What of the other target antigens identified? The only other yersinial antigen positively identified as a 19 kDa protein which has recently been shown to be the β-subunit of the bacterial urease (46). This highly cationic protein was previously shown to mediate arthritis when given intra-articularly to sensitized rats (47). Purification, N-terminal sequencing, and cloning have followed (48). It has been speculated that its involvement in arthritis might relate to its cationic nature which would allow it to be retained within cartilage matrix (49) and act as an antigenic reservoir to stimulate antigen-specific T cells in the joint. Interestingly, hsp 60 has been found complexed to the urease enzyme in *Helicobacter pylori* (50), but it is not known if such complexes are found in other gram-negative organisms.

B. *Chlamydia*-Induced ReA

Involvement of immune responses to chlamydial hsps in ReA has been postulated because of the evidence that hsp 60 is an important target of the T-cell response in an experimental model of trachoma (51,52). In recent experiments, we have for the first time demonstrated chlamydial hsp 60–specific clones within populations derived from ReA synovial fluid (R. Jecock, K. Deane, A. Hassell, J. Pearce, and H. Gaston, manuscript submitted). The frequency of such clones has not been established, but they were evident on screening a relatively small number of clones by T-cell immunblotting. Previous studies showed higher titers of antibody to hsp 60 in patients with ReA (53); increased humoral and cellular responses to hsp 60 have also been noted in patients with salpingitis (54,55), another chronic inflammatory sequel to chlamydial infection. It will be important to map epitopes recognized within chlamydial hsp 60 in order to assess the possibility of cross reaction with either human hsp or even hsp 60 from other bacteria. The former would be important in maintaining inflammation after clearance of chlamydial infection, whereas the latter could result in recrudescence of arthritis following infection with organisms other than chlamydiae.

We have recently defined a second chlamydial antigen recognized by synovial T cells; this is the 18-kDa histone-like protein (56). Synthetic peptides have been used to identify a DR1-restricted epitope, and it will now be possible to examine responses to this antigen in patients with uncomplicated chlamydial infection as compared with those developing reactive arthritis. In common with the 19-kDa yersinial urease β-subunit, this antigen is very cationic (reflecting its ability to bind DNA), and it thus may also be preferentially retained in the joint. A 30-kDa histone-like protein has also been described in *Chlamydia* (57), and this is a strong candidate for the target of *Chlamydia*-specific clones which have been shown to recognize a polypeptide of this size (58).

C. ReA Associated with Mycobacterial Infection (Poncet's Disease)

Mycobacterial hsp 60–specific clones were obtained from synovial fluid of a patient with clinical reactive arthritis but no evidence of infection with the usual triggering infections. The polyclonal synovial fluid T-cell response to hsp 60 was very vigorous, and there was no response to other bacteria. This case provided the first opportunity to test the possibility that mycobacterial hsp 60–specific clones might cross react with human hsp 60 (or another joint constituent). Thus, the minimal epitope for these clones was mapped; all clones isolated responded to amino acids 4–13 and were DR3 restricted (59). This is not in a conserved portion of the hsp 60 molecule and there was therefore no cross recognition of human hsp 60, nor of type II collagen or proteoglycan. This result in no way excludes the possibility that in other patients, perhaps those with more persistent disease, hsp 60–specific cells which are autoreactive play an important role in the pathogenesis. Other workers have mapped the DR3-restricted epitope to these residues and commented on the marked responses shown by $DR3^+$ individuals to this antigen (60). On this basis, we have postulated that $DR3^+$ individuals might be more prone to Poncet's disease because of hyperresponsiveness to hsp 60 (7). There is no evidence to confirm or refute this suggestion. In patients receiving BCG instilled into the bladder as immunotherapy for cancer, 44% of the arthritis cases have been $B27^+$ (9). It may be that $B27^+$ patients are more liable to arthritis when the BCG is encountered at a mucosal site, and that patients with Poncet's disease secondary to pulmonary or spinal tuberculosis will not necessarily be $B27^+$.

V. BACTERIAL ANTIGENS RECOGNIZED BY CD8$^+$ T CELLS FROM PATIENTS WITH REA

Since reactive arthritis is clearly associated with the class I MHC antigen, B27, it might be expected that the recognition of ReA-associated bacteria by CD8$^+$ T cells would form a large section of this chapter. Unfortunately, until recently, reports of such cells have been conspicuous by their absence. This may relate to technical difficulties in cloning human CD8$^+$ T cells, and also to the failure to design experiments which ensure delivery of bacterial antigenic peptides to a site where they can be transferred to newly formed class I MHC molecules. Hermann et al. have described *Yersinia*-specific B27-restricted CD8$^+$ clones, having used stimulation with B27-transfected L cells infected with *yersinia* prior to cloning (61). The yersinial antigens recognized by these clones remain to be defined; an additional clone recognized an antigen common to the *Salmonella*- and *Yersinia*-infected cells, suggesting involvement of a conserved antigen such as an hsp. Clearly the definition of the target antigens recognized by these cells is of great importance. Likewise, the antigens recognized by CD8$^+$ clones which lysed uninfected cells in a B27-restricted manner are of great interest, particularly if they can be shown to have any relationship to bacterial antigens.

VI. INFLUENCE OF THE LOCAL ENVIRONMENT ON TARGET ANTIGENS AND ITS POSSIBLE SIGNIFICANCE FOR PATHOLOGY IN REA

Hsps first attracted the attention of immunologists when these proteins were shown to be prominent targets of the immune response to several intracellular pathogens, particularly the facultatively intracellular mycobacteria. There have been several explanations as to why the immune system should recognize this category of antigen. One attractive idea is that the intracellular location imposes a physiological stress on the invader which responds by an upregulation of hsp. This has been clearly shown for salmonellae (62), and it appears that a somewhat similar process occurs for *Chlamydia trachomatis* (63). However, the concept that microbial response to "stress" can involve novel or upregulation of hsp synthesis now has a much wider significance with the evidence that mechanisms for "global" responses to a local environment can lead to significant alteration in bacterial polypeptide profiles. Such findings raise questions on how conditions in vitro can be manipulated to represent the in vivo milieu experienced by the pathogen. It is also beginning to appear that quite diverse bacteria can respond to environmental change by inducing proteins that have unexpected homologies with those in other bacteria and thus the potential for antigenic cross reaction.

A number of sensor systems modulating responses to environmental change are well documented. One common form in gram-negative bacteria involves an outer membrane-spanning sensor which can transfer signals to the genome via phosphorylation-induced changes in a transmitter protein (64). Rapid expression of the plasmid-encoded *spvB* gene following entry of salmonellae into host cells is one example of this type of system (65). More truly global mechanisms for change can in some cases arise from environmental shifts which affect the extent of supercoiling of genomic DNA. Alterations in DNA topology can be induced by changes in environmental osmolarity, anaerobiosis, temperature, and nutrient levels (66). For example, osmotically induced synthesis of proteins determining invasion in salmonellae and thermally dependent induction of Yop virulence proteins in *Y. enterocolitica* are mediated by topological changes (66). In the latter instance, the gene encoding YmoA, a DNA-binding protein, has been shown to have homologues in a number of genera of enterobacteria, and appears to represent a new family of regulatory proteins (67). The recent identification of a protein, BacA, induced during rhizobia invasion of plant tissue, which has significant similarity with an *E. coli* protein, Sbma, and with proteins in both gram-positive and gram-negative bacteria, lends further support to the view that regulatory proteins produced in response to environmental signals are common across a wide range of bacteria (68).

Chlamydiae are obligate intracellular parasites; experiments in vitro have shown that their development is strongly influenced by the nutritional environment, so that deprivation results in altered morphology and a decrease in infectivity (69). These experiments are relevant to the in vivo situation for two reasons. First, physiological levels of animo acids intracellularly are probably inadequate for optimal growth. Second, interferon-γ acts by inducing the ubiquitous enzyme indoleamine 2,3 dioxygenase whose effect is to deplete intracellular tryptophan, an amino acid which chlamydiae are unable to synthesize (63). Thus in vivo, particularly under immune attack, chlamydiae are likely to show many of the changes which can be induced in vitro by nutrient deprivation or interferon-gamma treatment. This probably explains the failure to cultivate chlamydiae from synovium even though nucleic acids can be demonstrated by PCR, since chlamydiae would likely to have assumed a noncultivable state. Such organisms also tend to have a large and sometimes bizarre morphology with absence of the central nucleoid seen in mature elementary bodies (69); some of the forms shown in synovium by immunofluorescence could correspond to these (11).

Possibly a major consequence of the noncultivable state of chlamydiae in vivo is alteration in polypeptide profile. Preliminary evidence indicates that synthesis of hsps is maintained, whereas that of the OMP1 and OMP2, major structural and antigenically dominant proteins in infectious forms, is reduced (63).

Increasing expression of hsp 60 has been reported for chlamydiae within U937 cells induced to differentiate to monocytes by phorbol ester (70). A clue to the possible involvement of sensor systems mediating changes in DNA topology is suggested by the absence of dense nucleoid in noncultivable forms (69). Two recently recognized 18- and 30-kDa histone-like proteins appear to induce DNA compaction and nucleoid formation in the later stages of the developmental cycle (56,57). Recognition of synovial T-cell specificities for the 18-kDa and possibly the 30-kDa proteins (see above) suggests that these may be upregulated or temporally altered in synthesis during nutrient deprivation.

It may thus be that hsps are only the first in a series of protein families to be recognized as modulated by intracellular organisms in response to changes in their environment. These proteins may be preferred targets of the immune response, and where they show homology to eukaryotic proteins (as hsps and histone-like proteins clearly do) they may present opportunities for cross reaction with self-proteins.

VII. CONCLUSIONS

It is not surprising that hsps have attracted the attention of immunologists interested in reactive arthritis, given their role in the immune response to infectious agents, particularly those which are intracellular and can establish persistent infection (e.g., *M. leprae*). It has been established that hsp 60 is a target antigen in ReA associated with both genitourinary and enteric infection. The relative importance of an immune response to this particular antigen in the pathogenesis of the arthritis has yet to be determined. Likewise, the attractive possibility that an immune response to the bacterial hsp results in cross recognition of self-antigens has not been fully established, and it is certainly possible that immune responses to bacterial antigens are solely responsible for driving reactive arthritis. Chronicity of disease would then depend entirely on persistence of infection. Perhaps the most valuable result of studying immune responses to hsps in ReA is the focus on the antigens presented by intracellular organisms to the immune system, and the consequences, in terms of antigen production, of the environment which the organism experiences and to which it adapts. Awareness of the nature and range of these adaptations seems increasingly likely to shape future research.

ACKNOWLEDGMENTS

This work has been funded by grants from the Wellcome Trust, the Arthritis and Rheumatism Council, and the Medical Research Council. We thank Rowena Jecock for critical review of the manuscript.

REFERENCES

1. A. Keat, Reiter's syndrome and reactive arthritis in perspective, *N. Engl. J. Med. 309*:1606 (1983).
2. A. Toivanen, and P. Toivanen, Reactive arthritis, CRC Press, Boca Raton, Florida, 1988, p. 186.
3. K. Aho, P. Ahvonen, A. Lassus, K. Sievers, and A. Tiilikainen, HL-A antigen 27 and reactive arthritis, *Lancet 2*:157 (1973).
4. A. Bradlow and A.G. Mowat, Reiter's syndrome in a 73-year-old man with bronchiectasis, *Scand. J. Rheumatol. 12*:207 (1983).
5. R. Saario and A. Toivanen, *Chlamydia pneumoniae* as a cause of reactive arthritis, *Br. J. Rheumatol. 32*:1112 (1993).
6. L. Dall, L. Long and J. Stanford, Poncet's disease: tuberculous rheumatism, *Rev. Infect. Dis. 11*:105 (1989).
7. T.R. Southwood and J.S.H. Gaston, The molecular basis of Poncet's disease?, *Br. J. Rheumatol. 29*:49 (1990).
8. R.A. Hughes, S.A. Allard, and R.N. Maini, Arthritis associated with adjuvant mycobacterial treatment for carcinoma of the bladder, *Ann. Rheum. Dis. 48*:432 (1988).
9. A.S.M. Jawad, L. Kahn, R.F.P. Copland, D.C. Henderson, and A.K. Abdul-ahad, Reactive arthritis associated with Bacillus Calmette-Guerin immunotherapy for carcinoma of the bladder: a report of two cases, *Br. J. Rheumatol. 32*:1018 (1993).
10. P. Goupille, D. Soutif, and J.P. Valat, Arthritis after Calmette-Guerin Bacillus immunotherapy for bladder cancer, *J. Rheumatol. 19*:1825 (1992).
11. A. Keat, B. Thomas, J. Dixey, M. Osborn, C. Sonnex, and D. Taylor-Robinson, *Chlamydia trachomatis* and reactive arthritis—the missing link, *Lancet 1*:72 (1987).
12. R.A. Hughes, E. Hyder, J.D. Treharne, and A.C.S. Keat, Intra-articular chlamydial antigen and inflammatory arthritis, *Q. J. Med. 80*:575 (1991).
13. R.D. Taylor, C. Gilroy, B. Thomas, and A. Keat, Detection of *Chlamydia trachomatis* DNA in joints of reactive arthritis patients by polymerase chain reaction, *Lancet 340*:81 (1992).
14. M.U. Rahman, M.A. Cheema, H.R. Schumacher, and A.P. Hudson, Molecular evidence for the presence of Chlamydia in the synovium of patients with Reiter's syndrome, *Arthritis Rheum. 35*:521 (1992).
15. K. Granfors, S. Jalkanen, R. von Essen, et al., Yersinia antigens in synovial fluid cells from patients with reactive arthritis, *N. Engl. J. Med. 320*:216 (1989).
16. K. Granfors, S. Jalkanen, A. Lindberg, et al., Salmonella lipopolysaccharide in synovial cells from patients with reactive arthritis, *Lancet 335*:685 (1990).
17. K. Granfors, S. Jalkanen, P. Toivanen, and A. Lindberg, Bacterial lipopolysaccharide in synovial fluid cells in Shigella triggered reactive arthritis, *J Rheumatol 19*:500 (1992).
18. M. Hammer, H. Zeidler, S. Klisma, and J. Heesemann, *Yersinia enterocolitica* in the synovial membrane of patients with *Yersinia*-induced arthritis, *Arthritis Rheum. 33*:1795 (1990).
19. Y.E. Johnston, P.H. Duray, A.C. Steere, et al., Lyme arthritis: spirochaetes found in synovial microangiopathic lesions, *Am. J. Pathol. 118*:26 (1986).

20. P.F. Life, N.J. Viner, P.A. Bacon, and J.S.H. Gaston, Synovial fluid antigen presenting cells unmask peripheral blood T cell responses to bacterial antigens in inflammatory arthritis, *Clin. Exp. Immunol. 79*:189 (1990).
21. R. Lahesmaa-Rantala, O. Lehtonen, K. Granfors, and A. Toivanen, Avidity of anti Yersinia antibodies in yersiniosis patients with and without Yersinia triggered reactive arthritis, *Arthritis Rheum. 30*:1176 (1987).
22. S. Nikkari, R. Merilahti-Palo, R. Saario, et al., Yersinia triggered reactive arthritis. Use of polymerase chain reaction and immunocytochemical staining in the detection of bacterial components from synovial specimens, *Arthritis Rheum. 35*:682 (1992).
23. K. Kapasi and R. Inman, HLA B27 expression modulates gram negative bacterial invasion into transfected L cells, *J. Immunol. 148:* 3554 (1992).
24. K. Granfors, M. Viljanen, A. Tiilikainen, and A. Toivanen, Persistence of IgM, IgG, and IgA antibodies to yersinia in Yersinia arthritis, *J. Infect. Dis. 141*:424 (1980).
25. K. Granfors and A. Toivanen, IgA anti yersinia antibodies in yersinia triggered reactive arthritis, *Ann. Rheum. Dis. 45*:561 (1986).
26. A. Jarry, N. Cerf-Bernussan, N. Brousse, F. Selz, and D. Guy-Grand, Subsets of CD3+ (T cell receptor alpha/beta or gamma/delta) and CD3-lymphocytes isolated from normal human gut epithelium displayed phenotypical features different from their counterparts in peripheral blood, *Eur. J. Immunol. 20*:1097 (1990).
27. B. Rocha, P. Vassalli, and D. Guy-Grand, The extrathymic T cell development pathway, *Immunol. Today 13*:449 (1992).
28. D.V. Havlir, J.J. Ellner, K.A. Chervenak, and W.H. Boom, Selective expansion of human gamma/delta T cells by monocytes infected with live mycobacterium tuberculosis, *J. Clin. Invest. 87*:729 (1991).
29. Hermann, A.W. Lohse, W.J. Mayet, et al., Stimulation of synovial fluid mononuclear cells with the human 65kDa heat shock protein or with live enterobacteria leads to a preferential expansion of TCR gamma/delta positive lymphocytes, *Clin. Exp. Immunol. 89*:427 (1992).
30. R. O'Brien and W. Born, Heat shock proteins as antigens for gamma delta T cells, *Semin. Immunol. 3*:81 (1991).
31. A. Haregewoin, G. Soman, R. Hom, and R. Finberg, Human gamma delta+ T cells respond to mycobacterial heat shock protein, *Nature 340*:309 (1989).
32. J. Holoshitz, F. Koning, J. Coligan, J. de Bruyn, and S. Strober, Isolation of CD4– CD8– mycobacteria reactive T lymphocyte clones from rheumatoid arthritis synovial fluid, *Nature 339*:226 (1989).
33. J. Strominger, The gamma delta T cell receptor and class Ib MHC related proteins: enigmatic molecules of immune recognition, *Cell 57*:895 (1989).
34. S. Porcelli, C.T. Morita, and M.B. Brenner, CD1b Restricts the response of human CD4-8– lymphocytes–T to a microbial antigen, *Nature 360*:593 (1992).
35. M. Breban, R.E. Hammer, J.A. Richardson, and J.D. Taurog, Transfer of the inflammatory disease of HLA-B27 transgenic rats by bone marrow engraftment, *J. Exp. Med. 178*:1607 (1993).
36. A. Townsend and H. Bodmer, Antigen recognition by class I-restricted lymphocytes, *Annu. Rev. Immunol. 7*:601 (1989).

37. L.M. Brunt, D.A. Portnoy, and E.R. Unanue, Presentation of *Listeria monocytogenes* to CD8+ T cells requires secretion of hemolysin and intracellular bacterial growth, *J. Immunol. 145*:3540 (1990).
38. J.D. Pfeifer, M.J. Wick, R.L. Roberts, K. Findlay, S.J. Normark, and C.V. Harding, Phagocytic processing of bacterial antigens for class-I MHC presentation to T-cells, *Nature 361*:359 (1993).
39. S. Chiplunkar, G. DeLibero, and S. Kaufmann, *Mycobacterium leprae* specific Lyt 2+ T lymphocytes with cytolytic activity, *Infect. Immun. 54*:793 ( ).
40. J.L. Flynn, M.M. Goldstein, K.J. Triebold, B. Koller, and B.R. Bloom, Major histocompatibility complex class I-restricted T cells are required for resistance to *Mycobacterium tuberculosis*, *Proc. Natl. Acad. Sci. USA 89*:12013 (1992).
41. N. Viner, L. Bailey, P. Life, P. Bacon, and J. Gaston, Isolation of Yersinia specific T cell clones from the synovial membrane and synovial fluid of a patient with reactive arthritis, *Arthritis Rheum. 34*:1151 (1991).
42. P.F. Life, E.O.E. Bassey, and J.S.H. Gaston, T-cell recognition of bacterial heat-shock proteins in inflammatory arthritis, *Immunol. Rev. 121*:113 (1991).
43. E. Hermann, A. Lohse, R. van der Zee, et al., Synovial fluid derived Yersinia reactive T cells responding to human 65 kDa heat shock protein and heat stressed antigen presenting cells, *Eur. J. Immunol. 21*:2139 (1991).
44. E. Hermann, W. Mayet, T. Poralla, K.H. Mayer zum Buschenfelde, and B. Fleischer, Salmonella reactive synovial fluid T cell clones in a patient with post infectious Salmonella arthritis, *Scand. J. Rheumatol. 19*:350 (1990).
45. P. Probst, E. Hermann, K.H. Mayer zum Buschenfelde, and B. Fleischer, Multiclonal synovial T-cell response to *Yersinia-enterocolitica* in reactive arthritis—the *Yersinia* 61-kDa heat-shock protein is not the major target antigen, *J. Infect. Dis. 167*:385 (1993).
46. P. Probst, E. Hermann, K.H. Mayer zum Buschenfelde, and B. Fleischer, Identification of the *Yersinia-enterocolitica* urease beta-subunit as a target antigen for human synovial T-lymphocytes in reactive arthritis, *Infect. Immun. 61*:4507 (1993).
47. A.K.H. Mertz, S.R. Batsford, E. Curschellas, M.J. Kist, and K.B. Gondolf, Cationic *Yersinia* antigen-induced chronic allergic arthritis in rats, *J. Clin. Invest. 87*:632 (1991).
48. M. Skurnik, S. Batsford, A. Mertz, E. Schlitz, and P. Toivanen, The putative arthritogenic catioic 19-kilodalton antigen of *Yersinia enterocolitica* is a urease beta subunit, *Infect. Immun. 67*:2498 (1993).
49. W.B. Van den Berg, L.D.A. Van de Putte, W.A. Zwarts, and L.A.B. Joosten, Electrical charge of the antigen determines intra-articular antigen handling and chronicity of arthritis in mice, *J. Clin. Invest. 74*:1850 (1984).
50. D.J. Evans, D.G. Evans, L. Engstrand, and D.Y. Graham, Urease-associated heat shock protein of *Helicobacter pylori*, *Infect. Immun. 67*:2125 (1992).
51. R. Morrison, R. Belland, K. Lyng, and H. Caldwell, Chlamydial disease pathogenesis. The 57 kD chlamydial hypersensitivity antigen is a stress response protein, *J. Exp. Med. 170*:1271 (1989).
52. R. Morrison, Chlamydial hsp60 and the immunopathogenesis of chlamydial disease, *Semin. Immunol. 3*:25 (1991).

53. R.D. Inman, M.E.A. Johnston, B. Chiu, J. Falk, and M. Petric, Immunochemical analysis of immune response to *Chlamydia trachomatis* in Reiter's syndrome and non-specific urethritis, *Clin. Exp. Immunol. 69*:246 (1987).
54. E. Wagar, J. Schachter, P. Bavoil, and R. Stephens, Differential human serologic response to two 60,000 molecular weight *Chlamydia trachomatis* antigens, *J. Infect. Dis. 162*:922 (1990).
55. S.S. Witkin, J. Jeremias, M. Toth, and W.J. Ledger, Cell-mediated immune response to the recombinant 57-kDa -heat-shock protein of *Chlamydia-trachomatis* in women with salpingitis, *J. Infect. Dis. 167*:1379 (1993).
56. T. Hackstadt, W. Behr, and Y. Ying, *Chlamydia trachomatis* developmentally regulated protein is homologous to eukaryotic histone H1, *Proc. Natl. Acad. Sci. USA 88*:3937 (1991).
57. T. Hackstadt, T.J. Brickman, C.E. Barry, and J. Sager, Diversity in the *Chlamydia-trachomatis* histone homologue Hc2, *Gene 132*:137 (1993).
58. A.B. Hassell, D.J. Reynolds, M. Deacon, J.S.H. Gaston, and J.H. Pearce, Identification of T-cell stimulatory antigens of *Chlamydia-trachomatis* using synovial fluid-derived T-cell clones, *Immunology 79*:513 (1993).
59. J.S.H. Gaston, P. Life, P. Jenner, M. Colston, and P. Bacon, Recognition of a mycobacteria specific epitope in the 65 kD heat shock protein by synovial fluid derived T cell clones, *J. Exp. Med. 171*:831 (1990).
60. W. Van Schooten, D. Elferink, J. Van Embden, D. Anderson, and R.R.P. de Vries, DR3 restricted T cells from different HLA DR3 positive individuals recognize the same peptide (amino acids 2–12) of the mycobacterial 65 kDa heat shock protein, *Eur. J. Immunol. 19*:2075 (1989).
61. E. Hermann, D.T.Y. Yu, K.H. Mayer zum Buschenfelde, and B. Fleischer, HLA-B27–restricted CD8 T-cells derived from synovial fluids of patients with reactive arthritis and ankylosing spondylitis, *Lancet 342*:646 (1993).
62. N.A. Buchmeier and F. Heffron, Induction of Salmonella stress proteins upon infection of macrophages, *Science 248*:730 (1990).
63. W.L. Beatty, G.I. Byrne, and R.P. Morrison, Morphologic and antigenic characterization of interferon gamma–mediated persistent *Chlamydia-trachomatis* infection in vitro. *Proc. Natl. Acad. Sci. USA 90*:3998 (1993).
64. J.B. Stock, A.J. Ninfa, and A.M. Stock, Protein phosphorylation and regulation of adaptive responses in bacteria, *Microbiol. Rev. 53*:450 (1989).
65. J. Fierer, L. Eckmann, F. Fang, C. Pfeifer, B.B. Finlay, and D.B. Guiney, Expression of the Salmonella virulence plasmid gene spv B in cultured macrophages and nonphagocytic cells, *Infect. Immun. 61*:5231 (1993).
66. C.J. Dorman and N. NiBhriain, DNA topology and bacterial virulence gene regulation, *Trends Microbiol. 1*:92 (1993).
67. A.V. Mikulskis and G.R. Cornelis, A new class of proteins regulating gene expression in enterobacteria, *Mol. Microbiol. 11*:77 (1994).
68. P. Yorgey and R. Kolter, A widely conserved developmental sensor in bacteria?, *Trends Genet. 9*:374 (1993).
69. A.M. Coles, D.J. Reynolds, A. Harper, A. Devitt, and J.H. Pearce, Low-nutrient

induction of abnormal chlamydial development—a novel component of chlamydial pathogenesis, *FEMS Microbiol. Lett. 106*:193 (1993).

70. K.H Kohler, E. Nettelnbreker, H. Holtmann, and H. Zeidler, Expression of the 60kD chlamydial heat shock protein during the culture of the monocytic cell line U937 infected with *Chlamydia trachomatis* serovar K, *Arthritis Rheum. 36*:S224 (1993).

7

Juvenile Chronic Arthritis and Heat Shock Proteins

E. R. de Graeff-Meeder, G. T. Rijkers, A. B. J. Prakken, W. Kuis, and B. J. M. Zegers
University Hospital for Children and Youth, Het Wilhelmina Kinderziekenhuis, Utrecht, The Netherlands

R. van der Zee and Willem van Eden
University of Utrecht, Utrecht, The Netherlands

I. INTRODUCTION

A. Juvenile Chronic Arthritis

Juvenile chronic arthritis (JCA) is defined as an arthritis in one or more joints in children under 16 years of age with a duration of at least 3 months and with exclusion of several other disorders (1). Arthritis can be defined as swelling or effusion or presence of two or more of the following symptoms: limitation of motion, tenderness or pain on motion, and increased temperature. According to the European League Against Rheumatism, JCA is subdivided into three onset types (2): polyarticular onset, oligoarticular onset, and systemic onset of disease (Table 1). Subtyping of JCA is performed after 3 months of disease.

JCA is defined as polyarticular when five or more joints are affected. This onset mode comprises about 40% of all JCA cases. In oligoarticular onset of JCA, less than four joints are affected, and this subtype occurs in about 50% of the cases. Systemic onset of JCA is characterized by spiking fever of unknown origin for at least 2 weeks in addition to systemic symptoms and rash at onset. In this subtype, found in 10% of the cases, overt arthritis may be absent for months or even years. The oligoarticular subtype of JCA can be further subdi-

Table 1 Classification of Types of Onset of JCA

	Polyarticular onset	Oligoarticular onset	Systemic onset
Frequency	40%	50%	10%
No. of joints involved	≥5	≤4	Variable
Age of onset peak	Childhood RF^-: 3 years RF^+: > 10 years	Childhood Type 1: 2 years Type 2: 10 years	Childhood 5 years
Sex ratio	RF^-: females >> males RF^+: females >>> males	Type 1: females >>> males Type 2: males << females	Females and males equally
Systemic involvement	Moderate	Not present	Prominent
Chronic uveitis	5%	20%	Rare
Rheumatoid factor	10%	Rare	Rare
Antinuclear antibodies	25–75%	Type 1: 75% Type 2: rare	0–10%
Prognosis	RF^-: moderate RF^+: poor	Type 1: good Type 2: moderate	Moderate to poor
Major complicatons	Erosions	Type 1: blindness Type 2: ankylosing spondylitis	Serositis Erosions

vided into two categories: type 1 and type 2. Type 1 shows an early onset and is predominantly found in girls where the disease is associated with the occurrence of chronic uveitis. Type 2 has a relative late onset, a male preponderance, and association with spondylarthropathies (see Table 1). Even within these two types of oligoarticular JCA a significant degree of heterogeneity exists. Clinical experience has learned that the different subtypes of JCA require distinct therapeutic strategies. Therapy is based largely on empirism. A major drawback in the development of specific and effective therapy has been lack of understanding of the etiology of JCA. It is only recently that new insights into the etiology of chronic arthritis (JCA and rheumatoid arthritis [RA]), reactive arthritis (ReA), and certain other systemic autoimmune diseases have developed.

B. Characterization of the Inflammatory Process in Juvenile Chronic Arthritis

The pathological features of the synovial membrane in JCA are similar to those described in adult rheumatoid arthritis (RA). The inflammatory process begins with congestion and edema which is followed by infiltration with in order of appearance polymorphonuclear leucocytes, small lymphocytes, and in advanced cases also with plasma cells and giant cells. Villous hyperplasia and hypertrophy of the synovial lining layer is present and vascular endothelial hyperplasia is often prominent. Generally, the inflammatory process is mild in comparison to findings in adult patients with RA. Only in a minority of children, pannus formation leading to progressive erosions and destruction of cartilage and bone is found. It has been suggested that the greater thickness of the cartilage in juveniles offers some form of protection against bone destruction. The histopathological findings do not allow to distinguish between the oligoarthritic subtype of JCA and the polyarthritic subtype. By means of immunohistological analysis, the inflamed synovium in JCA is generally characterized by a mild lymphoid infiltrate consisting mainly of activated HLA-DR$^+$ and CD4$^+$ T cells. Studies on the synovial fluid and synovial membranes of patients with RA showed that a high percentage of these CD4$^+$ T cells are CD29$^+$ (3,4). Although only a limited number of patients with JCA have been studied regarding this aspect, we found that the majority of CD4$^+$ synovial T cells in JCA is also CD29$^+$. Functionally, CD4$^+$CD29$^+$ T cells represent T memory cells; the CD29 molecule itself is the β1 chain of the very late antigen (VLA) integrin family (5). The predominance of CD29$^+$ T cells in synovial fluid could be either a direct reflection of the fact that this receptor is required for passage through blood vessels into the synovial compartment or the effect of local stimulation in the joint.

C. Heat Shock Proteins

Studies on adjuvant induced arthritis in Lewis rats showed that heat shock proteins (hsps) have a role in the disease (6). Hsps are immunodominant antigens of bacteria and parasites, including those bacteria that have been associated epidemiologically with chronic arthritis (7). However, from the experimental work with Lewis rats, we have learned that critical epitopes for protection against adjuvant arthritis are not necessarily immunodominant (8). Hsps are remarkably conserved in evolution, as would be expected in view of their critical roles in the maintenance of cell integrity (9). Striking is the extreme conservation of both amino acid sequence and function within a given hsp family (10). Comparison of the primary structure of hsps of different species shows regions of complete amino acid sequence identity (conserved sequences), as well as regions that have a species specific sequence (nonconserved sequences). Thus, every bacterial hsp will contain a number of epitopes that are shared with those of human hsps.

II. HEAT SHOCK PROTEINS AND JUVENILE CHRONIC ARTHRITIS

The involvement of hsps and immune reactivity to hsps in JCA is apparent from a number of observations.

A. Synovial Membranes

Initially expression of hsp 60 was studied in synovial membranes of rats with adjuvant arthritis and in adult patients with either reumatoid arthritis or osteoarthritis (11). Later on hsp 60 expression was also studied in patients with JCA (12). The monoclonal antibodies used in the first studies were raised in mice: ML-30 against hsp 60 of *Mycobacterium leprae*,TB-78 against hsp 60 of *M. tuberculosis*, and F-8 raised against the arthritogenic nonapeptide of *M. tuberculosis* (Table 2).

For all antibodies used, a similar staining pattern was found; that is, predominant staining of synovial lining ells, blood vessels, and macrophages. There was no correlation between the intensity of staining with the above-mentioned antibodies and the degree of lining cell proliferation or the density or location of lymphoid infiltration. In a second study, a monoclonal antibody to human hsp 60 was developed (LK-1) with a unique specificity for human hsp 60 and not reactive with the bacterial counterpart (12). Another antibody, LK-2, recognizes both the human and the bacterial hsp 60 (Table 2). With the availability of LK-1 and LK-2, the question whether epitopes of human hsp 60 or bacterial hsp 60 are expressed in synovial membranes of patients with JCA could be addressed. Expression of human hsp 60 ($LK1^{+}$) was found in the cytoplasm of

Table 2 Expression of Hsp 60 in Synovial Membranes of Rats and Humans

Monoclonal antibody	Ig class	Reactivity to epitope within aminoacid	Reference	Staining pattern
ML-30	IgG1	280–303 of mycobacterial hsp 60	13	Cytoplasmic staining of synovial lining and endothelial cells
TB-78	IgG1	170–234 of mycobacterial hsp 60	14	As with ML-30
F-8	IgM	180–188 of mycobacterial hsp 60	11,15	As with ML-30
LK-1	IgG1	383–447 of human hsp 60	12	As with ML-30
LK-2	IgG1	383–419 of human and bacterial hsp 60	12	As with ML-30

lining cells and endothelial cells and macrophages, all with a high level of expression of HLA-DR (11,16). The pattern of expression of human hsp 60 in synovial membranes of patients with JCA as detected by LK-1 parallelled exactly the staining pattern of the antibodies, ML-30, TB-78, and F-8 (unpublished observations by the authors). The cross-reactive LK-2 antibody also showed a similar staining pattern. With exception of F-8, it seems likely that cross reactivity with human hsp 60 was the basis for this result. The monoclonal antibody F-8 recognizes the 180–188 amino acid sequence of mycobacterial hsp 60, a sequence not present in human hsp 60. Overall it can be concluded that both endogenous hsp 60 and microbial hsp 60 may be expressed in synovial tissue of patients with JCA and that hsp 60 is localized mainly in synovial lining cells, endothelial cells of blood vessels, and macrophages.

B. Cell Surface Expression of Heat Shock Protein

Whether or not (processed) hsp 60 fragments are expressed on the cell surface, for example, of synovial lining cells or macrophages is a controversial issue due in part to the lack of suitable reagents. Class I–restricted $CD8^+$ T cells, raised against the bacterial hsp 60, recognize macrophages that were subjected to various stress stimuli, including interferon gamma activation and viral infection (17). The proliferative response of $\gamma\delta$ T cells to the Daudi B-cell line can be inhibited by a rabbit anti–hsp 60 antiserum (18). Ascites of ML-30, a monoclonal antibody specific for an epitope encoded by amino acids 280–303 of mycobacterial hsp 60 (13), shows surface staining of murine bone marrow–derived macrophages (19). However, purified monoclonal antibody or tissue culture derived ML-30 did neither show surface staining of macrophages nor a variety of murine and human myelomonocytic and B-cell lines (20,21). Our own observations (Fig. 1) indicate that on heat shock treatment, peripheral blood monocytes of normal individuals as well as patients with JCA express human hsp 60 determinants as evidenced by increased cell surface staining with LK-1. These results suggest that epitopes of human hsp 60 are present in inflamed synovial tissue of patients with JCA and have, when properly expressed, the potential to initiate a humoral and/or cellular immune response.

C. Juvenile Chronic Arthritis a T-Cell–Mediated Disease?

Even without knowledge of the nature and distribution of potential autoantigens that may be involved, there are various clinical arguments to consider JCA a T-cell–mediated autoimmune disease. The first argument in support of this is

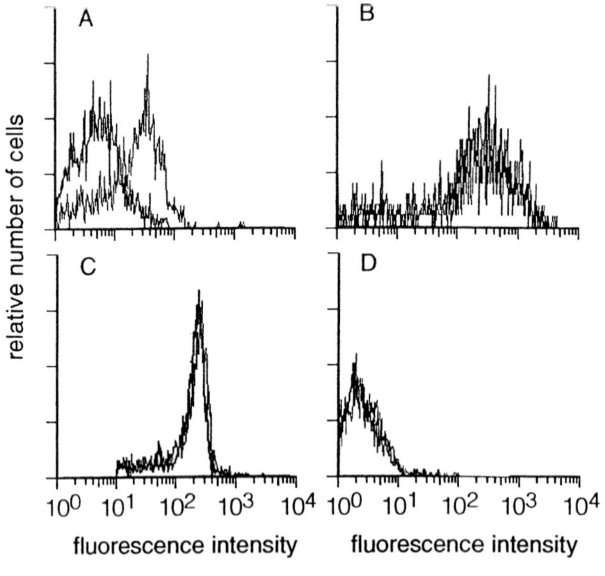

Figure 1 Induction of cell surface hsp 60 expression on human monocytes by heat shock treatment. Human peripheral blood mononuclear cells were incubated for 30 min at either 37°C (heavy tracing) or 41°C (light tracing), followed by 2 hr at room temperature. Cells were stained with biotinylated LK-1 (A), phycoerythrin conjugated anti–HLA-DR (B), or biotinylated R73 OX6, OX18, or OX19 (anti–rat leukocyte antigens; panel D). In a second step, cells were incubated with streptavidin-phycoerythrin (A and D) and fluoresceinated CD14 (pan-monocyte). Cells were analyzed on a FACStar Plus flowcytometer (Becton Dickinson); shown are histograms of $CD14^+$ gated monocytes. Note that apart from LK-1 (panel A), heat shock treatment does not change expression of HLA-DR (B) or CD14 (C), nor does it lead to binding of irrelevant MAbs (D).

the clinical observation that viral infections that cause suppression of cellular immunity, for example, measles (22–24) or cytomegalovirus infection (W. Kuis, personal communication)), may lead to remission of disease activity in patients with JCA. Furthermore, chronic arthritis is found in patients with B-cell deficiencies such as common variable immunodeficiency, X-linked agammaglobulinemia and selective immunoglobulin A (IgA) deficiency (25,26). The clinical features of the chronic arthritis in patients with B-cell defects are indistinguishable from the disease in (B-cell competent) patients with JCA. Because the etiology of JCA has not been established and may even be variable, it can not be excluded that the underlying cause for the chronic arthritis in both patient groups differs. In patients with primary T-cell deficiencies, chronic arthritis is not observed.

Evidence for the fact that endogenous hsp 60 may act as an autoantigen in arthritis in humans was provided by the observation that T cells from synovial fluid and peripheral blood of patients with JCA can be activated when cultured in vitro with recombinant human hsp 60 (27). In our hands, adult patients with RA, healthy children, nor healthy young adults show a T-lymphocyte proliferative response to human hsp 60 (27). This is in accordance with a recent study of Pope et al. (28), who found no positive response to human hsp 60 of peripheral blood mononuclear cells from 11 adult patients with RA and 8 patients with other forms of inflammatory synovitis. However, synovial fluid mononuclear cells from 3 of 11 RA patients did respond to human hsp 60 (28). Activation of peripheral blood lymphocytes of normal healthy individuals with killed *Mycobacterium tuberculosis* results in cytotoxic T-cell activity for autologous target cells primed with synthetic peptides containing homologous sequences of mycobacterial and human hsp 60 (29). This indicates that T cells with a specificity for epitopes of conserved sequences in hsp 60 are present in peripheral circulation of normal individuals. Apparently, owing to regulatory control exerted by suppressor T cells or otherwise, these T cells do not become activated under physiological conditions.

When peripheral blood T-cells of patients with JCA were cultured with either human hsp 60 or mycobacterial hsp 60, we found a statistically significant correlation between the stimulation indices ($r = 0.948$, $p < 0.02$; [27]), suggesting that the T-cell response to human hsp 60 is at least in part directed against conserved sequences within the hsp 60 molecule. In a number of patients, a proliferative T-cell response could also be induced by activation with mycobacterial hsp 70 or the mycobacterial hsp 60 nonapeptide 180–188. Whether the positive proliferative response to mycobacterial hsp 70 is due to recognition of conserved sequences present in the human hsp 70 as well is sofar unknown. We can only conclude that T lymphocytes of patients JCA recognize next to mycobacterial hsps autologous hsps also. Based on these results, it has become clear that an endogenous hsp can serve as an autoantigen and lead to substantial T-lymphocyte reactivity in patients with JCA.

III. HYPOTHESIS

Our data on hsp 60 expression in inflamed synovium and in vitro T-cell reactivity to human hsp 60 allow us to propose a sequence of events leading to chronic arthritis: (1) a trigger, for example, an infection, may lead to inflammation which leads to raised expression of (human) hsp in the synovial membranes; (2) human hsp 60 recognizing T lymphocytes will be activated; (3) T lymphocytes recognizing self–hsp 60 expressed at the site of inflammation may exert regulatory activities; and (4) such regulatory activities may lead to suppression of disease.

From studies in experimental animals, we have learned that a variety of stimuli (mycobacterial hsp 60, collagen type II, streptococcal cell wall, cartilage proteoglycans, pristane) can induce arthritis in genetically susceptible strains. For some arthritic diseases in humans, the genetic predisposition (HLA type, sex) and a clear association with microbial antigens are well known. For instance, in case of male HLA-B27$^+$ individuals, an infection of the gastrointestinal tract can easily lead to a reactive arthritis. In animals of various genetic backgrounds, both mice and rats, antigens such as collagen type II and mycobacterial hsp 60 have been shown to be critical in the induction of arthritis. Given our limited knowledge of the genetic make-up of arthritis-prone individuals and the fact that many arthritis-inducing antigens are unknown, it may be fair to conclude that the initiation of arthritis in humans and rodents may depend on similar pathogenic factors. The initial trigger leading to disease may vary from individual to individual, whereas in the progression of pathology many factors may be common.

In the model of adjuvant arthritis (AA), hsp 60 has been identified as a critical antigen in the induction of arthritis in Lewis rats. By immunization with recombinant hsp 60, the animals became resistant to subsequent induction of AA using whole mycobacteria. Mycobacterial hsp 60 was also found to protect against streptococcal cell wall–induced arthritis and pristane (oil)–induced arthritis. In other experimental models of arthritis (induction of arthritis with avridine, also called CP20961, and collagen type II), immunization with mycobacterial hsp 60 leads to a variable degree of protection from the disease. The fact that mycobacterial hsp 60 can prevent experimentally induced arthritis, also in models not induced with mycobacteria, suggests that selfhsp 60 is an autoantigen critically involved in downregulatory events involved in all forms of autoimmune arthritis in rodents.

Heat shock proteins are intracellular proteins having critical functions in the proteins housekeeping machinery of every living cell. Probably owing to such essential cellular functions, their evolutionary variation has remained markedly limited. Extensive amino acid sequence similarities exist between microbial and mammalian hsps. Based on this molecular mimicry induction of a response to microbial hsp 60 could lead to responses directed to self–hsp 60 in humans. The synthesis of hsps is raised when cells are subjected to stress.

At the onset of juvenile chronic arthritis, many triggering events could lead to an enhanced expression of hsps in synovial membranes. Apart from the well-known bacterial infections, viral infections could lead to the local synthesis of hsps as well. The association between persistent rubella virus infection and JCA is the best documented example so far. But other stresses, like ischemic processes, could result in overexpression in hsps in synovial membranes also. Given the finding that the normal T-cell repertoire includes receptors with a specificity for hsp 60 epitopes (29), any individual could, after an initial triggering event as discussed above, develop a chronic arthritis. Clearly, a variety of regulatory elements, both within and outside the immune system, prevent this from happening. One of such regulatory systems is at least in part formed by the highly polymorphic major histocompatibility complex (MHC). The capacity of different MHC molecules to present a given peptide for recognition by T-cell receptors does vary.

In humans, a difference was noticed in the capability to respond to human hsp 60 between children and adults. A significant and reliable response of mononuclear cells from both the peripheral blood and the synovial fluid to human hsp 60 was found in children with JCA. Probably with the exception of very early cases of reumatoid arthritis, proliferative responses to human hsp 60 are not seen in adults. From the comparison of patients in different subgroups of JCA, it became evident that patients with the oligoarticular onset of JCA (OA-JCA) the response is associated with the active phase of the disease. In case of remission of the disease, the response to human hsp 60 disappeared (Fig. 2). In the subgroups of polyarticular or systemic JCA, we documented less frequently positive responses to human hsp 60, and these responses did not show any correlation with the disease activity. Clinical relapse of arthritis patients with OA-JCA coincided with reappearance of anti–human hsp 60 reactivity. Furthermore, it was found that in vitro priming of nonresponder cells of patients OA-JCA in remission led to positive secondary in vitro responses only in the case of patients with OA-JCA and not in the case of other clinical subgroups (unpublished observations of (A.B.J. Prakken et al., unpublished observations). One could argue that the basis for recovery from the arthritis in patients with OA-JCA lies in the induction of peripheral tolerance based on specific T-cell responses directed at human hsp 60 in the context of a specific MHC make-up.

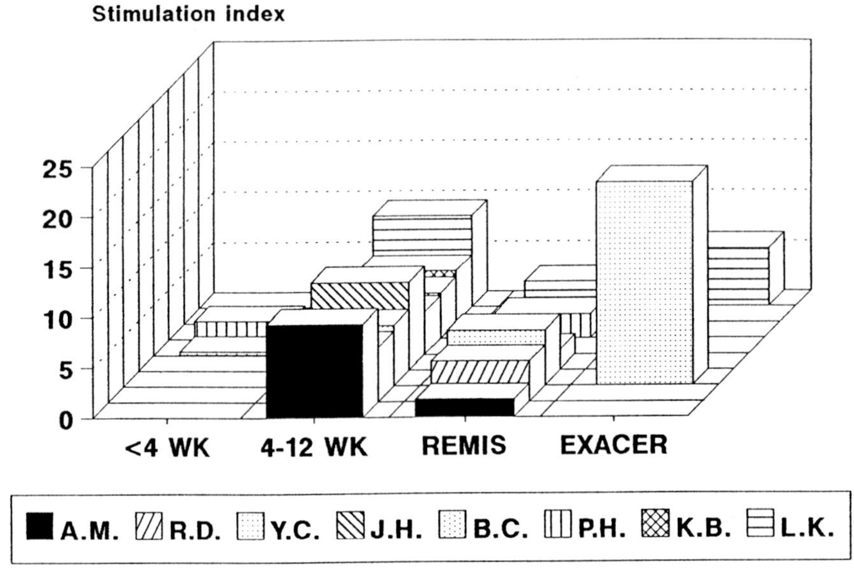

Figure 2 Relation between in vitro anti–hsp 60 T-cell response in time and clinical disease activity in eight patients with oligoarticular juvenile chronic arthritis (A.M., R.D., Y.C., J.H., B.C., P.H., K.B., L.K.). Peripheral blood mononuclear cells were collected at indicated intervals and cultured in vitro for 6 days with 10 μg/ml recombinant hsp 60. The degree of T-cell proliferation (measured as ^{3}H-thymidine incorporation in hsp 60 cultures relative to control cultures) is indicated as stimulation indices. Note that the T-cell response in the patients B.C. and P.H. at <4 weeks after the onset of the disease is not yet positive.

IV. CONCLUSIONS

In patients with JCA, a T-lymphocyte response to human hsp 60 as self-antigen occurs. Interestingly, these responses occur mainly in the subgroup with the best prognosis: the patients with OA-JCA. The presence of responses during the active phase of the disease, preceding remission, suggests that responses to self–hsp 60 may positively contribute to mechanisms leading to disease remission. In line with the AA model, this could be the basis for immunological intervention. Through immunization with peptides of the autoantigen hsp 60, peripheral tolerance could be induced. Identification of the relevant epitopes is the first step toward such a strategy.

REFERENCES

1. J.C. Jacobs, *The Differential Diagnosis of Arthritis in Childhood. Juvenile Rheumatoid Arthritis*, Springer-Verlag, New York, 1993.

2. P.H.N. Wood, Special meeting on nomenclature and classification of arthritis in children, *The Care of Rheumatic Children* (E. Munthe, ed.), Eular Publishers, Basle, 1978, pp. 47–50.
3. J.G. Hanley, D. Pledger, W. Parkhill, M. Roberts, and M. Gross, Phenotypic characteristics of dissociated mononuclear cells from rheumatoid synovial membrane, *J. Rheumatol. 17*:1274 (1990).
4. C. Pitzalis, G. Kingsley, J. Murphy, and G. Panayi, Abnormal distribution of the helper-inducer and suppressor-inducer T lymphocyte subsets in the rheumatoid joint, *Clin. Immunol. Immunopathol. 45*:252 (1987).
5. M.E. Hemler, VLA proteins in the integrin family: structure functions and their role on lymphocytes, *Annu. Rev. Immunol. 8*:365 (1990).
6. W. van Eden, Heat-shock proteins and the immune system, *Immunol. Rev. 121*:5 (1991).
7. D.B. Young, Stress protein as antigens during infection, *Stress Proteins and Inflammation* (C. Rice Evans et al., eds.), Richelieu Press, London, 1991.
8. S. Anderton, R. van der Zee, A. Noordzij, W. van Eden, Differential hsp65 T cell epitope recognition following Adjuvant Arthritis inducing or protective immunisation protocols, *J. Immunol. 152*:3656–3664 (1994).
9. S. Lindquist, The heat-shock response, *Annu. Rev. Biochem. 55*:1151 (1986).
10. R.A. Young, Stress proteins and immunology, *Annu. Rev. Immunol. 8*:401 (1990).
11. E.R. de Graeff-Meeder, M. Voorhorst, W. van Eden, H-J. Schuurman, J. Huber, D. Barkley, R.N. Maini, W. Kuis, G.T. Rijkers, and B.J.M. Zegers, Antibodies to the mycobacterial 65-kD heat shock protein are reactive with synovial tissue of adjuvant arthritic rats and patients with rheumatoid arthritis and osteoarthritis, *Am. J. Pathol. 137*:1013 (1990).
12. C.J.P. Boog, E.R. de Graeff-Meeder, M.A. Lucassen, R. van der Zee, M.M. Voorhorst-Ogink, P.J.S. van Kooten, H.J. Geuze, and W. van Eden, Two monoclonal antibodies generated against human hsp 60 show reactivity with synovial membranes of patients with juvenile chronic arthritis, *J. Exp. Med. 175*: 1805 (1992).
13. T.M. Buchanan, H. Nomaguchi, D.D. Anderson, R.A. Young, T.P. Gillis, W.J. Britton, J. Ivanyi, A.H.J. Kolk, O. Closs, B.R. Bloom, and V. Mehra, Characterization of antibody-reactive epitopes on the 65-kilodalton protein of Mycobacterium leprae, *Infect. Immun. 55*:1000 (1987).
14. A.R.M. Coates, B.W. Allen, J. Hewitt, J. Ivanyi, and D.A. Mitchison, Antigenic diversity of Mycobacterium tuberculosis and Mycobacterium bovis detected by monoclonal antibodies, *Lancet 2*:167 (1981).
15. W. van Eden, J.E.R. Thole, R. van der Zee, A. Noorzij, J.D.A. Embden, E.J. Hensen, and I.R. Cohen, Cloning of the mycobacterial epitope recognized by T lymphocytes in adjuvant arthritis, *Nature 331*:171 (1988).
16. A. Karlsson Parra, E. Dimeny, C. Juhlin, B. Fellstrom, and L. Klareskog, The anti-Leu4 (CD3) monoclonal antibody reacts with proximal tubular cells of the human kidney, *Sand. J. Immunol. 30*:719 (1989).
17. T. Koga, A. Wand Wurttenberger, J. DeBruyn, M.E. Munk, B. Schoel, and S.H. Kaufmann, T cells against a bacterial heat shock protein recognize stressed macrophages, *Science 245*;1112 (1989).
18. P. Fisch, M. Malkovsky, S. Kovats, E. Sturm, E. Braakman, B.S. Klein, S.D.

Voss, L.W. Morrissey, R. DeMars, W.J. Welch, R.L.H. Bolhuis, and P.M. Sondel, Recognition by human Vgamma9/Vdelta2 T cells of a GroEL homolog on Daudi Burkitt's lymphoma cells, *Science 250*:1269 (1990).
19. A. Wand Wurttenberger, B. Schoel, J. Ivanyi, and S.H. Kaufmann, Surface expression by mononuclear phagocytes of an epitope shared with mycobacterial heat shock protein 60, *Eur. J. Immunol. 21*:1089 (1991).
20. R. Kiessling, A. Gronberg, J. Ivanyi, K. Soderstrom, M. Ferm, S. Kleinau, E. Nilsson, and L. Klareskog, Role of hsp60 during autoimmune and bacterial inflammation, *Immunol. Rev. 121*:91 (1991).
21. S.H.E. Kaufmann, B. Schoel, J.D.A. Embden, T. Koga, A. Wand-Wurttenberger, M.E. Munk, and U. Steinhoff, Heat-shock protein 60: implications for pathogenesis of and protection against bacterial infections, *Immunol. Rev. 121*:67 (1991).
22. K. Yoshioka, H. Miyata, and S. Maki, Transient remission of juvenile rheumatoid arthritis after measles, *Acta Paediatr. Scand. 70*:419 (1981).
23. E. Simpanen, R. von Essen, and H. Isomaki, Remission of juvenile rheumatoid arthritis (Still's disease) after measles, *Lancet 2*:987 (1977).
24. G.F. Still, On a form of chronic joint disease in children, *Trans. R. Med. Chir. Soc. 80*:47 (1897).
25. R.E. Petty, J.T. Cassidy, and D.G. Tubergen, Association of arthritis with hypogammaglobulinemia, *Arthritis Rheum. 20*:441 (1977).
26. J.T. Cassidy, R.E. Petty, and D.B. Sullivan, Occurrence of selective IgA deficiency in children with juvenile rheumatoid arthritis, *Arthritis Rheum. 20*:181 (1977).
27. E.R. de Graeff-Meeder, R. van der Zee, G.T. Rijkers, H-J. Schuurman, W. Kuis, J. W. J. Bijlsma, B. J. M. Zegers, and W. van Eden, Recognition of human 60 kD heat shock protein by mononuclear cells from patients with juvenile chronic arthritis, *Lancet 337*: 1368 (1991).
28. R.M. Pope, R.M. Lovis, and R.S. Gupta, Activation of synovial fluid T lymphocytes by 60-kd heat-shock proteins in patients with inflammatory synovitis, *Arthritis Rheum. 35*:43 (1992).
29. M.E. Munk, B. Schoel, S. Modrow, R.W. Karr, R.A. Young, and S.H. Kaufmann, T lymphocytes from healthy individuals with specificity to self-epitopes shared by the mycobacterial and human 65-kilodalton heat shock protein, *J. Immunol. 143*: 2844 (1989).

8

T-Cell Responses to Heat Shock Proteins in Rheumatoid Arthritis

Harald G. Wiker, Morten Harboe, and Jacob B. Natvig
Institute of Immunology and Rheumatology, The National Hospital, University of Oslo, Oslo, Norway

I. INTRODUCTION

Rheumatoid arthritis (RA) is a chronic disabling disease characterized by tissue destruction in peripheral joints. Several characteristic features of RA are typically associated with autoimmune reactions, such as formation of rheumatoid factors and local generation of immune complexes with evidence for complement activation in the synovial fluid (SF). The local inflammatory responses primarily affect synovial tissue resulting in destruction of cartilage. An important clinical parameter of RA is thickening of synovial tissue due to an inflammatory response with infiltration of activated B and T lymphocytes, as well as professional antigen presenting cells like macrophages and dendritic cells (1). The inflammatory tissue may be highly organized, and lymphoid-like follicles or even germinal centers occur in the joints (2). There is also a marked HLA association in RA (3,4), and therefore the main current view is that RA is an autoimmune disease.

The B-cell responses in RA have been most extensively characterized and are typically autoimmune with formation of rheumatoid factors. Formation of these antibodies may be regulated by antigen-specific T cells, but little is known about the specificities of the cooperative interactions between B and T cells involved in rheumatoid factor production. An interesting possible function of B cells with membrane-bound rheumatoid factor is that they may capture immune complexes and present antigen within these complexes for T cells (5).

The specificities of T-cell responses in RA have been less well characterized and are less understood. No single autoimmune T-cell specificity has been sorted out comparable with rheumatoid factor formation on the B cell side. There is an active current search for antigenic epitopes and corresponding T cells with autoimmune specificities that might explain the local immune responses leading to tissue destruction.

A. Positively and Negatively Selected T Cells

A growing body of evidence supports the hypothesis that T-cell specificities are generated by a combination of positive and negative selection of the T-cell repertoire in the thymus. In the normal situation, there is a carefully balanced network of autoreactive T cells. In autoimmunity, this network may be distorted by foreign antigens cross-reacting with self. At present, we know of only a limited number of such cross reactive antigens. The heat shock proteins (hsps) are the most studied group of such molecules highly conserved during evolution. The mycobacterial hsp 65 antigen has been extensively studied in RA as part of the large group of mycobacterial antigens, whereas immune responses to other hsps have been less well studied. Data mainly on T-cell responses to hsp 65 in RA will therefore be considered here.

B. Cross-Reacting Antigens

The human homologue of the mycobacterial hsp 65 was initially characterized by Dudani and Gupta (6), and it is present in inflamed joints of RA patients (7). There are, however, several other human proteins with limited stretches of sequence homology with the mycobacterial hsp 65, which include the proteoglycan link protein (8) and human cytokeratin 1/2 (9). The most interesting is probably the homology with human lactoferrin showing seven tetrapeptide homologies (10). The tetrapeptide KDLL of mycobacterial hsp 65 may be of particular significance, since it is also found in the β chain of DR1, DR3, and DR4 (11). The hsp 70 antigen of mycobacteria (12) also has a homologue in eukaryotes (13). However, T-cell responses to hsp 70 have not been well studied in RA, but there are detailed studies of *Mycobacterium leprae*-vaccinated persons and patients with leprosy (14,15). The mycobacterial hsp 18 antigen is potentially quite interesting with respect to RA, because T cells restricted by HLA DR4,Dw4 have been isolated from *M. leprae*-vaccinated persons (16). This is the restriction element associated with RA (3,4). These observations have not been pursued thoroughly in RA but patients with juvenile rheumatoid arthritis (JRA) have strong antibodies to a *M. kansasii* 18-kDa antigen (17). It has not been established whether this antigen is hsp 18. The gene for hsp18 has been found in *M. leprae*, *M. avium*, *M. habana*, and *M. intracellulare* (18), and hsp

18 may thus be widely encountered. The gene has not been detected in *M. tuberculosis* (18).

II. ANTIGEN SPECIFICITIES OF T CELLS IN PATIENTS WITH RA

A. Antigen Specificities of Synovial Versus Peripheral T Cells in Patients with RA

It is generally believed that the inflammatory response in RA is antigen driven, and there has been a great interest in identifying particular antigens that stimulate T cells from rheumatoid synovial tissue. Relatively few have been demonstrated to induce significant responses. Self-antigens like immunoglobulin (IgG) and collagen would be of particular importance, but it has not been convincingly shown that these proteins can drive T-cell responses in RA (19). Among several microbial antigens tested, mycobacterial antigens have induced particularly strong responses of T cells from synovial tissue. This was first demonstrated by Abrahamsen et al. (20). The response to purified protein derivative (PPD), a precipitated fraction from autoclaved culture of *M. tuberculosis* (21), was significantly higher in synovial fluid than peripheral blood lymphocytes of eight patients with RA. In contrast, the responses to phytohemagglutinin (PHA) or pokeweed mitogen (PWM) showed the opposite pattern with the highest stimulation index in peripheral blood lymphocytes. Holoshitz et al. (22) and Res et al. (23) confirmed this finding using an acetone-precipitable fraction of *M. tuberculosis*, which also gave the strongest T-cell responses in cells from synovial tissue of patients with RA. This antigen fraction did not induce proliferation of SF cells from patients with degenerative joint disease (22). It is of interest to note that hsp 65 is present both in PPD (21,24) and the acetone-precipitable fraction (8).

Burmester et al. (25) showed that the same pattern with stronger responses of SF than peripheral blood (PB) cells could be repeated not only for hsp 65 but also for tetanus toxoid. T cells from other inflamed tissues, such as pleural exudates from patients not having tuberculosis, also react much more vigorously to hsp 65 than T cells from peripheral blood of the same patients (23). These findings suggest that the strong responsiveness is a general characteristic of inflamed tissue and that reactivity to hsp 65 is not a unique feature of inflamed RA tissue.

B. Antigen Specificities of Synovial T Cells in Patients with RA and Controls

There is general agreement that hsp 65 is only one of several mycobacterial antigens that stimulate SF T cells. Res et al. (23) demonstrated that in a panel of 15 T-cell clones from SF of patients with RA, 12 different mycobacterial antigenic specificities were represented. These findings correspond to the experience in our laboratory. Wilson et al. (26) raised seven *M. bovis* BCG-reactive clones from a patient with RA, being specific for at least four different proteins. The procedure for selection of T cells determines to a great extent the specificity repertoire of T-cell clones obtained. In general, it was necessary to use hsp 65 in the initial stimulation and screening to obtain clones with this specificity. Although careful frequency studies were not done, this indicates that hsp 65 specific T cells are relatively infrequent.

Gaston et al. (27) found that all of five patients with reactive arthritis responded to *M. leprae* hsp 65, whereas only 9 of 16 patients with RA showed positive responses. In addition, there were generally stronger responses in the group of patients with reactive arthritis. These data suggest that hsp 65 is not essential in the pathogenesis of RA. On the other hand, it may not be the number of responding cells that is essential but the specificity at epitope level of the responding cells. It is thus necessary to consider the fine specificities of the hsp 65-specific T-cells in RA versus other diseases.

C. Frequency of Responding Cells; Limiting Dilution Studies

Crick and Gatenby (28) did not detect any hsp 65-reactive cells by limiting dilution studies of lymphoid cells from patients with RA. This may be due to long disease duration in the patients of that study. Data from Holoshitz et al. (22) indicate that antimycobacterial responses are strongly dependant on disease duration. They found that patients with 1–10 years of disease duration had the strongest responses, whereas patients with shorter or longer lasting RA had weak responses.

Fischer et al. (29) found that the frequency of PPD-responsive T cells varied from 1 per 300 to 1 per 3000 in peripheral blood lymphocytes (PBLs) from healthy individuals. Hsp 65-responsive cells were represented approximately four times less frequently, varying from 1 per 3000 to 1 per 10,000. The frequencies of PPD-reactive and hsp 65-reactive peripheral blood and synovial fluid T cells in patients with RA did not differ significantly from these values. This observation apparently contradicts early studies showing that synovial T cells respond much more vigorously to PPD than PBLs (20,22). There are two possible explanations which unite these data. First, synovial fluid lymphocytes (SFLs) are preactivated and more alert resulting in a more rapid response. Second, and more likely however, SF T cells react more strongly and more rapidly, since they are supplied more efficient antigen presentation by dendritic cells occurring in higher numbers in SF than PB (1). In limiting dilution studies, this difference will not prevail, which would explain why similar frequencies of mycobacteria-specific T cells are present in PBLs and SFLs.

D. T Cells with Specificity for the Human Hsp 65 Homologue

No single T-cell specificity or subpopulation of T cells has been demonstrated to induce or maintain RA. T-cell responses to be considered as relevant for autoimmune diseases should be directed toward conserved epitopes on cross-reactive proteins. Lamb et al. (30) demonstrated the existence of cross-reactive T-cell clones from a patient with tuberculosis responding to the mycobacterial hsp 65, as well as the human homologue. As illustrated in Table 1, two T-cell clones (USSF1F11 and USSF2C6) from patients with RA have been mapped to react with conserved sequences, and one of these clones also responded to the relevant peptide of the human homologue. In cytotoxicity assays, these two clones were demonstrated to lyse target macrophages pulsed with the human hsp 65 homologue (31). Pope et al. (32) found that 5 (18%) of 28 patients with RA responded to the human protein, although the responses to the mycobacterial hsp 65 were stronger. Interestingly, the compiled data in Table 1 show that several T-cell clones have been mapped to nonconserved epitopes of hsp 65. These T-cell clones were mainly derived from other patient groups including leprosy, tuberculosis, and reactive arthritis, as well as vaccinated individuals. Relatively few T-cell clones have been mapped, so the significance of these data are uncertain, but it appears to be a tendency that T-cell clones reacting with cross-reactive epitopes were obtained more frequently from patients with RA. T cells reacting with mycobacteria-specific epitopes on hsp 65 in patients with RA probably reflect previous encounter with mycobacterial antigen.

III. CHARACTERISTICS OF THE RESPONDING T CELLS

A. T Cells with αβ Receptors

In an extensive study of patients with RA and control individuals, Jenkins et al. (33) found that peripheral blood lymphocytes of patients with RA and controls express very similar patterns of Vβ chains. There is, however, a skewed pattern of expressed Vβ families in the synovial tissues and synovial fluid T cells of patients with RA compared with the pattern expressed by their PBLs. The expression of Vβ6, Vβ14, and β15 was increased, whereas a reduced expression of Vβ4, Vβ5.1, Vβ10, Vβ16, and Vβ19 was observed (33). Others have also observed increased levels of Vβ6 (34) and Vβ14 (35,36) in RA. In a patient with early RA, isolated BCG-specific clones expressed mainly Vβ8 but also Vβ6 (26). As seen from Table 2, several hsp 65-specific T-cell clones from a patient with reactive arthritis expressed Vβ5, which has not been found to be associated with RA (33). The hsp 65-specific T-cell clones isolated from two patients with RA and listed in Table 2 expressed a heterogeneous pattern, but both patients had hsp 65-specific clones expressing Vβ6, which was found to be elevated in RA SF (33,34).

Table 1 Restriction Elements and Epitope Specificities for Mycobacteria-specific and Heat Shock–Specific T-Cell Clones in RA Compared with Other Disease States

Diagnosis	Clone designation[a]	HLA type	HLA restriction element	Antigen specificity	Epitope	Conserved residues	Ref.
RA	USSF1C6	DR4Dw4,DRw11, DQw7	DR	hsp 65	461–475	4/15	59
RA	USSF1F11	DR4Dw4,DRw11, DQw7	DR	hsp 65	251–265	12/15	59
RA	USSF2C6	DR4Dw4,DRw11, DQw7	DQA1*0301 DQB1*0301	hsp 65	241–255	9/15	59
RA	CF13/CFB39		DP4	hsp 65	456–466	5/10	60
ReA	MAW1.4	A1,B8,DR3	DR3	hsp 65	1–15	4/15	61
ReA	MAW1.5	A1,B8,DR3	DR3	hsp 65	1–15	4/15	61
M. leprae vaccinated	A7JM	A2,11;B5,12; DR1,2	DR1	hsp 65	211–225	3/15	62
M. leprae vaccinated	2/5A;2/4F;6/10F	A2,9;B15,40; DR4Dw4Dw14	Dr4Dw4	hsp 18	38–50	ND	16
M. leprae vaccinated	2/2F	A2,9;B7,35 DR2,4,Dw4	DR4Dw4	hsp 18	38–50	ND	16
Tuberculosis	K8AH	ND	DR2	hsp 65	231–245	4/15	62
Leprosy	R3F10	DR2,DR3	DQ	hsp 65	235–279	ND	63
Leprosy	R2E4	DR2,DR3	DR3	hsp 65	1–14	4/14	63
Leprosy	Br2-10	DR5,DRw6	DR5	hsp 65	17–61	ND	63
Leprosy	Br5-3	Dr5,DRw6	DR5	hsp 65	85–108	ND	63

[a]Clones designated with the same first two letters are from the same patient.
ND, not determined.

Table 2 Characteristics of Hsp 65–Specific T-Cell Clones

Diagnosis	Clone designation[a]	HLA restriction	Vα	Jα	Vβ	Dβ	Jβ	Ref.
ReA	MAW1.4	DR3	2.4	IGRJa01	5.2	ND	1.5	61
ReA	MAW1.5	DR3	2.4	IGRJa01	5.2	ND	1.5	61
ReA	MAW1.8	DR3	2.4	IGRJa01	5.2	ND	1.5	61
ReA	MAW10.15	DR3	w23.2	C	5.6	ND	1.3	61
ReA	MAW10.8a	DR3	w23.2	C	5.6	ND	1.3	61
ReA	MAW10.8b	DR3	2.5	CF13	5.1	ND	2.1	61
RA	CF13	DP4	1.1	CF13	6.7	ND	2.5	61
RA	CF519	DP4	1.4	AC17	22.3	ND	2.7	61
RA	CF520	DP4	1.4	AC17	22.3	ND	2.7	61
RA	CFB39	DP4	1.4	AC17	22.3	ND	2.7	61
RA	CFB42	DP4	1.4	0	12.2	ND	2.2	61
RA	USSF2C6	DQw7	ND	ND	6	1.1	2.3	59
RA	USSF1F11	DR	ND	ND	17	1.1	2.3	59
RA	USSF1C6	DR	ND	ND	1	1.1	2.2	59

[a]Clones designated with the same first two letters are from the same patient.
ND, not determined.

B. T Cells with $\gamma\delta$ Receptors

$\gamma\delta$ T cells may be of interest in RA, since some studies have found that these cells are enriched in SF compared with PB (37), although not confirmed by others (38). A large proportion of $\gamma\delta$ T cells from healthy individuals respond to stimulation with *M. tuberculosis* antigen. By limiting dilution of normal PBLs, it has been estimated that 1 out of 2– to 20 $\gamma\backslash\delta$ T cells proliferates on stimulation with heat-killed *M. tuberculosis* (39). This is a very high frequency, indicating oligoclonality of reactive cells, that a superantigen may be involved, or alternatively a mitogenic response. It has been established that there are hsp 65 reactive $\gamma\delta$ T cells in normal individuals (40) and patients with RA (31,41). By limiting dilution, 1 of 64 double-negative T cells responded to hsp 65 in normal persons (39). Considering the high proportion of *M. tuberculosis*–reactive $\gamma\delta$ T cells, it appears that hsp 65 is not responsible for stimulation of the majority of the $\gamma\delta$ T cells reacting with tubercle bacilli. On the other hand, the number of hsp 65–reactive $\gamma\delta$ T cells is high compared with the frequencies of hsp 65–reactive $\alpha\beta$ T cells (29). Interestingly, one mycobacterial hsp 65–reactive $\gamma\delta$ T-cell clone has been described to respond to the human homologue, suggesting that hsp 65–specific $\gamma\delta$ T cells may be involved in autoimmunity (42).

There is a highly restricted usage of $\gamma\delta$ V genes in T cells of healthy individuals. At birth, Vγ9 cells is only a minor population of the $\gamma\delta$ cells (43), increasing to more than 50% of the $\gamma\delta$ T cells in adults (44). There is one recent extensive study on Vγ usage in RA (45) which shows that, in contrast to the general pattern of Vβ usage (33), the pattern of Vγ usage is skewed in PB of patients with RA compared to healthy controls. Vγ9 was found to be less dominant in patients with RA than in the controls. There was an even more biased Vγ usage in SF of patients with RA toward Vγ 3/5 as detected by semiquantitative polymerase chain reaction (PCR). Further sequencing analysis showed that this selective expansion was limited to the Vγ3 subset (45). Stimulation of T cells with *M. tuberculosis* antigen induced an expansion of Vγ9/Vδ2 cells (46,47). It is possible that hsp 65–specific $\gamma\delta$ T cells also express Vγ9/Vδ2 (47), but the identity of the antigen that may stimulate the Vγ3 expanded population in RA SF has not been established. It has been suggested that a hitherto unknown antigen may be responsible for the Vγ3 bias. A high proportion of V$\delta1^+$ SF T cells has been found in patients with RA and JRA (48–50), which may be coincident with the Vγ3 bias. V$\delta1^+$ cells are infrequently represented among normal PBLs (51).

C. Cytokine Production of Hsp 65–Reactive T Cells in RA

$\alpha\beta$ T-cell clones isolated from SF of patients RA patients express mostly Th1 but also Th2 and mixed (Th0-like) cytokine patterns (52). Antigen-specific clones reacting with mycobacterial antigen tended to be of the Th1-like type, which is similar to the pattern observed in T cells from patients with tuberculosis and leprosy. Few hsp 65–reactive clones from patients with RA have been characterized with respect to cytokine secretion pattern. One hsp 65–reactive clone was Th1-like and another was of a mixed type secreting interferon gamma (IFN-γ), interleukin 10 (IL-10) and IL-4 but not IL-6 (52).

γδ T-cells from RA SF may also be divided into various subsets according to their cytokine profile. The distribution of Th1, Th2, and Th0-like clones was similar to that of αβ clones except that mean IL-4 levels were lower for γδ clones than αβ clones (53).

IV. PATHOGENETIC MECHANISMS

As indicated previously, it is generally believed that the local inflammatory and immune response in RA is antigen driven. Two main alternative explanations for this inflammatory response have been advocated. First, RA may be induced by a single antigenic epitope. Alternatively, RA may be induced by any epitope being cross-reactive with self. Between these two extremes, several intermediate alternatives may be postulated.

The strong association between RA and the epitope localized at residues 67–74 of the DRβ chain in several DR4 subtypes (3,4) indicates that there may be a specific antigenic peptide or a particular peptide motif interacting with this DR4 region which triggers RA. Most of the hsp 65–responsive T-cell clones were not DR4 restricted. On the other hand, two DR-restricted hsp 65–specific T-cell clones were isolated from a patient with RA (USSF1C6 and USSF1F11; see Table 1), but it was not established whether these cells were restricted by the particular "shared epitope" on DRβ. Presumably, a relevant antigen, for the shared epitope on DRβ, would be of external origin not being a self-antigen. However, several different autoreactive T-cell specificities have been observed in T cells from rheumatoid joints, which apparently contradicts the single-peptide motif hypothesis. An interesting explanation would be that different pathogenetic mechanisms are involved in the induction and maintanance of disease. RA may be induced by a single antigenic peptide motif, whereas several autoreactive T cells with different specificities may participate in the propagation and maintenance of the autoreactive inflammatory response.

During the course of the inductive primary inflammatory response, T cells with different specificities would be attracted to the inflamed joint. The assumption is that these T-cell specificities will participate together with the original inductive T-cell specificity and maintain the inflammatory response after the inductive antigenic stimulus has gone. It is of great interest to note that in proliferation assays, the antimycobacterial responses are quite weak in the first year after onset of RA, being stronger during the following 9 years (22). This would fit the suggestion that responses to mycobacterial antigen are not responsible for induction of RA, but that they may be relevant for maintanance of the disease. Several mycobacterial epitopes cross-react with self, and these epitopes are so far primarily detected on the heat shock proteins.

These assumptions are supported by a recent theory on mechanisms of autoimmunization (54) which has gained experimental support (55,56). Autoreactive T cells evading negative selection and thus reacting with the dominant self can trigger the autoimmune response if a cross-reactive foreign antigen is encountered. During the process, several additional autoreactive T cells are recruited. These are positively selected cells that react with the "cryptic self," and the recruitment of these cells is termed determinant spreading.

During development of the initial immune response, the whole repertoire of memory T cells would potentially be attracted to the local inflammatory site. However, certain T-cells subsets may be selectively attracted to particular tissues (57). On the other hand, there is no indication that the homing process is antigen specific. Mycobacterial antigens are potent immunogens, and they are almost ubiquitous. If not presented by infection, most people will be sensitized by naturally occurring mycobacteria in the environment or by the extensively applied BCG vaccine. This would explain why T cells isolated from chronic inflammatory tissues react vigorously to mycobacterial antigens. Furthermore, the wide repertoire of mycobacterial T-cell specificities observed in SF of patients with RA could be interpreted on this background.

Theoretically, autoreactive T cells may be enriched at the inflammatory site during the disease process owing to specific stimulation and proliferation of these cells, whereas irrelevant T cells either stay without excessive proliferation or leave for recirculation.

V. CONCLUSIONS

The role of hsp 65 in RA is uncertain. There is still a possibility that certain T cells specific for conserved sequences on hsp 65 may be essential for induction of RA, but it is more likely that these T cells play a role in the maintenance of disease together with a large repertoire of self-reactive T cells concentrated at the local site as a secondary phenomenon.

There are, however, several other hsps which have not been explored in any detail in RA. A peptide of hsp 18 may be of particular interest, because it has been demonstrated to induce DR4Dw4-restricted T-cell responses in healthy persons (16).

If it is possible to find a crucial peptide motif responsible for induction of RA, several strategies for immunomodulation would be available. These strategies include MHC blockade and T-cell receptor antagonism (58). One should aim at strategies that would interfere with the faulty T-cell network in a specific way so as to obtain downregulation of, or modification of, the phenotypic characteristics of the essential T-cell specificities.

REFERENCES

1. K. Waalen, Ø. Førre, J. Pahle, J.B. Natvig, and G.R. Burmester, Characteristics of human rheumatoid synovial and normal blood dendritic cells, *Scand. J. Immunol. 26*:525 (1987).
2. I. Randen, O.J. Mellbye, Ø. Førre, and J.B. Natvig, The identification of germinal centres and follicular dendritic cell networks in rheumatoid synovial tissue, *Scand. J. Immunol. 41*:481 (1995).
3. R.J. Winchester, The molecular basis of susceptibility to rheumatoid arthritis, *Adv. Immunol. 56*:289 (1994).
4. P.K. Gregersen, J. Silver, and R.J. Winchester, The shared epitope hypothesis. An approach to understanding the molecular genetics of susceptibility to rheumatoid arthritis, *Arthritis Rheum. 30*:1205 (1987).
5. E. Roosnek and A. Lancaveccia, Efficient and selective presentation of antigen-antibody complexes by rheumatoid factor B cells, *J. Exp. Med. 173*:487 (1991).
6. A.K. Dudani and R.S. Gupta, Immunological characterization of a human homolog of the 65-kilodalton mycobacterial antigen, *Infect. Immun. 57*:2786 (1989).
7. A. Karlson-Parra, K. Söderström, M. Ferm, J. Ivanyi, R. Kiessling, and L. Klareskog, Presence of human 65kD heat shock protein (hsp) in inflamed joints and subcutanous nodules of RA patients: a clue to pathogenicity of anti–65 kD hsp immunity, *Scand. J. Immunol. 31*:283 (1990).
8. W. Van Eden, J.E.R. Thole, R. Van der Zee, A. Noordzij, J.D.A. Van Embden, E. J. Hensen, and I. R. Cohen, Cloning of the mycobacterial epitope recognized by T lymphocytes in adjuvant arthritis, *Nature 331*:171 (1988).
9. A. Rambukkana, P.K. Das, S. Krieg, S. Yong, I.C. Le Poole, and J.D. Bos, Mycobacterial 65,000 MW heat-shock protein shares a carboxy-terminal epitope with human epidermal cytokeratin 1/2, *Immunology 77*:267 (1992).
10. N. Esaguy, A.P. Aguas, J.D.A. Van Embden, and M.T. Silva, Mycobacteria and human autoimmune disease: direct evidence of the cross-reactivity between human lactoferrin and the 65-kilodalton protein of tubercle and leprosy bacilli, *Infect. Immun. 59*:1117 (1991).
11. A.P. Aguas, N. Esaguy, C.E. Sunkel, and M.T. Silva, Crossreactivity and sequence homology between 65-kilodalton mycobacterial heat shock protein and human lactoferrin, transferrin, and DRβ subsets of major histocompatibility complex class II molecules, *Infect. Immun. 58*:1461 (1990).
12. R.B. Lathigra, D.B. Young, D. Sweetser, and R.A. Young, A gene from *Mycobacterium tuberculosis* which is homologous to the DnaJ heat shock protein of *E. coli*, *Nucleic Acids Res. 16*:1636 (1988).
13. R.J. Garsia, L. Hellqvist, R.J. Booth, A.J. Radford, W.J. Britton, L. Astbury, R.J. Trent, and A. Basten, Homology of the 70-kilodalton antigens from *Mycobacterium leprae* and *Mycobacterium bovis* with the *Mycobacterium tuberculosis* 71-kilodalton antigen and with the conserved heat shock protein 70 of eucaryotes, *Infect. Immun. 57*:204 (1989).

14. F. Oftung, A. Geluk, K.E.A. Lundin, R.H. Meloen, J.E.R. Thole, A.S. Mustafa, and T.H.M. Ottenhoff, Mapping of multiple HLA class II-restricted T-cell epitopes of the mycobacterial 70-kilodalton heat shock protein, *Infect. Immun. 62*:5411 (1994).
15. A.S. Mustafa, K.E.A. Lundin, and F. Oftung, Human T cells recognize mycobacterial heat shock proteins in the context of multiple HLA-DR molecules: studies with healthy subjects vaccinated with *Mycobacterium bovis* BCG and *Mycobacterium leprae*, *Infect. Immun. 61*:5294 (1993).
16. F. Oftung, T.M. Shinnick, A.S. Mustafa, K.E. A. Lundin, T. Godal, and A. Nerland, Heterogeneity among human T cell clones recognizing an HLA-DR4,Dw4–restricted epitope from the 18-kDa antigen of *Mycobacterium leprae* defined by synthetic peptides, *J. Immunol. 144*:1478 (1990).
17. M. Sioud, J. Kjeldsen-Kragh, A. Quayle, H.G. Wiker, D. Sørskaar, J.B. Natvig, and Ø. Førre, Immune responses to 18.6 and 30-kDa mycobacterial antigens in rheumatoid patients and V beta usage by specific synovial T cell lines and fresh T cells, *Scand. J. Immunol. 34*:803 (1991).
18. R.J. Booth, D.L. Williams, K.D. Moudgil, L.C. Noonan, P.M. Grandison, J.J. McKee, R.L. Prestidge, and J.D. Watson, Homologs of *Mycobacterium leprae* 18-kilodalton and *Mycobacterium tuberculosis* 19-kilodalton antigens in other mycobacteria, *Infect. Immun. 61*:1509 (1993).
19. J.B. Natvig, and A.J. Quayle, V-genes of T cell receptors in rheumatoid arthritis, *Proc. N.Y. Acad. Sci. 756*:138 (1995).
20. T.G. Abrahamsen, S.S. Frøland, and J.B. Natvig, In vitro mitogen stimulation of synovial fluid lymphocytes from rheumatoid arthritis and juvenile rheumatoid arthritis patients: dissociation between the response to antigens and polyclonal mitogens, *Scand. J. Immunol. 7*:81 (1978).
21. M. Harboe, Antigens of PPD, Old tuberculin, and autoclaved *Mycobacterium bovis* BCG studied by crossed immunoelectrophoresis, *Am. Rev. Respir. Dis. 124*:80 (1981).
22. J. Holoshitz, I. Drucker, A. Yaretzky, W. Van Eden, A. Klajman, Z. Lapidot, A. Frenkel, and I. Cohen, T lymphocytes of rheumatoid arthritis patients show augmented reactivity to a fraction of mycobacteria cross-reactive with cartilage, *Lancet 2*:305 (1986).
23. P.C.M. Res, D.L.M. Orsini, J.M. Van Laar, A.A.M. Janson, C. Abou-Zeid, and R. R. P. DeVries, Diversity in antigen recognition by *Mycobacterium tuberculosis*-reactive T cell clones from the synovial fluid of rheumatoid arthritis patients, *Eur. J. Immunol. 21*:1297 (1991).
24. J. De Bruyn, R. Bosmans, M. Turneer, M. Weckx, J. Nyabenda, J.-P. Van Vooren, P. Falmagne, H.G. Wiker, and M. Harboe, Purification, partial characterization, and identification of a skin-reactive protein antigen of *Mycobacterium bovix* BCG, *Infect. Immun. 55*:245 (1987).
25. G.R. Burmester, U. Altstidl, J.R. Kalden, and F. Emmrich, Stimulatory response towards the 65 kDa heat shock protein and other mycobacterial antigens in patients with rheumatoid arthritis, *J. Rheumatol. 18*:171 (1991).
26. K.B. Wilson, A.J. Quayle, S. Suleyman, J. Kjeldsen-Kragh, Ø. Førre, J.B. Natvig, and J.D. Capra, Heterogeneity of the TCR repertoire in synovial fluid T lympho-

cytes responding to BCG in a patient with early rheumatoid arthritis, *Scand. J. Immunol. 38*:102 (1993).
27. J.S.H. Gaston, P.F. Life, L.C. Bailey, and P.A. Bacon, In vitro responses to a 65-kilodalton mycobacterial protein by synovial T cells from inflammatory arthritis patients, *J. Immunol. 143*:2494 (1989).
28. F.D. Crick, and P.A. Gatenby, Limiting-dilution analysis of T cell reactivity to mycobacterial antigens in peripheral blood and synovium from rheumatoid arthritis patients, *Clin. Exp. Immunol. 88*:424 (1992).
29. H.P. Fischer, C.E.M. Sharrock, M.J. Colston, and G.S. Panayi, Limiting dilution analysis of proliferative T cell responses to mycobacterial 65-kDa heat-shock protein fails to show significant frequency differences between synovial and peripheral blood of patients with rheumatoid arthritis, *Eur. J. Immunol. 21*:2937 (1991).
30. J.R. Lamb, V. Bal, P. Mendez-Samperio, A. Mehlert, A. So, J. Rothbard, S. Jindal, R.A. Young, and D.B. Young, Stress proteins may provide a link between the immune response to infection and autoimmunity, *Int. Immunol. 1*:191 (1989).
31. S.G. Li, A.J. Quayle, Y. Shen, J. Kjeldsen-Kragh, F. Oftung, R.S. Gupta, J.B. Natvig and Ø.T. Førre, Mycobacteria and human heat shock protein–specific cytotoxic T lymphocytes in rheumatoid inflammation, *Arthritis Rheum. 35*:270 (1992).
32. R.M. Pope, R.M. Lovis, and R.S. Gupta, Activation of synovial fluid T lymphocytes by 60 kd heat shock proteins in patients with inflammatory synovitis, *Arthritis Rheum. 35*:43 (1992).
33. R.N. Jenkins, A. Nikaein, A. Zimmerman, K. Meek, and P.E. Lipsky, T cell receptor Vβ gene bias in rheumatoid arthritis, *J. Clin. Invest. 92*:2688 (1993).
34. C.M. Weyand, U. Oppitz, K. Hicok, and J.J. Goronzy, Selection of T cell receptor Vβ elements by HLA-DR determinants predisposing to rheumatoid arthritis, *Arthritis Rheum. 35*:990 (1992).
35. J.D. Capra and J.B. Natvig, Is there V region restriction in autoimmune disease?, *Immunologist 1*:16 (1993).
36. M.P. Davey, and D.D. Munkirs, Patterns of T-cell receptor variable β gene expression by synovial fluid and peripheral blood T-cells in rheumatoid arthritis, *Clin. Immunol. Immunopathol. 68*:79 (1993).
37. F.M. Brennan, M. Londei, A.M. Jackson, T. Hercend, M.B. Brenner, R.N. Maini, and M. Feldman, T cells expressing γδ chain receptors in rheumatoid arthritis, *J. Autoimmun. 1*:319 (1988).
38. J. Kjeldsen-Kragh, A. Quayle, C. Kalvenes, Ø. Førre, D. Sørskaar, O. Vinje, J. Thoen, and J.B. Natvig, Tγδ cells in juvenile rheumatoid arthritis and rheumatoid arthritis, *Scand. J. Immunol. 32*:651 (1990).
39. D. Kabelitz, A. Bender, S. Schondelmaier, B. Schoel, and S.H.E. Kaufmann, A large fraction of human peripheral blood γ/δ^+ T cells is activated by *Mycobacterium tuberculosis* but not by its 65-kD heat shock protein, *J. Exp. Med. 171*:667 (1990).
40. A. Haregewoin, G. Soman, R.C. Hom, and R.W. Finberg,, Human $\gamma\delta^+$ T cells respond to mycobacterial heat-shock protein, *Nature 340*:309 (1989).
41. J. Holoshitz, F. Koning, J.E. Coligan, J. De Bruyn, and S. Strober, Isolation of $CD4^-$ $CD8^-$ mycobacteria-reactive T lymphocyte clones from rheumatoid arthritis synovial fluid, *Nature 339*:226 (1989).

42. A. Haregewoin, B. Singh, R.S. Gupta, and R.W. Finberg, A mycobacterial heat-shock protein-responsive γ/δ T cell clone also responds to the homologous human heat shock protein: a possible link between infection and autoimmunity, *J. Infect. Dis. 163*:156 (1991).
43. G. Casorati, G. De Libero, A. Lanzavecchia, and N. Migone, Molecular analysis of human γ/δ^+ clones from thymus and peripheral blood, *J. Exp. Med. 170*:1521 (1989).
44. C. Bottino, G. Tambussi, S. Ferrini, E. Ciccone, P. Varese, M.C. Mingari, L. Moretta, and A. Moretta, Two subsets of human T lymphocytes expressing γ/δ antigen receptor are identifiable by monoclonal antibodies directed to two distinct molecular forms of the receptor, *J. Exp. Med. 168*:491 (1988).
45. Y. Kageyama, Y. Koide, S. Miyamoto, T. Inoue, and T.O. Yoshida, The biased Vγ gene usage in the synovial fluid of patients with rheumatoid arthritis, *Eur. J. Immunol. 24*:1122 (1994).
46. G. De Libero, G. Casorati, C. Giachino, C. Carbonara, N. Migone, P. Matzinger, and A. Lanzavecchia, Selection by two powerful antigens may account for the presence of the major population of human peripheral γ/δ T cells, *J. Exp. Med. 173*: 1311 (1991).
47. D. Kabelitz, A. Bender, T. Prospero, S. Wesselborg, O. Janssen, and K. Pechhold, The primary response of human γ/δ+ T cells to *Mycobacterium tuberculosis* is restricted to Vγ9-bearing cells, *J. Exp. Med. 173*:1331 (1991).
48. E.C. Keystone, C. Rittershaus, N. Wood, K.M. Snow, J. Flatow, J.C. Purvis, L. Poplonski, and P.C. Kung, Elevation of a γδ T cell subset in peripheral blood and synovial fluid of patients with rheumatoid arthritis, *Clin. Exp. Immunol. 84*:78 (1991).
49. M. Sioud, J. Kjeldsen-Kragh, A. Quayle, C. Kalvenes, K. Waalen, Ø. Førre, and J. B. Natvig, The Vδ gene usage by freshly isolated T lymphocytes from synovial fluids in rheumatoid synovitis: a preliminary report, *Scand. J. Immunol. 31*:415 (1990).
50. J. Kjeldsen-Kragh, A.J. Quayle, O. Vinje, J.B. Natvig, and Ø. Førre, A high proportion of the Vδ1+ synovial fluid γδ T cells in juvenile rheumatoid arthritis patients express the very early activation marker CD69, but carry the high molecular weight isoform of the leukocyte common antigen (CD45RA), *Clin. Exp. Immunol. 91*:202 (1993).
51. W. Haas, Gama/delta cells, *Ann. Rev. Immunol. 11*:637 (1993).
52. A.J. Quayle, P. Chomarat, P. Miossec, J. Kjeldsen-Kragh, Ø. Førre, and J. Natvig, Rheumatoid inflammatory T-cell clones express mostly Th1 but also Th2 and mixed (Th0-like) cytokine patterns, *Scand. J. Immunol. 38*:(1993).
53. P. Chomarat, J. Kjeldsen-Kragh, A.J. Quayle, J.B. Natvig, and P. Miossec, Different cytokine production profiles of γδ T cell clones: relation to inflammatory arthritis, *Eur. J. Immunol. 24*:2087 (1994).
54. E.E. Sercarz and S.K. Datta, Autoimmunity. Mechanisms of autoimmunization: perspective from the mid-90s, an editorial overview, *Curr. Opin. Immunol. 6*:875 (1994).
55. P.V. Lehmann, E.E. Sercarz, T. Forsthuber, C.M. Dayan, and G. Gammon, De-

terminant spreading and the dynamics of the autoimmune T cell repertoire, *Immunol. Today 14*:203 (1993).

56. P.V. Lehmann, T. Forsthuber, A. Miller, and E.E. Sercarz, Spreading of T-cell autoimmunity to cryptic determinants of an autoantigen, *Nature 358:*155 (1992).
57. N. Oppenheimer-Marks, and P.E. Lipsky, Transendothelial migration of T cells in chronic inflammation, *Immunologist 2*:58 (1994).
58. A. Franco, G.Y. Ishioka, L. Adorini, J. Alexander, J. Ruppert, K. Snoke, H.M. Grey, and A. Sette, MHC blockade and T-cell receptor antagonism, *Immunologist 2*:97 (1994).
59. J.A. Quayle, K.B. Wilson, S.G. Li, J. Kjeldsen-Kragh, F. Oftung, T. Shinnick, M. Sioud, Ø. Førre, J.D. Capra, and J.B. Natvig, Peptide recognition, T cell receptor usage and HLA restriction elements of human heat-shock protein (hsp) 60 and mycobacterial 65 kDa hsp-reactive T cell clones from rheumatoid synovial fluid, *Eur. J. Immunol. 22*:1315 (1992).
60. J. Henwood, J. Loveridge, J.I. Bell, and J.S.H. Gaston, Restricted T cell receptor expression by human T cell clones specific for mycobacterial 65-kDa heat shock protein: selective in vivo expansion of T cells bearing defined receptors, *Eur. J. Immunol. 23*:1256 (1993).
61. J.S.H. Gaston, P.F. Life, P.J. Jenner, M.J. Colston, and P.A. Bacon, Recognition of a mycobacteria-specific epitope in the 65-kD heat-shock protein by synovial fluid-derived T cell clones, *J. Exp. Med. 171*:831 (1990).
62. F. Oftung, A.S. Mustafa, T.M. Shinnick, R.A. Houghten, G. Kvalheim, M. Degre, K.E.A. Lundin, and T. Godal, Epitopes of the *Mycobacterium tuberculosis* 65 kilodalton protein antigen as recognized by human T cells, *J. Immunol. 141*:2749 (1988).
63. J.E.R. Thole, W.C.A. Van Schooten, W.J. Keulen, P.W.M. Hermans, A.A.M. Janson, R.R.P. De Vries, A.H.J. Kolk, and J.D.A. Van Embden, Use of recombinant antigen expressed in *Escherichia coli* K-12 to map B-cell and T-cell epitopes on the immunodominant 65-kilodalton protein of *Mycobacterium bovis* BCG, *Infect. Immun. 56*:1633 (1988).

9

Heat Shock Protein 60 Autoimmunity in Lyme Disease

Zhizhong Dai and Stanley Stein
Center for Advanced Biotechnology and Medicine, Piscataway, New Jersey and Robert Wood Johnson Medical School—UMDNJ New Brunswick, New Jersey

Stephanie Williams and Leonard H. Sigal
Robert Wood Johnson Medical School, New Brunswick, New Jersey

I. INTRODUCTION

For the past several years, we have been investigating the involvement of an autoimmune response in the pathogenesis of neurological features of Lyme disease (LD). Our recent findings implicate a role for autoreactive antibodies to the 60-kDa host heat shock protein (hsp 60) in this disease process. This chapter summarizes our discovery of a specific peptide sequence in hsp 60 as the molecular mimic of an immunodominant determinant in flagellin of *Borrelia burgdorferi* the etiological agent of Lyme disease.

II. HSP 60 IN AUTOIMMUNE RESPONSES

A variety of proteins are induced in all cells on stress, such as heat shock. This chapter will focus on one particular family of structurally conserved heat shock proteins, which includes GroEL in *Escherichia coli*, hsp 65 in *Mycobacterium bovis*, and hsp 60 in mammals. These proteins are termed chaperonins, corresponding to their role in assisting the folding and assembly of other proteins. Members of this family of proteins are highly homologous. For example, there is about 50% identity in amino acid sequence between human hsp 60 and *M. bovis* hsp 65 (1).

According to the molecular mimicry theory (2,3), it is possible that antibodies and/or T lymphocytes made in the immune response to a specific epitope found on a component of the pathogen may also bind to a structurally similar host determinant, thereby leading to autoimmune damage. This mimicry may be due to the presence of a common epitope in homologous proteins, such as the host and pathogen members of the hsp 60 family. Because of their high degree of homology across the broad stretches of phylogeny and their ubiquity, heat shock proteins represent a logical protein family in which to seek examples of molecular mimicry. As previously reviewed (4), it has been hypothesized that molecular mimicry between host and pathogen heat shock proteins constitutes an important physiological element of the protective immune response, which for unknown reasons may go awry in some individuals and may contribute to autoimmune disease. Another possibility, presented below, is molecular mimicry between unrelated pathogen and host proteins, one of which may be a heat shock protein. Sequence similarities between suspected autoimmune antigens and human hsp 60 have been demonstrated in several autoimmune diseases using an interesting but speculative database search paradigm (5).

One example of heat shock protein autoimmunity is the adjuvant arthritis induced by inoculation of mycobacteria into certain susceptible animals (7). A series of reports focused on a particular T-cell clone derived from rats with adjuvant arthritis that recognized a protein moiety of cartilage proteoglycan; the disease could be adoptively transferred by these cells (7–9). Furthermore, the cells strongly and specifically responded to a peptide comprising residues 180–188 of mycobacterial hsp 65 (10). Interestingly, administration of hsp 65 in oil did not induce arthritis but could provide resistance (11). Corresponding purported examples of arthritis in humans based on cross reactivity with bacterial heat shock proteins have been reviewed (12).

Another prominent example of heat shock protein involvement in autoimmunity is insulin-dependent diabetes mellitus (IDDM) of the nonobese diabetic (NOD) mouse strain. IDDM of NOD mice develops spontaneously (13). Appearance of both T-cell reactivity and antibodies that react with mycobacterial hsp 65 precedes overt diabetic symptoms (14,15). Infusion of T cells obtained from mice with advanced insulitis into prediabetic recipient mice produces severe disease within 1 week (14,15). Furthermore, injection of hsp 65 into 1-month-old mice induced diabetic changes well before disease would have manifested spontaneously (15). In contrast, when 1-month-old NOD mice were vaccinated with attenuated reactive T cells, resistance to both spontaneous and induced diabetes occurred (16). Finally, vaccination with a peptide comprising residues 437–460 of human hsp 60 could prevent diabetes in this model (17).

The earliest reactivity with an autoantigen in NOD mice (18,19) as well as in human IDDM (20,21) is believed to be with the 64-kDa protein. glutamatic acid decarboxylase (GAD). Autoimmunity to GAD may be a necessary but not a sufficient condition for disease progression. The initial insult of GAD immunoreactivity may lead to a cascade of reactivity with autoantigens, including hsp 60, thereby producing the diabetic state (19). Interestingly, sequence homology between GAD and hsp 60 has been noted (5), and the initial T-cell determinant of GAD (18) is within this region of homology.

In both IDDM of NOD mice and adjuvant arthritis, a correlation between autoimmune disease and members of the hsp 60 family has been suggested. We have hypothesized that an autoimmune response to hsp 60 underlies the neurological symptoms of chronic Lyme disease (22). Rather than molecular mimicry between pathogen and host heat shock proteins or between pathogen heat shock protein and an otherwise unrelated host protein, we have found that antibodies to the 41-kDa flagellin of *Borrelia burgdorferi* cross react with human hsp 60. Our studies supporting this hypothesis are now presented.

III. HEAT SHOCK PROTEIN ANTIGENICITY IN LYME DISEASE

GroEL homologues have been identified in *B. burgdorferi* as proteins with molecular weights of 60 and 66 kDa on gel electrophoresis, whereas the DnaK homologue has been identified as having a molecular weight of 72 kDa (23). On Western blot analysis, serum samples from patients with Lyme disease often react with both the 60- and 66-kDa proteins but not with the 72-kDa protein (23).

In another study (24), GroEL homologues in *B. burgdorferi* were identified as having molecular weights of 66 and 68 kDa, whereas DnaK homologues corresponded to 71 and 73 kDa. The 66- and 68-kDa proteins were further identified by aminoterminal sequencing, which showed similar and possibly identical sequences with one another, as well as strong homology with GroEL. In 9 of 12 sera tested by Western blot, strong reactivity to four antigens between 66 and 73 kDa was found; in contrast, no reactivity was found in seven control sera from healthy individuals. Another aspect of this study was the demonstration of peripheral blood mononuclear cell proliferation from five patients with Lyme disease in response to gel-purified 66-kDa antigen. Presumably, the 66- and 68-kDa antigens in this study (24) correspond to the 60- and 66-kDa antigens discussed above (22).

Research in another laboratory (25) also revealed that the 60-kDa antigen in *B. burgdorferi* is homologous to GroEL. Serum antibodies in patients with Lyme disease, but not control sera, were shown to bind to recombinant *B. burgdorferi* 60-kDa antigen prepared in *Escherichia coli* (26). Whereas one T-cell clone responsive to residues 260–274 of *B. burgdorferi* 60-kDa antigen was isolated from peripheral blood of a patient with Lyme disease, several other T-cell clones responded to other *B. burgdorferi* antigens. In a related study (27). a T-cell clone from a patient with Lyme disease recognized residues 37–54 of *B. burgdorferi* hsp 60. Neither T-cell clone responded to the corresponding sequences in the human and *E. coli* homologues (26,27).

A comparative survey for the presence of antibodies to *B. burgdorferi* and human hsp 60 was done by Western blot analysis (28). This study revealed that the *B. burgdorferi* hsp 60 homolog is an immunodominant protein in Lyme disease. Antibodies to *B.burgdorferi* 60 kDa (and *B. burgdorferi* 66 kDa) are present more frequently (91%) than to the next most immunodominant antigen, flagellin (81%). About 68% of the patients with Lyme disease also had antibodies to human hsp 60. However, a high incidence of immunoreactive antibodies to *B. burgdorferi* 60 and human hsp 60 was detected in patients with rheumatoid arthritis (67% and 73%, respectively), and there were even positive responses in healthy control individuals (17% and 25, respectively).

IV. A BRIEF REVIEW OF THE CLINICAL FEATURES OF LYME DISEASE

Lyme disease (LD) is a multisystem inflammatory disease caused by *B. burgdorferi*. As reviewed (29,30), the clinical features of the disease can be divided into three categories: early localized (31), early disseminated (32,33), and chronic or late disease (34,35). The immunology of Lyme disease has been reviewed previously (36,37).

A. Early Localized LD (Formerly Stage I)

The pathognomonic skin rash of LD, erythema migrans (EM), seen in about 50–70% of patients, usually begins as an erythematous macule or papule at the site of the (usually asymptomatic) tick bite. The lesion expands within a few days, often with central clearing. The organism can be grown from the lesion, especially from the outer margin of the expanding erythema. Secondary lesions are almost certainly due to hematogenous dissemination of the organism; that is, spirochetemia and not multiple bites. Early localized disease often includes nonspecific symptoms, described as a virus-like syndrome, thought to be due to the production and liberation of cytokines. Many patients with early disease are asymptomatic. Other findings, including pain on neck flexation, malar rash, conjunctivitis, erythematous throat, joint pains, and muscle tenderness, have been noted.

B. Early Disseminated LD (Formerly Stage II)

Approximately 8% of patients with untreated early localized LD progress to carditis, occurring within days of the onset of infection or as long as 9 months after the onset of EM. Lyme carditis causes fluctuating atrioventricular conduction defects, myopericarditis, and very mild congestive heart failure. The organism has been seen in myocardial biopsies and grown from one biopsy specimen. Approximately 10–15% of patients with untreated early localized disease will develop neurological features of early disseminated LD in the same time period as carditis. This neurological syndrome was first described over 70 years ago as tick-borne meningopolyneuritis or Bannwarth's syndrome. Headache, mild neck stiffness, and photophobia may occur. Fever may be mild or absent, fatigue and malaise are common, and mild encephalopathy, usually difficulty with memory and concentration and emotional lability, may be prominent features. Cranial neuropathies, often associated with a lymphocytic meningitis, are common, most often affecting the facial nerve (occasionally bilateral). Peripheral neuropathy and radiculoneuropathy may affect the limbs or trunk. A lymphocytic pleocytosis is typically found in the cerebrospinal fluid (CSF), especially in patients with meningitis plus radiculitis and there is concentration of antigen-specific T-cell and antibody reactivity in the CSF. The organism can occasionally be grown from the CSF, but there are no reports of *B. burgdorferi* being found in the biopsies of affected peripheral nerves.

C. Late or Chronic LD (Formerly Stage III)

Infection with *B. burgdorferi* can cause a variety of musculoskeletal complaints, including polyarthralgia, true inflammatory disease (poly-, oligo-, or monoarticular), tendonitis, bursitis, and fibromyalgia. Concentration of antigen-specific antibodies and T cells can be demonstrated in the synovial fluid. In the last decade, late neurological features of LD have been noted: progressive encephalopathy, polyneuritis, and mental and/or psychiatric changes. This syndrome is now called tertiary neuroborreliosis, which may occur as the first and only manifestation of previously latent infection. Late disease usually affects the central nervous system, in contrast with the primary peripheral nervous system disease of earlier neurological LD. Virtually all patients with tertiary neuroborreliosis are seropositive. Concentration of antigen-specific antibodies in the CSF can usually be demonstrated. The organism has been seen in brain biopsy and postmortem specimens but not in the peripheral nerve.

V. AUTOIMMUNITY IN CHRONIC LYME DISEASE

It is not clear if persistence of the organism is necessary for ongoing disease, since antibiotic therapy may not be as effective in late disease as it usually is at an early stage (38). Indeed, the presence of *B. burgdorferi* at local sites of inflammation is difficult to demonstrate (37). These findings are compatible with the premise that an autoimmune mechanism may be active in the pathogenesis of chronic Lyme disease.

Although an immune response to heat shock proteins in patients with Lyme disease has been demonstrated in the various studies cited above, there is no evidence that molecular mimicry between the homologous *B. burgdorferi* and host heat shock proteins leads to an autoimmune response. Three lines of evidence will be presented to demonstrate that neurological Lyme disease may be an autoimmune disease resulting from molecular mimicry between *B. burgdorferi* flagellin and human hsp 60. These include (1) immunostaining of nerve tissues and neuroblastoma cells, (2) identification of the cross-reacting epitopes, and (3) a proposed pathological mechanism of neuronal dysfunction.

A. Molecular Mimicry Between *B. burgdorferi* Flagellin and a Human Neuronal Protein

Based on a series of studies, we have concluded that molecular mimicry between *B. burgdorferi* 41-kDa flagellin and a human axonal protein is associated with neurological Lyme disease (39). Sera of patients with Lyme neurological disease were found to bind to normal human axons (39,40), as well as to cultured neural cells (41). This tissue and cell staining could be eliminated by preadsorption to whole *B. burgdorferi* or a flagellin-enriched preparation (39,40). In contrast, diabetic neuropathy and Guillain-Barré syndrome sera were also found to bind to axon, but this binding could not be removed by preadsorption with *B. burgdorferi* or *B. burgdorferi* flagellin(40).

Fortuitously, a monoclonal antibody, H9724, directed against an epitope unique to the flagellin of *B. burgdorferi* (42) was also found to stain axons and neural cells (40). Other antiflagellin antibodies, as well as antibodies against *B. burgdorferi* OspA and OspB (outer surface proteins), did not bind to axonal preparations (40). In another study, H9724, but not various other monoclonal antibodies, was shown to bind to antigens on peripheral and central nerves, as well as cardiac muscle, hepatocytes, and synovial cells (43).

B. Identification of the Lyme Disease–Specific Epitope in *B. burgdorferi* Flagellin

The epitope in flagellin recognized by H9724 was shown to be the same as that recognized by sera of patients with Lyme disease using multiple-peptide pin microsynthesis (44), as well as by recombinant DNA and standard peptide synthesis techniques (45). This epitope was reported to comprise residues 205–226 (44) or 213–224 (45). Independently, we found this same epitope using a peptide mapping procedure in which purified flagellin was chemically and proteolytically fragmented and the resultant peptides were identified by Western blotting and aminoterminal sequencing (46). Based on chemical synthesis and peptide dot blot analysis (47), we have defined the epitope to reside within the sequence,

VQEGVQQEGAQ (46). A semiquantitative comparison of H9724 binding to *B. burgdorferi* flagellin and two epitope peptides by dot blot analysis is shown in Figure 1. Based on multiple-peptide pin microsynthesis, we have suggested that there is a repeating epitope of homologous amino acids, VQEGVQ and QQEGAQ. Whereas flagellins from different bacteria are quite homologous, this particular epitope is unique to *B. burgdorferi* flagellin, since it is in a variable region of the protein.

C. Identification of the Host Cross-Reacting Protein

By Western blot analysis with H9724, it was possible to detect a 64-kDa protein in both neuroblastoma cells (Fig. 2) and human nerve (39). This protein could be immunoprecipitated from neuroblastoma cells and then detected on Western blot analysis with Lyme neurological disease sera (Fig. 3). Western blot analysis of numerous human and animal neuronal and non-neuronal tissues showed, essentially only one cross-reacting protein, in each case at the same molecular weight. Depending on the particular electrophoretic conditions, the molecular weight estimate varied between 60 and 64 kDa.

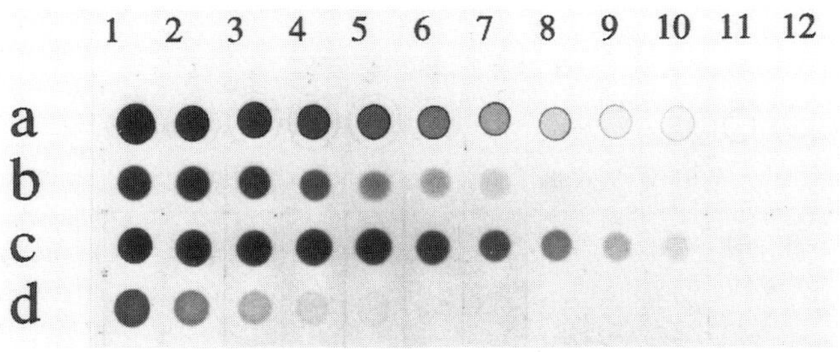

Figure 1 Dot blot analysis of epitope peptides. Peptides were synthesized by Fmoc chemistry, purified by reverse-phase HPLC, and analyzed for correctness of structure by mass spectrometry, aminoterminal sequencing, and amino acid compositional analysis. Peptides in twofold serial dilution were covalently attached to PVDF-AV membranes, according to a previously published procedure (48). The primary and secondary antibodies were H9724 and goat antimouse IgG-coupled alkaline phosphatase. Color development was with nitroblue tetrazolium and 5-bromo, 4-chloro, 3-indolyl phosphate. Samples are as follows: (a) *B. burgdorferi* flagellin purified by SDS-PAGE and diluted into 0.5% SDS/phosphate-buffered-saline, 4 ng in column 1; (b) 10-mer flagellin peptide, 215VQQEGA- AQQPA-224, 1000 ng in column 1; (c) 12-mer flagellin peptide, 213-EGVQQEGA- QQPA-224, 1000 ng in column 1; (d) hsp 60 13-mer peptide, 14-MLQGVDLLADAVA-26, 4000 ng in column 1. All peptides were diluted in 0.25 M sodium phosphate, pH7.5 containing 0.15 M sodium chloride, and 15% isopropanol. Qualitative comparisons may be made regarding the affinity of each peptide for H9724.

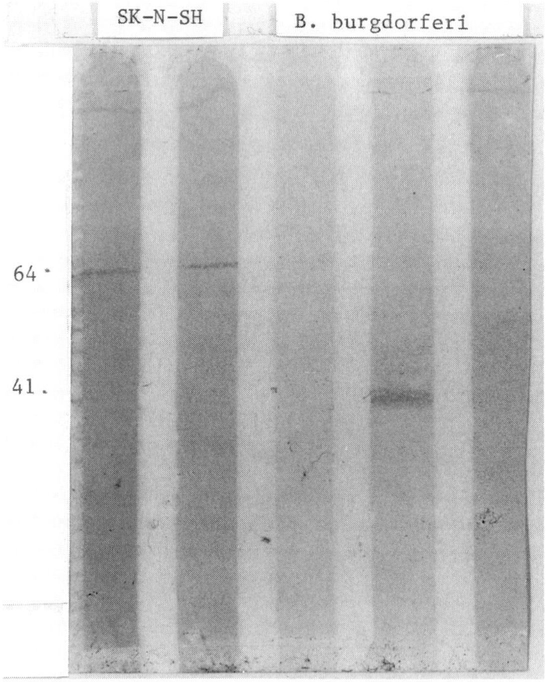

Figure 2 Western blot analysis showing proteins reactive with H9724 (from Ref. 39). Lanes 1 and 2 represent extracts from 1.7×10^6 and 1.4×10^7 SK-N-SH cells, respectively, grown to confluency, with H9724 as the primary antibody. Lane 3 is similar to lane 1, except the primary antibody was another anti-*B. burgdorferi* flagellin monoclonal antibody, 6TS. Lane 4 shows reactivity of H9724 toward purified *B. burgdorferi* flagellin, whereas normal mouse serum does not bind to flagellin (lane 5) nor to SK-N-SH proteins (not shown). Development of the blot was similar to that described in Figure 1.

For the purpose of protein isolation, we decided to use calf adrenal gland as the starting material (22). Frozen adrenal glands were trimmed, minced, and then homogenized in buffer containing a cocktail of protease inhibitors. After clarification by centrifugation, ammonium sulfate cuts were made. According to SDS-PAGE and Western blot analysis, the cross-reacting protein was most enriched in the 60% ammonium sulfate fraction. The protein was reconstituted in Tris buffer and resolved by reverse-phase high-performance liquid chromatography (HPLC) using a gradient of acetonitrile/propanol in aqueous 0.1% trifluoroacetic acid. Although recovery was poor, one late-eluting peak was found to include the cross-reacting protein, as determined by Western blotting.

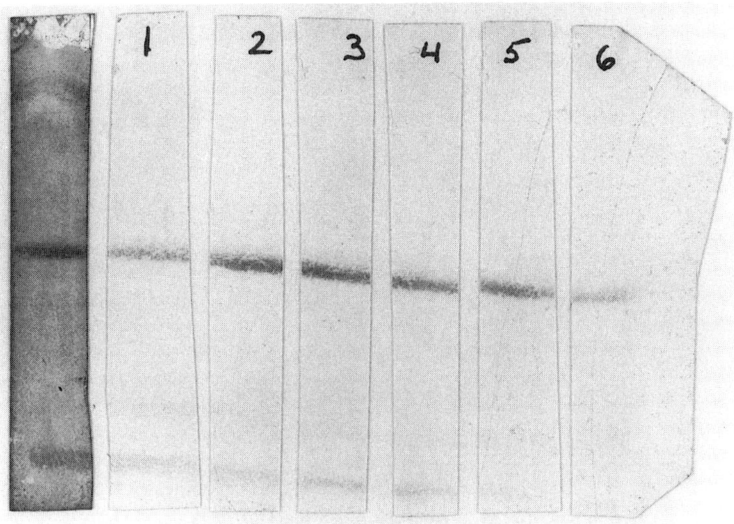

Figure 3 Western blot analysis showing cross reactivity of neuroblastoma cell–derived protein (from Ref. 39). Gammabind G-agarose beads were used to immunoprecipitate proteins from an extract of SK-N-SH cells in the presence of H9724. After appropriate incubations and washes, the beads were extracted with SDS sample buffer at 90°C. Following electrophoresis and transfer to a PVDF membrane, the same single band of protein was visualized using six different sera from patients with neuorlogical Lyme disease. Development of the blot was similar to that described in Figure 2, except for the use of goat antihuman immunoglobulin–coupled alkaline phosphatase.

A companion Coomassie blue-stained gel revealed a distinct band at the same migration position as the immunoreactive band (Fig. 4).

After electroblotting another gel onto a polyvinylidene difluoride (PVDF) membrane, the Coomassie blue-stained band was excised and subjected directly to automated Edman degradation in a gas-phase sequencer. It was possible to identify 13 residues from 16 cycles of degradation. By reference to the protein database, this partial sequence was found to match the aminoterminal sequence of mature chaperonin–heat shock protein 60. To confirm this result, an orthoganol method of protein purification was attempted. The 60% ammonium sulfate fraction was subjected to a combination of anion exchange (DEAE-Sepharose) chromatography and gel filtration (TSK 3000) HPLC. The most enriched fraction was then prepared for sequencing, as above, by SDS-PAGE and electroblotting onto a PVDF membrane. Now having a larger quantity of protein, it was possible to identify the 22 aminoterminal residues. Again, this sequence was identical to hsp 60.

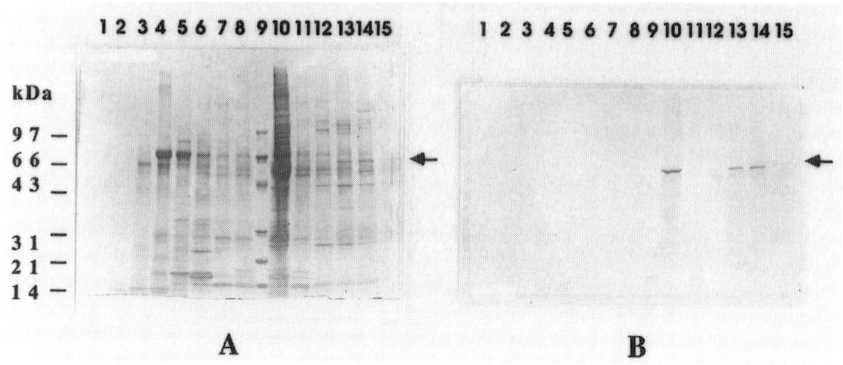

Figure 4 Monitoring of the purification of the cross-reacting protein using Western blot analysis (From Ref. 22). Solubilized proteins in the ammonium sulfate fraction were run in lane 10. Molecular weight markers were run in lane 9. Column fractions fom reverse-phae HPLC were run in the other lanes. The left panel depicts Coomassie blue staining of the gel, whereas the right panel is a Western blot of a duplicate gel developed with H9724 as the primary antibody in a manner similar to that in Figure 2. The cross-reacting band is seen in lanes 11–15, which correspond to fractions going across the final peak of protein in the reverse–phase chromatogram.

D. Identification of the Cross-Reacting Epitope

Recombinant hsp 60 of full mature length, as well as forms with various amino-terminal truncations, was kindly provided by Dr. R. S. Gupta (48). On Western blot analysis, only the full-length recombinant hsp 60 was detectable by H9724 (Fig 5). Since the protein with as small a deletion as the first 30 amino-terminal residues was found to be nonreactive, the epitope was presumed to be in this aminoteminal region. Comparison of the flagellin epitope with the amino-terminal of hsp 60 did not reveal any obvious sequence homology, although some similarity was observed. Two peptides comprising residues 2–12 and 14–26 of hsp 60 were synthesized, purified, and structurally characterized by mass spectrometry. On dot blot analysis, the first peptide was nonreactive, whereas the second peptide was bound by H9724 (see Fig. 1). Shortened versions of the second peptide were prepared by pin microsynthesis. ELISA analysis using H9724 provided identification of the minimal peptide epitope. Accordingly, the cross-reacting epitopes have been defined as:

B. burgdorferi: 211-V-Q-E-G-V-Q-216 216-Q-Q-E-G-A-Q-222; hsp 60: 14-M-L-Q-G-V-D-19, where the second repeating epitope of flagellin (216–222) is more reactive. The homology is not obvious, but the sequences are

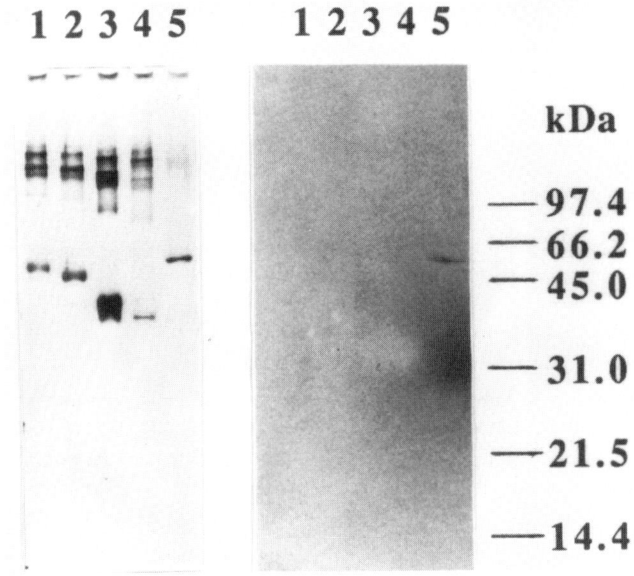

Figure 5 Analysis of recombinant forms of human hsp 60. Silver staining is shown in the left panel. H9724 was used as the primary antibody in the Western blot analysis shown on the right, and detection was as described in Figure 2. Lanes 1–4 are the four truncated recombinant hsp 60 forms, PKK13A, PKK13B, PKK13C, and PKK13D, respectively, whereas lane 5 is full-length, mature hsp 60.

suggestive of similarity in structure contributing to antibody recognition. As anticipated, this sequence in hsp 60 differs from the analogous region in *B burgdorferi* 60, which is 14-LLSGVE-19. Preliminary experiments using pin microsynthesis indicate the presence of antibodies reactive with this human hsp 60 epitope in some Lyme disease sera.

E. Proposed Biological Mechanism for Axonal Dysfunction

The neuropathy associated with chronic Lyme disease is an axonopathy rather than a demyelinating neuropathy (49). Our preliminary studies with human neuroblastoma cells (SK-N-SH), as well as other human and rodent nerve cell lines, have provided some interesting findings (41). In either serum-containing or serum-free media, the morphology of the cells will change in culture from the initial round form to an elongated, thin form. With further differentiation, each cell produces a long thin process, known as a neurite, which is structurally analogous to a normal nerve cell's axon. Reactivity of H9724 is seen in the neurite, neuritic bud, and cell body; occasionally, in those cells with very long neurites, the reactivity is seen preferentially in the neurite and bud.

Surprisingly, H9724 added to the culture medium was able to block the process of spontaneous differentiation and neurite outgrowth, such that treated cells were rounder, smaller, and did not produce either neuritic bud nor neurite. Other anti–*B. burgdorferi* antibodies, not directed against flagellin, of the same immunoglobulin (IgG) subclass did not mediate this effect. Differentiation resumed on removal of H9724 from the medium. Furthermore, H9724 did not interfere with the cells' ability to adhere to the culture plate. Differentiation induced by nerve growth factor (NGF) was also blocked by H9724. On the other hand, differentiation induced by dibutyryl cyclic adenosine monophosphate (cAMP), retinoic acid, or phorbol ester was not blocked by H9724. The simple explanation that H9724 might prevent binding of NGF to its receptor was ruled out: The amount of radiolabeled NGF that could be cross linked to surface receptors in cells treated with H9724 was not different from untreated cells (L.H.S., unpublished results).

Based on these findings, we propose the following scenario. Antibodies elicited against a particular epitope of *B. burgdorferi* flagellin are taken up by nerve cells and bind to intracellular hsp 60. This leads to inhibition of a previously unknown function of hsp 60 in axon maintenance, thereby causing dysfunction, not destruction, of the axons and the neuropathy of chronic Lyme disease. The plausibility of ready uptake of antibodies into neuronal cells is supported by the theory that humoral immunity may constitute an important arm of defense in these privileged cells (50). This concept is supported by the successful clearance of alphavirus in a SCID mouse model using hyperimmune serum but not with sensitized T lymphocytes (51). In the case of antiflagellin/anti–hsp 60 antibodies; however, this intracellular humoral immunity mechanism may cause axonal dysfunction owing to molecular mimicry. Gadjusek theorized that anti-axonal antibodies might cause axonopathy without frank necrosis (52), which is consistent with Lyme neurological disease (49).

VI. FUTURE STUDIES

Additional experiments are required to firmly establish our hypothesis on an anti–hsp 60 autoimmune mechanism in neurological Lyme disease. The monoclonal antibody H9724, which so remarkably duplicates the epitope specificity associated with Lyme disease antiflagellin serum antibodies, has proven to be a powerful tool in this research. However, human Lyme disease serum must be substituted for H9724 to prove crossreactivity with full-length recombinant hsp 60, as well as with the peptide representing the aminoterminal epitope. Similarly, the ability of anti–hsp 60 antibodies and Lyme disease serum antibodies to block neuritogenesis must be demonstrated. It should also be possible to prevent the antineuritogenesis activity of H9724 using peptides representing the epitopes of flagellin and hsp 60.

In any experiments with human serum or cerebrospinal fluid, we may have to first purify antiflagellin (or antiflagellin $peptide_{213-224}$) antibodies, such as by affinity chromatography, to clearly observe these phenomena. This is necessary, because a large portion of individuals with non–Lyme neurological disease or arthritis, as well as many healthy individuals, have anti–hsp 60 antibodies. Correlation between the presence of antibodies to the particular H9724-defined epitope on hsp 60 and the presence of neurological manifestations of Lyme disease must be proven. Finally, if this autoimmune mechanism of neurological pathogenesis is confirmed, a peptide representing the epitope sequence may eventually be used in a therapeutic strategy for Lyme disease based on induction of immune tolerance.

REFERENCES

1. S. Jindal, A.K. Dudani, C.B. Harley, B. Singh, and R.S. Gupta, Primary structure of a human mitochondrial protein homologous to bacterial and plant chaperonins and to hsp65 kD mycobacterial antigen, *Mol. Cell. Biol. 9*:2279 (1989).
2. R.T. Damian, Molecular mimicry: antigen sharing by parasite and host and its consequences, *Am. Naturalist 98:*129 (1989).
3. M.B.A. Oldstone, Molecular mimicry and autoimmune disease, *Cell 50*:819 (1987).
4. J.B. Winfield and W.N. Jarour, Stress proteins, autoimmunity, and autoimmune disease, Curr. *Topics Microbiol. Immunol. 167*:161 (1991).
5. D.B. Jones, A.F.W. Coulson, and G.W. Duff, Sequence homologies between hsp60 and autoantigens, *Immunol. Today 14*:115 (1993).
6. C.M. Pearson, Experimental models in rheumatoid disease, *Arthritis Rheum. 7*:80 (1964).
7. J. Holoshitz, A. Maitau, and I.R. Cohen, Arthritis induced in rats by cloned T lymphocytes responsive to mycobacteria but not to collagen type II, *J. Clin. Invest. 73*:211 (1984).
8. W. van Eden, J. Holoshitz, Z. Nevo, A. Frenkel, A. Klajman, and I.R. Cohen, Arthritis induced by a T lymphocyte clone that responds to *Mycobacterium tuberculosis* and to cartilage proteoglycans, *Proc. Natl. Acad. Sci. USA 82*:517 (1985).
9. I.R. Cohen, J. Holoshitz, W. van Eden, and A. Frenkel, T lymphocyte clones illuminate pathogenesis and effect therapy of experimental arthritis, *Arthritis Rheum. 28*:841 (1985).
10. W. van Eden, J. Thole, R. van der Zee, A. Noordzja, J.D.A. Embden, I.R. Cohen, and R.R.P. de Vries, Cloning of the mycobacterial epitope recognized by T lymphocytes in adjuvant arthritis, *Nature 331*:171 (1988).
11. M.E.J. Billingham, S. Carney, R. Butter, and M.J. Colston, A mycobacterial 65-kDa heat shock protein induces antigen-specific suppression of adjuvant arthritis, but is not itself arthritogenic, *J. Exp. Med. 171*:339 (1990).

12. W. van Eden, Heat shock proteins as immunogenic bacterial antigens with the potential to induce and regulate autoimmune arthritis, *Immunol. Rev. 121*:5 (1991).
13. L. Castano and G.S. Eisenbarth, Type-1 diabetes: a chronic autoimmune disease of human, mouse and rat, *Annu. Rev. Immunol. 8*:647 (1990).
14. I.R. Cohen, Type 1 insulin dependent diabetes mellitus, *Curr. Opin. Immunol. 1*:727 (1989).
15. I.R. Cohen, Autoimmunity to chaperonins in the pathogenesis of arthritis and diabetes, *Annu. Rev. Immunol. 9*:567 (1991).
16. I.R. Cohen, The physiological basis of T cell vaccination against autoimmune diseases, *Cold Spring Harbor Symp. Quant. Biol. LIV:879* (1989).
17. D. Elias, T. Reshef, O.S. Birk, R. van der Zee, M.D. Walker, and I.R. Cohen, Vaccination against autoimmune mouse diabetes with a T-cell epitope of the human 65-kDa heat shock protein, *Proc. Natl. Acad. Sci. USA 88*:3088 (1991).
18. D.L. Kaufman, M. Clare-Salzler, J. Tian, T. Forsthuber, G.S.P. Ting, P. Robinson, M.A. Atkinson, E.E. Sercarz, A.J. Tobin, and P.V. Lehmann, Spontaneous loss of T-cell tolerance to glutamic acid decarboxylase in murine insulin-dependent diabetes, *Nature 366*:69 (1993).
19. R. Tisch, X.-D. Yang, S.M. Singer, R.S. Liblau, L. Fugger, and H.O. McDevitt, Immune response to gluatmic acid decarboxylase correlates with insulitis in non-obese diabetic mice, *Nature 366*:72 (1993).
20. S. Baekkeskov, J.H. Nielsen, B. Marner, T. Bilde, J. Ludvigsson, and A. Lernmark, Autoantibodies in newly diagnosed diabetic children immunoprecipitate human pancreatic islet cell proteins, *Nature 298*:167 (1982).
21. S. Baekkeskov, H.-J. Aanstoot, S. Christgau, A.Reetz, M. Solimena, M. Cascalho, F. Folli, H. Richter-Olesen, and P.-D. Camilli, Identification of the 64K autoantigen in insulin-dependent diabetes as the GABA-synthesizing enzyme glutamic acid decarboxylase, *Nature 347*:151 (1990).
22. Z. Dai, H. Lackland, S. Stein, R. Radziewicz, S. Williams, and L.H. Sigal, Molecular mimicry in Lyme disease: monoclonal antibody H9724 to *B. burgdorferi* flagellin specifiically detects chaperonin-HSP60, *Biochim. Biophys. Acta 1181*:97 (1993).
23. M.M. Carriero, D.C. Laux, and D.R. Nelson, Characterization of the heat shock response and identification of heat shock protein antigens of *Borrelia burgdorferi*, *Infect. Immun. 58*:2186 (1990).
24. B.J. Luft, P.D. Gorevic, W. Jiang, P. Munoz, and R.J. Dattwyler, Immunologic and structural characterization of the dominant 66- to 73-kDa antigens of *Borrelia burgdorferi*, *J. Immunol. 146*:2776 (1991).
25. N.D. Mensi, D.R. Webb, C.W. Turck, and G. Peltz, Characterization of *Borrelia burgdorferi* proteins reactive with antibodies in Lyme arthritis synovial fluid, *Infect. Immun. 58*:2404 (1990).
26. M.-C. Shanafelt, P. Hindersson, C. Soderberg, N. Mensi, C.W. Turck, D. Webb, H. Yssel, and G. Peltz, T cell and antibody reactivity with the *Borrelia burgdorferi* 60-kDa heat shock protein in Lyme arthritis, *J. Immunol. 146*:3985 (1991).
27. R. Lahesmaa, M.-C. Shanafelt, A. Allsup, C. Soderberg, J. Anzola, V. Freitas, C. Turck, L. Steinman, and G. Peltz, Preferential usage of T cell antigen receptor V

region gene segment Vb5.1 by *Borellia burgdorferi* antigen-reactive T cell clones isolated from a patient with Lyme disease, *J. Immunol. 150*:4125 (1993).
28. L. Girouard, D.C. Laux, S. Jindal, and D.R. Nelson, Immune recognition of human hsp60 by Lyme disease patient sera. *Microb. Pathogen. 14*:287 (1993).
29. L.H. Sigal, Lyme-disease: a worldwide borreliosis, *Clin. Exp. Rheumatol. 6*:411 (1988).
30. A.C. Steere, Lyme disease, *N. Engl. J. Med. 321*:586 (1989).
31. A.C. Steere, N.H. Bartenhagen, J.E. Craft, et al., The early clinical manifestations of Lyme disease, *Ann. Intern. Med. 99*:76 (1983).
32. A.R. Pachner, and A.C. Steere, The triad of neurologic manifestations of Lyme disease: meningitis, cranial neuritis, and radiculoneuritis, *Neurology 35*:47 (1985).
33. L.H. Sigal, Severe complications of Lyme disease: recognition and management, *Management of the Critically Ill Patient with Rheumatic or Immunologic Illness*, Marcel Dekker, New York, 1994.
34. E.L. Logigian, R.F. Kaplan, and A.C. Steere, Chronic neurologic manifestations of Lyme disease, *N. Engl. J. Med. 323*:1438 (1990).
35. A.C. Steere, R.T. Schoen, and E. Taylor, The clinical evolution of Lyme arthritis, *Ann. Intern. Med. 107*:725 (1987).
36. A. Szczepanski and J.L. Benach, Lyme Borreliosis: host response to *Borrelia burgdorferi*, *Microbiol. Rev. 55*:21 (1991).
37. L.H. Sigal, Lyme disease, 1988: immunologic manifestations and possible immunopathogenic mechanisms, *Semin. Arthritis Rheum. 18*:151 (1989).
38. L.H. Sigal, Persisting complaints attributed to chronic Lyme disease: possible mechanisms and implications for management, *Am. J. Med.* (1994).
39. L.H. Sigal, Cross-reactivity between *Borrelia burgdorferi* flagellin and a human axonal 64,000 molecular weight protein, *J. Infect. Dis. 167*:1372 (1993).
40. L.H. Sigal, and A.H. Tatum, Lyme disease patients' serum contains IgM antibodies to *Borrelia burgdorferi* that cross-react with neuronal antigens, *Neurology 38*: 1439 (1988).
41. L.H. Sigal and S. Williams, Modification of neuroblastoma cells and other neural tumor cells growth and differentiation *in vitro* by an anti-*B. burgdorferi* flagellin monoclonal antibody (H9724): a functional correlate of the molecular mimicry, *Arthritis Rheum. 35*:S214 (1992).
42. A.G. Barbour, S.F. Hayes, R.A. Heiland, M.E. Schrumpf, and S.L. Tessier, A borrelia-specific monoclonal antibody binds to a flagellar epitope, *Infect. Immun. 52*:549 (1986).
43. E. Aberer, C. Brunner, G. Suchanek, H. Klade, A.G. Barbour, G. Stanek, and H. Lassmann, Molecular mimicry and Lyme borreliosis: a shared antigenic determinant between *Borrelia burgdorferi* and human tissue, *Ann. Neurol. 26*:732 (1989).
44. T. Schneider, R. Lange, W. Ronspeck, W. Weigelt, and H.W. Kolmel, Prognostic B-cell epitope on the flagellar protein of *Borrelia burgdorferi*, *Infect. Immun. 60*:316 (1992).
45. E. Fikrig, R. Berland, M. Chen, S. Williams, L.H. Sigal, and R. Flavell, Fine mapping of the serologic response to the *Borrelia burgdorferi* flagellin demonstrates an epitope common to neural tissue, *Proc. Natl. Acad. Sci USA 90*:183 (1993).

46. Z. Dai, *Definition of the Epitope on the 41-kDa Flagellin of Borrelia burgdorferi for a Monoclonal Antibody* H9724, and Identification of a H9724-*Reactive Protein from Calf Adrenal Gland*, Doctoral Thesis, University Microfilms International order #9320516, Ann Arbor, Michigan.
47. B. Canas, Z. Dai, H. Lackland, R. Poretz, and S. Stein, Covalent attachment of peptides to membranes for dot blot analysis of glycosylation sites and epitopes, *Anal. Biochem. 211*:179 (1993).
48. B. Singh and R.S. Gupta, Expression of human 60-kD heat shock protein (HSP60 or P1) in *Escherichia coli* and the development and characterization of corresponding monoclonal antibodies, *DNA Cell Biol. 11*:489 (1992).
49. J.H. Halperin, B.J. Luft, D.J. Volkman, and R.J. Dattwyler, Lyme neuroborreliosis: peripheral nervous system manifestations, *Brain 113*:1207 (1990).
50. R.H. Fabian, Uptake of antineuronal IgM by CNS neurons: comparison with antineuronal IgG, *Neurology 40*:419 (1990).
51. B. Levine, J.M. Hardwick, B.D. Trapp, T.O. Crawford, R.C. Bollinger, and D.E. Griffin, Antibody-mediated clearance of alphavirus infection from neurons, *Science 254*:856 (1991).
52. D.C. Gadjusek, Hypothesis: Interference with axonal transport of neurofilament as a common pathogenic mechanism in certain diseases of the central nervous system, *N. Engl. J. Med. 312*:714 (1985).

10

Stress Proteins in Behçet's Disease and Experimental Uveitis

T. Lehner, A. Childerstone, K. Pervin, A. Hasan, H. Direskeneli, M. R. Stanford, R. Whiston, E. Kasp, and D. C. Dumonde
United Medical and Dental School, Guy's and St. Thomas Hospitals, London, England

T. Shinnick
Hansen's Disease Laboratory, Atlanta, Georgia

R. van der Zee
Institute of Infectious Diseases and Immunology, Utrecht, The Netherlands

Y. Mizushima
St. Marianna University School of Medicine, Kawasaki, Japan

I. INTRODUCTION

Behçet's disease (BD) was first described by Hippocrates in ancient Greece, and the disease is named after Hulusi Behçet, a Turkish dermatologist. It is a multisystem disorder with oral and genital ulcers, uveitis, and cutaneous, arthritic, and neurological manifestations (1). The disease is most commonly found in Japan and probably in China and Korea. The prevalence of BD is also high in countries bordering the Mediterranean: Italy, Greece, Turkey, Israel, Egypt, Lebanon, Syria, Jordan, and Saudi Arabia, as well as Algeria and Tunisia. However, the prevalence of BD has been increasing in Europe owing to migration of Middle Eastern people. The cause of BD is unknown but HLA-B51 is significantly associated with the disease (1,2). Four hypotheses have been postulated in the etiology of BD: (1) autoimmunity or cross reactivity between microbial and oral mucosal antigens (3,4); (2) streptococcal infection causing the disease directly or by cross reactivity with the host tissues eliciting a damaging immune response (4–9); (3) herpes simplex virus type 1 modifying the immune responses (10–13); and (4) stress proteins of microbial origin cross reacting with the homologous human mitochondrial stress proteins and other host tissues (14,15).

The rationale for investigating stress proteins in the pathogenesis of BD is that the diverse microbial and possibly autoimmune manifestations might be accounted for by stress proteins (14). In addition, a significant increase in T-cell receptor γδ-positive cells have been found in BD (16,17). There is indeed evidence that a proportion of TCRγδ^{+} cells are sensitised to the 65-kDa hsp (18,19).

The objectives of this chapter are to review the evidence that the mycobacterial 65-kDa hsp and the homologous human mitochondrial 60-kDa hsp are involved in the pathogenesis of BD. The evidence is presented in four parts: (1) demonstration of the 65-kDa mycobacterial hsp in the selected strains of streptococci; (2) immunoglobulin A (IgA) and IgG antibodies to the 65-kDa hsp and to these streptococci are found in the sera of patients with BD; (3) four specific peptides determinants within the mycobacterial and homologous human hsp stimulate specific T-cell proliferative responses in patients with BD; (4) direct demonstration of the immunopathogenic potential of these four peptides is presented by the induction of uveitis in Lewis rats.

II. MATERIALS AND METHODS

A. Patients and Controls

The series of patients investigated consisted of 64 patients with BD using clinical criteria defined by the International study group for BD (20). Controls included patients with recurrent oral ulcers (n = 38) similar to those found in BD but without any extraoral manifestations (21); patients with rheumatoid arthritis (n = 41), recurrent herpetic infection of the lips (n = 8), and 53 healthy controls.

B. Streptococci, Hsp, Oral Mucosal Homogenate, and Synthetic Peptides

A number of *Streptococcus sanguis* serotypes and the uncommon serotypes found in BD are described in Table 1. We have also examined β-hemolytic streptococci of groups A and D and *S. salivarius, S. faecalis, S. mutans, S. sobrinus, S. mileri, S. mitis, S. mitior, and S. bovis* (Table 1). Recombinant 65-kDa heat shock protein derived from *Mycobacterium bovis* was prepared at the National Institute of Public Health and Environmental Protection, Bilthoven, The

Table 1 The Species, Types, and Sources of Streptococci, and Mr 65-kDa Band Produced by Western Blotting with Rabbit Antiserum

Streptococcus species	Type	Strain	Source	65-kDa band
S. sanguis	I	7863	NCTC[a]	+
S. sanguis	II	7864	NCTC	+++
S. sanguis I–II	7865	NCTC	-	
S. sanguis	–	11086	NCTC	+
S. sanguis (KTH-1)	–	49298	ATCC[b]	++
S. sanguis (KTH-2)	–	49296	ATCC[b]	+
S. sanguis (KTH-3)	–	49295	ATCC[b]	++
S. sanguis (KTH-4)	–	49297	ATCC[b]	+
S. sanguis ST3	–	–	ATCC[b]	+++
S. sanguis H83	–	–	ATCC[b]	+++
S. mutans	c	Guy's	J. Caldwell and T. Lehner	-
S. sobrinus	d	OMZ176	B. Guggenheim	+
S. milleri	–	10708	NCTC	±
S. mitis	–	OMZ100	B. Guggenheim	-
S. mitior	–	10712	NCTC	++
S. bovis	–	8177	NCTC	-
S. salivarius	–	8618	NCTC	-
S. faecalis	α	21C	NCTC	+++
S. faecalis	β	4949	NCTC	+++
S. pyogenes	Group A	8198	NCTC	+++

[a]National Culture Type Collection, Public Health Laboratories, London.
[b]From Professor Y. Mizushima (9), American Type Culture Collection.

Netherlands and was used at 10 μg/ml. Fetal oral mucosa was removed from an approximately 22-week-old fetus. The tissue was cut into small fragments, homogenized, and centrifuged as described previously (4). Both the supernatant and deposit were used for immunoblotting. A series of 29 overlapping synthetic peptides (15ers), derived from the sequence of the 65-kDa hsp were prepared at the Hansen Disease Laboratory, Centers for Disease Control, Atlanta, Georgia, and were used at 50 μg/ml. Human hsp peptides homologous to the selected mycobacterial peptides were also prepared and used at a concentration of 50 μg/ml (Table 2). Concanavalin A was used at 10 μg/ml (Sigma Chemical Co., Poole, Dorset, UK) as a positive control.

C. Antisera and Monoclonal Antibodies

Rabbit antiserum was prepared against the 65-kDa hsp. A number of monoclonal antibodies (MAbs) were obtained from several sources (14) to cover a large proportion of the 65-kDa hsp.

Table 2 Homologies Between Mycobacterial 65-kDa and the Human 60-kDa Hsp-Derived Peptides

Hsp	Peptide	Amino acid sequence	Identical residues	Identical and conserved residues
1. Mycobacterial	111–125	N P L G L K R G I E K A V E K		
		* * : : : * * : * * :		
Human	136–150	N P V E I R R G V M L A V D A	6/15	11/15
2. Mycobacterial	154–172	Q S I G D L I A F A M D K V G N E G V		
		* * : * * * * * * * *		
Human	179–197	K E I G N I I S D A M K K V G R K G V	10/19	11/19
3. Mycobacteria	219–233	L L V S S K V S T V K D L L P		
		* * * : * : : : *		
Human	244–258	L L S E K K I S S I Q S I V P	5/15	9/15
4. Mycobacterial	311–326	D L S L L G K A R K V V V T K D		
		* * * : * : * * * *		
Human	336–351	Q P H D L G K V G E V I V T K D	8/16	10/16
5. Mycobacterial	401–415	A K A A V E E G T V A G G G V		
		: * * * * * * * * * *		
Human	425–441	T R A A V E E G I V L G G G C	10/15	11/15

*Identical residues.
:Conserved residues.

D. Western Immunoblots and Radioimmunoassays

The techniques used have been described elsewhere (14). The antigens and antibodies were those described above. Human sera were used at a dilution of 1:10 for the Western blots, 1:20 and 1:40 for the radioassays. IgA and IgG antibodies were identified by means of ^{125}I-labeled affinity purified goat antihuman IgA or IgG antibodies (Sigma). Autoradiography was used in Western blots for 24 hr. The radioassay results were expressed as the mean percentage of radioactivity bound to the antigens (14).

E. ELISA

Plates were coated with predetermined optimal concentration of each of the peptides (at 10 μg/ml) and a random peptide (11er) as a control antigen (22). They were then incubated with doubling dilutions of serum samples (1:10, 1:20, etc.). Bound antibody was detected by incubation with goat antihuman IgG or IgA alkaline phosphatase conjugate (Sigma Fine Chemicals), followed by paranitrophenyl phosphate in diethanolamine buffer as the color reagent. The reaction was terminated with 3 M NaOH and the optical density (OD) measured at a wavelength of 405 nm. The results were expressed as the lowest dilution giving an OD of 0.15 units above the background sample, with no increase in OD seen with a random peptide of 11 residues.

F. Culture of Human Peripheral Blood Mononuclear Cells

Peripheral blood mononuclear cells (PBMCs) were separated from defibrinated blood samples from 134 subjects by lymphoprep density gradient centrifugation. In the initial studies, PBMCs (1×10^5 cells per well) from each subject were cultured for 6 days as described before (15). The results were assessed by calculating the net disintegrations per minute (dpm of cells with antigen − dpm of cells without antigen) and the stimulation index (SI), by the ratio of dpm of antigen stimulated to unstimulated cultures. The results were also expressed as the mean (±SEM) of triplicate cultures, which were considered to be significant if the SI was greater than 2 and the net dpm was greater than 500.

G. Separation of CD4- and CD8-Enriched Cells

Defibrinated blood samples from eight patients with BD were separated by Ficoll-Triosil density gradient centrifugation and monocyte-enriched cells were obtained by adherence to plastic dishes (Falcon Labware, Oxnard, CA). The nonadherent cells were separated into T- and B-enriched populations by rosetting with sheep red blood cells (SRBCs) and the CD4 cell subset was purified from the T cells by positive selection using a panning technique (15). The purity of the cell populations was assessed by indirect immunofluorescence, which showed that the positively selected CD4$^+$ cells had less than 5% CD8 contamination and the negatively selected CD8-enriched cells contained less than 10% CD4$^+$ cells.

H. Generation of Short-Term Cell Lines and T-Cell Epitope Mapping

Patients with BD who had shown significant proliferative responses to the 65-kDa hsp were selected for short-term cell lines (15). PBMCs were cultured at 2×10^5 cells per well in 96-well-round-bottomed plates in RPMI-1640 stimulated with 65-kDa HSP (10 μg/ml). On day 3, 100 μg/ml of the medium was replaced with medium containing 10% autologous serum, 10 units/ml of recombinant interleukin-2 (IL-2) and 10 units/ml of recombinant IL-4. On day 13, the cells were incubated (2×10^4) for 3 days with each of the 29 overlapping synthetic peptides (50 μg/ml) and 2×10^4 feeder cells, consisting of autologous PBMCs irradiated at 4000 rad. The cells were harvested on day 16 and [^{3}H]thymidine uptake was determined as for the cultures described above. The results were considered to be significant if the SI was greater than 3.

I. Immunization of Rats with Peptides

Mycobacterial and homologous human peptides which specifically stimulated T cells from patients with BD (15) were selected for the rat experiments (see Table 1). Lewis rats were injected with 500 μg of each of the five mycobacterial, five human homologous HSP peptides, and a control peptide (see Table 1), with enriched Freund's complete adjuvant into one hind foot pad, followed by intraperitoneal administration of 1×10^{10} heat-killed *Bordetella pertussis* (23). The rats were observed daily and the eyes were examined by slit lamp microscopy. The rats were killed either within 1–2 days of onset of uveitis or 28 days after immunization. The eyes were removed, fixed in formalin, processed, sectioned, and examined for histopathological changes.

J. Statistical Analysis

Antibody binding greater than the mean plus 2 standard deviation of healthy control sera was considered to be a significant increase in antibodies. Statistical analysis between the healthy controls, BD and the disease control groups was carried out by the X^2 test. The significance of the differences in the response of lymphocytes to 65-kDa hsp or synthetic peptides between patients with BD and healthy or disease controls was calculated using one-way analysis of variance for the 6-day lymphoproliferative assays and the X^2 analysis (with Yate's correction) or Fisher's exact test of probability for studies of short-term cell lines. Statistical differences between the selected mycobacterial synthetic peptides and their human homologues were compared using one-way analysis of variance and the paired *t*-test.

III. RESULTS

A. Immunoblot Analysis with the Antibodies of the 65-kDa Hsp, Streptococcal, and Oral Mucosal Antigens

A 65-kDa band was found by blotting with the polyclonal rabbit anti–mycobacterial hsp antiserum of *S. sanguis* NCTC 11086, KTH-1 to KTH-4, ST3, H83, and *S. pyogenes* but not *S. salivarius* (Fig. 1; Table 1). Fetal oral mucosa also showed a band of about 65-kDa with the antiserum. Examination of nine MAbs against the 65-kDa mycobacterial heat shock protein revealed that MAbs C1.1, II H9, and ML30 yielded 65-kDa bands with *S. sanguis* ST3 and H.83. These MAbs are directed against residues 88–123, 107–122, and 275–295, respectively, of the 65-kDa protein (24). *S. sanguis* KTH-1 reacted with MAb Y1.2, recognizing residues 11–27 (25), as well as MAbs C1.1 and ML30. *S. pyogenes* also reacted with MAb Y1.2, which is directed against residues 11–27, suggesting that the cross reactivities between the different species of streptococci invoked in the etiology of BD and the 65-kDa stress protein may occur in different parts of the protein. Although some of the other streptococci, such as *S. sanguis* KTH-2 and KTH-4, showed a 65-kDa band with the rabbit anti–65-kDa antiserum, this was not detected with any of the nine MAbs. These MAbs recognize only about 30% of the entire length of the 65-kDa heat shock protein, so that the streptococcal antigens may share some of the remaining 70% of the residues of the 65-kDa hsp and indeed other stress proteins.

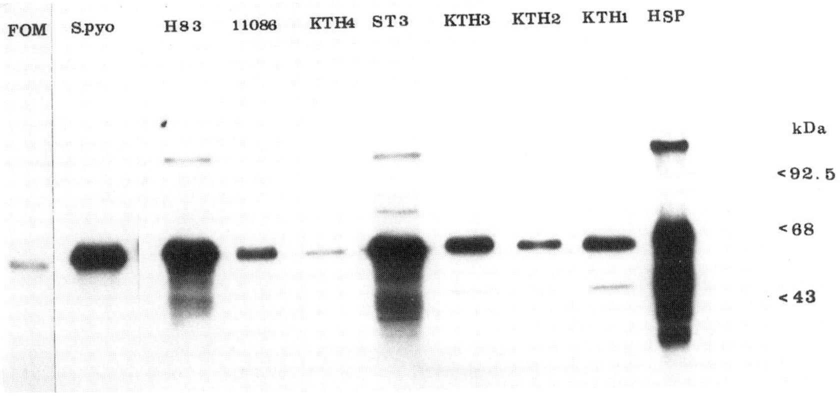

Figure 1 Immunoblot analysis with rabbit antiserum of the 65-kDa hsp and streptococcal and oral mucosal antigens. The slots were loaded with the 65-kDa hsp; *S. sanguis* KTH-1, KTH-2, KTH-3, KTH-4, 11086, H.83, and ST3; *S, pyogenes* 8198 (*S. pyogenes*), and fetal oral mucosal antigen (FOM). The numbers on the left indicate molecular masses in kilodaltons.

B. Immunoblot Analysis with Sera from Patients and a Selected Panel of Seven Antigens for IgA and IgG Antibodies

Immunoblots with sera from patients and controls carried out against a selected panel of antigens. Analysis of these results were frought with difficulties in view of the variations in the number of bands with different molecular masses of each antigen reacting with the different sera. Nevertheless, IgA antibodies to the mycobacterial 65-kDa hsp (Table 3) were found more frequently in sera from patients with BD (9 of 13) than in healthy controls (3 of 10; $X^2 = 3.4862$; $p < 0.05$). Significant increases in the 65- to 70-kDa bands with the uncommon serotypes of *S. sanguis* were found with KTH-1-3, testing serum IgA from patients with BD, and compared with control sera ($X^2 = 6.2947$; $p < 0.02$). However, significant differences in IgA antibodies were not detected with *S. sanguis* KTH-1, H83, or 11086 nor with *S. salivarius* and *S. pyogenes* (Table 3). A similar analysis of serum IgA from patients with recurrent oral ulcers (ROU) showed a significant increase in the 65- to 70-kDa bands only with *S. sanguis* KTH-3 ($X^2 = 5.7600$; $p < 0.02$). The results of immunoblot analyses for IgG antibodies were comparable with those for IgA (Table 3) except that the difference between the patients and controls failed to reach the 5% level of significance. Immunoblotting fetal oral mucosa showed a 43-kDa band with both IgA and IgG antibodies, and those were found more frequently with sera from patients with BD compared to controls, but the results failed to reach statistical significance (Table 3).

C. Antibodies to the 65-kDa Hsp

IgA antibody binding to the recombinant hsp greater than the mean plus 2 standard deviations of sera from healthy subjects (Table 4) was found in sera from 23.3% of patients with BD ($X^2 = 5.567$; $p < 0.02$). However, significant increases were also found in sera from patients with ROU (26.5%; $p < 0.01$) and rheumatoid arthritis (29.3%; $p < 0.01$) but not with those from recurrent herpes labialis. Significantly raised IgA antibodies were found in BD to the recombinant 65-kDa mycobacterial heat shock protein and to the soluble extracts of *S. sanguis* (ST3, KTH1, KTH-2, and KTH-3). IgA antibodies to *S. sanguis* KTH-1, KTH-2, and KTH-3 might be significantly elevated only in BD, although the proportions are rather low, reaching only 36.7, 22.6, and 30.6%, respectively (Table 4). The results obtained with IgG antibodies were less striking, and only sera from patients with BD showed elevated anti–65-kDa antibodies, reaching the 5% level of significance (Table 4). The four uncommon serotypes of *S. sanguis* failed to yield significant increases in IgG antibodies with sera from patients with BD.

Table 3 Western Blots of Mr 65–70kDa IgA and IgG Isotypes with Four Groups of Sera Using Nine Antigenic Preparations

		S. sanguis																	
		Hsp		KTH-1		KTH-2		KTH-3		H83		11086		*S. salivarius*		*S. pyogenes*		FOM[a]	
Sera	n	IgA	IgG	IgA	IgG	IgA	IgG	IgA	IgG	IgA	IgG	IgA	IgG	IgA	IgG	IgA	IgG	IgA[b]	IgG[b]
Controls	10	3	7	8/9	8/10	6	8	6	8	2	4	0	3	0	4	3	5	2/9	1/9
Behçet's disease	13	9	12	13	12	13	13	13	12	6	6	5	4	2	1	8	7	5/11	5/11
Recurrent oral ulcers	(6)	2	5	6	6	4	6	6	6	3	4	3	3	1	0	5	6	2/4	1/4

[a]Fetal oral mucosa.
[b]43-kDa band only.

Table 4 Radioassay for IgA Antibodies to *S. sanguis* in Human Sera: Number Greater than the Mean + 2 SD of the Controls (No.$^+$) and the Total Number (No.), and % Tested Are Indicated for Each Antigen

	65kD hsp			NCTC-11086			KTH-1			KTH-2			KTH-3			ST-3		
	No.$^+$	No.	%	No.$^+$	No.	%	No.$^+$	No.	%	No.$^+$	No.	%	No.$^+$	No.	%	No.$^+$	No.	%
1. Healthy controls	3	51	5.9	3	52	5.8	2	30	6.7	3	50	6.0	2	51	3.9	1	24	4.3
2. Recurrent herpes	0	8	0	0	8	0	0	6	0	0	8	0	0	7	0	4	5	80.0[c]
3. Recurrent oral ulcers	9	34	26.5[a]	0	32	0	0	24	0	3	24	12.5	3	31	9.7	7[a]	19	36.8
4. Behçet's disease	14	60	23.3[b]	5	61	8.2	11	30	36.7[a]	14	62	22.6[b]	19	62	30.6[c]	12	24	50.0[c]
5. Rheumatoid arthritis	12	41	29.3[a]	7	38	18.4	3	30	10.0	2	39	5.1	3	39	7.7[c]	12	24	50.0

[a] $p < 0.01$.
[b] $p < 0.02$.
[c] $p < 0.001$.

D. Antibodies to the Hsp Peptides

Significantly higher serum IgA and IgG antibody titers were found in patients with BD than controls against the mycobacterial peptides (111–125 and 311–326) ($p < 0.00001 - < 0.05$) and the human homologous peptides (136–150 and 336–351) ($p < 0.0001$–0.02) (22). Significant IgA and IgG antibodies were also detected to the mycobacterial peptide (154–172) and human peptide (244–258). It appears that the T- and B-cell epitopes in BD are overlapping and that the most significant B-cell epitopes are found within peptides (111–125 and 311–326) and their human homologues.

E. Proliferative Responses of PBMCs to Stimulation with the 65-kDa Hsp and 29 Overlapping Synthetic Peptides

Stimulation of PBMCs with the *M. bovis* 65-kDa hsp yielded a high SI with cells from patients with BD (10.9 ± 1.8), but this response was not significantly different when compared with the response of the three control groups (Fig. 2). The overlapping peptides (15ers) derived from the sequence of the hsp were then used to stimulate PBMCs and the mean SI ($\pm$SEM) are presented in Figure 2. Of the 29 peptides, used only 4 (shaded) showed a significant increase in SI ($p < 0.05 - < 0.005$) by the one-way analysis of variance. The dpm of cells stimulated by the four peptides were also significantly raised in BD as compared with the controls (data not presented). Peptide 401–415 showed a slightly higher SI in BD than the controls, but the 5% level of significance was not reached.

F. Proliferative Responses of Enriched CD4 and CD8 Cells to Stimulation with the 65-kDa Hsp and Selected Peptides

Enriched CD4 and CD8 cells from eight patients with BD were reconstituted with 10% adherent cells and stimulated with the selected hsp peptides. Significant SI (≥3) were found with the 65-kDa hsp stimulating CD4-enriched and CD8-enriched cells in all but one patient. In contrast, only enriched CD4 cells were stimulated with peptides 111–125, 219–233, and 311–325, but peptide 154–172 stimulated CD4-enriched (five of five) as well as CD8-enriched (three of five) cells.

G. Short-Term Cell Lines Established with the 65-kDa Hsp and Stimulated with the Selected Synthetic Peptides

The mononuclear cell cultures were stimulated with hsps for 13 days in order to expand the proportion of cells responding to the immunodominant peptides. The cell lines were then stimulated separately with the putative immunodominant T-cell peptides, a nonstimulating control peptide (21–35) and the 65-kDa hsp for 3 days. A SI of at least three times greater than the unstimulated cultures

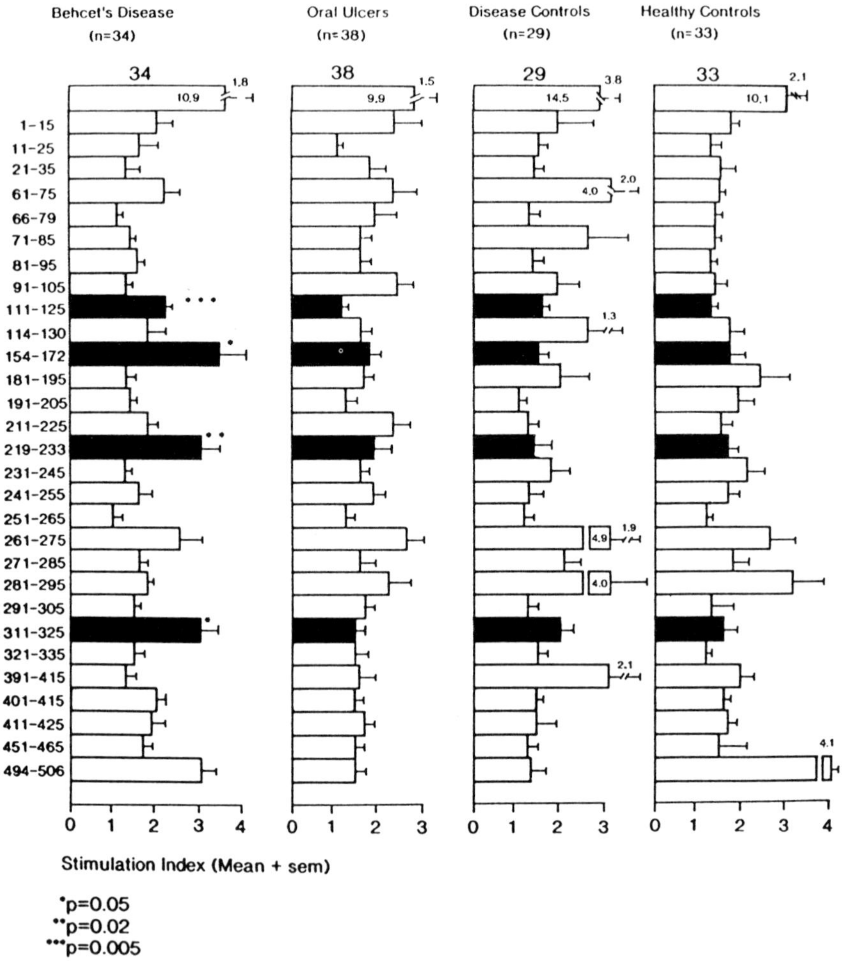

Figure 2 Mean stimulation indices (±sem) of peripheral blood mononuclear cells stimulated by 29 overlapping peptides (15ers) of the 65-kDa hsp in four groups of a total of 134 subjects.

was considered to be significant, and in order to enhance confidence in the statistical analysis only patients who yielded more than one significant cell line were analysed. The relative frequency of short-term cell lines (STCLs) responding to each peptide was expressed as the number of STCLs yielding a SI ≥ 3, divided by the number of STCLs examined, and multiplied by 100.

The hsp-stimulated cultures showed no significant difference in frequencies of 1061 STCLs between the four groups of subjects (Table 5). The control peptide (21–35) failed to stimulate lymphocytes from any of the groups of subjects. The four selected peptides were used to stimulate 4244 STCLs and showed greater frequencies with lymphocytes from patients with BD (8.6–10.0%) than those from healthy controls (0–0.6%), disease controls (0–1.5%), or patients with ROU (0-3.8%) (Table 5). χ^2 analysis of the lymphoproliferative responses stimulated by each of the four synthetic peptides showed significant differences between the four groups of subjects in the frequencies of STCLs ($p < 0.02$ – < 0.0001). Further analysis comparing BD with the healthy and disease controls showed significant differences with all peptides ($p = 0.001$–0.0001, but for peptide 219–233 $p = 0.025$). As ROU is found in all patients with BD, it was of special interest to find out if these T-cell determinants are specific to lymphocytes in BD or are shared with those in ROU. Indeed, significantly greater frequencies of STCLs were found with lymphocytes from patients with BD as compared with ROU with all but peptide 219–233 ($p < 0.001$ or < 0.02).

H. Analysis of the Frequencies of STCLs Stimulated with the Selected Peptides in the Ocular, Arthritic, and Mucocutaneous Types of BD

Further analysis of BD into ocular, arthritic, and mucocutaneous types revealed that lymphocytes from patients with the ocular type yielded the highest frequency of stimulation with the four hsp peptides (five of six patients), especially peptides 111–125 and 311–326. In the arthritic type of BD, lymphocytes from four of the six patients responded to these peptides, and there was no obvious difference in stimulation between the four peptides, although the highest frequency was elicited by peptide 154–172. Lymphocytes from only one patient with the mucocutaneous type responded to the peptides, with maximal frequency being elicited by peptide 219–233, although only three patients were analyzed in this group. If an arbitrary frequency of greater than 2.5% of STCLs is taken to be significant, then only 4 of the 15 patients with BD (or 4 of 12 with the ocular or arthritic type) responded to all four peptides. Single peptides stimulated STCLs in 12 of the 15 patients with BD or 9 of the 12 patients with the ocular or arthritic type of BD. Among the healthy controls, only 2 out of 16 gave a significant response to peptide 311–325 and 1 out of 16 to peptide 111–125; among the disease controls, 1 of 9 responded to peptide 111–125 and 2 of 9 to peptide 219–233. Patients with ROU responded significantly to peptide 219–233 (in 3 of 10 and to peptide 311–325 in 2 out of 10 patients).

Table 5 Relative Frequencies of 7353 Short-Term Cell Lines to the 65-kDa Hsp and Peptides with Stimulation Indices ≥ 3.0

Group	BD (n = 16) STCL (n = 2107) Frequency (%)	ROU (n = 10) STCL (1456) Frequency (%)	Disease Controls (n = 9) STCL (1326) Frequency (%)	Healthy Controls (n = 16) STCL (2464) Frequency (%)	Chi^2	Probability
65-kDa hsp	174/301 (58)	131/213 (62)	89/195 (46)	212/352 (60)	–	–
21–35	0/301	0/213	0/195	1/352 (0.3)	–	–
111–125	30/301 (10.0)[c]	4/213 (1.9)[e]	0/195	0/352	14.75	0.0001
154–172	28/301 (9.1)[c]	0/213[e]	0/195	0/352	26.68	0.0001
219–233	29/301 (9.6)[a]	6/213 (2.7)	3/195 (1.5)	0/352	6.26	0.0123
311–326	26/301 (8.6)[b]	8/213 (3.8)[d]	0/195	2/352 (0.6)	11.74	0.0006
401–415	4/201 (2.0)	0/178	1/156 (0.6)	0/352	–	–

Chi square analysis was first used to test the lymphoproliferative responses to each peptide given by the four groups of patients and controls (last two columns). Chi square with Yate's correction was then applied to find out if the responses by patients with BD were significantly different from healthy and disease controls ($^{a}p = 0.025$, $^{b}p = 0.001$, $^{c}p = 0.0001$). Fisher's exact test for small numbers was used to analyze the responses between patients with BD and those with ROU ($^{d}p = 0.016$, $^{e}p = 0.0007$).

I. Investigations of HLA Restriction in Stimulating STCLs by the Selected Peptides

HLA-B51 shows a significant association with BD (1,2). STCLs were analyzed in $B51^+$ and $B51^-$ patients with BD and in $B51^+$ healthy controls (Table 6). The frequency of STCLs with SI ≥ 3 was significantly greater ($p < 0.05$ or < 0.01) with lymphocytes from $B51^+$ patients with BD than those from $B51^+$ controls when stimulated with any of the four peptides. However, χ^2 analysis between $B51^+$ and $B51^-$ patients with BD showed a lower frequency of T-cell proliferative responses to peptides (154–172 and 311–326) in $B51^+$ patient, although the 5% level of significance was reached only with peptide 154–172. These results suggest that HLA-B51 is not associated with the proliferative responses of any of these peptides. PBMCs from all patients (n = 27) and controls (n = 38) were stimulated with the 65-kDa hsp, and significant proliferative responses were observed with lymphocytes from patients typed for DR2 to DR7 antigens or HLA A and B haplotypes (data not presented). The four selected peptides did not show restriction in stimulating STCLs derived from patients with HLA DR2-7 or HLA A or B haplotypes as compared with 11 disease and 38 healthy control groups.

J. Stimulation of PBMCs with the Homologous Human Mitochondrial 60-kDa Hsp

The selected mycobacterial peptides were compared with the homologous human hsp peptides in the 6-day proliferative assay. The human hsp peptides (179–197 and 336–351) gave similar results to the homologous mycobacterial peptides (154–172 and 311–326), respectively (Fig. 3). Somewhat higher stimulation indices resulted with the human hsp peptides (136–150 and 244–258) than with the homologous mycobacterial peptides (111–125 and 219–233), respectively. The results suggest that four human hsp peptides yielded more significant differences ($p = 0.003$–0.0006) than the homologous mycobacerial HSP peptides ($p < 0.03$ to $p = 0.0003$), with the exception of peptide 111–125, when patients with BD were compared with the disease and healthy control groups. All but one of the disease controls yielded negative lymphoproliferative responses with either the four human or four mycobacterial hsp peptides. Peptide 401–415 and the homologous human hsp peptide (425–441), as well as the control peptide (21–35; data not presented) failed to stimulate lymphocytes from the patients or controls. These results suggest that human hsp peptides may play a part in the immunopathogenesis of BD.

K. Ocular Changes in Rats

A total of 83 Lewis rats were used to determine if any of the eight selected mycobacterial or human hsp peptides can induce eye lesions. The results have

Table 6 Analysis of 1439 Short-Term Cell Lines Stimulated with the Mycobacterial 65-kDa Hsp Followed by One of the Five Peptides from B51$^+$ and B51$^-$ Patients with BD and B51$^+$ Healthy Controls

				Frequency (%) of STCLs stimulated with peptide				
Group	HLA-B	Disease	n	111–125	154–172	219–233	311–326	401–415
I	B51$^+$	BD	6	14/104(13.5)[b]	8/115(6.9)[a,c]	8/115 (6.9)[a]	8/115 (6.9)[a]	6/82 (7.3)[a]
II	B51$^+$	Controls	3	0/60	0/60	0/60	0/60	0/60
III	B51$^-$	BD	6	19/133 (14.3)	20/133 (15.0)	9/133 (6.8)	19/133 (14.3)	3/76 (3.9)

[a]$p < 0.005$ or [b]$p < 0.01$ by χ^2 analysis of the frequency of STCLs responding to a peptide with a SI ≥ 3 from B51$^+$ BD and B51$^+$ controls (groups I and II). [c]$p < 0.05$ by χ^2 analysis of B51$^+$ and B51$^-$ patients with BD (groups I and III).

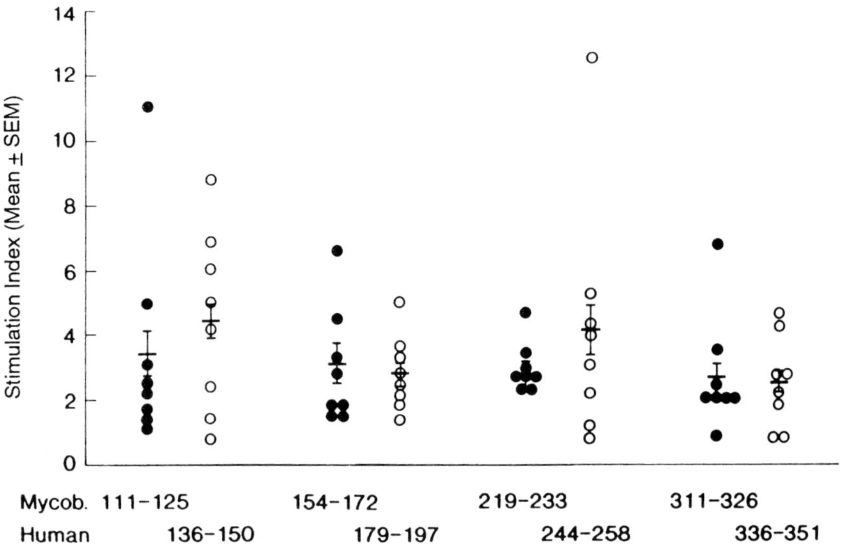

Figure 3 Comparison between mycobacterial and homologous human peptides stimulating lymphocytes from patients with Behçet's disease (n = 8); 23 of 27 controls yielded SI < 3.

shown (Table 7) that two of the mycobacterial peptides (111–125 and 311–326) and their human homologues (136–150 and 336–351) are capable of inducing vasodilation or hemorhage in the iris vessels (23). Histological examination revealed an inflammatory cell infiltration of the ciliary body, iris, vitreous, and retina, hemorrhage from the retinal vessels, and loss of photoreceptors. The control mycobacterial peptide (91–105) or human hsp peptide (116–130) did not induce clinical or histological changes in the eyes. Although the mycobacterial peptide (111–125) induced uveitis in a similar proportion of rats (70%) as the homologous human hsp-derived peptide (136–150; 64%), the other human hsp peptides appeared to be more uveitogenic than the homologous mycobacterial peptides (Table 7). Thus, peptide 311–326 induced uvetitis in 30–40% of rats as compared with the human hsp peptide (336–351), which induced uveitis in 75–92% of rats. Indeed, the remaining two mycobacterial peptides (154–172 and 219–233) failed to induce uveitis, unlike the corresponding human hsp peptides (179–197 and 244–258), which induced uveitis in one of four and three of four of the rats (23). The positive control (S antigen) induced uveitis in four of five rats, and the 65-kDa hsp also induced uveitis in three of four of the animals. It is noteworthy that the more severe histological changes were observed with peptides 111–125 and 336–351.

Table 7 The Effect of Administration of Synthetic Peptides Derived from the Sequences of the 65-kDa Mycobacterial and Homologous 60-kDa Human Hsps on the Eyes of Lewis Rats

Development of uveitis after injection of	
Mycobacterial peptide	Human peptide
111–125 (7/10)	136–150 (7/11)
154–172 (0/5)	179–197 (3/4)
219–233 (0/5)	244–258 (2/4)
311–326 (4/10)	336–351 (11/12)
91–105 (0/5)	116–130 (0/8)
S antigen (5/5)	CFA (0/10[a])

Vaccine: 500 μg peptide mixed with FCA administered into the foot pads followed by IP injection of *B. pertussis*. The results are expressed as number of rats with clinical and/or histological ocular lesions/number of rats vaccinated.
[a]One rat developed fibrinous exudate but not iridocyclitis.

IV. DISCUSSION

We have investigated in four stages the possibility that hsps are involved in the pathogenesis of BD. In stage 1, we have established that the 65-kDa hsp is found in the four species of streptococci, *S. pyogenes, S. sanguis, S. faecalis,* and *S. salivarius*, which have been implicated in the etiology of BD (8,9). In stage 2, IgA and IgG classes of antibodies to the 65-kDa hsp were found in the sera of patients with BD by immunoblotting and radioassay (14). Four B-cell epitopes were detected in sera from patients with BD, both against the mycobacterial and human homologous peptide sequences of hsp by ELISA (22). The B-cell epitopes overlapped with the T-cell epitopes, although the fine specificity of 15[er] peptides was not investigated. The most significant difference in IgA and IgG antibody titers between patients and controls were found with peptides 111–125, 136–150, and 311–326, and 336–351.

Earlier investigations of BD or ROU reported autoimmune responses to oral mucosal homogenates (3,4,26,27). In view of the high degree of homology between microbial and human cellular heat shock proteins (28), we examined the possibility that the autoimmune findings might be interpreted on the basis of a cross-reacting determinant. Indeed, a 65-kDa band was found with the polyclonal anti–65-kDa antiserum, and MAbs Y1.2, C1.1, ML30, and IV D8 showed one or two discrete bands of about 100 or 190 kDa with the fetal mucosa. It is noteworthy that MAb Y1.2 shows about 60% cross reactivity between the human mitochondrial P1 antigen and the 65-kDa stress proteins (29), and MAb ML30 recognizes human mitochondria (30). These findings raise the possibility that molecular cross reactivity between microbial stress proteins and fetal oral mucosal antigens may account for the earlier reports in the literature of autoimmune responses in this disease (3,4,6,31).

In stage 3, T-cell epitopes of the 65-kDa hsp were mapped in patients with BD by stimulating T cells with the overlapping synthetic peptides derived from the sequences of the *Mycobacterium tuberculosis* 65-kDa hsp (15). Significant lymphoproliferative responses were stimulated with the four hsp peptides in BD as compared with the related disease (recurrent oral ulcers), unrelated disease, and healthy controls ($p < 0.05$–0.005). In order to assess the relative frequency of sensitised lymphocytes by these peptides, 7353 STCLs were generated from the lymphocytes of patients and controls. Peptides 111–125, 154–172, and 311–325 ($p < 0.001$) and peptide 219–233 ($p < 0.02$) yielded significantly greater frequency of STCLs in BD than in healthy and disease controls. All but peptide 154–172 stimulated only the $CD4^+$ subset of T cells. There was no evidence that reactivity to the selected peptides is restricted by DR2–DR7 antigens. HLA-B51 is significantly associated with BD, but there was no evidence that B51 was a restricting element when $B51^+$ patients were compared with $B51^-$ patients with BD and with $B51^+$ healthy control subjects. A comparative investigation between the corresponding mycobacterial and human hsp peptides showed similar or higher lymphoproliferative responses by the human peptides compared with the mycobacterial peptides.

These findings suggest that four overlapping T- and B-cell epitopes have been mapped in mycobacterial and homologous human hsps in patients with BD, and these do not overlap with hsp peptides mapped for adjuvant, reactive, juvenile, or rheumatoid arthritis (32–35). Whether these epitopes, within the two polypeptides of 215 residues (111–325 and 136–351), may function as discontinuous epitopes depends on the conformational structure of the hsps, which needs to be determined. The possibility was examined that the four peptide determinants within the 65-kDa hsp might be involved in the pathogenesis of BD. Whereas the high microbial load and associated stress proteins found in oral ulceration of BD may initiate an immune response to these conserved epitopes, expression of autoreactive T-cell clones might be stimulated by immunodominant T-cell epitopes of endogenous hsps, which may induce immunopathological changes.

In stage 4, mycobacterial and four human hsp peptides were tested directly in Lewis rats in order to find out if any of these peptides can induce uveitis. The experimental evidence in Lewis rats has identified at least two mycobacterial and the corresponding human hsp peptides in the development of uveitis. The peptides synthesized from the human hsp sequence induced a higher proportion of rats to develop uveitis than the corresponding mycobacterial peptides. The most uveitogenic peptide (336–351) induced the lesions in a similar proportion of rats to those elicited by the S antigen, although the histopathological changes were less severe. We suggest that stress proteins might be involved in the etiology of BD, in view of the high microbial load of both gram-positive and gram-negative organisms that colonize the oral mucosa. Indeed, oral ulceration is usually the earliest and most consistent feature of BD, and some of the streptococci or their breakdown products might find their way into the lamina propria through the ulcerated mucosa, or they are carried by the large number of macrophages and Langerhans cells at that site and initiate an immune response (14,36). It is significant that the proportion of $\gamma\delta^+$ T cells is increased in BD (16,17) and that $\gamma\delta^+$ T cells may be sensitized to the 65-kDa hsp (18,19). We suggest that if an immunodominant T-cell epitope within the hsp were to stimulate expansion of the corresponding autoreactive clone (such as peptide 111–125 or 311–326), this may react with the host cell endogenous hsp peptides (136–150 or 336–351) generated by the stress response to infection and lead to autoimmune disease.

These results suggest that specific T- and B-cell epitopes identified by stimulating human T cells and in human sera can induce at least one manifestation of the disease found in patients from whom the T cells were isolated. Furthermore, two mycobacterial and the homologous human hsp peptides stimulated specifically human T-cell proliferative responses and induced uveitis in rats. This appears to be unique to hsps, and the potential therapeutic implications will now be explored by inducing oral or nasal tolerance with the synthetic peptides.

V. SUMMARY

A variety of streptococcal species have been implicated in the etiology of BD, and a hypothesis was postulated that stress proteins might account for these findings. Indeed, antibodies to the mycobacterial 65-kDa hsp reacted with a number of serotypes of *S. sanguis* and *S. pyogenes*. Immunoblot analyses revealed a 65- to 70-kDa band with *S. sanguis* more frequently in sera from patients with BD than those from controls ($p < 0.02$). Serum antibodies were then studied by an immunoassay and this revealed significantly raised IgA antibodies to the recombinant 65-kDa mycobacterial hsp and to soluble protein extracts of *S. sanguis* in sera from patients with BD.

T- and B-cell epitopes of the 65-kD mycobacterial and the 60-kD human mitochondrial hsp were mapped in patients with BD. Significant lymphoproliferative responses were stimulated with four mycobacterial hsp peptides in BD as compared with disease and heathy controls. In order to assess the relative frequency of sensitized lymphocytes by these peptides, 7353 STCLs were generated from the lymphocytes of patients and controls. Peptides 111–125, 154–172, and 311–325 ($p < 0.001$) and peptide 219–233 ($p < 0.02$) yielded significantly greater frequency of STCLs in BD than in controls. There was no evidence to suggest that reactivity to the selected peptides is restricted by HLA-B51 or DR antigens. A comparative investigation showed similar or higher lymphoproliferative responses stimulated by the homologous human peptides (136–150, 179–197, 244–258, and 336–351) as compared with the mycobacterial peptides. Four B-cell epitopes were mapped and these overlapped with the four T-cell epitopes. The peptides were then tested for their potential to induce uveitis in Lewis rats by injecting the peptides with enriched complete Freund's adjuvant. Peptides 336–351, 136–150, 111–125, and 311–325, in a descending order of frequency, induced uveitis in rats, with an inflammatory cell infiltration of the iris, ciliary body, vitreous, and retina. The results suggest that these peptide determinants within the 65-kDa hsp might be involved in the pathogenesis of BD.

REFERENCES

1. T. Lehner and J.R. Batchelor, Classification and an immunogenetic basis of Behcet's syndrome, *Behcet's Syndrome; Clinical and Immunological Features (*T. Lehner and C.G. Barnes, eds.). Academic Press, London, 1979, p. 13.
2. S. Ohno, E. Nakayama, S. Suguira, K. Itakura, K. Aoki, and M. Aizawa, Specific histocompatibility antigens associated with Behcet's disease, *J. Rheumatol. 3*:1 (1975).
3. Y. Oshima, T. Shimizu, R. Yokohari, T. Matsumoto, K. Kano, T. Kagami, and H. Hagaya, Clinical studies on Behcet's syndrome, *Ann. Rheum. Dis. 22*:36 (1963).
4. T. Lehner, Recurrent aphthous ulceration and autoimmunity, *Lancet ii*1154–1155.
5. E.A. Graykowski, M.F. Barile, W.B. Lee, and H.R. Stanley, Recurrent aphthous stomatis. Clinical therapeutic, histopathologic and hypersensitivity aspects, *J.A.M.A. 196*:637 (1966).
6. O. Donatski, An immunofluorescence study on the cross-reaction between Strep. 2A and human oral mucosa, *Scand. J. Dent. Res. 83*:111 (1975).
7. J.M.A. Wilton and T. Lehner, An investigation into the antigenic relationship between oral bacteria and oral mucosa. (abstr.), *J. Dent. Res. 47*:1001 (1968).
8. F. Kaneko, Y. Takahashi, Y. Muramatsu, and Y. Miura, Immunological studies on aphthous ulcer and erythema nodosum-like eruptions in Behcet's disease, *Br. J. Dermatol. 113*:303 (1985).
9. Y. Mizushima, Skin hypersensitivity to streptococcal antigens and the induction of systemic symptoms by the antigens in Behcet's disease—a multicentre study, *J. Rheumatol. 16*:506 (1989).
10. A.M. Denman, P.J. Fialkow P.J., B.K. Pelton, A.C. Salo, and D.J. Appleford, Attempts to establish a viral aetiology for Behcet's syndrome, T. Lehner and C.G. Barnes, eds.), Behcet's syndrome, Academic Press, London, 1979, pp. 91–105.
11. R.P. Eglin, T. Lehner, and J.H. Subak-Shape, Detection of RNA complimentary to herpes-simplex virus in mononuclear cells from patients with Behcet's syndrome and recurrent oral ulcers, *Lancet 2*:1356 (1982).
12. W.A. Bonass, J.A. Bird-Stewart, M.A. Chamberlain, and J.W. Halliburton, Mo-

lecular studies in Behcet's syndrome, *Recent Advances in Behcet's Disease*, (T. Lehner and C.G. Barnes, eds.), International Congress and Symposium Series, Royal Society of Medicine, *103*:37 (1986).
13. M. Studd, D.J. McCance, and T. Lehner, Detection of HSV-1 DNA in patients with Behcet's syndrome and in patients with recurrent oral ulcers by the polymerase chain reaction, *J. Med. Microbiol. 34*:39 (1991).
14. T. Lehner, E. Lavery, R. Smith, R. van der Zee, Y. Mizushima, and T. Shinnick, Association between the 65-kilodalton heat shock protein, *Streptococcus sanguis*, and the corresponding antibodies in Behcet's disease, *Infect. Immun. 59*:1434 (1991).
15. K. Pervin, A. Childerstone, T. Shinnick, Y. Mizushima, R. van der Zee, A. Hasan, R. Vaughan, and T. Lehner, T cell epitope expression of mycobacterial and homologous human 65-kilodalton heat shock protein peptides in short term cell lines from patients with Behcet's disease, *J. Immunol. 151*:2273 (1993).
16. F. Fortune, J. Walker, and T. Lehner, The expression of $\gamma\delta$ T cell receptor and the prevalence of primed, activated and IgA-bound T cells in Behcet's syndrome, *Clin. Exp. Immunol. 82*:326 (1990).
17. Y. Suzuki, K. Hoshi, T. Matsuda, and Y. Mizushima, Increased peripheral blood $\gamma\delta$ T cells and natural killer cells in Behcet's disease, *J. Rheumatol. 16*:588 (1992).
18. R.L. O'Brien, M.P. Happ, A. Dalbas, E. Palmer, R. Kubo, and W.K. Born, Stimulation of a major subset of lymphocytes expressing T cell receptor $\gamma\delta$ by an antigen derived from Mycobacterium tuberculosis, *Cell 57*:667 (1989).
19. A.G. Haregewoin, C. Hom, and R.W. Finberg, Human $\gamma\delta$ T cells respond to mycobacterial heat shock proteins, *Nature 340*:309 (1989).
20. International Study Group for Behcet's Disease, Criteria for diagnosis of Behcet's disease, *Lancet 335*:1078 (1990).
21. T. Lehner, Autoimmunity in oral diseases with special reference to recurrent oral ulceration, *Proc. R. Soc. Med. 61*:515 (1968).
22. H. Direskeneli, T. Shinnick, A. Hasan, F. Fortune, Y. Mizushima, R. van der Zee, and T. Lehner, Submitted for publication.
23. M.R. Stanford, E. Kasp, E. Whiston, D.C. Dumonde, K. Pervin, A. Hasan, S. Todryk, Y. Mizushima, and T. Lehner, Heat shock protein peptides reactive in patients with Behcet's disease are uveitogenic in Lewis rats, *Clin. Exp. Immunol. 97*:226 (1994).
24. D.C. Anderson, M.E. Barry, and T.M. Buchanan, Exact definition of species-specific and cross-reactive epitopes of the 65-kilodalton protein of *Mycobacterium leprae* using synthetic peptides, *J. Immunol. 141*:607 (1988).
25. V. Mehra, D. Sweetser, and R.A. Young, Efficient mapping of protein antigenic determinants, *Proc. Natl. Acad. Sci. USA 83*:7013 (1986).
26. A.I. Dolby, Recurrent aphthous ulceration: effect of sera and peripheral blood upon epithelial tissue culture cells, *Immunology 17*:709 (1969).
27. R. Rogers, W. Sams, and R. Shorter, Lymphocytotoxicity in recurrent aphthous stomatitis, *Arch. Dermatol. 109*:361 (1974).
28. S. Jindal, A.K. Dudani, B. Singh, B., C.B. Harley, and R.S. Gupta, Primary structure of a human mictochondrial protein homologous to the bacterial and plant chaperonins and to the 65-kilodalton mycobacterial antigen, *J. Immunol. 9*:2279 (1989).

29. A.K. Dudani, and R.S. Gupta, Immunological characterization of a human homolog of the 65-kilodalton mycobacterial antigen, *Infect. Immun. 57*:2786 (1989).
30. D. Evans, P. Norton, and J. Ivanyi, Distribution in tissue sections of the human groEL stress-protein homologue, *APMIS 98*:437 (1989).
31. T. Lehner, Characterization of mucosal antibodies in recurrent aphthous ulceration and Behcet's syndrome, *Arch. Oral. Biol. 14*:843 (1969).
32. W. Van Eden, E.R. Thole, R. Van der Zee, A. Noordzij, J.D.A. van Embden, E.J. Hensen, and I.R. Cohen, Cloning of the mycobacterial epitope recognized by T lymphocytes in adjuvant arthritis, *Nature 331*:171 (1988).
33. J. Holoshitz, F. Koning, J.E. Coligan, J. De Bruyn, and S. Strober, Isolation of CD4-CD8$^-$ mycobacteria-reactive T lymphocyte clones from rheumatoid arthritis synovial fluid, *Nature 339*:226 (1989).
34. J.S.H. Gaston, P.F. Life, P.J. Jenner, M.J. Colston, and P.A. Bacon, Recognition of a mycobacteria-specific epitope in the 65kD heat shock protein by synovial fluid-derived T cell clones, *J. Exp. Med. 171*:831 (1990).
35. A.J. Quayle, K.B. Wilson, S.G. Li, J. Kjeldsen-Kragh, F. Oftung, T. Shinnick, M. Sioud, O. Forre, J.D. Caora, and J.B. Natvig, *Eur. J. Immunol. 22*:1315 (1992).
36. L.W. Poulter and T. Lehner, Immunohistology of oral lesions from patients with recurrent oral ulcers and Behcet's syndrome, *Clin. Exp. Immunol. 78*:189 (1989).

11

Stress Proteins in Systemic Lupus Erythematosus

John B. Winfield and Wael N. Jarjour
University of North Carolina, Chapel Hill, North Carolina

I. INTRODUCTION

Systemic lupus erythematosus (SLE) is a chronic systemic autoimmune disease of unknown etiology that primarily affects females during their child-bearing years. Genetic and environmental factors contribute to the development of this disorder, which is characterized by endogenous T- and B-cell activation, the appearance of multiple autoantibodies to self-constituents, immune complex formation, and inflammatory tissue injury in various organ systems. Consideration of stress proteins and the stress response as factors of possible etiopathogenetic significance in SLE came after the initial stress protein forays in adjuvant arthritis and rheumatoid arthritis (1), and this derives from the fortuitous detection of autoantibodies to heat shock proteins (hsps) 90 and 70 in SLE sera (2,3). Investigation during the past 5 years lies principally in three areas: (1) stress protein expression in and on cells of the immune system; (2) further definition of antibodies and autoantibodies to stress proteins; and (3) studies of a polymorphism of the hsp 70-2 gene, which is located in the class III region of the major histocompatibility complex (MHC). This chapter discusses the possible relevance of recent information in these areas to the pathogenesis of SLE.

II. STRESS PROTEIN EXPRESSION IN SLE

A. Intracellular Expression of Stress Proteins

It is well recognized that many stimuli other than heat, such as drugs, hormones, certain prostaglandins, mitogens, and cytokines, can induce the synthesis of stress proteins in human cells. For example, synthesis of hsp 90 is dramatically, albeit transiently, increased in T cells following stimulation with mitogens (4), and the stress response may be part of a negative-feedback loop that downregulates the expression of genes for interleukin-1 β (IL-1β) and perhaps other cytokines (5). Because SLE is inherently a disease of endogenous lymphocyte activation, with increased numbers of immunoglobulin (Ig)–secreting B cells and spontaneously activated T cells, reports of increased expression of certain stress proteins in peripheral blood mononuclear cells (PBMCs) of patients with SLE are not surprising.

The initial observation in this regard—that specific incorporation of [^{35}S]methionine by hsp 70 and hsp 85 (hsp90) was increased in SLE PBMCs relative to levels in control cells (6)—has been extended to analyses of this phenomenon with respect to disease activity status and relationship to clinical manifestations (7–9). Thus, transcription of hsp 70 genes (8), but not ubiquitin genes (10), was found to be increased in PBMCs from patients with active SLE. Elevated levels of hsp 90 and hsp 72, but not hsp 60 or hsp 73, were detected in PBMCs from a subset of patients with SLE by densitometric scanning of Western blots probed with anti–stress protein monoclonal antibodies (7,9). Of special interest here was an apparent association of increased hsp 90, but not hsp 72, expression with active SLE in serial studies of individual patients, particularly those exhibiting neuropsychiatric or cardiorespiratory manifestations. The discoordinate elevation of different stress proteins in these studies, and the lack of an association of high hsp 90 levels with other indices of disease activity, for example, erythrocyte sedimentation rates and anti-DNA titers, raise the possibility that expression of hsp 90 in SLE PBMCs may be an independent marker of disease activity and/or clinical subset in this disorder (9). Further investigation will be required to establish the significance of these findings with respect to disease pathogenesis in this disorder.

B. Cell Surface Expression of Stress Proteins

Although primarily intracellular, certain data suggest that stress proteins, or close homologues thereof, may be expressed on plasma membranes of lymphocytes and other cells. For example, a hsp 70–related peptide-binding protein that appears to act in antigen presentation (11,12) has been detected on the surface of mouse B cells. Similarly, our laboratory has described a GroEL-related ~77-kDa protein on the surface membrane of γδ cells, which is not present on T cells with αβ receptors (13). In an extension of this latter finding, we used monospecific antibodies as probes to define the surface localization of several large stress proteins (hsp 60, hsp 70, hsp 90, and hsp 110), or cross-reactive molecules thereof, on various types of resting, heat-shocked, or activated lymphocytes and phagocytic cells from normal peripheral blood or other non-SLE sources (W.N.J., et al., unpublished observations). B-cell lines, T cells with γδ receptors, and mitogen-activated peripheral T cells were shown to react with anti–hsp 60 in indirect immunofluorescence and complement-dependent microcytotoxicity assays. Mitogen-activated T cells, but not resting T cells or other cell types, also reacted with anti–hsp 90. Surface expression of proteins related to hsp 110 or members of the hsp 70 family of stress proteins on T cells, B cells, and several monomyelocytic cell lines was not observed. In contrast to our preliminary data suggesting that the monomyelocytic line HL60 expressed hsp 90 after heat shock or stimulation with phorbol esters (14), positive anti–hsp 90 staining of this cell line and several other phagocytic cell lines was not demonstrable when cell death was limited to 5% or less.

The issue of cell surface expression of hsp 90 on PBMCs from patients with SLE or other autoimmune diseases has been approached by Erkeller-Yuksel and colleagues (15). Using AC88, an IgG1 monoclonal antibody raised against hsp 90 isolated from *Achyla ambisexualis* that recognizes human hsp 90, and flow cytometry, ~2% of PBMCs from 62 patients with SLE exhibited immunofluorescent staining, a significantly higher value than the 0.4% staining seen with PBMCs from 42 control subjects. Anti–hsp 90 surface staining was particularly prominent in patients with active SLE, but it was not linked to expression of lymphocyte activation markers. Disease control patients with Sjögren's syndrome, systemic sclerosis, rheumatoid arthritis, and polymyositis/dermatomyositis were only occasionally positive in this investigation.

Excluding molecules with structural characteristics of integral membrane proteins (16,17), stress proteins (or peptides thereof) could be expressed on the cell surface by three mechanisms: (1) presentation of stress protein peptides by MHC class I, class II, class Ib molecules; (2) translocation of stress proteins to the cell surface as they chaperon nascent integral membrane proteins or damaged intracellular proteins to the plasma membrane; and (3) secondary binding of released or secreted extracellular stress proteins to surface membrane proteins. Regardless of the mechanism(s) by which stress proteins (or their epitopes) get to the cell surface, the fact that they do broadens their potential significance vis-à-vis repertoire selection, cellular activation, immune surveillance, regulation of lymphocyte growth and differentiation, autoimmunity, and perhaps autoimmune disease (18).

III. ANTIBODIES AND AUTOANTIBODIES TO STRESS PROTEINS

A. Antibodies to Microbial Stress Proteins

Antibodies to microbial hsp 60 and/or hsp 70 have been demonstrated in patients with various rheumatic and autoimmune diseases, including rheumatoid arthritis, SLE, Sjögren's syndrome, systemic sclerosis, mixed connective tissue disease, and ankylosing spondylitis, to name a few (19–26). In many of these reports, however, it is unclear whether detection of antibodies to microbial stress proteins in patient sera represents a true increase in specific antibodies to agents of putative etiological significance or whether their detection simply reflects a polyclonal increase in immunoglobulin levels. Indeed, a careful analysis using ELISA techniques found some IgG and/or IgM antibodies to mycobacterial hsp 60 in essentially all sera from patients with rheumatoid arthritis and various other rheumatic diseases *and* in all normal sera; moreover, patients and controls did not differ significantly in the levels of antibodies to this antigen (27).

B. Autoantibodies to Stress Proteins

Only limited data are available concerning the development of true autoantibodies to stress proteins; that is, antibodies that react with human antigens. Unconfirmed reports suggest that patients with SLE exhibit autoantibodies to ubiquitin and a charged synthetic octapeptide of ubiquinated histone H2A (28,29). Using immunoblotting techniques, our group described autoantibodies to hsp 90 and to the constitutively expressed 73-kDa member of the hsp 70 family in a minority of patients with SLE (2,3). Others also have reported autoantibodies to human hsp 70 in rheumatoid arthritis, SLE, and ankylosing spondylitis (24). Our initial studies have been extended in a more comprehensive survey of 268 patients with SLE or various other rheumatic diseases, inflammatory bowel disease, and several autoimmune skin diseases (30). Autoantibodies to human hsp 60, hsp 73, and hsp 90 were absent or rare in Sjögren's syndrome, rheumatoid arthritis, ankylosing spondylitis, Reiter's syndrome, and in control sera and infrequent ($<$20% of sera positive) in SLE and progressive systemic sclerosis. On the other hand, $>$20% of patient sera exhibited IgM and/or IgG autoantibodies to hsp 60 in mixed connective tissue disease, polymyositis/dermatomyositis, psoriatic arthritis, inflammatory bowel disease, epidermolysis bullosa acquisita, and bullous pemphigoid and to hsp 73 in Lyme disease and ulcerative colitis. In a similar study that used ELISA techniques rather than immunoblotting, some IgG and IgM autoantibodies to human hsp 60 (recombinant P1 protein) were detected in nearly all sera from patients with SLE, various other rheumatic disease, and normal subjects (27). Here the majority of sera from patients with SLE or Sjögren's syndrome had significantly higher levels of IgG antibodies to human hsp 60 than to mycobacterial hsp 60. KindÅs-Mügge and colleagues (31) also observed a similar (~20–30% frequency of IgG and IgM antibodies to hsp 72 and hsp 73 in patients with SLE and control subjects, with 2 of 47 patients with SLE exhibiting rather higher levels of IgM antibodies. Both of these latter studies probably are consistent with our own if one considers that the immunoblotting conditions that we used excluded the very low levels of anti–stress protein autoantibodies, that, like rheumatoid factor, may be present in essentially all human sera. In any case, it is clear that patients with autoimmune disease do not develop high-affinity/high-titer autoantibodies to stress proteins of the kind that are commonly seen to DNA and certain other nuclear antigens in SLE.

The ultimate significance of autoantibodies to stress proteins remains to be determined. Precise information concerning their diagnostic specificity, linkage with other autoantibody systems, relationship to disease activity status, and reactive epitopes is not available. Although anti–stress protein autoantibodies clearly do not appear to be disease-specific "markers," and their relatively low concentration and prevalence argues against a role in disease pathogenesis, these issues should remain open until examined more directly in acutely ill or early-onset patients. For example, certain stress proteins are spontaneously shed or selectively released (16,17,32) and theoretically could form immune complexes of potential significance in tissue injury. Perhaps of more interest is the possibility that stress proteins, such as hsp 70 or ubiquitin, may be ingredients in the "immunogenic particle" concept (33) of the origin of antinuclear and other autoantibodies. Thus, hsp 70 associates with proteins in the nucleus and nucleolus in a cell cycle–dependent fashion (34), migrates among different cell compartments, and binds to a variety of nuclear and cytoplasmic proteins in virus-infected cells and in cells exposed to other stressful stimuli (35–37). Consistent with this idea are recent data from Furakawa and colleagues demonstrating that induction of hsp 72 synthesis in cultured human keratinocytes by ultraviolet light or a cytotoxic prostaglandin is accompanied by increased binding of anti-U1RNP and anti–SS-A/Ro to cultured keratinocytes (38).

IV. POLYMORPHISM OF THE HSP 70-2 GENE AS A SUSCEPTIBILITY MARKER IN SLE

Analysis of overlapping cosmid clones has enabled the identification and precise mapping of three hsp 70 genes (hsp 70-1, hsp 70-2, and hsp 70-hom) within the class III region of the human MHC (39). Hsp 70-1 and hsp 70-2 encode an identical protein, whereas hsp 70-hom encodes a previously unknown protein. These genes map close to each other, about 90 kb telomeric to the C2 gene and about 280 kb centromeric to the tumor necrosis factor α (TFN-α) gene. A restriction fragment length polymorphism (RFLP) of hsp 70 gene(s) in the MHC has been described using an hsp 70 probe, ph 2.3, and the restriction enzyme *Pst*I (40). The two alleles, represented by a 9.0- or an 8.5-kb fragment, occur with frequencies of 0.62 and 0.38, respectively. Because a member of the hsp 70 family is involved in the handling and presentation of antigen peptides to the T-cell receptor (11,12), we hypothesized that hsp 70 gene expression could play a role in the development of autoimmunity in SLE. Indeed, several groups have reported an increase in the frequency of the *Pst*I 8.5-kb allele of the hsp 70-2 gene associated with a C4A deletion in type I diabetes and Graves' disease (41,42). In addition, Pociot et al. (43) have observed decreased hsp 70-2 mRNA expression in individuals who are homozygotes for the *pst*I 8.5-kb allele as compared with homozygotes for the *pst*I 9.0-kb allele.

Using pH 2.3 and *Pst*1 in studies of hsp 70 gene polymorphism in SLE, genomic DNA from 49 individuals with SLE and 45 healthy controls with no family history of autoimmune disease have been analyzed. As expected, we observed two alleles, 8.5 and 9.0 kb, resulting in three genotypes (9.0/9.0, 9.0/8.5, and 8.5/8.5). There was a statistically significant association between the absence of the 9.0-kb allele and SLE ($p = 0.0474$). Conversely, the presence of 8.5-kb allele was associated with SLE ($p = 0.039$), and its allelic frequency was increased in patients with SLE relative to the control subjects ($p = 0.0066$). When the data were analyzed with respect to race, the increase in the 8.5 allele was associated with SLE in African-Americans ($p = 0.042$) but not whites. Because multiple associations with specific MHC genes have been described in SLE, we next examined the extended haplotype of our patient and control populations. Linkage of the hsp 70-2 polymorphism with C4A deletion, DR3, or an NCOI polymorphism of TNF-α was not demonstrable. Taken together, these data suggest that a polymorphism of the hsp 70-2 gene in the class III region of the MHC is a new susceptibility marker for SLE, particularly in African-Americans.

V. CONCLUSIONS

1. Because SLE is characterized by endogenous lymphocyte activation, the increased expression of certain stress proteins in peripheral blood mononuclear cells of patients with this disorder may simply reflect cellular activation. However, the demonstration of elevated levels of hsp 90 in lymphocytes from a clinically discrete subset of patients with active disease raises the possibility of a special role for this stress protein in pathogenesis.
2. Increasing evidence suggests that certain stress proteins can, under certain circumstances, be expressed on the surface of cells. Although the significance of such observations remains to be determined, surface expression of stress proteins could render cells susceptible to specific immunological attack by autoreactive T cells or anti–stress protein autoantibodies.
3. Autoantibodies to stress proteins, especially hsp 90 and certain members of the hsp 70 family, occasionally are elevated in patients with SLE. However, sensitive assay procedures detect such autoantibodies in roughly equivalent frequency and titer in both patients and normal individuals, suggesting that autoantibodies to stress proteins are a type of natural antibody. The intriguing idea that

stress proteins could contribute to the formation of immunogenic particles in the development of high titer antinuclear antibodies merits further study.

4. A polymorphism of the hsp 70-2 gene in the class III region of the MHC may be a new susceptibility marker for SLE in African-Americans.

REFERENCES

1. J.B. Winfield, Stress proteins, arthritis, and autoimmunity, *Arthritis Rheum. 32*: 1497 (1989).
2. S. Minota, S. Koyasu, I. Yahara, and J.B. Winfield, Autoantibodies to the heat-shock protein hsp90 in systemic lupus erythematosus, *J. Clin. Invest. 81*:106 (1988).
3. S. Minota, B. Cameron, W.J. Welch, and J.B. Winfield, Autoantibodies to the constitutive 73-kD member of the hsp70 family of heat shock proteins in systemic lupus erythematosus, *J. Exp. Med. 168*:1475 (1988).
4. D.K. Ferris, A. Harel Bellan, R.I. Morimoto, W.J. Welch, and W.L. Farrar, Mitogen and lymphokine stimulation of heat shock proteins in T lymphocytes, *Proc. Natl. Acad. Sci. USA 85*:3850 (1988).
5. J.A. Schmidt and E. Abdulla, Down-regulation of IL-1β biosynthesis by inducers of the heat-shock response, *J. Immunol. 141*:2027 (1988).
6. Y. Deguchi, S. Negoro, and S. Kishimoto, Heat-shock protein synthesis by human peripheral mononuclear cells from SLE patients, *Biochem. Biophys. Res. Commun. 148*:1063 (1987).
7. P.M. Norton, D.A. Isenberg, and D.S. Latchman, Elevated levels of the 90 kd heat shock protein in a proportion of SLE patients with active disease, *J. Autoimmun. 2*:187 (1989).
8. Y. Deguchi and S. Kishimoto, Enhanced expression of the heat shock protein gene in peripheral blood mononuclear cells of patients with active systemic lupus erythematosus, *Ann. Rheum. Dis. 49*:893 (1990).
9. V.B. Dhillon, S. McCallum, P. Norton, B.M. Twomey, F. Erkeller Yuksel, P. Lydyard, D.A. Isenberg, and D.S. Latchman, Differential heat shock protein overexpression and its clinical relevance in systemic lupus erythematosus, *Ann. Rheum. Dis. 52*:436 (1993).
10. B. Twomey, S. McCallum, V. Dhillon, D. Isenberg, and D. Latchman, Transcription of the genes encoding the small heat shock protein ubiquitin is unchanged in patients with systemic lupus erythematosus, *Autoimmunity 13*:197 (1992).
11. E.K. Lakey, E. Margoliash, and S.K. Pierce, Identification of a peptide binding protein that plays a role in antigen presentation, *Proc. Natl. Acad. Sci. USA 84*:1659 (1987).
12. A. Vanbuskirk, B.L. Crump, E. Margoliash, and S.K. Pierce, A peptide binding protein having a role in antigen presentation is a member of the HSP70 heat shock family, *J. Exp. Med. 170*:1799 (1989).
13. W. Jarjour, L.A. Mizzen, W.J. Welch, S. Denning, M. Shaw, T. Mimura, B.F. Haynes, and J.B. Winfield, Constitutive expression of a groEL-related protein on the surface of human gamma/delta cells, *J. Exp. Med. 172*:1857 (1990).
14. W. Jarjour, V. Tsai, V. Woods, W. Welch, S. Pierce, M. Shaw, H. Mehta, W.

Dillmann, N. Zvaifler, and J. Winfield, Cell surface expression of heat shock proteins, *Arthritis Rheum. 32*:S44 (1989).
15. F.M. Erkeller Yuksel, D.A. Isenberg, V.B. Dhillon, D.S. Latchman, and P.M. Lydyard, Surface expression of heat shock protein 90 by blood mononuclear cells from patients with systemic lupus erythematosus, *J. Autoimmun. 5*:803 (1992).
16. P.J. McCormick, B.J. Keys, C. Pucci, and A.J.T. Millis, Human fibroblast-conditioned medium contains a 100k dalton glucose-regulated cell surface protein, *Cell 18*:173 (1979).
17. P.J. McCormick, A.J.T. Millis, and B. Babiarz, Distribution of a 100k dalton glucose-regulated cell surface protein in mammalian cell cultures and sectioned tissues, *Exp. Cell Res. 138*:63 (1982).
18. W. Born, M.P. Happ, A. Dallas, C. Reardon, R. Kubo, T. Shinnick, P. Brennan, and R. O'Brien, Recognition of heat shock proteins and gamma/delta cell function, *Immunol. Today 11*:40 (1990).
19. G.M. Bahr, G.A.W. Rook, M. Al-Saffar, J. van Embden, J.L. Stanford, and K. Behbehani, Antibody levels to mycobacteria in relation to HLA type: evidence for non–HLA-linked high levels of antibody to the 65 kD heat shock protein of *M. bovis* in rheumatoid arthritis, *Clin. Exp. Immunol. 74*:211 (1988).
20. G. Tsoulfa, G.A. Rook, J.D. van Embden, D.B. Young, A. Mehlert, D.A. Isenberg, F.C. Hay, and P.M. Lydyard, Raised serum IgG and IgA antibodies to mycobacterial antigens in rheumatoid arthritis, *Ann. Rheum. Dis. 48*:118 (1988).
21. I.L. McLean, J.R. Archer, M.I.D. Cawley, F.S. Pegley, B.L. Kidd, and P.W. Thompson, Specific antibody response to the mycobacterial 65 kDa stress protein in ankylosing spondylitis and rheumatoid arthritis, *Br. J. Rheumatol. 29*:426 (1990).
22. H.H. Handley, S. Nonaka, R.S. Gupta, and J.H. Vaughan, Autoimmune responses to the human chaperonin, HSP63 (abstr.), *Arthritis Rheum. 33 (Suppl.)*:S46 (1990).
23. G. Tsoulfa, G.A. Rook, J.D. van Embden, D.B. Young, A. Mehlert, D.A. Isenberg, F.C. Hay, and P.M. Lydyard, Raised serum IgG and IgA antibodies to mycobacterial antigens in rheumatoid arthritis, *Ann. Rheum. Dis. 48*:118 (1989).
24. G. Tsoulfa, G.A.W. Rook, G.M. Bahr, M.A. Sattar, K. Behbehani, D.B. Young, A. Mehlert, J.D.A. Van-Embden, F.C. Hay, D.A. Isenberg, and P.M. Lydyard, Elevated IgG antibody levels to the mycobacterial 65-kDa heat shock protein are characteristic of patients with rheumatoid arthritis, *Scand. J. Immunol. 30*:519 (1989).
25. J. Panchapekesan, M. Daglis, and P. Gatenby, Antibodies to 65 kDa and 70 kDa heat shock proteins in rheumatoid arthritis and systemic lupus erythematosus, *Immunol. Cell Biol. 70*:295 (1992).
26. M.G. Danieli, M. Candela, A.M. Ricciatti, R. Reginelli, G. Danieli, I.R. Cohen, and A. Gabrielli, Antibodies to mycobacterial 65 kDa heat shock protein in systemic sclerosis (scleroderma), *J. Autoimmun. 5*:443 (1992).
27. H.H. Handley, S. Nonaka, R.S. Gupta, and J.H. Vaughan, Autoimmune responses to the human chaperonin, HSP63 (abstr.), *Arthritis Rheum. 33 (Suppl.)*:S46 (1990).
28. S. Muller, J.P. Briand, and M.H. Van Regenmortel, Presence of antibodies to ubiquitin during the autoimmune response associated with systemic lupus erythematosus, *Proc. Natl. Acad. Sci. USA 85*:8176 (1988).

29. S. Plaué, S. Muller, and M.H.V. Van Regenmortel, A branched, synthetic octapeptide of ubiquitinated histone H2A as target of autoantibodies, *J. Exp. Med. 169*:1607 (1990).
30. W.N. Jarjour, B.D. Jeffries, J.S. Davies, IV, W.J. Welch, T. Mimura, and J.B. Winfield, Autoantibodies to stress proteins: a survey of various rheumatic and inflammatory diseases, *Arthritis Rheum. 34*:1133 (1991).
31. I. Kindas Mugge, G. Steiner, and J.S. Smolen, Similar frequency of autoantibodies against 70-kD class heat-shock proteins in healthy subjects and systemic lupus erythematosus patients, *Clin. Exp. Immunol. 92*:46 (1993).
32. L.E. Hightower and P.T. Jr. Guidon, Selective release from cultured mammalian cells of heat-shock (stress) proteins that resemble glia-axon transfer proteins, *J. Cell Physiol. 138*:257 (1989).
33. E.M. Tan, E.K. Chan, K.F. Sullivan, and R.L. Rubin, Antinuclear antibodies (ANAs): diagnostically specific immune markers and clues toward the understanding of systemic autoimmunity, *Clin. Immunol. Immunopathol. 47*:121 (1988).
34. K.L. Milarski, W.J. Welch, and R.I. Morimoto, Cell cycle-dependent association of HSP70 with specific cellular proteins, *J. Cell Biol. 108*:413 (1989).
35. W.J. Welch and J.P. Suhan, Cellular and biochemical events in mammalian cells during and after recovery from physiological stress, *J. Cell Biol. 103*:2035 (1986).
36. N.B. La Thangue and D.S. Latchman, A cellular protein related to heat-shock protein 90 accumulates during herpes simplex virus infection and is overexpressed in transformed cells, *Exp. Cell Res. 178*:169 (1988).
37. P.C. Collins and L.E. Hightower, Newcastle disease virus stimulates the cellular accumulation of stress (heat shock) mRNAs and proteins, *J. Virol. 44*:703 (1982).
38. F. Furukawa, K. Ikai, N. Matsuyoshi, K. Shimizu, and S. Imamura, Relationship between heat shock protein induction and the binding of antibodies to the extractable nuclear antigens on cultured human keratinocytes, *J. Invest. Dermatol. 101*:191 (1993).
39. C.M. Milner and R.D. Campbell, Structure and expression of the three MHC-linked *HSP70* genes, *Immunogenetics 32*:242 (1990).
40. A.M. Goate, D.N. Cooper, C. Hall, T.K.C. Leung, E. Solomon, and L. Lim, Localization of a human heat shock HSP70 gene sequence to chromosome 6 and detection of two other loci by somatic-cell hybrid and restriction fragment length polymorphism analysis, *Hum. Genet. 75*:123 (1987).
41. N.J. Caplen, A. Patel, A. Millward, R.D. Campbell, S. Ratanachaiyavong, F.S. Wong, and A.G. Demaine, Complement C4 and heat shock protein 70 (HSP70) genotypes and type I diabetes mellitus, *Immunogenetics 32*:427 (1990).
42. S. Ratanachaiyavong, A. G. Demaine, R.D. Campbell, and A.M. McGregor, Heat shock protein 70 (HSP70) and complement C4 genotypes in patients with hyperthyroid Graves' disease, *Clin. Exp. Immunol. 84*:48 (1991).
43. F. Pociot, K.S. Ronningen, and J. Nerup, Polymorphic analysis of the human MHC-linked heat shock protein 70 (HSP70-2) and HSP70-Hom genes in insulin-dependent diabetes mellitus (IDDM), *Scand. J Immunol. 38*:491 (1993).

12

Expression of and Immune Response to Heat Shock Protein 65 in Crohn's Disease

Willy E. Peetermans
University Hospitals, K. U. Leuven, Leuven, Belgium

I. INTRODUCTION

A. Characteristics

Crohn's disease is an inflammatory bowel disease of unknown etiology. The diagnosis of Crohn's disease is based on clinical, pathological, radiological, and endoscopic features. The syndrome exhibits the classic pathological characteristics of chronic inflammation; that is, an infiltration of macrophages and lymphocytes, including large numbers of plasma cells. In the active stage of Crohn's disease, there is an acute inflammatory component as well; that is, an influx of neutrophils and monocytes into the inflamed mucosa (1,2).

B. Pathogenesis

The pathogenesis of Crohn's disease remains unknown. Research focuses at the possible initiating events, the proinflammatory mediators, and deficient downregulatory mechanisms of intestinal inflammation. The role of the mucosal immune system is thought to be of major importance in this respect (3). There is accumulating evidence that a microbial factor may be involved in the initiation or amplification of the intestinal inflammation in Crohn's disease. In an animal model, nonsteroidal anti-inflammatory drugs (NSAIDs) induce an intestinal ulceration and an enhanced permeability of the intestinal wall similar to that found in Crohn's disease. NSAID-induced intestinal inflammatory lesions do not develop readely in germ-free animals and are greatly reduced by the administration of antibiotics to conventional animals (4,5). In human Crohn's disease, antibiotics also favorably influence signs and symptoms of active inflammatory bowel disease. Furthermore, the recurrence of Crohn's disease in the neoterminal ileum after curative surgical resection of the diseased segments does not occur when the neoterminal ileum is excluded from the passage of feces, but it does occur within 6 months after reanastomosis allowing the passage of the fecal stream, which contains high numbers of microbial and dietary antigens (6). The search for a microbial trigger of Crohn's disease is further enhanced by epidemiological studies that show clustering of Crohn's disease cases (7). Much interest has been attributed to the possible role of atypical mycobacteria in the pathogenesis of Crohn's disease, because *Mycobacterium paratuberculosis* causes Johne's disease, a chronic ileitis and colitis in ruminants that has some similarities with Crohn's disease in humans. Many patients with Crohn's disease have antibodies against *M. paratuberculosis*, but there is a considerable overlap with controls (8) and other studies do not confirm these results (9). A slow-growing *M. paratuberculosis* has been isolated from intestinal tissue samples of a minority of patients with Crohn's disease, but it is also seldomly found in normal controls (10). Using DNA-based diagnostic tools, *M. paratuberculosis* has been demonstrated in intestinal tissue samples of 65% of patients with Crohn's disease and 12.5% of the controls (11,12). Various other mycobacterial DNA sequences are found in these tissue samples as well (13,14), and these data still await confirmation in other laboratories.

The key question in the pathogenesis of Crohn's disease is whether the inflammatory lesions reflect an appropriate response to a persistent abnormal stimulus (e.g., a luminal microbial agent) or an inappropriate and prolonged inflammatory reaction initiated by a transient intestinal stimulus (e.g., a bacterial or food antigen) and is perpetuated by an aberrant regulation of the intestinal immune system. In healthy intestinal mucosa, there is always some degree of chronic inflammation, often referred to as controlled inflammation (3). Since the mucosa of the ileum and colon is constantly in contact with a broad range of microbial agents, it must display tolerance on the one hand and remain capable of mounting an adequate and protective immune response in case of hostile invasion of pathogens on the other hand. In Crohn's disease, alterations in the regulatory mechanisms of this state of immune tolerance and readiness have been observed. These alterations include changes in the relative number of macrophages and T- and B-cell populations and an increased production of immunoglobulins (3,15). The inability of the gut-associated lymphoepithelial tissue of patients with Crohn's disease to stimulate $CD8^+$ suppressor T lymphocytes and the overwhelming activation of mucosal $CD4^+$ helper T lymphocytes may be involved in the sustained intestinal inflammation (16–18). An enhanced production of interferon gamma (IFN-γ) by activated T lymphocytes and an enhanced secretion of the proinflammatory cytokines tumor necrosis factor α (TNF-α), interleukin-1β (IL-1β), and interleukin-6 (IL-6) by lamina propria mononuclear cells from patients with Crohn's disease reflect the state of inflammation at the mucosal level. These cytokines may be important mediators of tissue injury as well and possible targets of antiinflammatory therapy (3,18–20).

Hsp 65 is an attractive antigen to study in Crohn's disease, since it is ubiquitously present in the fecal bacterial and mycobacterial flora. Moreover, the hypothesis of molecular mimicry between (myco)bacterial and human hsp 65 homologues may provide an explanation for the possible link between an immune response to colonization or infection with microbial pathogens and autoimmune inflammatory disorders (21–23). In the literature, there are conflicting reports on the possible role of hsp 65 in Crohn's disease. Raised immunoglobulin G (IgG) antibodies against mycobacterial hsp 65 are detected in half of the patients with Crohn's disease, whereas antibodies against the human hsp 65 counterpart are not increased in one study (24). To which extent the treatment with steroids abrogated the production of antibodies against self–hsp 65 in these patients remains unclarified. Significantly increased serum IgA antimycobacterial hsp 65 antibody titers are reported in patients with Crohn's disease, indicating that a mucosal challenge may have provoked the humoral immune response. Serum IgG anti–hsp 65 antibodies do not differ between patients and controls in that study (25). Furthermore, there is one report that shows that T lymphocytes taken from the inflamed mucosa of patients with Crohn's disease exhibit an increased proliferative response to stimulation with hsp 65 or other antigens in vitro, whereas T lymphocytes from uninflamed mucosa are unresponsive. The peripheral blood lymphocytes of these patients with Crohn's disease also have a raised proliferation index when assayed with hsp 65 as the antigenic stimulus. There are, however, no healthy controls included in that study (26). Taken together with the supposed availability of hsp 65 from the abundantly present microorganisms and the easy access of luminal antigens to reach the mucosal immune system owing to an enhanced permeability of the intestinal wall, these data warrant further investigation of the cellular and humoral immune response to hsp 65 in patients with Crohn's disease. To which extent hsp 65 is expressed in intestinal tissue samples of these patients and whether this hsp 65 is of (myco)-bacterial or human origin is also explored.

II. MATERIALS AND METHODS

A. Immunohistochemistry

Surgical biopsy specimens were obtained from patients with Crohn's disease as well as from controls operated on because of malignancy. Multiple transmural biopsies were taken from diseased areas as well as from areas that appeared macroscopically normal. In control patients, biopsies were taken at random in areas that appeared normal on macroscopic examination. One specimen was fixed in Bouin's fluid and stained with hematoxylin/eosin. The specimens for immunohistochemistry were snap frozen in liquid nitrogen–precooled isopentane and stored at −70°C until further analysis. Five-micrometer cryostat sections were dried overnight at 20°C and fixed in acetone for 10 min. Rehydrated sections were incubated with antibodies for 30 min. A three-step immune peroxidase technique was applied. Endogenous peroxidase was blocked with 0.3% H_2O_2 in methanol. Antimycobacterial hsp 65 antibodies, used in this study, were the monoclonal anti–hsp 65 antibodies LK-1 and LK-2 (gift from Dr. R. van der Zee and Dr. W. van Eden, Department of Immunology and Infectious Diseases, Faculty of Veterinary Medicine, University of Utrecht, The Netherlands) (27). LK-1 has unique specificity for the mammalian hsp 65 homologue without cross reactivity with the bacterial counterpart. LK-2 recognizes both mammalian and bacterial hsp 65 homologues. Double staining for hsp 65 and B7 expression was assayed with a combination of the peroxidase technique (brown reaction product) and the alkaline phosphatase antialkaline phosphatase technique (blue reaction product). Endogenous alkaline phosphatase was blocked by levamisole. The expression of the B7 was assessed by using the monoclonal antibody B7-24 (gift from Dr. M. de Boer, Innogenetics NV, Division of Immunology, Gent, Belgium) (28,29). The biopsy specimens were independently analyzed by three observers. Staining positivity was scored on a subjective scale whereby o indicated absence of staining, + weak, ++ moderate, +++ strong, and ++++ very strong staining positivity.

B. Peripheral Blood Lymphocyte Stimulation Assay

Peripheral blood lymphocytes of patients with Crohn's disease and healthy controls were isolated from heparinized blood on a Ficoll-Hypaque gradient (Pharmacia LKB, Uppsala, Sweden). The cells were washed and resuspended in RPMI-1640 medium (Gibco, Paisley, Scotland) supplemented with penicillin, streptomycin, glutamine, and 5% autologous plasma. Cells were cultured at a concentration of 1×10^6 cells/ml in 96-well round-bottom culture plates (Falcon, Lincoln Park, NY). For stimulation with mycobacterial hsp 65 (gift from Dr. J.D.A. van Embden, National Institute of Public Health and Environmental Hygiene, Bilthoven, The Netherlands), a final concentration of 1 and 10 μg/ml was used. The following other antigens and mitogen were also used for stimulation, each at an indicated final concentration that has been shown in previous experiments to induce optimal T-cell proliferation: streptokinase-streptodornase 100 IU/ml (Lederle, Belgium), *Candida albicans* 0.5 μg/ml (Haarlem Allergenen Laboratorium, The Netherlands), tuberculin 0.4 μg/ml, sensitin 0.4 μg/ml (Statens Serum Institute, Copenhagen, Danmark), and pokeweed mitogen 0.5 μg/ml (Calbiochem, La Jolla, CA). Cells were cultured in a 5% CO_2 humidified atmosphere for 6 days. Eight hours after a 1-μCi [^{3}H]thymidine pulse (Amersham, Buckinghamshire, UK), cells were harvested and processed for determination of [^{3}H]thymidine incorporation in a liquid scintillation counter. The proliferative response was expressed as the stimulation index (SI), which was calculated as the ratio between the mean counts per minute (cpm) of quadruplicate cultures of mitogen- or antigen-stimulated cells divided by the mean cpm of cells in medium only.

C. ELISA for Detection of Anti–Hsp 65 Antibodies

Nunc Immunoplates (Gibco, Paisley, Scotland) with 96 flat-bottom wells were coated with hsp 65 by incubation of a 10 μg/ml solution in bicarbonate buffer (pH 9.6) for 2 hr at 37°C. One hundred microliter aliquots of twofold serum dilutions from 1/25 to 1/800 were added in duplicate to the wells and incubated for 1 hr at 37°C. ELISA was performed using incubation for 1 hr at 37°C of a 100-μl volume of peroxidase-conjugated goat-antihuman IgG (diluted 1/4000) (Kirkegaard and Perry Laboratories, Gaitersburg, MD) and 2.2′-azino-di-(3-ethyl-benzthiazoline sulfonat) as substrate for the colorimetric reaction. Optical density was read at 405 nm, and the results are expressed in arbitrary units (AU), which were calculated from a standard for which a strongly reactive serum sample was used and to which a concentration of 1000 AU/ml was attributed.

D. Statistical Analysis

Data from patients and controls were compared by the Student's *t*-test (two-tail, unpaired). Correlations were studied by the Spearman rank correlation analysis. Level of significance was set at 0.05.

III. RESULTS

A. Immunohistochemistry

Nine patients with Crohn's disease and seven controls were examined. Routine histology of hematoxylin/eosin–stained sections confirmed the diagnosis of Crohn's disease. Hsp 65 was expressed in a scattered pattern in the mucosa and, to a lesser degree, in the submucosa and tunica musculosa of tissue specimens (both ileum and colon) from patients with Crohn's disease (Fig. 1a,b). The degree of hsp 65 expression correlated with the severity of transmural inflammation. In tissue segments from patients with Crohn's disease that were normal on macroscopy and routine histology, expression of hsp 65 was detected as well. In controls, hsp 65 expression was absent or very scarce and staining positivity was much less intense than in patients with Crohn's disease (Fig. 1c).

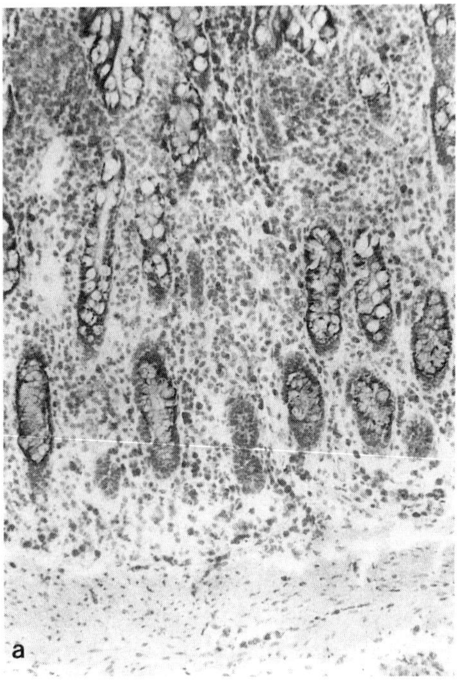

Figure 1 Expression of hsp 65 and B7 in intestinal tissue biopsies. (a and b) Biopsy specimen from a patient with Crohn's disease. Hsp 65 staining with LK1 monoclonal antibody. (c) Biopsy specimen from a control patient. Hsp 65 staining with LK1 monoclonal antibody. (d) Biopsy specimen from a patient with Crohn's disease. Double staining for B7 (B7-24 monoclonal antibody; blue reaction product) and hsp 65 (LK1 monoclonal antibody; brown reaction product).

Comparison between staining with LK-1 and LK-2 in neighboring tissue sections revealed that staining positivities were observed in similar cells and in a similar distribution pattern, indicating that the human homologue was expressed.

Double staining for B7 and hsp 65 showed that in tissue specimens from patients with Crohn's disease, all hsp 65–positive cells were also B7 positive and that most, but not all, B7-positive cells also expressed hsp 65 (Fig. 1D).

B. Antibodies Against Hsp 65 and Peripheral Blood Lymphocyte Stimulation Assay

Forty-five patients and 15 controls were included in the study. Four patients were excluded because of steroid or cyclosporin A treatment and one control was

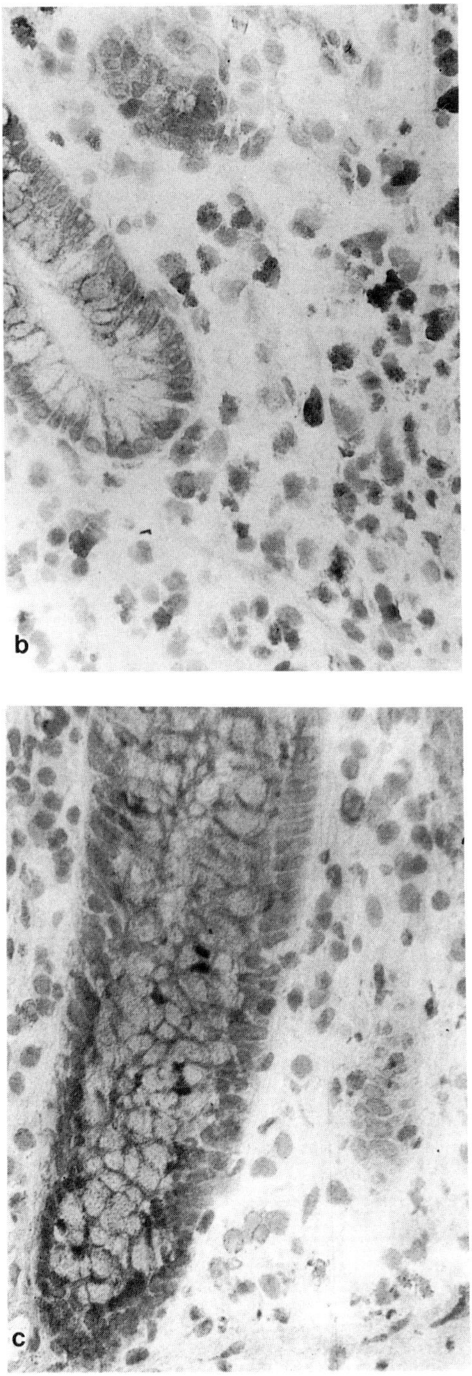

Figure 1 Continued

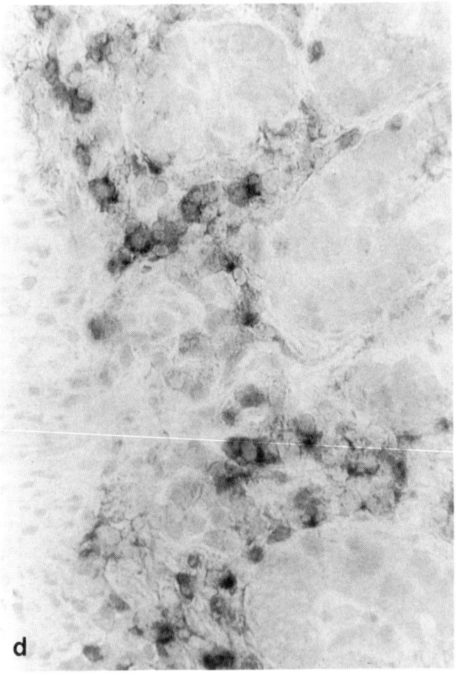

Figure 1 Continued

refused because of prior tuberculosis. There was no difference in age or gender between patients and controls (p $>$ 0.1, χ^2). The patient population was representative for the whole spectrum of Crohn's disease activity. The concentrations of antibodies against hsp 65 were 129 $\pm$ 195 for patients and 55 $\pm$ 55 for controls (mean $\pm$ SD; $p < 0.05$) (Fig. 2). There was a considerable overlap between both groups, but 45% of the patients with Crohn's disease had anti–hsp 65 antibody titers that exceeded the 90% confidence limit of the antibody titers of the controls.

Proliferation of peripheral blood lymphocytes in patients with Crohn's disease on stimulation with hsp 65 in vitro was observed. The peripheral blood lymphocyte stimulation index for hsp 65 was, however, significantly lower ($p < 0.05$) for patients than for controls. The SI for the other antigens or mitogen did not differ significantly between patients and controls (Table 1). There was no correlation between SI for hsp 65 and SI for tuberculin or sensitin, indicating that the response to hsp 65 was not a nonspecific bystander phenomenon of a response to mycobacterial antigens. In patients with Crohn's disease,

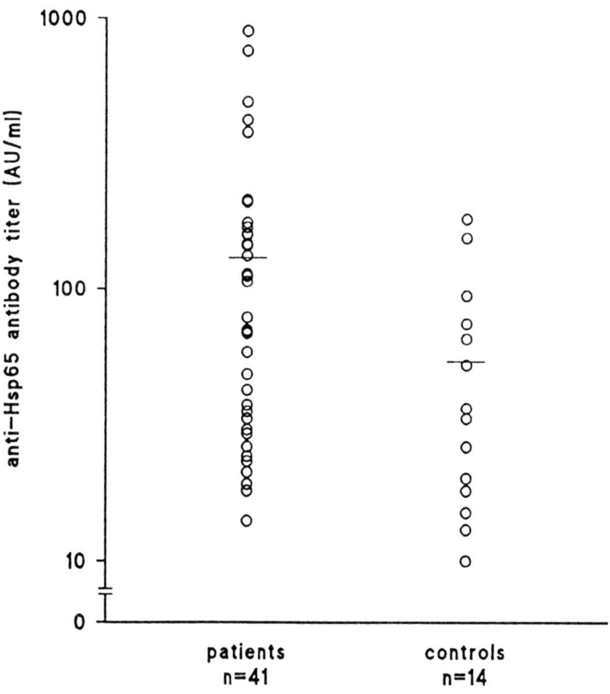

Figure 2 Titer of anti–hsp 65 antibodies (AU/ml) in patients with Crohn's disease and controls. The horizontal line indicates the mean titer.

Table 1 Peripheral Blood Lymphocyte Stimulation Assay

	S.I.	
Stimulation	patients (n = 41)	controls (n = 14)
Pokeweed mitogen	116 ± 98	101 ± 79
Streptokinase-streptodornase	82 ± 85	101 ± 117
Candida albicans	149 ± 128	126 ± 65
Tuberculin	14 ± 32	24 ± 37
Sensitin	6 ± 7	14 ± 13
Hsp 65 (1 μg/ml)	5 ± 6	12 ± 11[a]
Hsp 65 (10 μg/ml)	13 ± 113	29 ± 23[a]

Peripheral blood lymphocytes were stimulated in vitro with hsp 65 and a number of other antigens or a mitogen as indicated. The results are expressed as the mean ± SD stimulation index. [a] A significant difference ($p < 0.05$) between patients and controls.

there was a negative correlation between the proliferative T-cell response to hsp 65 and the level of anti–hsp 65 antibodies ($r = -0.44$; $p < 0.01$).

IV. DISCUSSION

The main conclusions from this study are that the human hsp 65 homologue is strongly expressed by B7-positive antigen-presenting cells in colon and ileum tissue sections from patients with Crohn's disease and that the level of anti–hsp 65 antibodies is significantly higher in patients with Crohn's disease than in controls, whereas the peripheral blood lymphocyte proliferative response to stimulation with hsp 65 in vitro is impaired in patients with Crohn's disease compared with controls. Taken together with the negative correlation between the level of anti–hsp 65 antibodies and the lymphocyte SI for hsp 65, these data suggest that a hsp 65–mediated immune response is generated in patients with Crohn's disease. Whether this hsp 65–specific immune response is involved in the pathogenesis of Crohn's disease remains an attractive hypothesis for which several observations and suggestions provide support. In the present study, it is demonstrated that the human hsp 65 homologue that is strongly expressed in the intestinal mucosa of patients with Crohn's disease, is presented by B7-positive mononuclear cells. The expression of the B7 glycoprotein on antigen-presenting cells correlates with their capacity to prime T lymphocytes. Moreover, the receptor-ligand interaction between B7 on the antigen-presenting cells and CD28 on T lymphocytes during antigen presentation is shown to be involved in the reversal of an anergic state (30). The observation that hsp 65 is presented by B7-positive cells therefore strongly suggests that sensitization to hsp 65 may develop and contribute to the loss of immune tolerance and controlled inflammation in the intestinal mucosa of patients with Crohn's disease. This activation of mucosal T-cell immunity is thought to be crucial in the development of Crohn's disease (3,15). Since self–hsp 65 is expressed in the gut resection specimens of fully established and long-lasting Crohn's disease, this inflammatory disorder is to be considered as an autoimmune disease rather than an infectious disease, at least in its chronic phase. This does, however, not exclude that initially exposure to (myco)bacterial hsp 65 from the fecal flora triggers a hsp 65–mediated immune response. The enhanced permeability of the intestinal wall that is observed in patients with Crohn's disease and that is shown to predispose to recurrence of the intestinal inflammatory lesions (31) facilitates the leakage of this antigen from the lumen to the immune reactive cells of the gut-associated lymphoepithelial tissue. The exposure of these cells to this (myco)-bacterial hsp 65 may then induce a humoral and cellular immune response that also recognizes cross-reactive epitopes on self–hsp 65.

In an earlier study, we demonstrated that exposure of human monocytes and macrophages from healthy volunteers to hsp 65 directly stimulates the release of the proinflammatory cytokines TNF-α and IL-1β. These cytokine responses are obtained in the absence of previously sensitized T lymphocytes (32). This is in agreement with another report that exposure of the monocytic leukemia cell line THP-1 to hsp 65 induces the release of TNF-α, IL-6, and IL-8 (33). Moreover, cytokines such as IFN-γ and TNF-α are shown to induce an increased production of self–hsp 65 in these cells (34). These results indicate that hsp 65 directly mediates an inflammatory reaction that may lead to an upregulation of self–hsp 65 homologues in the inflammatory lesions. This may provide a stimulus for proliferation of autoreactive T lymphocytes. This hypothesis is supported by the observation that αβ T lymphocytes from draining lymph nodes of a sterile inflammation display a proliferative response on stimulation in vitro with mycobacterial, mammalian, and human hsp 65 homologues (35). The direct proinflammatory effects of hsp 65 suggest that besides the supposed role of hsp 65 is autoimmune inflammatory disorders that is based on a cross-reactive immune response due to molecular mimicry between prokaryotic and self–hsp 65 homologues, other pathways may be involved in the pathogenesis of hsp 65–related inflammatory disorders as well.

The increased anti–hsp 65 antibody titers observed in this study are in agreement with earlier reports (24) and may be due to a mucosal challenge with either prokaryotic or self–hsp 65. Priming for hsp 65 occurs early in life, for example, by vaccination with the whole-cell pertussis vaccine (36). The anti–hsp 65 antibodies generated by this vaccine cross react with the *Escherichia coli* and human hsp 65 homologues. The enhanced permeability of the intestinal wall for luminal antigens as well as the upregulation of self–hsp 65 in the inflammatory lesions provide an explanation for the raised anti–hsp 65 antibody titers in patients with Crohn's disease.

In the present study, proliferation of peripheral blood lymphocytes of patients with Crohn's disease and healthy controls on in vitro stimulation with hsp 65 is observed. As reported earlier, T cells of PPD-negative individuals with no history of tuberculosis or BCG vaccination may display a proliferative response to hsp 65 (37). This is probably due to priming early in childhood and boosting by recurrent exposure to hsp 65 from commensal or pathogenic (myco)-bacteria. The proliferative peripheral blood lymphocyte response to hsp 65 is lower in patients with Crohn's disease than in controls. This impaired response to hsp 65 in patients with Crohn's disease is not due to an overall T-cell suppression, since the SI for the other antigens, including tuberculin and sensitin, does not differ between patients and controls. Moreover, there is no correlation between the SI for hsp 65 and the SI for tuberculin or sensitin, indicating that the observed effect is hsp 65 specific. A possible explanation for the impaired peripheral blood lymphocyte response to hsp 65 is that hsp 65–reactive T lymphocytes are trapped at the site of inflammation leading to a reduced number of hsp 65–reactive cells in the circulating pool. Similar findings are reported for other hsp 65–related disorders. Peripheral blood lymphocytes of patients with long-lasting rheumatoid arthritis do not display an enhanced proliferative response to hsp 65, whereas peripheral blood lymphocytes from patients who recently developed rheumatoid arthritis and most patients with juvenile chronic arthritis yield a raised SI for hsp 65. Moreover, synovial fluid lymphocytes from patients with long-lasting rheumatoid arthritis have a much higher SI than peripheral blood lymphocytes (38,39).

V. CONCLUSIONS

Our study provides evidence that a hsp 65–mediated immune response is mounted in patients with Crohn's disease. Taken together with the abundant expression of hsp 65 by B7-positive antigen-presenting cells in the intestinal mucosa of patients with Crohn's disease, these data support the hypothesis that hsp 65 may be involved in the initiation and maintenance of inflammation in this disease. Whether extraintestinal ("reactive") symptoms of Crohn's disease such as arthritis, uveitis, and erythema nodosum have a hsp 65–reactive etiology still is a provocative question.

ACKNOWLEDGMENTS

We greatly acknowledge Dr. R. van der Zee and Dr. W. van Eden for providing the monoclonal antibodies LK-1 and LK-2; Dr. J.D.A. van Embden for providing the recombinant mycobacterial hsp 65 and Dr. M. de Boer for providing the B7-24 monoclonal antibody. The production and distribution of recombinant hsp 65 are supported by the UNDP/World Bank/WHO special programme for research and training in tropical diseases. Part of the present work was supported by a grant of the National Fund for Scientific Research (NFWO), Brussels, Belgium. This work was done in collaboration with Geert R. D'haens, Paul Rutgeerts, and Jan L. Ceuppens from the Department of Internal Medicine and Karel Geboes from Department of Pathology, University Hospitals, K.U. Leuven, Leuven, Belgium.

REFERENCES

1. D.K. Podolsky, Inflammatory bowel disease (first of two parts), *N. Engl. J. Med. 325*:928 (1991).
2. D.K. Podolsky, Inflammatory bowel disease (second of two parts), *N. Engl. J. Med. 325*:1008 (1991).

3. S. Schreiber, A. Raedler, W.F. Stenson, et al., The role of the mucosal immune system in inflammatory bowel disease, *Gastroenterol. Clin. North Am. 21*:451 (1992).
4. T.H. Kent, R.M. Cardelli, and F.W. Stamler, Small intestinal ulcers and intestinal flora in rats given indomethacin, *Am. J. Pathol. 54*:237 (1969).
5. I. Bjarnason and T.J. Peters, Intestinal permeability, non-steroidal anti-inflammatory drug enteropathy and inflammatory bowel disease: an overview, *Gut 30*:252 (1989).
6. P. Rutgeerts, K. Geboes, M. Peeters, et al., Effect of faecal stream diversion on recurrence of Crohn's disease in the neoterminal ileum, *Lancet 338*:771 (1991).
7. J. Hermon-Taylor, Causation of Crohn's disease: the impact of clusters, *Gastroenterology 104*:643 (1993).
8. A. Elsaghier, C. Prantera, C. Moreno, et al., Antibodies to Mycobacterium paratuberculosis-specific protein antigens in Crohn's disease, *Clin. Exp. Immunology 90*:503 (1992).
9. K.J. Stainsby, J.R. Lowes, R.N. Allan, et al., Antibodies to Mycobacterium paratuberculosis and nine species of environmental mycobacteria in Crohn's disease and control subjects, *Gut 34*:371 (1993).
10. G. Gitnick, J. Collins, B. Beaman, et al., Preliminary report on isolation of Mycobacteria from patients with Crohn's disease, *Dig. Dis. Sci. 34*:925 (1989).
11. M.T. Moss, E.P. Green, M.L. Tizard, et al., Specific detection of Mycobacterium paratuberculosis by DNA hybridisation with a fragment of the insertion element IS900, *Gut 32*:395 (1991).
12. J.D. Sanderson, M.T. Moss, M.L.V. Tizard, et al., Mycobacterium paratuberculosis DNA in Crohn's disease tissue, *Gut 33*:890 (1992).
13. S. Wall, Z.M. Kunze, S. Saboor, et al., Identification of spheroplast-like agents isolated from tissues of patients with Crohn's disease and control tissues by polymerase chain reaction, *J. Clin. Microbiol. 31*:1241 (1993).
14. H.H. Yohsimura, D.Y. Graham, M.K. Estes, et al., Investigation of association of mycobacteria with inflammatory bowel disease by nucleic acid hybridization, *J. Clin. Microbiol. 25*:45 (1987).
15. R.P. MacDermott and W.F. Stenson, Alterations of the immune system in ulcerative colitis and Crohn's disease, *Adv. Immunol. 42*:285 (1988).
16. L. Mayer and D. Eisenhardt, Lack of induction of suppressor T cells by intestinal epithelial cells from patients with inflammatory bowel disease, *J. Clin. Invest. 86*:1255 (1990).
17. M. Zeitz, W.C. Greene, N.J. Peffer, et al., Lymphocytes isolated from the intestinal lamina propria of normal nonhuman primates have increased expression of genes associated with T-cell activation, *Gastroenterology 94*:647 (1988).
18. E. Breese, C.P. Braegger, C.J. Corrigan, et al., Interleukin-2- and interferon-γ-secreting T cells in normal and diseased human intestinal mucosa, *Immunology 78*:127 (1993).
19. H.C. Reinecker, M. Steffen, T. Witthoeft, et al., Enhanced secretion of tumour necrosis factor-alpha, IL-6 and IL-1β by isolated lamina propria mononuclear cells from patients with ulcerative colitis and Crohn's disease, *Clin. Exp. Immunol. 94*:174 (1993).

20. K.W. Beagley and C.O. Elson, Cells and cytokines in mucosal immunity and inflammation, *Gastroenterol. Clin. North Am. 21*:347 (1992).
21. J.R. Lamb, V. Bal, P. Mendez-Samperio, et al., Stress proteins may provide a link between the immune response to infection and autoimmunity, *Int. Immunol. 1*:191 (1989).
22. J.B. Winfield and W.N. Jarjour, Stress proteins, autoimmunity and autoimmune disease, *Curr. Top. Microbiol. Immunol. 167*:161 (1991).
23. P. Res, J. Thole and R. de Vries, Heat-shock proteins and autoimmunity in humans, *Springer Semin. Immunopathol. 13*:81 (1991).
24. A. Elsaghier, C. Prantera, G. Bothamley, et al., Disease association of antibodies to human and mycobacterial hsp70 and hsp60 stress proteins, *Clin. Exp. Immunol. 89*:305 (1992).
25. T.J.R. Stevens, V.R. Winrow, D. R. Blake, et al., Circulating antibodies to heat-shock protein 60 in Crohn's disease and ulcerative colitis, *Clin. Exp. Immunol. 90*:271 (1992).
26. U. Pirzer, A. Schönhaar, B. Fleischer, et al., Reactivity of infiltrating T lymphocytes with microbial antigens in Crohn's disease, *Lancet 338*:1238 (1991).
27. C.J.P. Boog, E.R. de Graeff-Meeder, M.A. Lucassen, et al., Two monoclonal antibodies generated against human Hsp60 show reactivity with synovial membranes of patients with juvenile chronic arthritis, *J. Exp. Med. 175*:1805 (1992).
28. P. Vandenberghe, J. Delabie, M. de Boer, et al., In situ expression of B7/BB1 on antigen-presenting cells and activated B cells: an immunohistochemical study, *Int. Immunol. 5*:317 (1993).
29. M. de Boer, P. Parren, J. Dove, et al., Functional characterization of a novel anti-B7 monoclonal antibody, Eur. J. Immunol. 22:3071 (1992).
30. P.S. Linsley and J.A. Ledbetter, The role of the CD28 receptor during T cell responses to antigen, *Annu. Rev. Immunol. 11*:191 (1993).
31. J. Wyatt, J. Vogelsang, W. Hübl, et al., Intestinal permeability and the prediction of relapse in Crohn's disease, *Lancet 341*:1437 (1993).
32. W.E. Peetermans, C.J.I. Raats, J.A.M. Langermans, et al., Mycobacterial heat-shock proteins 65 induces proinflammatory cytokines, but does not activate human mononuclear phagocytes, *Scand. J. Immunol. 39*: 613 (1994).
33. J.S. Friedland, R. Shattock, D.G. Remick, et al., Mycobacterial 65-kD heat-shock protein induces release of proinflammatory cytokines from human monocyt cells, *Clin. Exp. Immunol. 91*:58 (1993).
34. M.T. Ferm, K. Söderström, S. Jindal, et al., Induction of human Hsp60 expression in monocytic cell lines, *Int. Immunol. 4*:305 (1992).
35. S.M. Anderton, R. Van der Zee, and J.A. Goodacre, Inflammation activates self Hsp 60–specific T-cells, *Eur. J. Immunol. 23*:33 (1993).
36. G. Del Giudice, A. Gervaix, P. Costantino, et al., Priming to heat-shock protein in infants vaccinated against pertussis, *J. Immunol. 150*:2025 (1993).
37. M.E. Munk, B. Schoel, and S.H.E. Kaufmann, T cell responses of normal individuals towards recombinant protein antigens of Mycobacterium tuberculosis, *Eur. J. Immunol. 18*:1835 (1988).

38. P.C.M. Res, C.G. Schaar, F.C. Breedveld, et al., Synovial fluid T cell reactivity against 65 kD heat-shock protein of mycobacteria in early chronic arthritis, *Lancet* *2*:478 (1988).
39. E.R. Degraeff-Meeder, R. van der Zee, G.T. Rijkers, et al., Recognition of human 65 kD heat-shock protein by mononuclear cells from patients with juvenile chronic arthritis, *Lancet 337*:1368 (1991).

13

Stress Proteins in Multiple Sclerosis and Other Central Nervous System Diseases

Gary Birnbaum
University of Minnesota, School of Medicine, Minneapolis, Minnesota

I. INTRODUCTION

Descriptions of stress proteins and their functions are set forth elsewhere in this book. For the purposes of this chapter, I have categorized stress proteins in the nervous system in two ways:

1. Those expressed in the central nervous system (CNS) under normal conditions
2. Those present, or increased, in the abnormal CNS

Defining the roles of stress proteins in the normal and abnormal CNS is complicated by the large numbers of different cell types present in this organ system. Neurons, astrocytes, oligodendrocytes, microglial cells, endothelial cells, and choroidal cells are sufficiently different from one another, both functionally and metabolically, to preclude extrapolation of results from one cell type to the other. Superimposed on this complexity are the significant regional differences that exist within the CNS as they relate to differences in function and differences in the patterns of constitutive stress protein expression. With these caveats in mind, I will proceed.

II. STRESS PROTEINS IN THE NORMAL CNS

Most cells within the normal CNS express constitutive forms of stress proteins. This was demonstrated in our laboratory using a panel of 20 monoclonal antibodies with specificities either for the 65-kDa stress protein of mycobacteria, the human 70/72–kDa stress protein, or the human 60-kDa stress protein. A rabbit polyvalent antibody to the *Escherichia coli* stress proteins was also used. These antibodies were used to immunocytochemically stain normal CNS (1). Patterns of staining varied with the particular antibody. Most monoclonals to the 65-kDa mycobacterial stress protein stained neuronal, glial, and microglial cell bodies. Similar staining patterns were seen with the polyvalent rabbit antiserum. Little, if any, staining was noted of oligodendrocytes with any of the monoclonals tested. Three monoclonals bound to normal central and peripheral nervous system myelin. One such monoclonal was IH9, specific for an epitope of the *Mycobacterium leprae* 65-kDa stress protein. Another monoclonal was II-13, specific for an epitope on the human 60-kDa stress protein. The third monoclonal was N27F3-4, specific for the both the constitutive and inducible form of the human 70/72–kDa stress protein. We could not determine, using our immunocytochemical method, whether these monoclonals bound to myelin epitopes that cross reacted with stress proteins or with stress proteins per se within the myelin sheaths.

III. STRESS PROTEINS DURING DEVELOPMENT OF THE NERVOUS SYSTEM

Increasing evidence suggests that stress proteins may play an important role during nervous system development. This was demonstrated in both drosophila and mammalian systems (2–5). In drosophila, (2,3), mRNA for heat shock proteins (hsps) 23 and 26 were constitutively expressed during particular stages of larval development, mainly in neurocytes and thoracic ganglion cells. In rat embryos (4,5), severe heat shock resulted in marked deformities in the forebrain and eyes. Less severe heat shock resulted in lengthening of neuroectoderm cell cycle times. Both these phenomena were associated with increased levels of mRNA of hsps 27, 71, 73, and 88. Different types of hsp mRNA were induced during different times in the cell cycles of the neuroectoderm, suggesting that stress proteins may play a role in regulating cell cycle during the ontogeny of the CNS.

IV. STRESS PROTEINS AS MARKERS OF NERVOUS SYSTEM INJURY

Stress protein expression has been used as a marker to define the extent of injury caused by particular experimental or natural disease processes. For example, it was used to define regions of ischemia (3,6–8), hyperthermia (8–12), and trauma (8,13) in several experimental systems. It was also used to identify injured cells in patients with Alzheimer's disease (10,14–16) and other neurodegenerative diseases (10,14,16). Expression of stress proteins in affected cells frequently occurred in the absence of other anatomical changes, suggesting that stress protein expression may be a sensitive marker of cell injury. The functions of stress proteins in injured nervous system are not clear, but they may play a protective role. For example, Lowenstein and coworkers (17) noted that induction of hsp 72 in cultured rat cerebellar granular cells protected them from the damaging effects of the excitotoxin glutamate.

V. STRESS PROTEINS AS TARGETS OF IMMUNE RESPONSES

Two characteristics of stress proteins greatly increase the likelihood that these proteins may become autoantigens involved either in initiating or perpetuating autoimmune diseases. First is their great phylogenetic conservation across major species barriers from prokaryotes to mammals. The second is their immunodominance; that is, their ability to evoke strong immune responses during the course of infections with a large number of infectious agents (detailed elsewhere in this book). Thus, exposure to an infectious agent, in a genetically susceptible host, at the appropriate time during the host's development could result in an immune response to the infecting agent's stress proteins that either cross reacts with normal host stress proteins or cross reacts with organ-specific proteins, resulting in an autoimmune disease. Several laboratories tested the feasibility of this hypothesis in the human disease multiple sclerosis and in its animal model, experimental autoimmune encephalomyelitis (EAE).

A. Multiple Sclerosis

Multiple sclerosis (MS) is a chronic, often progressive, inflammatory demyelinating disease of the CNS. Large numbers of immunocompetent cells are present in brain and spinal fluid and concentrations of immunoglobulins also are increased in these compartments. The target of the inflammatory response is presumed to be the oligodendrocyte-myelin complex. As a result, there is extensive, multifocal demyelination, little in the way of remyelination, and frequent astrocytic proliferation.

The evidence that MS is an autoimmune disease, or for that matter, an immunologically mediated disease, is circumstantial. Numerous laboratories tried to find immune responses to myelin and other antigens that were unique to patients with MS, but none succeeded. Persons with MS do have increased immune responses to a variety of myelin antigens, including myelin basic protein, proteolipid protein, and myelin/oligodendrocyte glycoprotein (18–24), and several of these are potent autoantigens capable of inducing EAE (25,26). However, it is not clear whether the heightened immune responses to these antigens are primary or secondary to the disease process.

A well-documented clinical phenomenon in MS is the association between disease activity and infections (27; personal observations). The study by Sibley and associates showed a clear association between antecedent infections, usually viral, and attacks or worsening of disease. Although antecedent symptoms of MS will worsen during the acute phase of an infection, especially if there is an associated fever, true disease exacerbations occur 3–14 days after the acute infection during recovery.

There are several mechanisms whereby infections could cause exacerbations of an inflammatory, possibly autoimmune, disease:

1. Infections result in increased concentrations of circulating cytokines. Some of these, such as tumor necrosis factor or interferon gamma, in addition to having direct effects on oligodendrocytes, could alter the blood brain-barrier, allowing easier entry of myelin-specific immunocompetent cells and/or antibodies. Cytokines could also nonspecifically activate antimyelin T cells resident within the CNS of patients with MS.
2. Infections result in immune responses to the infectious organism's stress proteins. These immune responses could cross react with phylogenetically conserved stress proteins expressed by oligodendrocytes in areas of inflammation.
3. Immune responses to an infectious agent's stress proteins could cross react with myelin antigens, resulting in myelin destruction.
4. A combination of above effects may be important.

Different approaches were used to determine whether stress proteins, or immune responses to stress proteins, are involved in the pathogenesis of MS. These included studies on

1. Expression of stress proteins in the brains of persons with MS
2. Distribution of $\gamma\delta$ T lymphocytes in MS brains
3. Immune responses to stress proteins in patients with MS
4. Studies on cross reactivity between stress proteins and CNS myelin

1. *Expression of Stress Proteins in MS CNS*

Several laboratories studied expression of stress proteins in areas of MS demyelination. Selmaj and his associates (28,29) demonstrated expression of hsp 65 in immature oligodendrocytes at the edges of MS plaques. There also was constitutive expression of hsp 72 in MS and non–MS brain, especially in astrocytes. Wucherpfenning et al. (30) demonstrated expression of hsp 60 in foamy macrophages at the edges of acute plaques and expression of hsp 90 in reactive astrocytes. As noted above, we utilized a panel of 20 monoclonal and polyvalent antibodies to human and mycobacterial stress proteins to immunocytochemically stain fixed, frozen sections of normal and MS brains (1). Patterns of staining varied among antibodies. Some antibodies stained cell bodies of neurons and astrocytes, others stained neurofilaments, and still others stained normal myelin. No increased staining was noted at the edges of either acute or chronic plaques and demyelinated areas did not stain at all.

Using a different approach, a number of laboratories studied the expression of stress proteins by glial cells in vitro. Again, results varied with the techniques used and the antibodies used for the assays. Selmaj et al. (29), Freedman et al. (31), and Satoh et al. (32) identified oligodendrocytes as cells expressing stress proteins. Selmaj et al. detected constitutive hsp 65 expression in oligodendrocytes but not astrocytes, yet noted some hsp 70/72 expression in astrocytes. Satoh et al. detected constitutive expression of hsp 60 in cultured murine oligodendrocytes and "marginally detectable" levels in most astrocytes. Expression of hsp 60 was increased in oligodendrocytes following heat stress. Freedman et al. detected constitutive expression of hsp 60 and hsp 70/72 expression in oligodendrocytes with increased expression of hsp 70/72 following heat stress. No astrocytes expressed hsp 70/72, and only a small percentage of astrocytes expressed hsp 60 proteins. In contrast, Marini and coworkers (11), using tissue cultures of rat astrocytes and neurons, demonstrated heat-inducible expression of hsp 70 stress proteins predominantly in astrocytes, but oligodendrocytes may have been absent from these cultures. Using purified cultures of rat astrocytes, Dwyer et al. (33) demonstrated that such cells synthesized stress proteins of 30–34, 68, 70, 89, and 97 kDa and that exposure to heat readily induced expression of hsp 65.

Since MS is an inflammatory disease, concentrations of cytokines are increased in MS brain (34,35). De Souza et al. (36) studied the effects of cytokines on the expression of stress proteins in mixed cultures of human glial cells (33, 34). A mixture of cytokines induced expression of hsp 72, predominantly in oligodendrocytes. The specific cytokines involved in this induction were interleukin-1 (IL-1), interferon gamma (INF-γ), and tumor necrosis factor alpha (TNF-α). Birnbaum et al. (manuscript in preparation) prepared cultures of purified murine astrocytes and exposed them to heat shock with or without exposure to a mixture of cytokines. Cytokines alone induced small amounts of hsp 70/72. However, the combination of heat shock and cytokine exposure augmented expression of hsp 70/72 2- to 11-fold. The particular cytokines involved in this were INF-γ and TNF-α.

It is apparent from the above data, depending on the specificities of the antibodies used and the particular methods, results can vary widely. Nevertheless, there is good evidence that some antibodies to stress proteins bind to normal myelin, whereas others bind to oligodendrocytes or astrocytes. In addition, the inflammatory milieu within MS brains increases the expression of stress proteins within glia, increasing the possibility that immune responses to infectious agents' stress proteins could cross react with their human homologues.

2. *The Distribution of γ/δ T Cells in MS Brains*

Although the vast majority of T cells responding to stress proteins express antigen-specific receptors composed of α and β chains, their numbers comprise only a small fraction of the total α/β–expressing T-cell pool. In contrast is the population of T cells expressing antigen-specific receptors composed of γ and δ chains. These cells comprise only about 1–5% of the total T-cell pool, yet a majority of this cells respond to stress proteins (37–41). Investigators, therefore, studied the association between areas of MS demyelination, expression of stress proteins, and the distribution of γ/δ T cells. Their assumption was that the presence of γ/δ T cells could indicate the presence of an immune response to stress proteins in that region.

Selmaj et al. (28) and Wucherpfenning et al. (30) found accumulations of γ/δ T cells in areas of MS demyelination. Selmaj et al. found them in regions of chronic demyelination in proximity to immature oligodendrocytes that expressed hsp 65. Wucherpfenning et al. found γ/δ T cells in regions of acute demyelination. The 60- and 90-kDa stress proteins were found in regions of normal myelin but were "overexpressed" in acute MS plaques. Cells expressing these stress proteins were foamy macrophages and reactive astrocytes, respectively. On the basis of sequence analyses, Wucherpfenning postulated there were oligoclonal expansions of γ/δ cells in MS brains, but since the repertoire of T-cell receptors in this population is already restricted, these data do not necessarily indicate reactivity to a single antigen. Hvas and coworkers (42) looked for the presence of γ/δ cells in MS brains utilizing the technique of polymerase chain reaction amplification of T-cell receptor mRNA. They found γ/δ messages in all of their 12 MS brains, in one of their 10 nonneurological disease controls, and in 2 of their three neurological controls. Sequence analyses of the MS brain–derived cDNA failed to show evidence of clonal expansions of particular populations of γ/δ cells. Two groups studied γ/δ T cells in MS and control spinal fluids. Shimonkevitz et al. (43) found increased numbers of γ/δ cells in the spinal fluid of patients with MS with recent onset of disease but not in patients with chronic MS or in disease controls. Sequence analyses of the junctional regions of these cells' antigen receptors suggested an oligoclonal expansion of this population, perhaps in response to a particular antigen. Perella et al. (44) found equivalent numbers of γ/δ T cells in MS and other neurological disease (OND) spinal fluids, and did not think that γ/δ cells had a unique role to play in the pathogenesis of MS.

Since γ/δ T cells are present in brains of MS and other neurological disease brains and spinal fluids, their presence per se is not disease specific. However, their presence does suggest an in situ immune response to stress proteins. There may be disease specificity in these responses, since it is possible that the antigens stimulating these cells may be different in persons with MS than in persons with OND.

3. *Immune Responses to Stress Proteins in Patients with MS and Controls*

Work in this area can be divided into studies involving cellular immune responses and those involving humoral responses to stress proteins.

Two groups that described cellular immune responses to heat shock proteins in persons with MS and OND are Salvetti et al. (45) and our laboratory (46). Salvetti et al. studied peripheral blood T-cell proliferative responses to recombinant hsp 65 and hsp 70 from *M. bovis* in 31 persons with MS, 19 individuals with OND, and 19 normal controls. Proliferative responses to hsp 70 were significantly more frequent in persons with MS compared with OND and healthy controls. Responses to hsp 65 were equivalent in the three groups. Lines of T cells were established from 10 patients with MS and 12 healthy controls using PPD as the antigen. Again, hsp 70–reactive lines were significantly more common in patients with MS than in healthy controls. Interestingly, cytofluorometric analyses of PPD-responsive lines revealed that only a minority of responding cells expressed γ/δ T-cell receptors. Our laboratory studied T-cell proliferative responses to mycobacterial stress proteins, tetanus toxoid, and recombinant hsp 65 from *M. leprae*. The T cells were concurrently collected from the peripheral bloods and spinal fluids of 20 persons with MS and 9 persons with inflammatory neurological diseases other than MS. Cells were cultured in vitro and stimulated with the above antigens. Significantly increased spinal fluid lymphocyte proliferative responses to mycobacterial sonicate, relative to responses from paired peripheral blood lymphocytes, were present in 14 of the 20 specimens from patients with MS compared with 2 of 9 specimens from patients with other neurological diseases ($p < 0.025$). Spinal fluid lymphocytes also responded to tetanus toxoid, but differences between blood and spinal fluid were not statistically significant. Lymphocytes from one patient with MS responded only to recombinant hsp 65. When patients with MS were classified according to duration of disease, 9 of 10 with duration less than 2 years had spinal fluid T cells responding to *M. tuberculosis* compared with 5 of 10 with disease longer than 2 years ($p < 0.012$). These data supplement the observations of Shimonkevitz et al. (43), who described increased numbers of activated γ/δ cells predominantly in spinal fluids from persons with recent-onset MS.

Additional data in support of the hypothesis that immune responses to stress proteins play a role in autoimmune demyelination are the observations of Mor et al. (47). These investigators prepared T cells from the spinal cords, blood, spleen, and lymph nodes of rats during the acute phase of EAE or during recovery from EAE. Using limiting dilution analyses, they determined the frequency of T-cell responses to the myelin protein, myelin basic protein (MBP), and recombinant hsp 65 and hsp 70. As expected, responses to MBP were enriched in the spinal cords of rats during and after acute EAE. However, there was also enrichment of T cells responsive to hsp 65. T-cell lines established from spinal cord lymphocytes responded to MBP, hsp 65, and hsp 70. When EAE was induced from an anti-MBP–responsive T-cell line, similar patterns of enrichment for MBP and hsp 65–reactive T cells were noted, indicating that responses to stress proteins occurred in the absence of exposure to adjuvant mycobacteria.

Studies of humoral immune responses to stress proteins in MS are mainly the work of in our laboratory (48,49) and Freedman et al. (50). We used immunoblots to detect antibodies to native and recombinant mycobacterial stress proteins and to bacterial stress proteins in spinal fluids and paired sera from persons with MS and OND. Antibodies to many stress proteins, including those of the 60-kDa and 70-kDa families, were present in CSF and sera from all patient groups. Patterns of antibodies varied between CSF and sera and between patients, but no disease-distinctive pattern was seen. When anti–stress protein antibodies were analyzed for isotypes, patients with MS had higher concentrations of anti–stress protein immunoglobulin A (IgA) antibodies than did patients with OND. This suggested an in situ synthesis of such antibodies within the CNS in persons with MS. Freedman et al. studied antibody concentrations to recombinant hsp 60 using an ELISA. Titers of antibodies in MS spinal fluids were significantly higher than those seen in persons with OND, and the higher titers correlated with the presence of oligoclonal bands in MS CSF but not OND CSF.

4. *Studies on Cross Reactivity Between Stress Proteins and CNS Myelin*

As noted above, certain anti–stress antibodies stain normal myelin. We studied this phenomenon using immunoblots (Birnbaum, et al., submitted for publication). A panel of 20 anti–stress protein antibodies was assayed for the ability to bind to myelin proteins separated on SDS-PAGE. Three monoclonal antibodies specific for either mycobacterial or human stress proteins stained bands of normal myelin proteins. One murine monoclonal to *M. leprae* hsp 65, IIH9, stained a 44- to 46-kDa doublet. This doublet was the same size as the myelin protein 2′, 3′-cyclic 3′-nucleotide phosphodiesterase (CNP). To study this observation further, purified CNP was used in immunoblots. IIH9 bound to this protein in a pattern identical to that seen with whole myelin. A nonapeptide region of sequence homology was identified between the epitope of hsp 65 recognized by IIH9 and CNP. This peptide was synthesized and used in immunoblots. IIH9 strongly bound to this peptide, proving that this region of sequence homology was responsible for the observed cross-reactivity.

In a more circumspect manner, data from several other laboratories suggest that cross reactivity occurs between stress proteins and myelin. In 1975, Wisniewski and Bloom (51) noted primary demyelination in the brains of guinea sensitized to tuberculin, who were then subsequently challenged with PPD intracranially. The investigators interpreted these data to indicate that demyelination was the result of a passive bystander effect induced by the localized delayed hypersensitivity response. In retrospect, an alternative explanation is that demyelination occurred because of a specific cross reactivity between myelin and PPD. Additional observations that support this alternative conclusion come from several different laboratories (52–55), all of them demonstrating that exposure of EAE susceptible animals to either *Bordetella pertussis* or *M. tuberculosis* rendered them highly resistant to subsequent development of EAE.

It thus appears that at both the humoral and cellular level there is cross reactivity between infectious agents' stress proteins and normal myelin components.

VI. SUMMARY

Stress or heat shock proteins are constitutively expressed in normal central nervous system tissues in a variety of cell types (oligodendrocytes, astrocytes, and neurons). Their function is uncertain, but they may be critical during nervous system development and may protect cells from various stresses, such as hypoxia, anoxia, and excessive excitatory stimulation.Increased amounts of stress proteins are expressed in various cells of the CNS during acute toxic-metabolic states and in more chronic degenerative diseases. Increased expression of stress proteins may constitute a sensitive marker of cell injury.

Antibodies to mycobacterial stress proteins bind to normal human myelin and to oligodendrocytes in regions of MS demyelination. Some proteins in myelin that cross react with stress proteins are not human homologues. In at least one instance, it is an endogenous myelin protein, namely CNP. Cellular immune responses to stress proteins occur with increased frequency and magnitude in persons with MS, especially those with recent onset of disease. In addition, there are populations of T cells expressing γ/δ T cells in the brains and spinal fluids of persons with MS. These data support the hypothesis that an immune response to an infectious agent's stress proteins could result in a cross-reactive immune response to CNS myelin, and result in demyelination. This may be an especially important mechanism in MS cases of recent onset.

REFERENCES

1. G. Birnbaum, H.B. Clark, and D. Psihos, Expression of heat shock proteins in the brains of patients with multiple sclerosis and other neurological diseases, *Ann. Neurol. 30*:305 (1991).
2. R. Marin, J.P. Valet, and R.M. Tanguay, hsp23 and hsp26 exhibit distinct spatial and temporal patterns of constitutive expression of Drosophila adults, *Dev. Genet. 14*:69 (1993).
3. M. Chopp, The roles of heat shock proteins and immediate early genes in central nervous system normal function and pathology. *Curr. Opin. Neurol. Neurosurg. 6*:6 (1993).
4. D.A. Walsh, K. Li, J. Speirs, C.E. Crowther, and M.J. Edwards, Regulation of the inducible heat shock 71 genes in early neural development of cultured rat embryos, *Teratology 40*:321 (1989).
5. D. Walsh, K. Li, J. Wass, A. Dolnikov, F. Zeng, L. Zhe, and M. Edwards, Heat-shock gene expression and cell cycle changes during mammalian embryonic development, *Dev. Genet. 14*:127 (1993).
6. K. Vass, W.J. Welch, and T.S.J. Nowak, Localization of 70-kDa stress protein induction in gerbil brain after ischemia, *Acta Neuropathol (Berl). 77*:128 (1988).
7. D.M. Ferriero, H.Q. Soberano, R.P. Simon, and F.R. Sharp, Hypoxia-ischemia induces heat shock protein-like (HSP72) immunoreactivity in neonatal rat brain, *Brain Res. Dev. Brain Res. 53*:145 (1990).
8. J.R. Brown, Induction of heat shock (stress) genes in the mammalian brain by hyperthermia and other traumatic events: a current perspective, *J. Neurosci. Res. 27*:247 (1990).
9. I.R. Brown, and S.J. Rush, Expression of heat shock genes (hsp70) in the mammalian brain: distinguishing constitutively expressed and hyperthermia-inducible mRNA species, *J. Neurosci. Res. 25*:14 (1990).
10. P.J. Harrison, A.W. Procter, T. Exworthy, G.W. Roberts, A. Najlerahim, A.J. Barton, and R.C. Pearson, Heat shock protein (hsx70) mRNA expression in human brain: effects of neurodegenerative disease and agonal state, *Neuropathol. Appl. Neurobiol. 19*:10 (1993).
11. A.M. Marini, M. Kozuka, R.H. Lipsky, and T.S.J. Nowak, 70-Kilodalton heat shock protein induction in cerebellar astrocytes and cerebellar granule cells in vitro: comparison with immunocytochemical localization after hyperthermia in vivo, *J. Neurochem. 54*:1509 (1990).
12. G.K. Sprang and I.R. Brown, Selective induction of a heat shock gene in fibre tracts and cerebellar neurons of the rabbit brain detected by in situ hybridization, *Brain Res. 427*:89 (1987).
13. Z.Y. Xue and R.M. Grossfeld, Stress protein synthesis and accumulation after traumatic injury of crayfish CNS. *Neurochem. Res. 18*:209 (1993).
14. Y. Namba, M. Tomonaga, K. Ohtsuka, M. Oda, K. Ikeda, HSP 70 is associated with abnormal cytoplasmic inclusions characteristic of neurodegenerative diseases, *No To Shinkei 43*:57 (1991).
15. J.E. Hamos, B. Oblas, S.D. Pulaski, W.J. Welch, D.G. Bole, and D.A. Drachman, Expression of heat shock proteins in Alzheimer's disease, *Neurology 41*:345 (1991).

16. S. Cisse, G. Perry, R.G. Lacoste, T. Cabana, and D. Gauvreru, Immunochemical identification of ubiquitin and heat-shock proteins in corpora amylacea from normal aged and Alzheimer's disease brains, *Acta Neuropathol. (Berl). 85*:233 (1993).
17. D.H. Lowenstein, P.H. Chan, and M.F. Miles, The stress protein response in cultured neurons: characterization and evidence for a protective role in excitotoxicity, *Neuron 7*:1053 (1991).
18. M. Allegretta, J.A. Nicklas, S. Sriram, and R.J. Albertini, T cells responsive to myelin basic protein in patients with multiple sclerosis, *Science 247*:718 (1990).
19. C.N. Baxevanis, G.J. Reclos, C. Servis, E. Anastasopoulos, P. Arsenis, A. Katsiyiannis, N. Matikas, J.D. Lambris, and M. Papamichail, Peptides of myelin basic protein stimulate T lymphocytes from patients with multiple sclerosis, *J. Neuroimmunol. 22*:23 (1989).
20. D.A. Hafler, D.S. Benjamin, J. Burks, and H.L. Weiner, Myelin basic protein and proteolipid protein reactivity of brain- and cerebrospinal fluid-derived T cell clones in multiple sclerosis and postinfectious encephalomyelitis, *J. Immunol. 139*:68 (1987).
21. D. Johnson, D.A. Hafler, R.J. Fallis, M.B. Lees, R.O. Brady, R.H. Quarles, and H.L. Weiner, Cell-mediated immunity to myelin-associated glycoprotein, proteolipid protein, and myelin basic protein in multiple sclerosis, *J. Neuroimmunol. 13*:99 (1986).
22. R. Liblau, L.E. Tournier, J. Maciazek, G. Dumas, O. Siffert, G. Hashim, and M.A. Bach, T cell response to myelin basic protein epitopes in multiple sclerosis patients and healthy subjects, *Eur. J. Immunol. 21*:1391 (1991).
23. R.P. Lisak and B. Zweiman, In vitro cell-mediated immunity of cerebrospinal-fluid lymphocytes to myelin basic protein in primary demyelinating diseases, *N. Engl. J. Med. 297*:850 (1991).
24. M. Pette, K. Fujita, B. Kitze, J.N. Whitaker, E. Albert, L. Kappos, and H. Wekerle, Myelin basic protein-specific T lymphocyte lines from MS patients and healthy individuals, *Neurology 40*:1770 (1990).
25. R.B. Fritz, C.H. Chou, and D.E. McFarlin, Relapsing murine experimental allergic encephalomyelitis induced by myelin basic protein, *J. Immunol. 130*:1024 (1983).
26. V.K. Tuohy, R.A. Sobel, and M.B. Lees, Myelin proteolipid protein-induced experimental allergic encephalomyelitis. Variations of disease expression in different strains of mice, *J. Immunol. 140*:1868 (1988).
27. W.A. Sibley, C.R. Bamford, and K. Clark, Clinical viral infections and multiple sclerosis, *Lancet. 1*:1313 (1985).
28. K. Selmaj, C.F. Brosnan, and C.S. Raine, Colocalization of lymphocytes bearing gamma delta T-cell receptor and heat shock protein hsp65+ oligodendrocytes in multiple sclerosis, *Proc. Natl. Acad. Sci. USA 88*:6452 (1991).
29. K. Selmaj, C.F. Brosnan, and C.S. Raine, Expression of heat shock protein–65 by oligodendrocytes in vivo and in vitro: implications for multiple sclerosis, *Neurology 42*:795 (1992).
30. K.W. Wucherpfennig, J. Newcombe, H. Li, C. Keddy, M.L. Cuzner, and D.A. Hafler, Gamma delta T-cell receptor repertoire in acute multiple sclerosis lesions, *Proc. Natl. Acad. Sci. USA 89*:4588 (1992).

31. M.S. Freedman, N.N. Buu, T.C. Ruijs, K. Williams, and J.P. Antel, Differential expression of heat shock proteins by human glial cells, *J. Neuroimmunol. 41*:231 (1992).
32. J. Satoh, H. Nomaguchi, and T. Tabira, Constitutive expression of 65-kDa heat shock protein (HSP65)–like immunoreactivity in cultured mouse oligodendrocytes, *Brain Res. 595*:281 (1992).
33. B.E. Dwyer, R.N. Nishimura, J. de Velbis, and K.B. Clegg, Regulation of heat shock protein synthesis in rat astrocytes, *J. Neurosci. Res. 28*:352 (1991).
34. K. Selmaj, C.S. Raine, B. Cannella, and C.F. Brosnan, Identification of lymphotoxin and tumor necrosis factor in multiple sclerosis lesions, *J. Clin. Invest. 87*:949 (1991).
35. F.M. Hofman, D.R. Hinton, K. Johnson, and J.E. Merrill, Tumor necrosis factor identified in multiple sclerosis brain, *J. Exp. Med. 170*:607 (1989).
36. S.D. DeSouza, J.P. Antel, and M.S. Freedman, Cytokine induction of heat shock protein expression in human oligodendrocytes: an interleukin-1–mediated mechanism, *J. Neuroimmunol. 50*:17 (1994).
37. W. Born, M.P. Happ, A. Dallas, C. Reardon, R. Kubo, T. Shinnick, P. Brennan, R. OBrien, Recognition of heat shock proteins and gamma delta cell function, *Immunol Today 11*:40 (1990).
38. J.A. Bluestone, R.Q. Cron, T.A. Barrett, B. Houlden, A.I. Sperling, A. Dent, S. Hedrick, B. Rellahan, and L.A. Matis, Repertoire development and ligand specificity of murine TCR gd cells, *Immunol. Rev. 120*:5 (1991).
39. A. Haregewoin, G. Soman, R.C. Hom, and R.W. Finberg, Human gamma delta+ T cells respond to mycobacterial heat-shock protein, *Nature 340*:309 (1989).
40. D. Kabelitz, A. Bender, S. Schondelmaier, B. Schoel, and S.H. Kaufmann, A large fraction of human peripheral blood gamma/delta + T cells is activated by Mycobacterium tuberculosis but not by its 65-kD heat shock protein, *J. Exp. Med. 171*: 667 (1990).
41. K. Pfeffer, B. Schoel, H. Gulle, S.H. Kaufmann, and H. Wagner, Human gamma/ delta T cells responding to mycobacteria, *Behring Inst. Mitt. 88*:36 (1991).
42. J. Hvas, J.R. Oksenberg, R. Fernando, L. Steinman, and C.C. Bernard, Gamma delta T cell receptor repertoire in brain lesions of patients with multiple sclerosis, *J. Neuroimmunol. 46*:225 (1993).
43. R. Shimonkevitz, C. Colburn, J.A. Burnham, R.S. Murray, and B.L. Kotzin, Clonal expansions of activated gamma/delta T cells in recent-onset multiple sclerosis, *Proc. Natl. Acad. Sci. USA 90*:923 (1993).
44. O. Perrella, P.B. Carrieri, M.D. De, and G.A. Buscaino, Markers of activated T lymphocytes and T cell receptor gamma/delta + in patients with multiple sclerosis, *Eur. Neurol. 33*:152 (1993).
45. M. Salvetti, C. Buttinelli, G. Ristori, M. Carbonari, M. Cherchi, M. Fiorelli, M.G. Grasso, L. Toma, and C. Pozzilli, T-lymphocyte reactivity to the recombinant mycobacterial 65- and 70-kDa heat shock proteins in multiple sclerosis, *J. Autoimmun. 5*:691 (1992).
46. G. Birnbaum, L. Kotilinek, and L. Albrecht, Spinal fluid lymphocytes from a subgroup of multiple sclerosis patients respond to mycobacterial antigens, *Ann. Neurol. 34*:18 (1993).

47. F. Mor and I.R. Cohen, T cells in the lesion of experimental autoimmune encephalomyelitis. Enrichment for reactivities to myelin basic protein and to heat shock proteins, *J. Clin. Invest. 90*:2447 (1992).
48. G. Birnbaum and P. Schlievert, Antibodies to mycobacterial antigens in the spinal fluids and blood from patients with multiple sclerosis and other neurologic diseases, *Neurology 42*:247 (1992).
49. G. Birnbaum, and L. Kotilinek, Antibodies to 70-kd heat shock protein are present in CSF and sera from patients with multiple sclerosis, 45th Annual Meeting of the American Academy of Neurology, New York, NY, April.
50. M.S. Freedman, S. Prabhakar, R. Gupta, E. Kurien, and J.P. Antel, Cerebrospinal fluid immunoreactivity to heat shock protein is elevated in multiple sclerosis, *Ann. Neurol. 32*:259 (1992).
51. H.M. Wiesniewski and B.R. Bloom, Primary demyelination as a nonspecific consequence of a cell-mediated immune reaction, *J. Exp. Med. 141*:346 (1975).
52. A.A. Vandenbark, D.R. Burger, and R.M. Vetto, Cell-mediated immunity in experimental allergic encephalomyelitis: cross-reactivity between myelin basic protein and mycobacteria antigens, *Proc. Soc. Exp. Biol. Med. 148*:1223 (1975).
53. M. Mostarica-Stojkovic, S. Vukmanovic, M. Petrovic, Z. Ramic, and M.L. Lukic, Dissection of the adjuvant and suppressive effects of mycobacteria in experimental allergic encephalomyelitis production, *Int. Arch. Allery Appl. Immunol. 85*:82 (1988).
54. K. Hempel, A. Freitag, B. Freitag, B. Endres, B. Mai, and G. Liebaldt, Unresponsiveness to experimental allergic encephalomyelitis in Lewis rats pretreated with complete Freund's adjuvant, *Int. Arch. Allergy Appl. Immunol. 76*:193 (1985).
55. D. Lehmann and A. Ben Nun, Bacterial agents proposed against autoimmune disease. I. Mice pre-exposed to Bordetella pertussis or Mycobacterium tuberculosis are highly refractory to induction of experimental autoimmune encephalomyelitis, *J. Autoimmun. 5*:675 (1992).

14

Heat Shock Protein 60 and Insulin-Dependent Diabetes Mellitus

David B. Jones and Nigel W. Armstrong
University of Liverpool, Liverpool, England

I. INTRODUCTION

Insulin-dependent diabetes mellitus (IDDM) is the result of the complete loss of the insulin-producing pancreatic islet β cells (1). Gepts initially identified lymphocytic infiltration around the islets (insulitis) in 1965 (2). Bottazzo subsequently demonstrated that the serum of patients with polyendocrine autoimmune syndrome contained antibodies which reacted with islets (3). This finding was extended to include newly diagnosed patients with IDDM (4), establishing IDDM as an autoimmune disease. Two animal models of spontaneous IDDM have also been studied: the biobreeding (BB) rat (5), and the nonobese diabetic (NOD) mouse (6). Only the NOD mouse and human IDDM are considered in this chapter. There appear to be two components of the development of IDDM. First, a genetic predisposition which is strongly linked to the major histocompatibility complex (MHC) haplotypes DR3 and DR4 and to specific DQA and DQB alleles confers an increased risk for the development of IDDM (7). However, there is almost certainly an environmental agent(s), not yet identified conclusively, which converts the genetic predisposition to clinical disease (8). This is supported in studies of monozygotic twins, in which the concordance rate for IDDM is less than 50% (9).

In this chapter, the antigenic targets and effector mechanisms of the autoimmune attack on the β cells are described, with particular reference to the stress protein heat shock protein 60 (hsp 60) as an autoantigen.

II. MECHANISMS OF β-CELL DESTRUCTION IN IDDM

Several effector mechanisms of β-cell destruction in IDDM have been proposed. These include humoral immunity, direct T-cell involvement, and cytokine-mediated β-cell lysis.

A. Autoantibodies

Although autoantibodies to several β-cell antigens have been extensively described (10), it is now generally accepted that IDDM is not predominantly an autoantibody-mediated disease (1,11). Disease transfer in the animal models cannot be mediated by autoantibody alone. The autoantibodies described include islet cell antibodies (ICAs), (3,12), which react with intracellular components, complement-fixing antibodies (13), and islet cell surface antibodies (ICSAs) (14), which may be targeted against a cell surface glycoprotein, although there is also evidence that they may represent a high titer subset of ICAs. In addition to these uncharacterised targets, several β-cell autoantigens were initially identified on the basis of antibody reactivity, including glutamic acid decarboxylase, (GAD 65), hsp 60, and carboxypeptidase H. Although probably not of etiological significance, they have proved useful in clinical practice, in the identification of prediabetic individuals, and in the confirmation of a diagnosis of IDDM (15).

B. $CD4^+$ T Cells as Direct Effectors of β-Cell Lysis

Immunohistochemical examination of the pancreata of patients with IDDM who died early in the course of the disease revealed that the β cells aberrantly expressed class II MHC molecules (16). This led to the hypothesis that class II MHC-restricted autoreactive $CD4^+$ T cells (which have been identified against many islet cell proteins) could act as the effector cells, recognizing antigenic peptides derived from a β-cell–specific antigen. It was considered that these $CD4^+$ T cells would not have been subject to thymic tolerance mechanisms (17). This view has the attraction of unifying the known genetic susceptibility conferred by class II MHC haplotypes with experimental data concerning β-cell antigen–specific $CD4^+$ autoreactive T cells.

A transgenic mouse model in which allogeneic MHC class II expression is limited to β cells using a rat insulin promoter did not support this hypothesis (18). The mice expressed allogeneic class II MHC on the β cells but did not develop insulitis or IDDM. Potentially autoreactive $CD4^+$ T cells were rendered anergic. However, this model is an incomplete test of the hypothesis, as the β cells did not express the necessary costimulator molecules (19), such as B7 (CD80) or B7-2, and MHC class II–associated invariant chain (Ii) (20) was not included in the transgenic model.

Recent evidence in a triple transgenic model shows that expression of the B7 molecule on the β cells was critical in the activation of autoreactive T cells (21). The model used the rat insulin promoter to direct expression of a viral protein, lymphocytic choriomeningitis virus (LCMV) cell surface glycoprotein, and B7 to the β cells, and a transgenic T-cell receptor (TCR) which skews the T-cell repertoire exclusively to express a TCR specific for an epitope of the viral glycoprotein. Although the double transgenic TCR/viral protein mice required external activation in the form of infection by the LCMV (22), the triple transgenic mice spontaneously develop IDDM.

C. CD8$^+$ T Cells as Effectors

Two alternative mechanisms require consideration. The first is that the β cells are destroyed by antigen-specific MHC class I–restricted CD8$^+$ T cells. Evidence which supports this hypothesis includes the histology of the human diabetic pancreata both at postmortem (16) and in several series of grafts between genetically identical but IDDM discordant twins (23). Approximately 80% of the leukocyte infiltrate is composed of CD8$^+$ T cells.

Several studies in the NOD mouse demonstrate that IDDM can only be transferred to syngeneic disease-free littermates using both the Lyt2 (CD8) and L3T4 (CD4) T-cell subset (24). Transgenic mice with a β_2-microglobulin knockout have been crossed with IDDM-susceptible NOD mice (25). The β_2 null mutation leads to a complete absence of class I MHC expression, and CD8$^+$ T cells and is thus a useful model to assess the involvement of CD8$^+$ T cells in the autoimmune process. The NOD/β_2–/– mice have no insulitis and do not develop IDDM.

Other studies have shown that CD4$^+$ T-cell clones which respond to uncharacterized islet proteins are able to transfer the disease into CD8$^+$ T-cell depleted (26), SCID/NOD mice (27), or athymic nude/NOD mice (28) (that have no endogenous T-cell population), which argues against a role for CD8$^+$ T cells even in the effector phase.

D. CD4$^+$ T Cells and Macrophages: Autoantigen Presentation

The evidence that the development of IDDM in the NOD mouse requires macrophages is contradictory. Although macrophage depletion using silica completely prevents diabetes (29), liposome depletion does not (30). However, the importance of CD4$^+$ T cells is uniformly accepted. Transgenic mice with an Aβ null phenotype, which leads to an absence of class II MHC expression and the CD4$^+$ T-cell subset, crossed onto a NOD mice background have no insulitis and no diabetes (31). Histological analysis of the development of insulitis in NOD mice shows that in the early infiltrate, CD4$^+$ T-cells predominate, and that CD8$^+$

T cells become more common only in late insulitis (32). The same sort of longitudinal studies are clearly not possible in human IDDM, but as IDDM is a chronic disease, different areas of the pancreas at a single point in time appear to show that early and late stages of infiltration can be distinguished in the same patient. Some studies have proposed that macrophages or Langerhans cells are the earliest type of infiltrating cell in human IDDM (33), although this remains controversial.

The interaction between autoreactive $CD4^+$ T cells and professional antigen-presenting cells, presumed to be macrophages and/or Langerhans cells, requires that the autoantigen is first available for presentation. In the case of hsp 60, it is not clear whether spontaneous release from the normal location in the β cell of the insulin secretory granule occurs or presentation follows an initial β-cell insult, although redistribution of hsp 60 into the cytoplasm has been observed (34).

E. Cytokines in IDDM: Direct Effectors or Promoters of Cell-Mediated Immunity or Autoantigen Expression

Several lines of evidence suggest that cytokines have an important role in the destruction of the β cells in IDDM either by direct cytotoxicity or by a cytokine imbalance promoting other effector mechanisms. Interleukin-1β (IL-1β) is currently the main candidate for a direct effector (33). The relative sensitivity of islet cells to the cytotoxic effects of IL-1β and the high molar excess of the natural receptor antagonist required to inhibit the cytokine have been demonstrated in vitro. However, in vivo assays have shown that IL-1β administered systemically reduces the incidence of diabetes in the NOD mouse; this contrasting result is thought to be the due to corticosteroid induction effecting immunosuppression.

The role of TH1 and TH2 subsets of $CD4^+$ T cells has recently become the focus of intense investigation. T cells of the TH1 subset express IL-2, interferon gamma (IFN-γ), and tumor necrosis factor α (TNF-α), cytokines whose expression enhances the activation of macrophages, cytotoxic T cells, and natural killer cells. Conversely, TH2 $CD4^+$ T cells express IL-4, IL-10, and transforming growth factor β. In the NOD mouse, several studies have shown that the systemic administration of TH1 cytokines significantly increases insulitis and IDDM, whereas TH2 cytokines reduce the incidence of diabetes (35). Transgenic mouse models in which constitutive β-cell expression of cytokines is mediated by the rat insulin promoter generally support this view; IL-2 and TNF-α transgenic mice both develop characteristic insulitis, although β-cell destruction is only observed in mice made homozygous for the IL-2 transgene.

Our own unpublished data suggest that heat shock protein production is increased in vitro in purified human islets by several cytokines, whereas under the same conditions, GAD 65 expression appeared to be downregulated.

III. HSP 60 AS A β-CELL ANTIGEN IN IDDM

Hsp 60, previously referred to as hsp 65, is the human equivalent of the *Escherichia coli* GroEL protein and mycobacterial hsp 65 (36). It is a member of the chaperonin family of proteins, whose diverse functions include protein folding and cellular protection (37). The amino acid sequences are highly conserved between species, with greater than 65% homology between the human and mycobacterial forms (36). The hsp 65 of lower organisms is considered to be immunodominant. Following infection or immunization with mycobacteria, for example, over 40% of reactive T-cell clones respond to this single protein (38).

Hsp 60 has been implicated in several autoimmune diseases (39) because of its immunodominance and the high degree of homology between mammalian and microbial hsp 60/65. This hypothesis is referred to as molecular mimicry (40), and it incorporates immune cross reactivity between environmental microorganisms and mammalian stress proteins expressed by the target cell.

The role of hsp 60 in the autoimmune destruction of the pancreatic islet β cells in IDDM has been investigated in both the NOD mouse and human IDDM.

A. Heat Shock Proteins in the NOD Mouse

1. *Immunostimulation Reduces the Incidence of IDDM in the NOD Mouse*

Several studies have established that the incidence of IDDM in the NOD mouse can be reduced by exposure to mycobacterial preparations including complete Freund's adjuvant (killed *Mycobacterium tuberculosis* in oil emulsion) (41–43) and the bacille Calmette-Guérin vaccine (BCG, live attenuated *Mycobacterium bovis*) (44). The mode of action is unclear, but hypotheses have included nonspecific immune stimulation, activation of suppressor lymphocytes or macrophages (45), alteration of cytokine profiles from TH1 to TH2 (46), or possibly by the common presence of mycobacterial hsp 65 in the administered antigen.

2. *M. bovis Hsp 65 Cross-Reactive Antigen Identified as Target of Cellular Immune Response*

Support for this last hypothesis came in a report from Elias and coworkers which suggested that the pathogenesis of IDDM in the NOD mouse was associated with a T-cell and antibody reactivity to an antigen which was cross reactive with mycobacterial hsp 65 (47). These investigators reported that the primary event appeared to be the release into the serum of the hsp 65 cross-reactive antigen, which preceded the onset of diabetes and the other autoimmune phenomena. It was implied that the hsp 65 cross-reactive antigen was a β-cell protein. The onset of insulitis was associated with the spontaneous development of T-cell responses to hsp 65. This group reported a reduction in the incidence of IDDM in mice treated at the age of 30 days with a preparation of mycobacterial hsp 65 in saline.

At this age, insulitis is frequently observed, but clinical diabetes does not usually occur before 120 days. T-cell clones were generated against mycobacterial hsp 65, and transfer of a single hsp-65–reactive $CD4^+$ T-cell clone to young prediabetic NOD mice accelerated the onset of IDDM suggesting that the clone may have elicited an autoimmune response against a host protein. Although the implication was clearly that a heat shock protein was the target of the T-cell clones, it was postulated that the specific β-cell destruction observed in IDDM could result from the expression of a β-cell stress protein or that an unrelated protein shared sequence homology with the mycobacterial hsp 65. Immunization with mycobacterial hsp 65 in incomplete Freund's adjuvant (ICFA) lead to transient insulitis, but the incidence of IDDM was also reduced compared with the control population. This unexpected result will be discussed later.

3. *$CD4^+$ T Cells Responsive to an Immunodominant Epitope of Human Hsp 60 Transfer IDDM*

In a later paper, Elias et al. repeated a number of elements of the experiments using human hsp 60 in place of the mycobacterial hsp 60 (48). The major difference was that immunization with hsp 60 did not reduce the incidence of IDDM but instead rapidly induced hyperglycaemia. An epitope of human hsp 60 was also identified which stimulated a T-cell response in NOD mouse. This peptide, p277, comprised amino acids 437–460 and differs by only one amino acid from mouse hsp 60, but it contains more limited sequence homology with mycobacterial hsp 65. Using the T-cell clones generated against mycobacterial hsp 65, it was found that in several cases the responses to human hsp 60 were stronger, implying that this antigen was more closely related to the β-cell target protein than mycobacterial hsp 65. Three T-cell clones responded to the peptide p277, and only these clones produced insulitis and hyperglycemia in transfer experiments. The other T-cell clones, which responded more strongly to mycobacterial hsp 65 than human hsp 60 were not able to transfer insulitis. This important peptide could be used in a nonimmunogenic form to downregulate the immune response to hsp 60 in the NOD mouse, and T-cell clones recognizing this peptide could induce insulitis and hyperglycemia. Further, the same clones, when attenuated by gamma irradiation, could be used as therapeutic vaccines. This paper is of interest as it clearly shows that autoreactive T-cell clones which recognize a peptide derived from the ubiquitous hsp 60 are able to exert a β-cell–specific effect. Although the protein is expressed in most or all the cells in the body, it is only the β cells that are destroyed by the p277 reactive T-cell clones.

4. *Difference Between Mycobacterial Hsp 65 and Human Hsp 60 Immunization*

Immunization with human hsp 60 in ICFA led to hyperglycemia, whereas mycobacterial hsp 65 under the same conditions reduced the later incidence of IDDM.

The mechanism of action was unclear at the time of publication of the papers describing hsp 60 as an autoantigen in NOD mouse IDDM (47,48). It was well established that the administration of the native peptide in a nonimmunogenic form (either supraoptimal or oral route) is able to anergize the responding T-cell population, but immunization in adjuvant would have been expected to stimulate a response. However, it was recently shown that presentation of an altered peptide ligand could induce anergy in T cells even when administered in a system in which a response would normally be stimulated (49). The mycobacterial homologue of the mouse/human p277 peptide may therefore constitute an altered ligand which is able to tolerize the p277-responsive $CD4^+$ T-cell population even when administered in IFA.

5. *Disease Transfer to NON.H-2^{NOD} Mice*

The p277-responsive T-cell clones were also transferred to NON.H-2^{NOD} mice (48). This strain of mice does not develop spontaneous IDDM but shares a common MHC class II genotype. These mice, however, developed IDDM after administration of this T-cell clone. Importantly, although these mice do not spontaneously develop IDDM, immunization with human hsp 60 resulted in hyperglycemia and insulitis. Thus, activation of a population of T-cell clones which react with a host heat shock protein is the key event in the autoimmune process. In the NOD mice, this event occurs spontaneously, but in the related NON mice, the T-cell clones must be activated experimentally by immunization. At the time this paper was published, the concept of TH1 and TH2 subsets of $CD4^+$ T cells was not well established. It would be of interest to examine whether the T-cell clones were of a TH1 or TH2 phenotype.

6. *Treatment of Established IDDM in NOD Mice with p277*

More recently, the peptide p277 has been used to treat NOD mice with established insulitis in 84-, 105-, and 119-day-old mice (50). Of 30 mice treated with the control peptide, bovine serum albumin (BSA), 28 had died by 280 days. Conversely, none of 21 mice treated before 105 days with a 50-μg dose of peptide p277 died, although many exhibited hyperglycemia. Similar results were also obtained in a larger series of 157 NOD mice: Total mortality in the peptide-treated group was 3% compared with 84% in the control BSA group.

Together these reports clearly identify hsp 60 as a key autoantigen in the etiology of IDDM in the NOD mouse.

B. Hsp 60 in Human IDDM

1. *Cellular Stress Induces Increased Hsp 60 Expression in Rat Insulinoma Cells and Purified Human Islets*

Our initial experiments with a rat insulinoma (RIN) cell line showed that hsp 60 was expressed at a low level in unstressed cell. Cellular stress induced by hyperthermic incubation or by treatment with the cytokines interleukin-1β (IL-1β), IFN-γ, or TNF-α led to increased synthesis of hsp 60 following a prolonged recovery phase (Fig. 1) (51).

More recently, we have used purified human islet tissue in the same system. Hsp 60 expression was considerably higher in the unstressed islets than the insulinoma cell line, probably as a consequence of the procedure used to remove the islets from the pancreas, but again, a qualitative increase in expression of hsp 60 was observed. The expression of GAD 65 appeared to be reduced under the same conditions.

2. *Antibody Reactivity to 64-kDa Antigen is Reduced by Pretreatment with Anti-Hsp 60 Antibody*

The earliest evidence for involvement of hsp 60 in human IDDM came from our work using stressed RIN cells as a target (51). These experiments demonstrated that expression of a ~64-kDa protein was induced and that antibodies present in the serum of patients with insulin-dependent diabetes bound to this region in Western blots. Further, the binding of anti-64-kDa antibodies in the serum of newly diagnosed patients with IDDM was inhibited by the preincubation of the cell suspensions with antibodies against mycobacterial hsp 65. These experiments suggested that the 64-kDa antigen, a common target of hum-

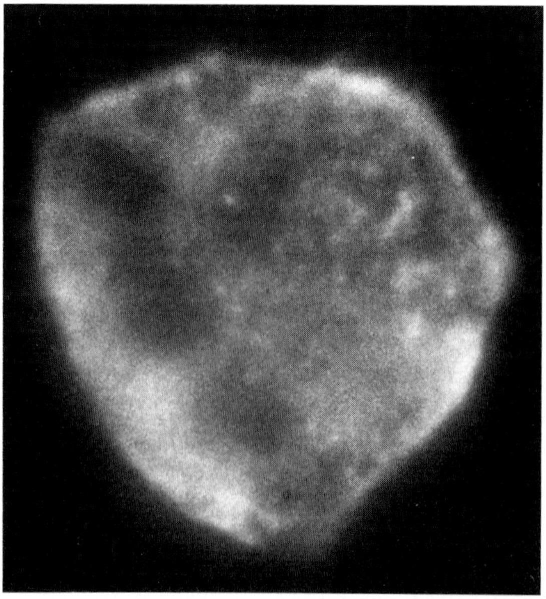

Figure 1 Expression of hsp 60 in heat shocked RIN cells. (From Ref. 51.).

oral immunity in IDDM, may be a heat shock protein or was an unrelated antigen with sequence homology to a heat shock protein.

This study was performed before the identity of the 64-kDa antigen was known. The 64-kDa antigen was subsequently demonstrated to be the 64-kDa islet isoform of GAD 65. However, several studies have established that 64-kDa antibody reactivity may not be composed only of GAD 65–reactive immunoglobulin: We have proposed that hsp 60 autoreactivity may constitute a significant fraction of 64-kDa reactivity in some subjects (52).

These investigations (51) were not supported by studies by Atkinson and coworkers (53). Using a similar model, they found that hyperthermic incubation of islets stimulated the expression of heat shock proteins with M_rs of 72, 75, and 90-kDa. The 60-kDa heat shock protein was constitutively expressed, and it was not upregulated by hyperthermic stress in their system. In addition, Western blot analysis with 64-kDa antibody–positive individuals did not identify hsp 60 as a target. However, our initial observations were dependent on a prolonge recovery time of the heat shocked RIN cells, and the data obtained by Atkinson and coworkers did not include an recovery phase, suggesting that the time scale of production of an islet-specific hsp 60 homologue is not immediate and may follow a cellular response and recovery.

3. *Hsp 60 Antibody Reactivity Is Significantly Higher in IDDM Subjects*

More recently, we have studied the humoral autoreactivity to hsp 60 in IDDM using recombinant human hsp 60. A total of 167 sera from 26 cases newly diagnosed IDDM, 45 long-standing cases of IDDM, 46 non–insulin-dependent diabetic patients (NIDDM), and 50 control subjects were assayed in an ELISA for antibodies to human and mycobacterial hsp 60. Significantly higher reactivity was found both in newly diagnosed patients and in those with long-standing IDDM than in those with NIDDM and control subjects (Table 1). The demonstration of increased autoantibody reactivity to human hsp 60 in IDDM subjects has established hsp 60 as a potential autoantigen in IDDM. However, the earlier discussion of effector mechanisms clearly shows that $CD4^+$ T-cell reactivity against an antigen is more etiologically important.

4. *Proliferative T-Cell Responses to Hsp 60 in IDDM*

We have therefore measured $CD4^+$ T-cell responses to human hsp 60 in IDDM subjects in an in vitro T-cell proliferation assay (Fig. 2). In this assay, autoreactive $CD4^+$ T cells are stimulated by antigen, and the proliferative response is measured on the basis of thymidine incorporation into the dividing antigen-responsive clones. A positive proliferative response is defined by a stimulation index greater than 3, (where the stimulation index = proliferation with antigen/proliferation with medium alone). This cut-off is widely used, and it represents

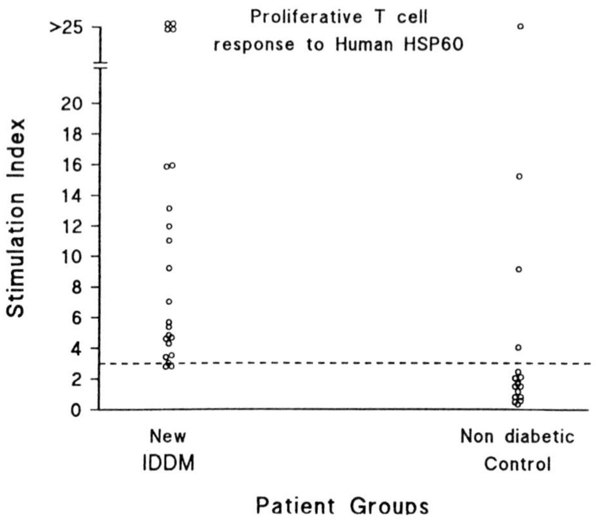

Figure 2 T-cell proliferative responses to hsp 60 in IDDM.

the mean plus 3 SD of the wells containing medium alone. In a series of 22 patients with IDDM, 20 showed positive proliferative responses to recombinant human hsp 60. Conversely, only 4 of 17 nondiabetic subjects showed proliferative responses. The difference in both frequency of positive response and magnitude of response were significant between the two groups.

Table 1 Serum Antibody reactivity to human and mycobacterial hsp 60 in newly diagnosed and long-standing IDDM, NIDDM, and Nondiabetic control subjects

	Subject groups			
Hsp type	Newly diagnosed IDDM	Long duration	NIDDM	Nondiabetic control
Human hsp 60				
median	0.108	0.178	0.092	0.057
range	0.026–0.306	0.037–0.368	0.010–0.373	0.010–0.137
significance (p)	$p < 0.02$	$p < 0.001$	ns	–
Mycobacterial hsp 65				
median	0.106	0.367	0.150	0.066
range	0.023–0.720	0.029–0.610	0.032–0.687	0.10–0.268
significance	$p < 0.04$	$p < 0.001$	ns	–

ns, not significant.

In the most recent study of p277 in the NOD mouse, Cohen and Elias outlined preliminary data which showed that in human IDDM, the human homologue of the p277 peptide appeared to be a common T-cell epitope. In their experimental system, 20 of 23 newly diagnosed patients with IDDM showed proliferation in response to this peptide (50).

Hsp 60 has not been studied as extensively as GAD 65 in human IDDM. Nonetheless, the emerging pattern is that hsp 60 appears to be a common target of the cellular immune response in IDDM, and that the dominant T-cell epitope may be conserved with the NOD mouse model. It is to be hoped that the therapeutic effects of p277 tolerization in this animal model will be extended to the treatment of human disease. The recent paper by Shehadeh et al. may be of particular interest in this respect (54). In a small series of newly diagnosed IDDM subjects, BCG immunization appeared to influence the 'honeymoon period' frequently observed following clinical diagnosis of IDDM, with a significantly increased frequency of subjects remaining insulin independent 3–6 months after treatment. However, this study showed a lower than expected frequency of the subjects treated with placebo in remission at these time points; more extensive studies are clearly required.

C. A Wider Role for Hsp 60 in Autoimmunity

1. *Sequence Similarities Between Tissue-Specific Autoantigens and Hsp 60*

We have also previously described a wide range of sequence similarities between human hsp 60 and other autoantigens (39). Many of the tissue-specific autoantigens are intracellular enzymes and others are structural proteins or membrane constituents. A comprehensive search of the protein sequence database identified extensive areas of amino acid similarity between a wide range of established autoantigens which are targets of an autoimmune response in a single disease and regions of human heat shock protein 60 (Table 2).

The expression of hsp 60 is increased under conditions of stress in many tissues. It has been proposed that a physiological function of this stress protein may be to alert the immune system to the presence of damaged cells and to function as a target for their removal (55). In autoimmune disease, the initial insult against the target cell may activate a population of cross-reactive T cells, which are then able to react with the homologous sequences on the tissue-specific antigens. As the potential epitopes exhibit a variety of motifs, it would be expected that different MHC alleles would present alternative hsp 60 epitopes in the target cells, and thus susceptibility to specific autoimmune diseases would be associated with the expression of distinct MHC haplotypes. In the specific case of IDDM, the amino acid similarity between hsp 60 and the tissue-specific

Table 2 Regions of Similarity Between Human Hsp 60 and autoantigens

Hsp 60 region	Known autoantigen	Autoantigen region	Similarity score	Disease
1–7	Myelperoxidase	594–610	60	Glomerulonephritis
7–21	Cytochrome P-450	359–372	64	Chronic active hepatitis
7–21	17α-Hydroxylase	359–372	64	Addison's disease
12–32	Pyruvate dehydrogenase	18–36	62	Primary biliary cirrhosis
37–49	KU autoimmune antigen	81–93	63	Scleroderma
38–53	NADH dehydroreductase	45–63	62	Primary biliary cirrhosis
40–50	Cerebellar degeneration associated antigen	93–103	60	Paraneoplastic cerebellar degeneration
57–68	Cytochrome P-450	402–412	59	Chronic active hepatitis
65–75	Thyroglobulin	393–403	58	Hashimoto's thyroiditis
86–100	Cardiac myosin heavy chain	516–530	62	Coxsackie myocarditis
101–123	Cytokeratin	83–105	58	Rheumatoid arthritis
108–117	DNA-binding protein	73–82	58	Systemic lupus erythematosus
152–167	Bullous pemphigoid antigen	308–323	59	Pemphigoid
152–175	Neurofilament triplet M protein	727–749	67	Multiple sclerosis
202–221	Laminin β_2 chain	1354–1375	61	Basement membrane
290–300	21-Hydroxylase	61–71	53	Addison's disease
328–339	Myelin-associated protein	286–297	60	Multiple sclerosis
329–345	Cytokeratin	34–50	62	Rheumatoid arthritis
354–374	Laminin β_2 chain	205–225	60	Goodpasture's syndrome
384–396	Cardiac myosin heavy chain	136–148	65	Coxsackie myocarditis
391–405	Glutamic acid decarboxylase	520–534	56	Insulin-dependent diabetes
432–450	Dihydrolipoamide dehydrogenase	70–88	57	Primary biliary cirrhosis
451–470	Cardiac myosin heavy chain	253–272	64	Coxsackie myocarditis
453–472	KU autoimmune antigen	192–212	60	Scleroderma
468–480	Acetylcholine receptor	133–145	63	Myasthenia gravis
492–502	KU autoimmune antigen	449–459	57	Scleroderma
552–572	Hsp 90	633–653	60	Systemic lupus erythematosus
562–571	Cytokeratin	545–554	81	Rheumatoid arthritis

Source: From Ref. 39.

autoantigen GAD 65 covers a sequence of 15 residues, with 6 identities and 6 conserved substitutions (Table 3).

The identification of the primary epitope of GAD 65 in the NOD mouse provides support for this hypothesis (56). The initial T-cell proliferative response to a series of synthetic peptides which spanned the GAD 65 sequence was obtained with two overlapping peptides, with the conserved region corresponding to the sequence similarity with hsp 60 (Table 4). It would be of interest to determine whether this GAD 65 peptide–responsive T-cell population is also able to proliferate to the homologous hsp 60 peptide.

2. *Common Epitope Motifs Between β-Cell Autoantigens and Environmental Agents*

In IDDM, epitope cross reactivity with an environmental agent which initiates β cell stress and hsp 60 upregulation may also be important. In a similar database search, we examined sequence homologies between hsp 60 and viral proteins. A further region of sequence homology between hsp 60 and GAD 65 was identified, which also showed a conserved motif with the nucleocapsid protein of coxsackie B4 virus. The homology between GAD 65 and the coxsackie sequence had been noted previously (57): however, the additional involvement of an hsp 60 epitope may also be of interest.

The epitope identified has been investigated as a target of both the humoral and cellular immune response in IDDM. Significantly increased antibody reactivity was identified to both the GAD 65 (PEVKEK) and hsp 60 (VEVNEK) peptides. In addition, the T-cell proliferative responses to extended PEVKEK peptides were significantly higher in a limited series of newly diagnosed patients with IDDM. However, the frequency of a positive response, with a stimulation index greater than 3, was not significantly higher, so the statistically significant difference reflects only weak proliferation to this peptide. This region does appear to be a key secondary epitope in the NOD mouse; in this animal model, the activation of the proliferative response is spontaneous, and it remains to be determined whether coxsackie infection or spontaneous loss of tolerance leads to activation of the cross-reactive epitope in human IDDM.

D. Genetic Location of Hsp 60: Similarity to NOD IDD5 Susceptibility Region

The genomic location of human hsp 60 has also attracted considerable attention. The sequencing and cloning of this gene by Gupta's group have been followed by gene mapping. Using a biotinylation method, we have identified several apparent pseudogenes of the human hsp 60 in metaphase preparation of the human chromosomes (58). Five binding sites of the human hsp 60 gene were located, including one at chromosome 2q32, which is the human equivalent of

Table 3 T-Cell Epitope of GAD 65 in NOD Mice Showing Homology to Hsp 60

Antigen	Amino acid region	Sequence														
Human hsp 60	391–405	N	E	R	L	A	K	L	S	D	G	V	A	V	L	K
		_	*	*	*	*	*	*	_	_	*	*	*	*	_	*
M. bovis hsp 65	366–380	Q	E	R	L	A	K	L	A	G	G	V	A	V	I	K
		_	*	*	_	_	_	*	*				_	*	_	*
Human GAD 65	520–534	E	E	R	M	S	R	L	S	K	V	A	P	V	I	K

*Identity.
-Conserved substitution.

the NOD mouse diabetes-susceptibility region known as IDD5. This gene was located close to a mycobacterial response gene, and it was suggested that this therefore may be a location for a human hsp 60 gene (possibly pancreatic β-cell specific), which may have influenced the susceptibility to IDDM. However, other work has suggested that this gene may be a nitrous oxide–regulating gene (59), and it would therefore function as a susceptibility gene presumably through modification of the effect of oxidants on β cells during the process of immune activation.

IV. ROLE FOR OTHER STRESS PROTEINS IN THE ETIOLOGY OF IDDM

Expression of several species of heat shock protein is induced in pancreatic islet β cells by cytokine or hyperthermic stress, including species with M_rs of 70 and 90 kDa. Hsp 70 was considered to be of particular interest, and it has been investigated as a potential autoantigen in IDDM (53). The gene is located in the class III region of the MHC (60), which is adjacent to the major susceptibility locus for IDDM. However, there are no reports of antibody or T-cell reactivity against hsp 70 in IDDM, and tolerization to hsp 70 does not prevent insulitis or IDDM in the NOD mouse. No linkage has been found between a particular allele of hsp 70 and the susceptibility to IDDM in humans (61), and the general consensus is that there is no role for hsp 70 in the etiology of IDDM. To date, there are no published studies of T-cell reactivity to any stress protein other than hsp 60 in IDDM.

V. OTHER β-CELL AUTOANTIGENS

In addition to hsp 60, several other target autoantigens have been identified in the NOD mouse and human IDDM. These include insulin (62), GAD 65 (63),

Table 4 Sequence Similarities Between Hsp 60 and autoantigens

Antigen	Amino acid region	Sequences/homologies
Thyroglobulin	393–403	E K R W A S P R V A R
		* * – * * – * – –
Hsp 60	65–75	E Q S W G S P K V T K
DNA-binding protein	73–82	E A G E A T T T T T
		* * * – – * * * – *
Hsp 60	108–117	E A G D G T T T A T
Cytokeratin	545–555	G G M G G G L G G G
		* * * * * * – * * *
Hsp 60	562–571	G G M G G G M G G G
KU autoantigen	193–212	P L K G I T E Q Q K E G L E I V K M V M
		– * * – * * * – * * – * –
Hsp 60	451–473	S L T P A N E D Q K I G I E I I K R T L
GAD	520–534	E E R M S R L S K V A P V I K
		– * * – – – * * – * – *
Hsp 60	391–405	N E R L A K L S D G V A V L K

*Identity.
⁻Conserved substitution.
Source: From Ref. 39.

carboxypeptidase H (64), a 38-kDa insulin secretory granule antigen (65), and a 69-kDa antigen which exhibited sequence similarity with bovine serum albumin (66). More recently, systematic examination of islet cell or insulinoma cell protein extracts has shown that T-cell reactivity develops first against other as yet uncharacterised autoantigens (67,68).

A. 64-kDa Antigen/GAD 65

Most attention has focused on a 64-kDa antigen that was shown through immunoprecipitation of islet cell extracts with diabetic sera to be a common target of a humoral response (69). A component of the 64-kDa antigen was finally demonstrated to be GAD 65 (70), the major islet isoform of the gamma aminobutyric acid (GABA) shunt enzyme, which is also expressed in several other tissues, including brain, stomach, testis, and, in humans, in the islet α and δ cells in addition to the β cell (71).

Humoral autoimmunity to GAD 65 was initially identified in newly diagnosed patients with IDDM (69) and in prediabetic individuals (72). More recent publications have demonstrated a proliferative T-cell response to GAD 65 both in NOD mouse (56,73) and in human IDDM (74). In the NOD mouse, $CD4^+$ T-cell response to GAD 65 have been suggested to be the earliest event in the autoimmune cascade, although T-lymphocyte responses against other antigens, including hsp 60 can also be demonstrated simultaneously. The administration of GAD 65 in a nonimmunogenic form to NOD mice resulted in tolerization to GAD 65 and the prevention of IDDM in a significant proportion of the population. In addition, this resulted in tolerization to other β-cell autoantigens.

Harrison et al. analyzed the humoral and cellular autoreactivity to the core region of GAD 67 (75), which is the major central nervous system isoform and is highly homologous to the same region of GAD 65. It was shown that most of the prediabetic patients fitted into one of two distinct profiles with either low antibody reactivity and high T-cell proliferation or high antibody reactivity and low T-cell proliferation. The development of IDDM was accelerated in those subjects with high T-cell proliferation and low antibody reactivity. It was suggested that these findings may have reflected a TH1/TH2 split in the breakdown of tolerance to the β cell autoantigen, with the TH1 phenotype resulting in a more rapid progression to IDDM.

VI. CONCLUSIONS

In this chapter, the potential mechanisms of β-cell destruction in IDDM have been outlined, and evidence that supports hsp 60 as a key autoantigen in the NOD mouse and human IDDM reviewed. IDDM is a complex multifactorial disease, in which both genetic susceptibility and environmental agents each have a distinct role. The expression of hsp 60 in islets is known to be altered by cellular stress. Cohen and Birk have argued that hsp 60 may be an important target of the immune response in IDDM, bridging the gap between the environment and the autoimmune attack on the islet β cells (76). We have proposed that at the molecular level this may be reflected in a T-cell response directed against an environmental agent leading to a cross-reactive immune activation to a shared epitope of hsp 60 and subsequently to a relatively tissue specific protein. In the specific case of IDDM, the sequence of events would be initiation by coxsackie viral infection through the VEVNEK/PEVKEK conserved sequence.

Ultimately, an understanding of the immunological mechanisms that lead to the initiation and propagation of islet β-cell autoimmunity may lead to the development of novel immunotherapies that may prevent or halt the autoimmune destruction of the β cells.

REFERENCES

1. G.S. Eisenbarth, Type I diabetes mellitus. A chronic autoimmune disease, *N. Engl. J. Med. 314*:1360 (1986).
2. W. Gepts, Pathologic anatomy of the pancreas in juvenile diabetes mellitus, *Diabetes 14*:619 (1990).
3. G.F. Bottazzo, A. Florin-Christensen, and D. Doniach, Islet cell antibodies in diabetes mellitus with autoimmune polyendocrine deficiencies, *Lancet 1279* (1974).
4. A.C. MacCuish, E.E.W. Barnes, W.J. Irvine, and L.J.B. Duncan, Antibodies to pancreatic islet cells in insulin-dependent diabetes with coexistent autoimmune disease, *Lancet 2*:1529 (1974).
5. J.F. Bach, Mechanisms of autoimmunity in insulin dependent diabetes mellitus, *Clin. Exp. Immunol. 72*:1 (1988).
6. P. Pozzilli, A. Signore, A.J.K. Williams, and P.E. Beales, NOD mouse colonies around the world—recent facts and figures, *Immunol. Today 14*:193 (1993).
7. J.A. Todd, J.I. Bell, and H.O. McDevitt, HLA-DQB gene contributes to susceptibility and resistance to insulin dependent diabetes mellitus, *Nature 329*:599 (1987).
8. D.R. Gamble, The epidemiology of insulin dependent diabetes with particular reference to virus infection to its aetiology, *Epidemiol. Rev. 2*:49 (1980).
9. A.H. Barnett, L. Eff, R.D.G. Leslie, and D.A. Pyke, Diabetes in identical twins: a study of 200 pairs. *Diabetologia 20*:87 (1981).
10. L.C. Harrison, Islet cell antigens in insulin dependent diabetes - Pandora's box revisited. *Immunol. Today 13*:348 (1992).
11. D. Faustman, Mechanisms of autoimmunity in type I diabetes, *J. Clin. Immunol. 13*:1 (1993).
12. P.G. Colman, R.C. Nayak, I.L. Campbell, and G.S. Eisenbarth, Binding of 'cytoplasmic' islet cell antibodies is blocked by human pancreatic glycolipid extracts, *Diabetes 37*:645 (1988).
13. G.F. Bottazzo, A.N. Gorsuch, B.M. Dean, A.G. Cudworth, and D. Doniach,

Complement-fixing antibodies in type I diabetes: possible monitors of active beta cell damage, *Lancet 1*:668 (1980).
14. A. Lermark, Z.R. Freeman, C. Hofman, et al. Islet cell surface antibodies in juvenile diabetes mellitus, *N. Engl. J. Med. 299*:375 (1978).
15. E. Bonifacio, P. Bingley, M. Shattock, et al., Quantification of islet cell antibodies and prediction of insulin dependent diabetes mellitus, *Diab. Metab. Rev. 3*:893 (1990).
16. G.F. Botazzo, B.M. Dean, J.M. McNally, E. H. Mackay, P.G.F. Swift, and D.R. Gamble, *In situ* characterisation of autoimmune phenomena and expression of HLA molecules in the pancreas in diabetic insulitis, *N. Engl. J. Med. 313*:353 (1985).
17. G.F. Botazzo, Beta cell damage in diabetic insulitis: are we approaching the solution?, *Diabetologia 26*:241 (1984).
18. D. Lo, L.C. Burkly, G. Widera, et al., Diabetes and tolerance in transgenic mice expressing class II MHC molecules in pancreatic islet beta cells, *Cell 53*;159 (1988).
19. J.P. Allison, CD28-B7 interactions in T-cell activation, *Curr. Opin. Immunol. 6*:414 (1994).
20. A.J. Sant and J. Miller, MHC class II antigen processing: biology of invariant chains, *Curr. Opin. Immunol. 6*:57 (1994).
21. D.M.Harlan, H. Hengartner, M.L. Huang, et al., Mice expressing both B7-1 and viral glycoprotein on pancreatic beta cells along with glycoprotein specific transgenic T cells develop diabetes due to a breakdown of T lymphocyte unresponsiveness, *Proc. Natl. Acad. Sci. USA 91*:3137 (1994).
22. M.B.A. Oldstone, M. Nerenberg, P. Southern, J. Price and H. Lewicke, Virus infection triggers insulin dependent diabetes mellitus in a transgenic model: role of anti-self (virus) immune response, *Cell 65*:319 (1991).
23. R.K. Sibley and D.E.R. Sutherland, Pancreas transplantation: an immunohistologic and histopathologic examination of 100 grafts, *Am. J. Pathol. 128*:151 (1987).
24. A. Bendelac, C. Carnaud, C. Boitard, and J.F. Bach, Syngeneic transfer of autoimmune diabetes from diabetic NOD mice to healthy neonates: requirement for both $L3T4^+$ and $Lyt2^+$ T cells, *J. Exp. Med. 166*:823 (1987).
25. J. Katz, C. Benoist, and D. Mathis, Major histocompatibility complex class I molecules are required for the development of insulitis in the non-obese diabetic mouse, *Eur. J. Immunol.* (1993).
26. B.J. Bradley, K. Haskins, F.G. Larosa, and K.J. Lafferty, CD8 T cells are not required for islet destruction induced by a CD4+ T cell clone, *Diabetes 41*:1603 (1992).
27. S.W. Christianson, L.D. Scultz and E.H. Leiter, Adoptive transfer of diabetes into immunodeficient NOD-scid/scid mice- relative contribution of CD4+ and CD8+ T cells from diabetic versus prediabetic NOD.NON-thy-1a donors, *Diabetes 42*:44 (1993).
28. M. Matsumoto, H. Yagi, K. Kunimoto, and J. Kawaguchi, Transfer of autoimmune diabetes from diabetic NOD mice to NOD athymic nude mice- the roles of T cell subsets in the pathogenesis, *Cell Immunol. 148*:189 (1993).
29. B. Charlton, A. Bacelj, and T.E. Mandel, Administration of silica particles or anti-Lyt2 antibody prevents beta cell destruction in NOD mice given cyclophosphamide, *Diabetes 37*:930 (1988).

30. G. Kraal, G. Martens, and K. Kuystermans, Systemic elimination of macrophages using liposomes does not prevent the induction of type I diabetes, *Int. Arch. Allergy Immunol. 100*:115 (1993).
31. J. Katz, C. Benoist, and D. Mathis, Major histocompatibility complex class I molecules are required for the development of insulitis in non-obese diabetic mice, *Eur. J. Immunol. 23*:3358 (1993).
32. M. Nagata and J.W. Yoon, Studies on autoimmunity for T cell mediated beta-cell destruction—distinct difference in beta-cell destruction between CD4+ and CD8+ T cell clones derived from lymphocytes infiltrating the islets of NOD mice, *Diabetes 41*:998 (1992).
33. T. Mandrup-Poulsen, U. Zumsteg, J. Reimers, et al., Involvement of interleukin-1 and interleukin-1 antagonist in pancreatic b-cell destruction in IDDM, *Cytokine 5*:185 (1993).
34. K. Brudzynski, Insulitis caused redistribution of heat shock protein 60 inside beta cells correlates with induction of hsp60 autoantibodies, *Diabetes 42*:908 (1993).
35. M.J. Rappaport, A. Jaramilo, D. Zipris, et al., IL-4 reverses T-cell proliferative unresponsiveness and prevents the onset of diabetes in NOD mice, *J. Exp. Med. 178*:87 (1993).
36. S. Jindal, A.K. Dudani, B. Singh, C.B. Harley, and R.S. Gupta, Primary structure of a human mitochondrial protein homologous to the bacterial and plant chaperonins and to the 65-kilodalton Mycobacterial antigen, *Mol. Cell. Biol. 9*:2279 (1989).
37. R.J. Ellis, Proteins as molecular chaperones, *Nature 328*:378 (1987).
38. S.H.E. Kaufmann, U. Vath, J.E.R. Thole, et al. Enumeration of T cells reactive with Mycobacterium Tuberculosis organisms and specific for the recombinant mycobacterial 64kD protein. *Eur. J. immunol 17*, 351 (1987).
39. D.B. Jones, A.F.W.Coulson, and G.W. Duff, Sequence homologies between hsp60 and autoantigens, *Immunol Today 14*:115 (1993).
40. M.B. Oldstone, Molecular mimicry and autoimmune disease, *Cell 50*:819 (1987).
41. M.W.J. Sadelain, H.-J. Qin, J. Lauzon, and B. Singh, Prevention of autoimmune diabetes in NOD mice by adjuvant immunotherapy, *Diabetes 39*:583 (1990).
42. M.F. McInerey, S.B. Pek, and D.W. Thomas, Prevention of insulitis and diabetes onset by treatment with complete Freund's adjuvant in NOD mice, *Diabetes 40*:715 (1991).
43. H.-Y. Qin, M.W.Y. Sadelain, C. Hitchon, J. Lauzon, and B. Singh, Complete Freund's adjuvant induced T cells prevent the development and adoptive transfer of diabetes in nonobese diabetic mice, *J. Immunol. 150*:2072 (1993).
44. M. Harada, Y. Kishimoto, and S. Makino, Prevention of overt diabetes and insulitis in NOD mice by a single BCG vaccination, *Diabetes Res. Clin. Pract. 8*:85 (1990).
45. H. Yagi, M. Matsumoto, S. Suzuki, R. Misaki, R. Suzuki, S. Makino, and M. Harada, possible mechanism of the preventive effect of BCG against diabetes mellitus in the NOD mouse. I. Generation of suppressor macrophages in spleen cells of BCG vaccinated mice. *Cell Immunol. 138*:130 (1991).
46. N.N. Shehadeh, F. Larosa, and K.J. Lafferty, Altered cytokine activity in adjuvant inhibition of autoimmune diabetes, *J. Autoimmune 6*:293 (1993).
47. D. Elias, D. Markovits, T. Reshef, R. van der Zee, and I.R. Cohen, Induction and

therapy of autoimmune diabetes in the non-obese diabetic (NOD/It) mouse by a 65kD heat shock protein, *Proc. Natl. Acad. Sci. USA 87*:1576 (1990).
48. D. Elias, T. Reshef, O.S. Birk, R. van der Zee, M.D. Walker, and I.R. Cohen, Vaccination against autoimmune mouse diabetes with a T cell epitope of the human 65kD heat shock protein. *Proc. Natl. Acad. Sci. USA 88*:3088 (1991).
49. J. Sloan-Lancaster, B.D. Evavold, and P.M. Allen, Induction of T cell anergy by altered T cell receptor ligand on live antigen presenting cells, *Nature 363*:156 (1993).
50. D. Elias and I.R. Cohen, Peptide therapy for diabetes in NOD mice, *Lancet 343*:704 (1994).
51. D.B Jones, N.R. Hunter, and G.W. Duff, Heat shock protein 65 as a beta cell antigen of insulin dependent diabetes, *Lancet 335*:583 (1990).
52. D.B. Jones and G.W. Duff, Is there no role for heat shock protein in diabetes? *Lancet 337*:115 (1991).
53. M.A. Atkinson, L.A. Holmes, D.W. Scharp, P.E. Lacy, and N.K. McLaren, No evidence for serological autoimmunity to islet cell heat shock proteins in insulin dependent diabetes, *J. Clin. Invest. 87*:721 (1991).
54. N. Shehadeh, F. Calcarino, B.J. Bradley, I. Bruchlim, P. Vardi, and K.J. Lafferty, Effect of adjuvant therapy on development of diabetes in mouse and man, *Lancet 343*:706 (1994).
55. S.H.E. Kaufmann, Heat shock proteins and the immune response, *Immunol. Today 11*:129 (1990).
56. D.L. Kaufman, M. Clare-Salzler, J. Tian, et al., Spontaneous loss of T-cell tolerance to glutamic acid decarboxylase in murine insulin-dependent diabetes, *Nature 366*:69 (1993).
57. D.L. Kaufman, M.G. Erlander, M. Clare-Salzler, N.K. Atkinson, N.K. McLaren, and A.J. Tobin, Autoimmunity to two forms of glutamate decarboxylase in insulin dependent diabetes mellitus, *J. Clin. Invest. 89*:283 (1992).
58. D. Bowen Jones, D.E. Fantes, and R.S. Gupta, Diabetes and heat shock protein, *Nature 355*:119 (1992).
59. D. Latchmann, No diabetes link to hsp65? *Nature 356*:114 (1992).
60. C.A. Sargent, I. Dunham, J. Trowsdale, and R.D. Campbell, Human major histocompatibility complex contains genes for the major heat shock protein hsp70, *Proc. Natl. Acad. Sci. USA 86*:1968 (1989).
61. A. Pugliese, Z.L. Awdeh, A. Galluzzo, E.J. Yunis, C.A. Alper, and G.S. Eisenbarth, No independent association between hsp70 gene polymorphism and IDDM, *Diabetes 41*:788 (1992).
62. J.P. Palmer, C.M. Asplin, P. Clemons, et al., Insulin autoantibodies in insulin dependent diabetes before insulin treatment, *Science 222*:1337 (1983).
63. M.A. Atkinson, D.L. Kaufman, L. Campbell, et al., Response of peripheral blood mononuclear cells to glutamate decarboxylase in insulin dependent diabetes, *Lancet 339*:458 (1992).
64. L. Castano, E. Russo, L. Zhou, and M.A. Lipes, and G.S. Eisenbarth, Identification and cloning of a granule autoantigen (carboxypeptidase H) associated with type I diabetes, *J. Clin. Endocrinol. Metab. 73*:1197 (1991).

65. B.O. Roep, A.A. Kallan, W.L.W. Hazenbos, et al., T cell reactivity to 38kD insulin secretory granule protein in patients with recent onset type I diabetes, *Lancet 337*:1439 (1991).
66. J. Karjalainen, J.M. Martin, M. Knip, et al., Bovine serum albumin peptide as a possible trigger of insulin dependent diabetes mellitus, *N. Eng. J. Med. 327*:302 (1992).
67. C. Gelber, L. Paborsky, S. Singer, et al., Isolation of nonobese diabetic mouse T cells that recognise novel autoantigens involved in the early events of diabetes, *Diabetes 43*:33 (1994).
68. B. Bergman and K. haskins, Islet specific T cell clones from the NOD mouse respond to beta-granule antigen, *Diabetes 43*:197 (1994).
69. S. Baekkeskov, J.H. Nielsen, B. Marner, T. Bilde, J. Ludvigsson, and A. Lernmark, Autoantibodies in newly diagnosed diabetic children immunoprecipitate human pancreatic islet cell proteins, *Nature 298*:167 (1982).
70. S. Baekkeskov, H. Jan-Aanstoot, S. Christgau, et al., Identification of the 64K antigen of insulin dependent diabetes as the GABA synthesising enzyme glutamic acid decarboxylase, *Nature 347*:151 (1990).
71. J.S. Petersen, S. Russel, M.O. Marshall, et al., Differential expression of glutamic acid decarboxylase in rat and human islets, *Diabetes 42*:484 (1993).
72. S. Baekkeskov, M. Landin-Olsson, J.K. Kristensen, et al., Antibodies to a 64,000 M_r human islet cell antigen precede the clinical onset of insulin dependent diabetes, *J. Clin. Invest. 79*:926 (1987).
73. R. Tisch, X.D. Yang, S.M. Singer, R.S. Liblav, L. Fuggar, and McDevitt, Immune response to glutamic acid decarboxylase correlates with insulitis in non-obese diabetic mice, *Nature 366*:72 (1993).
74. M.A. Atkinson, D.L. Kaufman, L. Campbell, et al., Response of peripheral blood mononuclear cells to glutamate decarboxylase in insulin dependent diabetes, *Lancet 339*:458 (1992).
75. L.C. Harrison, M.C. Honeyman, H.J. DeAizpuria, et al., Inverse relation between humoral and cellular immunity to glutamic acid decarobxylas in subjects at risk of insulin dependent diabetes, *Lancet 341*:1365 (1993).
76. O.S. Birk and I.R. Cohen, T cell autoimmunity in type I diabetes mellitus, *Curr. Opin Immunol 5*:903 (1993).

15

Protection Against Tumors by Stress Protein Gene Transfer

Katalin V. Lukacs, Douglas B. Lowrie, and M. Joseph Colston
Laboratory of Leprosy and Mycobacterial Research, National Institute for Medical Research, London, England

I. INTRODUCTION

During the course of our studies on the interaction of the immune system with mycobacterial stress proteins, we found that a tumor cell line which had been transfected with the mycobacterial heat shock protein 65 (hsp 65) gene lost its ability to form tumors while at the same time immunization with the genetically modified cells protected the host against the otherwise lethal parent tumor (1). These observations suggest that there could be a role for stress proteins in manipulating the tumorigenicity of tumor cells and the host's response to those cells. In this chapter, we review the basis of immunological recognition of tumor cells, describe the experiments by which we were able to demonstrate the effects of hsp 65 gene transfer, and discuss the potential applications of such an approach.

One of the most important developments in cancer research in the past decade has been the demonstration that cancer is a multistep genetic disorder. Activation of dominant oncogenes and inactivation of tumor-suppressor genes by mutation or rearrangement provide the molecular basis of the malignant transformation resulting in two phenomena shared by malignant tumor cells:

1. Uncontrolled cell growth (tumorigenicity)
2. Expression of tumor-associated structures, encoded by the altered genes, which can be recognized by the immune system (immunogenicity)

A. Tumor-Associated Antigens

The presence of tumor-associated structures recognized by the immune system (tumor-associated antigens) was demonstrated in animal tumors long ago. When mice were freed by surgery from tumors induced by large doses of carcinogenic agents and subsequently injected with cells from the same tumor, they could eliminate otherwise lethal tumor loads (2). Although spontaneous human cancers have been generally considered to be weakly immunogenic or not immunogenic at all, in the past 5 years, significant progress has been made in identifying structures on the tumor cell surface which can serve as targets for the immune system.

The products of oncogenes and tumor-suppressor genes modified by mutations or rearrangements form one large group of antigens. For example, T cells specific for BCR-abl, mutated ras, and p53 proteins have been found in various murine and human tumors (3,4).

In many virus-associated tumors, the immune response is targeted against viral gene products, such as the Epstein-Barr nuclear antigen in some lymphomas and the human papilloma viruses E6 and E7 in cervical cancer (5,6). Recently, the first genes (MAGE-1,-2, -3) coding for antigens on a human tumor have been identified in melanoma. In addition to melanoma, the members of the MAGE gene family are expressed in many tumors of several types such as head and neck carcinoma, lung cancer, and breast carcinoma but not in normal adult tissues except for testes. They are readily recognized by T cells, providing a potential target for immunotherapy (7).

Despite the presence of T cells capable of recognising tumor antigens, the tumor cells often continue to proliferate. Attempts to increase the immune response by coinjection of tumor cells with bacterial adjuvants such as BCG or *Corynebacterium parvum* have been used to increase the immunogenicity and enhance tumor rejection.

B. Enhancing the Immune Response by Genetic Manipulation

The major purpose of genetic manipulation is to convert the weekly immunogenic tumor that elicits only a minor response into strongly immunogenic one that can provoke tumor rejection. Three main approaches have produced promising preliminary results.

1. It is well accepted that the major histocompatibility complex (MHC) class I and class II antigens are essential for antigen presentation. Transfer of genes encoding these molecules into mouse tumors decreased their oncogenic potential and increased their immunogenic recognition (8). The great diversity of the "outbred population" and the large number of different MHC loci involved in antigen presentation may limit this approach in humans.

2. Gene transfer of B7, a major costimulatory molecule required for the activation of T cells, has also been shown to increase tumor rejection in mice (9). It is possible, however, that an additional "strong" tumor antigen is required for this effect. In humans, the efficiency of B7 gene transfer remains to be determined.

3. Cytokines can significantly increase the antitumor immune response in clinical applications but have significant systemic side effects and short half-lives. Insertion of cytokine genes into cells of the immune response targeting the tumor (such as tumor-infiltrating lymphocytes) or into the tumor cells themselves results in the production of cytokines locally at very high concentration with no systemic side effects. In experimental models, insertion of genes encoding several cytokines, including interleukin-1 (IL-1), IL-2, IL-4, IL-7, and IL-12, interferon gamma (IFN-γ), tumor necrosis factor (TNF), macrophage and granulocyte colony-stimulating factors, have been applied with varying effects on both tumorigenicity and immunogenicity (reviewed refs. 10 and 11). In most cases, a complex antitumor response is induced. CD8 cells seem to be essential for most cytokines to exert their effect, but CD4 cells, natural killer (NK) cells, macrophages, neutrophils, and eosinophils can also participate depending on the cytokine gene used. Surprisingly, Hock at al. have observed that tumor cells engineered to produce IL-2, IL-4, IL-7, TNF, or INF-γ are no more immunogenic than parental cells mixed with a nonspecific adjuvant, *Corynebacterium parvum* (12). Several clinical trials are planned or are currently underway using various cytokine genes, which will determine if patients with cancer can benefit from such an approach.

II. STRESS PROTEIN GENE TRANSFER INTO TUMOR CELLS

Work over the past decade has demonstrated the importance of heat shock proteins in the immune response to many infectious agents. Heat shock proteins have also been implicated in a variety of autoimmune conditions and situations where inappropriate immune responses are occurring, such as the immunopathologically mediated nerve damage seen in leprosy (13) and in thryoid follicles of patients with Graves' disease (14). Among the explanations for such findings was the possibility that high levels of heat shock proteins in a particular cell or tissue could be increasing or altering processing and/or presentation of self-molecules such that they were much more likely to be recognized by the immune system. This hypothesis was tested by transfecting the macrophage cell line J774 with the mycobacterial hsp-65 gene; J774 cells are weakly immunogenic tumor cells of BALB/c origin and are also derived from antigen-presenting cells, and hence they provide an ideal model for testing the hypothesis.

A. Transfection of J774 Cells with Hsp 65

The gene encoding the *Mycobacterium leprae* 65 hsp was cloned into a retroviral shuttle vector pZIPNeoSV(x) to give pZIPML65 (Fig. 1). This was transduced into the virus-packaging cell psi-CRE, which was then used as a source of virus to transfect the J774 cells (1,15). Expression of hsp 65 by the transfected cells was demonstrated by Northern blotting analysis for mRNA and by FACS analysis using the anti–hsp 65 antibody IIIC8 (Fig. 2). In most experiments, either untransfected (J774) cells or cells transfected with vector alone (J774-vector) were used as controls.

B. Hsp-Transfected J774 Cells Lose Tumorigenicity

When mice received an intraperitoneal injection of 10^6 J774 cells they developed large tumors within 3 weeks. These were highly malignant lymphoreticular neoplasms with several metastases in the spleen, liver, kidney, mesentery, and lung. At the same time, J774 cells transfected with hsp 65 were not able to produce tumors (Fig. 3; Table 1). The lack of tumor formation was demonstrated in immunocompetent and T-cell–deficient athymic nude mice, indicating that expression of hsp 65 results in a genuine loss of tumorigenicity.

Lost tumorigenicity could be the result of the interaction of hsp 65 with oncogenes or tumor-suppressor genes or their protein products. Tumor-suppressor genes are critical sites for DNA damage, because they normally function as physiological barriers against clonal expansion. Mutations in the p53 gene

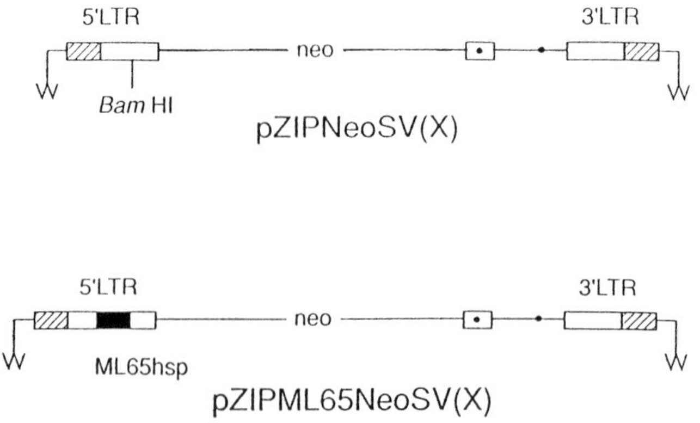

Figure 1 The J774 tumor cells were transfected by a retroviral shuttle vector, which was constructed by cloning 3.6-kb DNA including the *M. leprae* hsp 65 gene into pZIPNeoSV(X).

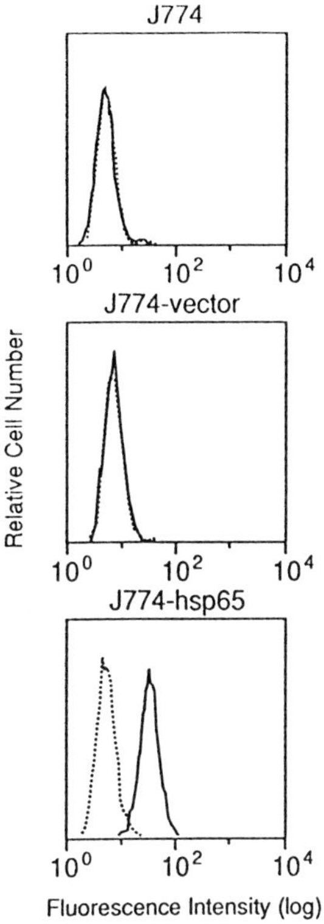

Figure 2 Expression of hsp 65 on the surface of hsp-transfected J774 cells assessed by FACS analysis. Control and transfected cells were stained by a FITC-conjugated monoclonal antibody IIIC8, which is specific for the *M. leprae* hsp 65.

resulting in dysfunctional p53 tumor-suppressor protein is the most frequently observed genetic lesion in malignant tumors; it is mutated in about half of almost all types of cancers (16).

Hsp-transfected and control tumor cells were examined for the expression of p53 tumor-suppressor protein. We have found elevated expression of p53 tumor-suppressor protein in hsp transfected J774 cells with two p53-specific monoclonal antibodies provided by D. Lane (data not shown). These findings suggest that hsp 65 may exert, or may partially exert, its effect through the p53 tumor-suppressor protein. The increased chaperone activity of the hsp 65–transfected cells could result in proper folding and conformation of the ineffective, mutant protein, thereby correcting its loss of tumor-suppressor function.

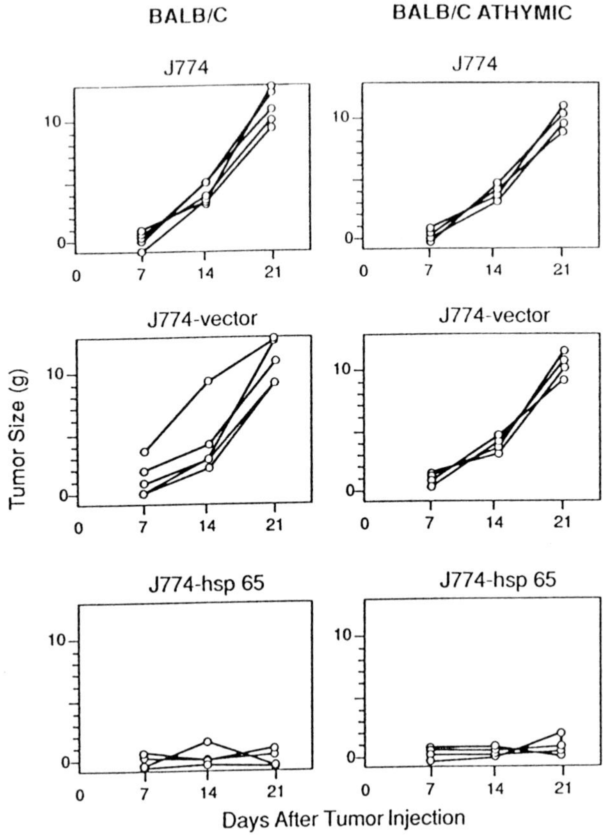

Figure 3 Growth of transfected tumor cells in BALB/c and T-cell–deficient athymic BALB/c mice. Mice were injected intraperitoneally with 10^6 tumor cells and killed after various time periods. Tumor size was calculated by weighing the primary tumor and abdominal organs of the test mice and subtracting the weight of abdominal organs of five age-matched normal controls.

C. Mice Immunized with Hsp 65–Transfected J774 Cells Are Protected Against Challenge with the Parent Tumor Cells

Mice were immunized with four intraperitoneal injections of 10^6 J774–hsp 65 cells at weekly intervals and then challenged 1 month later with untransfected J774 cells (Fig. 4; Table 2). The controls, unimmunized mice, developed progressive, metastasizing tumors which weighed approximately 10 g 3 weeks after inoculation. The immunized animals, however, were all completely protected against the challenge and failed to develop tumors even following several months

Table 1 The Tumorigenicity of Transfected and Parent Reticulum Sarcoma Cells in BALB/c,BALB/c Athymic and CBA Mice

		Tumor incidence		
Injected tumor cells		BALB/c	Athymic	CBA
J774	10^5	4/5	–	–
	10^6	9/10	5/5	0/10
	10^7	5/5	–	–
J774-vector	10^5	3/5	–	–
	10^6	14/15	5/5	–
J774-hsp 65	10^6	0/10	0/10	0/10
	10^7	1/10	–	–
	5×10^7	0/5	–	–

Mice were injected intraperitoneally with the indicated number of tumor cells. Tumor cells were either the parent cell line (J774), the parent cell line transfected with the mycobacterial hsp 65 gene (J774–hsp 65), or the parent cell line transfected with the vector alone (J774-vector). Tumor incidence was determined by necropsy and histological examination 21 days after injection of tumor cells.
Source: From Ref. 1.

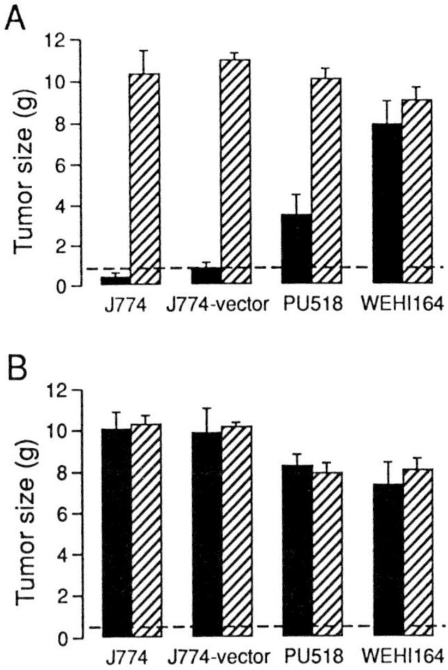

Figure 4 Growth of tumor cells in normal (A) and athymic (B) BALB/c mice that had been immunized with J774–hsp 65 cells. The results show tumor size 21 days after the challenge with tumor cells of differing origin in immunized (solid columns) and non-immune (hatched columns) mice. (From Ref. 1.)

Table 2 Effect of Immunization with Transfected Tumor Cells (J774–hsp 65) on Subsequent Challenge with Reticulum Sarcoma Cells

		Tumor incidence	
Immunization	Challenge	BALB/c	BALB/c athymic
J774–hsp 65	J774	0/10	5/5
None	J774	9/10	5/5
J774–hsp 65	J774-vector	0/15	5/5
None	J774-vector	10/10	5/5
J774–hsp 65	Pu518	2/10	5/5
None	Pu518	5/5	4/4
J774–hsp 65	Wehi 164	2/10	5/5
None	Wehi 164	5/5	4/4

Mice, either normal or athymic BALB/c, were immunized with four intraperitoneal injections of 10^6 J774–hsp 65 cells given at weekly intervals. On day 28, the mice were challenged intraperitoneally with 10^6 reticulum sacoma cells of differing origins.
Source: From Ref. 1.

of observation. To investigate the specificity of the protective effect, mice immunized with J774–hsp 65 cells were challenged with two other cell lines: Pu518, which is an unrelated, spontaneous reticulum sarcoma, and Wehi 164, which is derived from a methylcholanthrene–induced BALB/c sarcoma. The results show a significant level of protection against Pu518 cells but not Wehi 164 cells.

Our results suggested that expression of the hsp 65 protein greatly increased the immunological recognition of the untransfected cells. If we used irradiated J774 cells for immunization rather than transfected cells, we found that approximately 50% of the mice were protected on subsequent challenge. This would indicate that J774 cells do have tumor-associated antigens; however, the co-expression of the 65-kDa molecule provides greatly enhanced recognition of the tumor-associated antigen(s) either by virtue of its own highly immunogenic nature or by altering the efficiency of processing/presentation of the tumor antigen(s).

D. Mice Immunized with Hsp 65–Transfected J774 Cells Have T Cells Which Are Cytotoxic to the Parent Tumor Cells

In order to confirm that vaccinated mice were capable of mounting a cytotoxic response against the parent tumor cells, we then investigated cytotoxicity in vitro using a ^{51}Cr release assay (17). Both CD4 and CD8 T cells cultured from the spleens of J774-hsp–vaccinated mice were highly cytotoxic for the parent J774 cells (Fig. 5). Levels of cytotoxicity against other tumor cells (Pu518 and Wehi164) were significantly lower or absent altogether, reflecting the results of the in vivo protection experiments. These experiments also confirmed that in addition to the CD8 cytotoxic lymphocytes, CD4 T cells are also involved in the recognition and elimination of hsp-expressing J774 cells (18).

Thus, we could clearly demonstrate that expression of the mycobacterial hsp 65 gene greatly enhanced the recognition of native structures on the parent tumor cells, and this was reflected both in terms of the in vivo protection response as well as in vitro cytotoxicity. We do not know at this stage the nature of the molecule(s) which is recognized. There have been several reports that heat shock proteins themselves may be tumor-associated antigens (19–22). Hsps, including gp 96, hsp 90, and hsp 70, extracted from Met-A, a methyl cholanthrene-induced sarcoma can protect mice against subsequent challenge with sarcoma cells from the same tumor. Recent data, however, suggest that small peptides associated with the hsps are critical for this effect (22). After the elu-

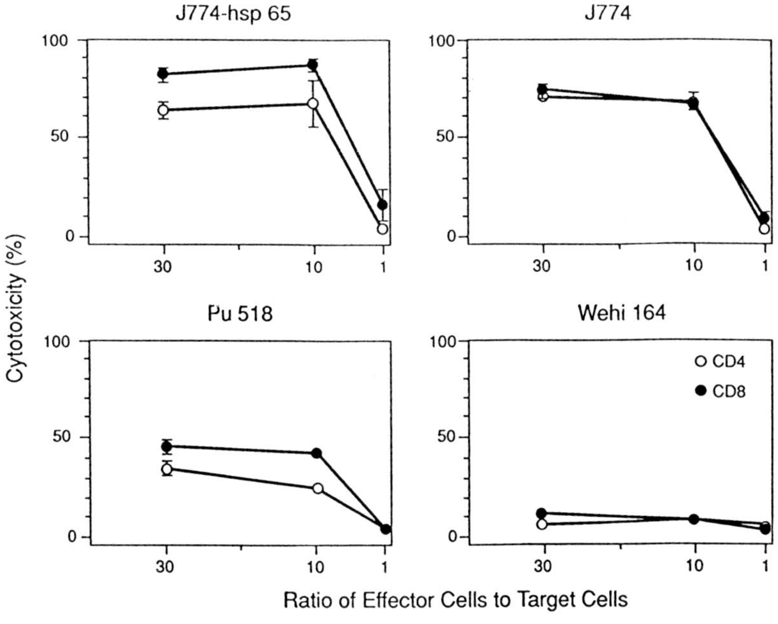

Figure 5 In vitro cytotoxic activity of splenocytes from mice immunized with J774-hsp 65 cells. Splenocytes from mice immunized with four intraperitoneal injections of J774-hsp 65 cells at weekly intervals were cultured for 6 days with irradiated J774-hsp 65 stimulator cells and tested in vitro in ^{51}Cr release assay. Purified populations of CD4 (open circles) and CD8 (closed circles) T cells were tested on various target cells.

tion of tumor-specific peptides, hsp 70 lost its protective effect and behaved as hsp 70 extracted from normal tissues. These data indicate that hsps are carriers of tumor-specific peptides rather than being the tumor-specific antigens themselves.

On the other hand, Ahsan and his colleagues showed (23) that immunization with the *M. bovis* BCG 64-kDa surface protein, which shares antigenic determinants with several mouse tumor cell lines, including Met-A, CT-26, and line 10 hepatoma, protects mice against these tumors. They also showed that monoclonal antibodies developed against the *M. leprae* 65-kDa heat shock protein could bind to these tumors and also to the BCG 64-kDa surface protein, suggesting that the 64-kDa stress protein is related to the tumor-specific antigen. It is not easy to imagine, however, how the highly conserved hsps could provide the diversity of individual tumor antigens highly specific for each tumor. It is worth noting that those tumors against which protection has been demonstrated by the BCG 64-kDa protein are all chemically induced tumors. It is possible that large doses of carcinogenic agents induce increased levels of host stress protein expression in the transformed cells, which cross react with the mycobacterial hsp 65; immune response generated by the mycobacterial hsp could eliminate tumors expressing high levels of cross reactive hsps behaving as shared tumor-associated antigens in these tumors. In our experiments, two spontaneous sarcomas (J774 and Pu518) showed no surface epitopes cross reactive with *M. leprae* hsp 65, and no protection was generated against these tumors by immunization with recombinant *M. leprae* hsp 65, indicating that hsps might play different roles in tumors of different origin.

Alternatively, rather than acting as a tumor-associated antigen itself, the 65-kDa molecule through its highly immunogenic nature could provide greatly enhanced, associated recognition of other tumor-associated antigen(s). Purified protein derivative (PPD), which contains the mycobacterial 65-kDa hsp, and the recombinant mycobacterial hsp protein have previously been shown to enhance immune responses to an associated but unrelated antigen (24–27).

Finally, heat shock proteins are chaperonins and are thought to be involved in the transport and folding of protein antigens during processing; thus, the increased chaperone activity of the hsp 65–transfected cells could have resulted in an increased efficiency of tumor-associated antigen processing and presentation and hence an increase in immunogenicity.

III. POSSIBLE APPLICATIONS FOR HSP 65 GENE TRANSFER IN THE TREATMENT OF TUMORS

The results with the J774 cells show that transfection of malignant tumor cells with the bacterial hsp 65 gene can abolish their ability to form tumors and re-

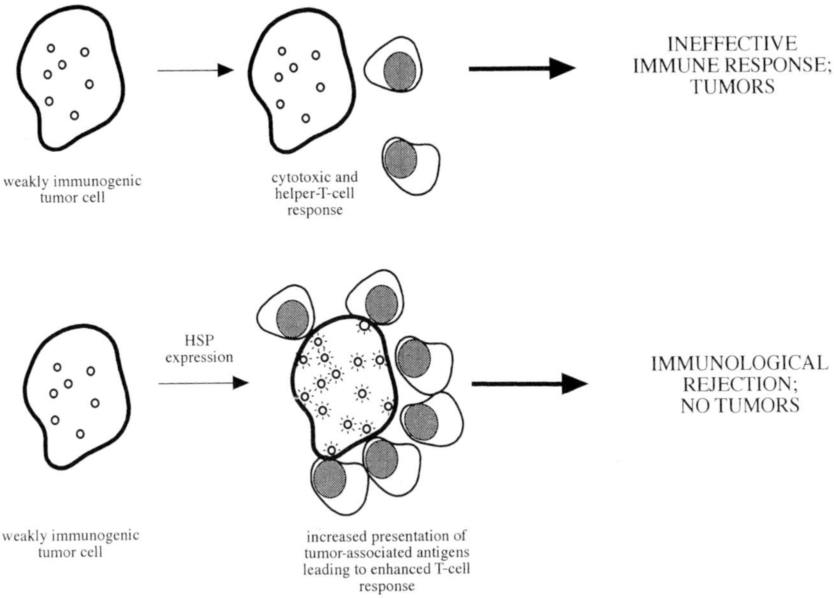

Figure 6 Possible explanation of increase in immunogenicity of tumour cells expressing the hsp 65 gene. Expression of hsp 65 results in more effective presentation of tumor-associated antigens, leading to a more effective immune response.

sult in the induction of immunity against the parent tumor cells. If this approach can be applied to reduce the tumorigenicity and increase the immunogenicity of other cells, then one could envisage its application in the treatment of tumors using either ex vivo or in vivo gene therapy with the hsp 65 gene. In the ex vivo application, tumor cells obtained from biopsies could be transduced in vitro with the hsp gene and reinjected into the host. Cytotoxic T cells induced by the hsp-expressing tumor cells could then recognize and lyse the untransduced parent tumor cells. Alternatively, plasmid DNA encoding the hsp 65 gene could be injected into the tumor in vivo with dual effect. Those tumor cells which express the gene in vivo would lose their tumorigenic potential, and by generating cytotoxic T lymphocytes able to lyse both transduced and untransduced tumor cells, the tumor could be eliminated.

Clearly, many questions remain to be answered before such an approach could be applied in humans. However, the results with the murine J774 cells clearly demonstrate the role of heat shock proteins in the augmentation of immune responses.

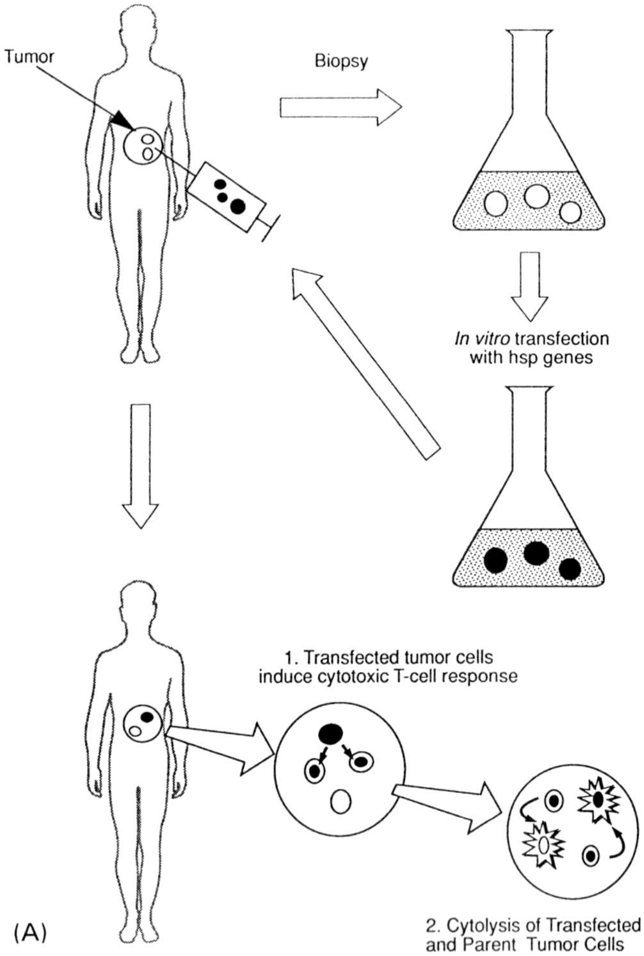

Figure 7 Possible in vivo gene therapy with hsp 65 gene. Ex vivo gene therapy (A) could be carried out with tumor cells obtained from biopsies, transduced with the hsp 65 gene in vitro, and reinjected into the patient. Cytotoxic T cells induced by the transfected tumor cells could eliminate the tumor. Alternatively, DNA encoding the hsp 65 gene could be injected into the tumor in vivo (B). Those cells which become transduced and express the hsp 65 protein would lose their tumorigenic potential and induce cytotoxic T-cell response effective against both the transduced and untransduced tumor cell, resulting in the elimination of the tumor.

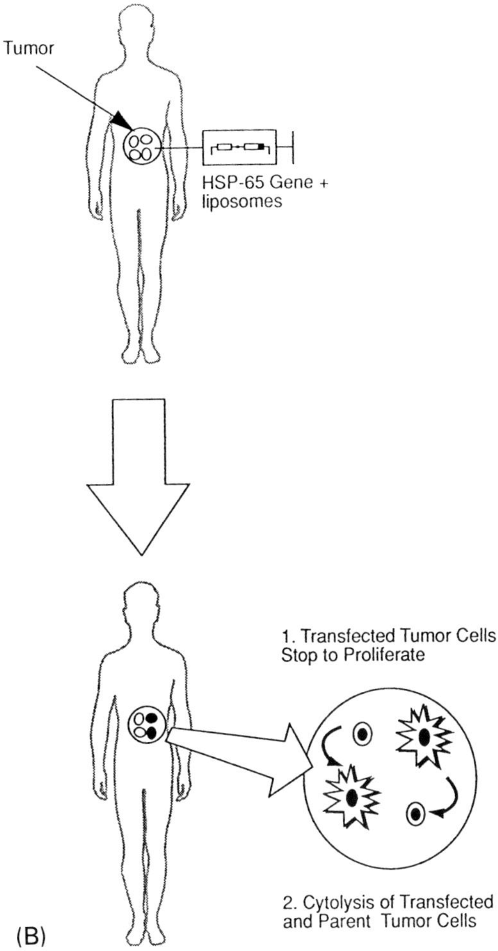

Figure 7 Continued.

REFERENCES

1. K.V. Lukacs, D.B. Lowrie, R.W. Stokes, and M.J. Colston, Tumor cells transfected with a bacterial heat-shock gene lose tumorigenicity and induce protection against tumors, *J. Exp. Med. 178*:343 (1993).
2. G. Klein, H.O. Sjogren, E. Klein, and K.E. Hellstrom, Demonstration of resistance against methylcholanthrene-induced sarcomas in the primary autochthonous host, *Cancer Res. 20*:1561 (1960).
3. D.J. Peace, W. Chen, H. Nelson, and M.A. Cheever, T cell recognition of transforming proteins encoded by mutated *ras* proto-oncogenes, *J. Immunol. 146(6)*:2059 (1991).

4. W. Chen, D.J Peace, D.K. Rovira, S-G. You, and M.A. Cheever, T-cell immunity to the joining region of p210$^{BCR\text{-}ABL}$ protein, *Proc. Natl. Acad. Sci. USA 89*: 1468 (1992).
5. G. Pallesen, S.J. Hamilton-Dutoit, M. Rowe, and L.S. Young, Expression of Epstein-Barr virus latent gene products in tumor cells of Hodgkins disease, *Lancet 337*:320 (1991).
6. L. Chen, E.K. Thomas, S-L. Hu, I. Hellstrom, and K.E. Hellstrom, Human papillomavirus type 16 nucleoprotein E7 is a tumor rejection antigen, *Proc. Natl. Acad. Sci. USA 88*:110 (1991).
7. T. Boon, Toward a genetic analysis of tumor rejection antigens, *Adv. Cancer Res. 58*:177 (1992).
8. R.F.L. James, S. Edwards, K.M. Hui, P.D. Bassett, and F. Grosveld, The effect of class II gene transfection on the tumorgenicity of the H-2k–negative mouse leukaemia cell line K36.16, *Immunology 72*:213 (1991).
9. S. E. Townsend and J.P. Allison, Tumor rejection after direct costimulation of CD8+ cells by B7-transfected melanoma cells, *Science 259*:368 (1993).
10. A. Gutierrez, N.R. Lemoine, and K. Sikora, Gene therapy for cancer, *Lancet 339*:715 (1992).
11. M.P. Columbo and G. Forni, Cytokine gene transfer in tumor inhibition and tumor therapy: where are we now? *Immunol. Today 15(2)*:48 (1994).
12. H. Hock, M. Dorsch, U. Kunzendorf, K. Uberla, Z. Qin, T. Diamantstein, and T. Blankenstein, Vaccinations with tumor cells genetically engineered to produce different cytokines: effectively not superior to a classical adjuvant, *Cancer Res. 53*:714 (1993).
13. U. Steinhoff, B. Schoel, and S.H. Kaufmann, Lysis of interferon-gamma activated Schwann cell by cross-reactive CD8+ alpha/beta T cells with specificity for mycobacterial 65 kd heat shock protein, *Int. Immunol. 2*:279 (1990).
14. R.S. Bahn and A.E. Heufelder, Pathogenesis of Graves' ophthalmopathy, *N. Engl. J. Med. 329*:1468 (1993).
15. C.L. Silva, A. Palacois, M.J. Colston, and D.B. Lowrie, *Mycobacterium leprae* 65hsp antigen expressed from a retroviral vector in a macrophage cell line is presented to T cells in association with MHC class II in addition to MHC class I, *Microb. Pathogen. 12*:27 (1992).
16. M. Hollstein, D. Sidransky, B. Vogelstein, and C.C. Harris, p53 mutations in human cancer, *Science 253*:49 1991.
17. C.L. Silva, K. Lukacs, and D.B. Lowrie, Major histocompatibility complex nonrestricted presentation to CD4 T lymphocytes of Mycobacterium leprae heat-shock protein 65 antigen by macrophages transfected with the mycobacterial gene, *Immunology 78*:35 (1993).
18. K. Lukacs and R.J. Kurlander, MHC-unrestricted transfer of antilisterial immunity by freshly isolated immune CD8 spleen cells, *J. Imunol. 143*:3731 (1989).
19. S.J. Ulrich, E.A. Robinson, L.W. Law, M. Millingham, and E. Appella, A mouse tumor-specific transplantation antigen is a heat shock-related protein, *Proc. Natl. Acad. Sci. USA 83*:3121 (1986).
20. M.A. Palladino Jr, P.K. Srivastava, H.F. Oettgen, and A.B. DeLeo, Expression

of a shared tumor–specific antigen by two chemically induced BALB/c sarcomas, *Cancer Res. 47*:5074 (1987).

21. Y. Tamura, N. Tsuboi, and K. Kikuchi, 70 kDa heat shock cognate protein is a transformation-associated antigen and a possible target for the host's anti-tumor immunity, *J. Immunol. 151*:5516 (1993).
22. H. Udono and P.K. Srivastava, Heat shock protein 70 associated peptides elicit specific cancer immunity, *J. Exp. Med. 178*:1391 (1993).
23. C.R. Ahsan, J. Sasaki, and H.Nomaguchi, The 65-kDa stress protein: is it related to the tumor specific antigen? *Immunol. Lett. 35*:291 (1993).
24. A. Vyakarnman and P.J. Lachmann, The killing of tumor cell targets coupled to tuberculin (PPD) by human and murine PPD-reactive T helper clones. II. Major histocompatibility complex restriction of killing. *Scand. J. Immunol. 27*:347 (1988).
25. A. Vyakarnam, P.J. Lachmann, and D.Y. Sia, The killing of tumor cell targets coupled tuberculin (PPD) by human and murine PPD-reactive T helper clones. I. PPD specificity of killing, *Scand. J. Immunol. 27*:337 (1988).
26. A.R. Lussow, C. Barrois, J. Van Embden, R van der Zee, A. S. Verdini, A. Pessi, J.A. Louis, P-H Lambert, and G. Del Giudice, Mycobacterial heat-shock proteins as carrier molecules, *Eur. J. Immunol. 21*:2297 (1991).
27. C. Barrios, A.R. Lussow, J. Van Embden, R. van der Zee, Rappuoli, P. Constantino, J.A. Louis, P.H. Lambert, and G. Del Giudice, Mycobacterial heat-shock proteins as carrier molecules. II. The use of the 70-kDa mycobacterial heat-shock protein as carrier for conjugated vaccines can circumvent the need for adjuvants and Bacillus Calmette Guerin priming, *Eur. J. Immunol. 22*:1365 (1992).

16

Immune Responses to Stress Proteins in Mycobacterial Infections

Juraj Ivanyi
MRC Clinical Sciences Centre, Royal Postgraduate Medical School, London, England

Pamela M. Norton
Institute for Animal Health, Compton, Newbury, England

Goro Matsuzaki
Medical Institute of Bioregulation, Kyushu University, Fukuoka, Japan

I. INTRODUCTION

Immune responses to mycobacterial heat shock proteins (hsps) have been of central interest in the study of the immunological significance of stress proteins for the past 7 years (1). There are several reasons which accounted for this special attention: (a) Pronounced immunogenicity of hsp 65 has been claimed from the abundance of monoclonal antibodies and on the grounds that the frequency of hsp 65–reactive T cells amounted to 20% of all antimycobacterial specificities (2): although obtained merely following immunization with killed *Mycobacterium tuberculosis* in complete Freund's adjuvant, the interpretation of this finding was sometimes overgeneralized also for tuberculous infections. (b) It has been thought that pathogenic mycobacteria replicating intracellularly would respond to oxidative stimuli in the cytoplasmic environment by enhanced hsp expression, resulting in amplified immunogenicity and possibly constituting also a prerequisite for intracellular parasitism. (c) Infection with tubercle bacilli was considered to represent a stress stimulus, leading to enhanced synthesis of host stress proteins and consequently to influence the antigen processing within infected cells. (d) In view of the broad B- and T-cell cross reactivity of the mycobacterial hsp 65 with bacterial and eukaryotic species, it has been speculated that T cells to conserved epitopes may lead to autoimmune pathology, whereas immunity directed against the polymorphic species-specific epi--topes could be protective (1). (e) Several studies using a purified recombinant preparation of the hsp 65 protein (3) implicated the role of mycobacterial infection without due consideration, that the used mycobacterial protein served merely as a "probe," with structure shared with several homologous microbial and eukaryotic proteins. Thus, initial interest in the special role of mycobacterial hsp 65–reactive T cells in experimental models for arthritis and in rheumatoid arthritis are discussed in another chapter. However, recent experiments demonstrated (a) a wide variety of mycobacterial specificities in the synovial fluid (4), and (b) a paucity of hsp 65–reactive cells in the synovium of patients who were not BCG vaccinated (5). This chapter attempts to review the experimental data so far obtained with a focus on the immunogenicity of stress proteins during infections with pathogenic mycobacteria.

II. ANTIGENIC STRUCTURE

A. Hsp 65

The mycobacterial hsp 65 is highly homologous with the sequence of GroEL from *Escherichia coli* and a 58- to 60-kDa mammalian constituent (5). It accounts for 1% of total protein of *M. tuberculosis* at 37°C, rising to 11% between 42 and 45°C (6). The conserved and variable regions and consequently the respective cross-reactive and species-specific linear T and B epitopes are distributed throughout the protein sequence. A large number of linear B epitopes have been identified by monoclonal antibodies (MAbs) (3). These epitopes are mostly shared between all mycobacterial species, some are distributed even broader to the other bacterial genera, and a few had been found to be specific for either tubercle bacilli or *M. leprae* (Table 1). MAb ML30, cross reactive with human and other mammalian homologues (7), has been localized to bind the peptide sequence 285–295 (8). Of the two MAbs which bind selectively to tubercle bacilli but very few other species, TB78 binds to a conformational epitope, whereas DC16 (9) binds to the linear peptide sequence 128–138. One *M. leprae*-specific linear epitope, IIIE9, resides in the sequence 428–434 (10)

Antigenic mimicry has been represented on the grounds of cross reactivity and/or sequence homology between hsp 65 and mammalian or prokaryotic proteins or unrelated origin (Table 2). Such cross reactivity was demonstrable for the link protein from proteoglycan, human lactoferrin, transferrin, HLA-DR molecules, and human epidermal cytokeratin 1/2 (11–14) and for the secreted antigen 85 complex of mycobacteria (15). The possible functional implications

Table 1 Species-Specific or Selective Epitopes of Hsp 65

Sequence	Code	Characteristics, Specificity	Reference
128–138	DC16	MAb: *M. tuberculosis* selective, linear	9
171–234	TB78	MAb: *M. tuberculosis* selective, conformational	10
231–245	K8AH	T clone: *M. tuberculosis* specific human	45
285–296	ML30	MAb: cross reactive: mammalian hsp 60	7
428–434	IIIE9	MAb: *M. leprae* specific	8

from these cases of antigenic mimicry for the host-parasite relationships and for autoimmune diseases are at present unknown.

A second gene (Cnp60-1) of 61% sequence homology with hsp 65 has been identified in an operon together with the hsp 10 gene (16,17); its immunological function is yet to be evaluated, since it carries practically none of the known B and T epitopes.

B. Hsp 10

The sequence of hsp 10 is widely conserved between several organisms. The *M. tuberculosis* and *M. leprae* genes bear 90% identity and have 44% identity with GroES of *E. coli* (18,19). Although GroES functions as a chaperonin for transport and folding of molecules across membranes in *E. coli*, its role in mycobacteria is poorly defined. Hsp 10 has been found to be present in mycobacterial culture filtrate during the initial 72 hr of culture (20); since a signal peptide is not contained in the sequence, the mechanism of secretion remains undefined. Species-specific epitopes are recognized by monoclonal antibodies

Table 2 Mimicry of Hsp 65

Sequence	Detection	Mimicry with unrelated proteins:	Reference
180–188	T clones	Link protein from proteoglycan, Epstein-Barr virus, hepatitis A virus, lamine	11
229–231 KDLL	Sequence	DR1,2,3 lactoferrin, transferrin	12
277–288	Sequence	Gliadin	Unpublished
?	MAb	Epidermal cells	13
425–540	MAbs	Cytokeratin 1/2	14
479–540 303–424	MAbs	Fibronectin-binding Ag 85 complex,	15

on hsp 10 from *M. tuberculosis* (IT10) and from *M. leprae* (ML06/ML10). This indicates that despite the highly conserved structure, at least some B-cell epitopes are determined by polymorphic sequences. However, these monoclonal antibody–defined epitopes have been found poorly or nonimmunogenic in patients with tuberculosis and lepromatous leprosy.

C. Hsp 71

The sequence identity is 83% between *M. tuberculosis* and *M. leprae*, 56% with DnaK from *E. coli*, and 47% with the hsp 70 mammalian counterpart (21). Unlike hsp 65, where the conserved and polymorphic residues are interspersed throughout the sequence, the hsp 71 amino terminal major part of the molecule is of highly conserved structure, whereas the divergent carboxyterminal 100 amino acids have only 17% conserved sequence between the various species. The protein comprises less than 0.5% at 37°C and rises to almost 20% at 48°C (when hsp 65 synthesis is already switched off) in *M. tuberculosis* (6).

Monoclonal antibodies (L7, L27 B7, B8) bind to most mycobacterial species but neither to *E. coli* nor to mammalian hsp 70. They have all been mapped to the carboxyterminal portion of the molecule using a series of truncated recombinants where they recognize predominantly conformational determinants, whereas B8 showed weak binding to peptide 498–515 (22,23). Scanning of overlapping peptides from the carboxyterminal portion of the molecule with sera from patients with tuberculosis identified four distinct epitope cores: 545–550, 558–564, 571–576, and 591–596 (24). All these epitopic regions differ from the *E. coli* and human sequence and the last three differ also between *M. tuberculosis* and *M. leprae*. Predictive epitope evaluation (Proteus Biotechnology Ltd., Manchester, UK) allocated the highest score to a single stretch of residues 565–591.

D. 18-kDa α-Crystallin

This protein is a member of a family of low molecular weight hsp α-crystallins which has only about 30% of sequence conservation. The 18-kDa protein initially found in *M. leprae*, responds to heat shock in *M. simiae (habana)* serovar 1 (25). It has also been found to be present in the *M. avium* complex (containing two genes) and several other species of mycobacteria but not in *M. tuberculosis* (26). The failure to find homologous gene in *M. tuberculosis* is of concern in view of cross reactivity, shown most explicitly by cross stimulation of *M. leprae*–18-kDa–primed T cells (27). A tentative explanation could be that the cross-reactive molecule in *M. tuberculosis* is the 16-kDa α-crystallin, although it contains only two short isolated stretches with sequence homology (28,29). Immunization with the whole molecule and with synthetic peptides revealed the presence of several linear epitopes expressed on the whole molecule and also cryptic epitopes expressed only by peptides (30).

E. 16-kDa α-Crystallin

This protein of 144 residues (originally designated as 14 kDa) also has about 30% sequence homology with the same family of low molecular weight heat shock proteins which has been discussed above, but evidence that it functions as a stress protein has so far not been obtained (28,29). The protein is found in the external part of the cell wall, but it is not secreted. Two linear epitopes (sequences 31–40 and 61–70) and one conformational *M. tuberculosis* complex-specific epitope (TB68) have been identified with MAbs (31).

III. T-CELL RESPONSES

A. CD4 αβ T Cells

1. *Murine Repertoire*

Following infection with *M. tuberculosis* or *M. bovis* BCG, T-cell proliferation interferon gamma (INF-γ) and other cytokine responses to hsp 65 and hsp 16 have been reported (32,33). The magnitude of hsp responses was similar to that found for other protein antigens. The hsp 65 epitope specificity of T cells from draining lymph nodes of primed mice was examined in respect to seven selected peptides (34). Immunodominant as sell as cryptic epitopes had been identified. Interestingly, the response to both immunodominant epitopes (153–171 and 180–196) was found to be H-2 permissive, whereas responses to cryptic peptide epitopes were found to be H-2 restricted. T cells stimulated with peptide 153–171 recognize also GroEL (hsp 60 from *E. coli*); hence implicating the potential role of sensitization with the gut flora. As antigen-presenting molecules, both IA and IE were shown to be involved in respect to peptide 65–85. The H-2-permissive responsiveness is not unique to hsp antigens and was observed also with other mycobacterial antigens.

The role of non–H-2 genetic control of the T-cell responses of 18-kDa antigen in the proliferation, but not in the delayed-type hypersensitivity (DTH) assay, was indicated by the low responsiveness of primed C57B/10J, and high response of BALB.B (both $H\text{-}2^b$) lymph node cells (27,35,36). BALB/cJ strain T cells recognized the most potent immunodominant epitope within residues 106–125, whereas residues 1–20, 31–50, and 121–140 stimulated in the B10.BR strain. It is of interest that epitope 106–125 is adjacent to the epitope of the MAb L5 (110–115) and overlapping with a conserved portion of sequence of this protein.

Analysis of 16-kDa epitopes with primed lymph node cells demonstrated four immunodominant epitopes and a multitude of cryptic epitopes localized throughout the sequence (37). However, only two of the immunodominant epitopes (sequences 21–40 and 71–91), both of which are *M. tuberculosis* complex specific, have been recognized by T cells from *M. tuberculosis*-infected mice. The mechanisms which lead to the focusing of the T-cell repertoire during tuberculous infection of mice has not been identified, but it could be explained on the assumption that infected macrophages have a more pronounced proteolytic degradation of this antigen. Furthermore, since both epitopes which remain immunogenic following infection bind to hsp 70 (38), it is possible that complexing with this chaperone may confer selective protection to the peptides during antigen processing (39).

2. *Human Repertoire*

Mycobacterial hsp antigens are recognized by CD4 $\alpha\beta$ T-cell lines of predominantly TH1 phenotype in the context of multiple HLA-DR haplotypes (40,41). Irrespective of cytokine profile or epitope specificity, T cells can lyse BCG-infected or activsated monocytes (42–44). One epitope of hsp 65 with *M. tuberculosis complex specificity* has been localized to the peptide sequence 231–245 using an HLA-DR2–restricted CD4 cell clone (45). The lack of reaction with the homologous peptide from the hsp 65 of *M. leprae* is based on differences in single residues at two positions (G $\Rightarrow$ Q and P $\Rightarrow$ S), both of which are involved in binding to the antigen-presenting DR2 molecule and can potentially change the conformation of this peptide. Recognition of a DR3-restricted hsp 65 peptide 2–12 by the T-cell receptors of different V_β families was examined using 15 clones (46). Although $V_\beta 5$ was used by 14 clones, this could not be interpreted as general dominance in the repertoire, since T cells responding to phytohemagglutinin (PHA) also used the same T-cell receptor (TCR) family expression. T clones used in the two quoted studies were derived from the blood of healthy sensitized or arthritic donors, and it is not known if the repertoire of patients with tuberculosis or leprosy has a different representation of T-cell specificities.

T-cell clones reacting against epitopes shared between mycobacterial and mammalian hsp 60 have been obtained from the peripheral blood of healthy individuals or induced by local inflammation (47–49). It is not known, at present, whether this apparent autoimmune T-cell reaction plays a pathogenic role in the natural development of mycobacterial or autoimmune disease. In contrast, a protective effect of hsp-65–targeted autoimmune CD4 T cells against certain tumors in the natural immune surveillance against transformed autologous cells was suggested on the basis of suppression of in vivo growth of high hsp 65–expressing fibrosarcoma cells (50). The recognition of certain hsp 65 by antibodies is open to even more practicable therapeutic application, as demonstrated by the antitumor action of MAb ML30 labeled with the immunotoxin saporin (51).

Mycobacterial hsp 10 was found to be stimulatory skin DTH reactions (52,53) and of in vitro responses of T cells from the peripheral blood of sensitized healthy subjects and patients with tuberculoid leprosy (19,54). T-cell clones, both specific for *M. leprae* and cross reactive, have been mapped to a 27-mer aminoterminal peptide (19), and one epitope within the sequence 11–25 was identified with pleural fluid T cells (55).

Hsp 70 stimulates proliferative T-cell responses in the majority of sensitised healthy persons, patients with tuberculosis as well as tuberculoid leprosy, and T-cell precursor frequencies have been found correlated with those which recognize hsp 65 (56,57). Analysis of epitope specificity, first with truncated gene products, localized the epitopes to a fragment adjacent to the carboxyterminal, the most divergent part of the molecule. Using synthetic peptides, the position of several potential epitopes was allocated to the region 278–502. It is of interest that hsp 71 can also be recognized in the context of HLA DRw53, which is frequent in populations of tropical countries (41).

Proliferative responses to the 18-kDa antigen of *M. leprae* have been demonstrable in 93% of leprosy contacts but only in 50% of patients with tuberculoid leprosy and even less in European PPD-positive donors (58). T-cell clones from healthy sensitized individuals against the 18-kDa antigen of *M. leprae* have been found to be restricted by HLA DR4.Dw4 while all mapping to peptide 38–50 (59), whereas DR1 individuals recognize a different, yet unidentified, epitope (41). It is not known if recognition of these epitopes is altered in any way in patients with leprosy, but a follow-up study of leprosy contacts during lepromin conversion showed that the early T-cell repertoire was directed neither to the 18-kDa nor to the hsp 65 proteins but predominantly against the fibronectin-binding antigen 85 (60).

B. CD8 T Cells

Cloned murine cytotoxic cells of CD8 phenotype, restricted by H-2 class I molecules, were reported to be cytolytic against target cells expressing both mycobacterial and autologous hsp 65 molecules (61). The stimulatory autologous macrophages needed preincubation with INF-γ or cytomegaloviral infection, which presumably led to enhanced expression of the target hsp 60 molecule. In another context, CD8 T cells purified from the spleens of *M. vaccae*-primed mice have been found active as "suppressor" cells (62). Suppressive activity, without H-2 restriction specificity, was demonstrable following the stimulation of cells with the hsp 65 antigen and attributed to competition for interleukin-2 (IL-2), because the target cells were represented in the test by IL-2–dependent CD4 T cells.

C. γδ T Cells

Increase of γδ T cells have been reported in normal and athymic mice immunized with *M. tuberculosis* in incomplete Freund's adjuvant (63,64). γδ T cells were also induced by PPD-containing aerosol in resident pulmonary lymphocytes (65). Increase of γδ T cells has also been reported in *M. bovis* BCG-infected mice (66,67). Although *M. tuberculosis* has been the antigenic stimulus in these murine studies and also for human Vγ9/Vγ2 T cells, representing a major population in normal individuals (68), the exact antigenic specificity in these experiments has not been determined.

Experiments using T cell hybridomas have shown that hsp 65 is recognized without preimmunization by γδ T cells mainly Vγ1 and Vδ6 from the newborn thymus or from the spleen and liver of adult mice (69,70). The Vγ1 family is dominantly expressed in peripheral lymphoid tissues of unimmunized mice and most of the hybridomas (62% Vγ1-bearing cells) recognized hsp 65, suggesting that a large γδ T-cell repertoire specific to hsp 60 exists in normal animals. The specificity of γδ T-cell recognition against hsp 65 was shown to be determined predominantly by the Vγ1 chain rather than the δ chain. Furthermore, CDR3 sequence of γ chain (Vγ1–Jγ1 junction) is not important in recognition of hsp 65, since there is no consensus sequence in γ chain CDR3 sequence expressed by hsp 65–recognizing γδ T-cell hybridomas. The hsp 65–specific γδ T cells have spontaneous reactivity and recognize oligopeptides corresponding to amino acids 180–196 and to lesser extent to its homologous peptide of mammalian hsp 60. Recent studies identified the minimal stimulatory 7-mer peptide FGLQLEL, which turned out to be mycobacteria specific (71). The authors classify this response as a separate type with intermediary features between superantigenic and conventional antigenic responsiveness.

A second family of murine γδ T cells against hsp 60 resides in Vδ5 cells, expanded by syngeneic mixed lymphocyte culture or PPD from BALB/c mice (72). The Vδ5 chain carries the invariant junctional sequence (called BALB/c invariant delta, BID), and a BID-bearing γδ T-cell hybridoma recognized irradiated BALB/c–derived splenic and MethA fibrosarcoma-stimulator cells. The self-recognition of the BID-bearing γδ T-cell hybridoma was blocked by addition of anti–hsp 65 MAb ML30(7) which cross reacts with murine hsp 60, suggesting that BID-bearing γδ T cells have the ML30-binding 285–296 sequence specificity expressed on the cell surface (Table 3).

Supporting this hypothesis, ML30 MAb prepared from serum free medium-stained cell surface of a population of MethA fibrosarcoma and irradiated spleen cells of BALB/c mice (50; Matsuzaki and Kobayashi unpublished observations). A similar specificity of γδ T cells against a cell surface epitope detected by anti–hsp 60 antibodies was also reported for human cells (73).

IV. ANTIBODY LEVELS

A. Murine Models

1. *Genetic Control*

The genetic control over antibody levels to hsp antigens depends on the mode of immunization and was shown to involve both H-2 and non–H-2 genes. Immunization with hsp 65 in adjuvants, however, revealed H-2 control in respect

Table 3 Detection of TCR Gene Products in T Cells Stimulated by Syngeneic MLR Using Dot Blot Analysis

		ML30 MAb in culture	
TCR gene	Family	absent	present
Vγ	½	+++	+
	4	++	+
	5	–	–
	6	+	–
	7	–	–
Vδ–Jδ1	1	++	+
	2	–	–
	3	–	–
	4	+	+
	5	+++	–
	6	++	+
	7	+	+

Cell culture for 5 days. PCR positive following 20(+++), 25(++), 30(+) cycles or negative (–) at 30 cycles.
Source: Data from Ref. 72.

of antibody levels to a single epitope. Antibody levels to the TB78 epitope of hsp 65 were found to be H-2 restricted, but with a different haplotype hierarchy in H-2 congenic strains with different background genes (74). Antibody levels to the 16-kDa antigen are also H-2 controlled, but they display the unusual dominant low-response phenotype in F1 hybrid animals.

Following infection of mice with *M. tuberculosis*, C57BI strains, irrespective of their H-2 haplotype, are low responders, whereas all BALB background strains are high responders to both hsp 65 and hsp 71 (75). These genetic differences in antibody levels are demonstrable following tuberculous infection but not after immunization with the hsp antigens in adjuvant. Antibody levels to both hsp 65 and hsp 71 antigens have been found high in (C57BI × BALB)F1 mice and they segregate in the C57BI backcross population independently for the hsp 65 and hsp 71 specificities (Table 4). The high frequency of low-responder animals indicates low penetration of these genes, perhaps due to unknown environmental factors which are common for both antibody specificities. Although these preliminary data indicate a role for one or a few yet unidentified genes, the molecular gene targets and their mechanism of action are not understood. Non–H-2 genes were shown to control also the response to individual linear peptide epitopes of hsp 65 of *M. leprae* (76); in this study, antisera from several inbred strains of mice reacted with peptides of the sequences 51–71, with

Table 4 Segregation of Antibody Responsiveness to Hsp Antigens in Backcross Mice

Responders to		Hsp 71 negative	Hsp 71 positive
Hsp 65	Negative	135	48
	Positive	18	22

Backcross mice (n = 223): (BALB/c × C57BI) × C57BI were infected with H37Rv and bled after 10 weeks. Antibody levels to purified recombinant mycobacterial hsp 65 (Ma6A) and hsp 71 (both from WHO-TDR Bank) were tested by ELISA. Positive serum titers > 1:500.

476–495 coinciding with the binding of MAb ML30 and with the 153–171 immunodominant T-cell epitope (75).

2. *Carrier Function of Hsps*

Both hsp 65 and hsp 71 recombinant proteins from *M. bovis* have an intriguing capacity to act as adjuvant-free carriers for the stimulation of antibody responses to otherwise nonimmunogenic peptide and oligosaccharide haptens (77). The Hsp 65–peptide conjugate became immunogenic only in mice which had previously been infected with live BCG, presumably as a priming stimulus for hsp 65 T cells. In contrast, two injections of the hsp 71–peptide conjugate without adjuvant led to equal levels of antipeptide antibodies in BCG-primed and BCG-nonprimed animals. However, T-carrier effects were not influenced by the level of anticarrier antibodies. The authors considered that hsp 71 was more potent due to either (1) natural environmental priming or (2) chaperone function during the processing of antigenic peptides; therefore, the carrier effect was achieved by a different mechanism from that which operated with hsp 65.

B. Human Diseases

Serological diagnosis of tuberculosis and leprosy require antigens of high species specificity and immunogenicity. Since stress proteins have mainly shared sequences, at least between pathogenic and several environmental mycobacteria, their serodiagnostic potential hinges on the question of whether the immunogenic epitopes will be determined by the conserved cross-reactive regions or by the polymorphic species-specific sequences. Indeed, hsp 65 is not useful for the diagnosis of tuberculosis, because antibodies to the whole molecule have high background levels, whereas the *M. tuberculosis*-specific TB78 epitope is poorly immunogenic (78). However, antibody levels to mycobacterial hsp 60 have been found elevated in the sera of patients with Crohn's disease; thus, supporting its mycobacterial etiopathogenesis (79). Interestingly, antibodies from patients with Crohn's disease did not react with human hsp 60, in striking contrast to ulcerative colitis, where the sera reacted with human but not with mycobacterial hsp 65.

Studies of antibody levels to hsp 71 have indicated serodiagnostic potentials in both tuberculosis and leprosy. Antibodies from sera of patients with lepromatous leprosy bound predominantly to the carboxyterminal recombinant fragments which contain the polymorphic sequences (22). Although the immunogenicity of the carboxyterminal part of mycobacterial hsp 71 and of other infectious pathogens is diagnostically favorable, sera from patients with leprosy and tuberculosis remain mutually cross reactive. Another attribute of hsp 71 is the elevated antibody levels even in patients with smear-negative tuberculosis, although the persisting titers in self-healed or past-treated tuberculosis and in lung fibrosis are diagnostically detrimental (80). However, it remains of interest to learn whether the hsp 71 immunogenicity is due to enhanced expression of hsp 71 in the infected macrophages. However, the specificity may be hard to improve using carboxyterminal peptides, because the bulk of antibodies seem to be directed to conformational rather than linear epitopes in leprosy (23) as well as in tuberculosis (unpublished data). Finally, antibody levels to the TB68 epitope of the 16-kDa antigen have been found elevated in about 70% of patients with smear-positive tuberculosis (78) and serological positivity in TB meningitis, childhood TB, and in family or professional contacts of patients with TB suggest that the 16-kDa protein is selectively immunogenic in the early stages of tuberculous infection (81,82).

V. EXPRESSION OF HOST STRESS PROTEINS

Mammalian cells normally express immunogenic hsp peptides which are recognized in MHC-restricted manner by autoreactive CD4 T cells under conditions of syngeneic mixed lymphocyte culture (83). The expression of mammalian hsp molecules in infected macrophages is of keen interest in the light of the report that enhanced hsp 60 expression closely correlated with the host protective immunity following immunization with *Toxoplasma gondii* (84). In the case of mycobacterial infection, it was suggested that macrophages producing reactive oxygen metabolites may protect themselves from damage by enhanced synthesis of stress proteins (85). However, hsp 70 synthesis in human B cells was found to be stimulated following heat shock but not by peroxide treatment (86); moreover, heat stress failed to protect the cells from the adverse effects of peroxide treatment on antigen processing. So far, only one study has reported an increased production of the heat-inducible, but not of the constitutive, form of hsp 70 in monkey Schwann cells following infection with *M. leprae* (87).

Recently, we examined the expression of hsp 70 in murine macrophages following infection with *M. bovis* BCG and other mycobacteria. The results showed that phagocytosis of BCG stimulated the hsp 70 production in macrophages which had matured from bone marrow cells for 7 days but not in those macrophages which matured for 10 days (Fig. 1, lanes b and h). Furthermore, *M. avium, M. cheloniae,* or opsonized zymosan failed to stimulate the hsp 70 production (Fig. 1, lanes a, c, d, and i). The BCG-stimulated hsp 70 production was weaker than that produced by heat (Fig. 1, lanes f and j) and did not appear to be an obligatory event following infection. The hsp 70 response seemed to depend critically on the stage of cell maturation and probably on some other yet undefined characteristics of the infected cells. Differences in the level of constitutive IL-1 production, depending on the stage of macrophage differentiation, may be one regulatory factor, but BCG stimulated consistently produced IL-1 production even in the absence of demonstrable hsp 70 synthesis (Fig. 1, lane h). The stimulation of IL-1 and other cytokines has previously been attributed to the action of lipoarabinomannan derived from *M. tuberculosis* (88). Alternatively, the possible role of BCG–derived hsp 65 needs to be considered in the light of the demonstration that lipopolysaccharide-depleted recombinant

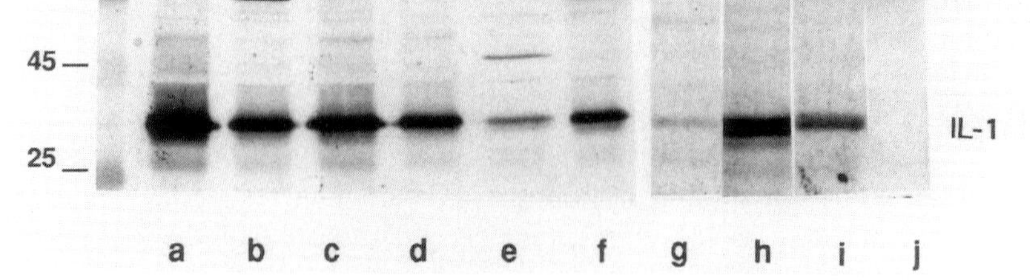

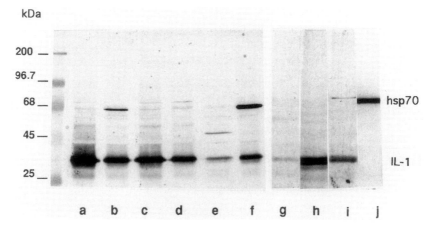

Figure 1 Hsp 70 and IL-1 production by mouse microphages. Bone marrow cells from BALB/c mice have been matured in vitro for 7 (lanes a–f) or 10 days (lanes g and j), followed by incubation for 1 hr with *M. avium* (lane a) BCG (lanes b and h), *M. cheloniae* (lane c), opsonized zymnosan (lanes d and i) medium (lanes 3 and g), or heat shock 45°C for 30 min (lanes f and j). Cells were washed, incubated for 20 hr, and cell extracts were analyzed by Western blot. Each strip was developed with antibodies against hsp 70 (RPN 1197, Amersham), producing a band of corresponding molecular weight and subsequently against IL-1β (S329 BM, NBSB, Potters Bar), producing a band of 33 kDa.

mycobacterial hsp 65 stimulated tumor necrosis factor (TNF), IL-6, and IL-8 cytokine mRNA in the THP-1 human monocyte cell line (89).

The variable occurrence of enhanced hsp 71 synthesis following BCG infection at some, but not all, stages of macrophage differentiation in our experiments may be due to the particular mechanism of mycobacterial entry into macrophages via complement receptors CR1 or CR3 which do not stimulate the production of reactive oxygen metabolites (90); the latter can be efficiently inactivated by mycobacterial constituents, such as superoxide dismutase or lipoarabinomannan. Indeed, recent studies suggest that BCG can merely amplify the action of oxidase stimulants phorbol myristate acetate or zymosan but does not act alone (91).

VI. CONCLUSIONS

1. The magnitude of response and cytokine profiles of hsp 65–reactive T cells in mice following infection with pathogenic mycobacteria have been found to be generally similar to responses against other mycobacterial proteins. Evidence for the regulation of immunogenicity for T cells by either bacterial or host cell stress response during infection has so far not been obtained.

2. T cells which carry either the $\alpha\beta$ (both CD4 and CD8) or $\gamma\delta$ receptor recognize antigenic determinants common for mycobacterial and homologous mammalian hsp 65/60 in healthy humans and mice. The function of these T cells may be either the beneficial clearing of aberrant host cells or they might be involved in autoimmune pathology (92).

3. Both hsp 65 and hsp 71 can act as carriers which convey immunogenicity to various haptens and are therefore relevant for vaccine development (77). This adjuvant function is attractive by its apparent lack of toxicity, but it has the undesirable element of autoimmunization; a possible role of hsp-mediated immunomodulation during infection with pathogenic mycobacteria has not been determined.

4. The magnitude of the antibody response to hsp 65 and hsp 71 (but not to the other mycobacterial antigens) is strongly influenced by non–H-2 genetic factors (75). This genetic control is expressed in *M. tuberculosis*-infected mice, but it is overridden by immunization with stress proteins in adjuvants. However, it is not known whether the genetic dependence results from either parasite or host cell stress responses.

5. The changes in the expression of host stress proteins following intracellular parasitism of pathogenic mycobacteria remain of considerable interest. So far, only hsp 70 synthesis has been monitored in infected cells, but the exact conditions under which hsp synthesis is elevated require better characterization; furthermore, changes in intracellular distribution will need to be monitored in view of the report that hsp molecules normally expressed intracellularly become expressed on the cell surface following HTLV-1 infection (93).

6. Stress induced overexpression of antigens within the infected host cells corresponding to that which exists for *Salmonella typhimurium* and *Shigella flexneri* (94,95) has so far not been documented for pathogenic mycobacteria; experimental work along these lines may advance following recent progress in detecting the mRNA of hsp 71 of *M. leprae* in vivo (96).

7. It had originally been proposed that mycobacterial hsp proteins of conserved structure could mimic the intracellular function of homologous eukaryotic proteins within infected cells (81); it was speculated at the same time that mycobacterial hsp "decoys" could cause a defect in host defense reactions and hence constitute a basis for mycobacterial virulence. This theory can derive support from recent evidence that both mammalian and mycobacterial hsp 71 can bind certain peptide epitopes (38) which could impart a chaperoning role in antigen presentation (39). This concept of "heterologous chaperoning" may have wider implications for the pathogenesis of bacterial infections with intracellular pathogens.

REFERENCES

1. D.B. Young, A. Mehlert, V. Bal, P. Mendez-Semperio, J. Ivanyi, and J.R. Lamb, Stress proteins and the immune response to mycobacteria—antigens as virulence factors?, *Antonie. van Leeuwenhoek J. Microbiol. 54*:431 (1988).
2. S.H.E. Kaufmann, U. Vath, J.E.R. Rhole, J.D.A. van Embden, and F. Emmrich, Enumeration of T cells reactive with *Mycobacterium tuberculosis* organisms and specific for recombinant mycobacterial 64-kDa protein, *Eur. J. Immunol. 17*:351 (1987).
3. J.E.R. Thole and R. Van der Zee, The 65 kD antigen:molecular studies on a ubiquitous antigen, *Molecular Biology of the Mycobacteria* (J. McFadden, ed.) 37–67, Surrey University Press, London, 1990, p. 30.
4. P.C.M. Res, D.L.M. Orsini, J.M. Van Laar, A.A.M. Janson, C. Abou-Zeid, and R.R.P. De Vries, Diversity in antigen recognition by *Mycobacterium tuberculosis*-reactive T cell clones from the synovial fluid of rheumatoid arthritis patients, *Eur. J. Immunol. 21*:1297 (1991).
5. F.D. Crick and P.A. Gatenby, Limiting dilution analysis of T cell reactivity to mycobacterial antigens in peripheral blood and synovium from rheumatoid arthritis patients, *Clin. Exp. Immunol. 88*:424 (1992).
6. D.B. Young and T.R. Garbe, Heat shock proteins and antigens of *Mycobacterium tuberculosis, Infect. Immun. 59*:3086 (1991).
7. D.J. Evans, P. Norton, and J. Ivanyi, Distribution in tissue section of the human groEL stress-protein homologue, *APMIS 98*:437 (1990).
8. D.C. Anderson, M.E. Barry, and T.M. Buchanan, Exact definition of species-specific and cross-reactive epitopes of the 65-kilodalton protein of *Mycobacterium leprae* using synthetic peptides, *J. Immunol. 14*:607 (1988).
9. A.H. Hajeer, J. Worthington, K. Morgan, and R.M. Bernstein, Monoclonal anti-

body epitopes of mycobacterial 65-kD heat-shock protein defined by epitope scanning, *Clin. Exp. Immunol.* *89*:115 (1992).
10. H.D. Engers, B.R. Bloom, and T. Godal, Monoclonal antibodies against mycobacterial antigens, *Immunol. Today* *6*:345 (1985).
11. W. van Eden, E.J.M. Hogervorst, R. van der Zee, J.D.A. van Embden, E.J. Hensen, and I.R. Cohen, The mycobacterial 65 kD heat-shock protein and autoimmune arthritis, *Rheumatol. Int.* *9*:187 (1989).
12. A.P. Aguas, N. Esaguy, C.E. Sunkel, and M.T. Silva, Cross-reactivity and sequence homology between the 65-kilodalton mycobacterial heat shock protein and human lactoferrin, transferrin, and DR_β subsets of major histocompatibility complex class II molecules, *Infect. Immun.* *58*:1461 (1990).
13. Th. W. Van Den Akker, B. Naafs, A.H.J. Kolk, E. De Glopper-Van der Veer, R.A.M. Chin, A. Lien, and Th. Van Joost, Similarity between mycobacterial and human epidermal antigens, *Br. J. Dermatol.* *127*:352 (1992).
14. A. Rambukkana, P.K. Das, S. Krieg, S. Yong, I.C. Le Poole, and J.D. Bos, Mycobacterial 65,000 MW heat-shock protein shares a carboxy-terminal epitope with human epidermal cytokeratin 1/2. *Immunology* *77*:267 (1992).
15. A. Rambukkana, P.K. Das, J.D. Burggraaf, W.R. Faber, P. Teeling, S. Krieg, J.E.R. Thole, and M. Harboe, Identification and characterization of epitopes shared between the mycobacterial 65-kilodalton heat shock protein and the actively secreted antigen 85 complex: their in situ expression on the cell wall surface of *Mycobacterium leprae, Infect. Immun.* *60*:4517 (1992).
16. T.F. Rinke de Wit, S. Bekelie, A. Osland, T.L. Miko, P.W.M. Hermans, D. van Sooligen, J.-W. Drijfhout, R. Schoningh, A.A.M. Janson, and J.E.R. Thole, Mycobacteria contain two GroEL genes: the second *Mycobacterium leprae* GroEL gene is arranged in an operon with groES, *Mol. Microbiol.* *6*:1995 (1992).
17. T.H. Kong, A.R.M. Coates, P.D. Butcher, C.J. Hickman and T.M. Shinnick, *Mycobacterium tuberculosis* expresses two chaperonin-60 homologs, *Proc. Natl. Acad. Sci. USA* *90*:2608 (1993).
18. P.N. Baird, L.M.C. Hall, and A.R.M. Coates, Cloning and sequence analysis of the 10 kDa antigen gene of *Mycobacterium tuberculosis, J. Gen. Microbiol.* *135*:931 (1989).
19. V. Mehra, B.R. Bloom, A.C. Bajardi, C.L. Grisso, P.A. Sieling, D. Alland, J. Convit, X. Fan, S. W. Hunter, P.J. Brennan, T.H. Rea, and R.L. Modlin, A major T cell antigen of *Mycobacterium leprae* is a 10-kD heat-shock cognate protein, *J. Exp. Med.* *175*:275 (1992).
20. C. Abou-Zeid, I. Smith, J.M. Grange, T.L. Ratliff, J. Steele, and G.A. W. Rook, The secreted antigens of *M. tuberculosis* and their relationship to those recognised by the available antibodies, *J. Gen. Microbiol.* *134*:531 (1988).
21. K.R. McKenzie, E. Adams, W.J. Britton, R.J. Garsia, and A. Basten, Sequence and immunogenicity of the 70-kDa heat shock protein of *Mycobacterium leprae, J. Immunol.* *147*:312 (1991).
22. M.P. Davenport, K.R. McKenzie, A. Basten, and W.J. Britton, The variable C-terminal region of the *Mycobacterium leprae* 70-kilodalton heat shock protein is the target for humoral immune responses, *Infect. Immun.* *60*:1170 (1992).

23. P.W. Peake, W.J. Britton, M.P. Davenport, P.W. Roche, and K.R. McKenzie, Analysis of B-cell epitopes in the variable C-terminal region of the *Mycobacterium leprae* 70-kilodon heat shock protein, *Infect. Immun. 61*:135 (1993).
24. A. Elsaghier, R. Lathigra, and J. Ivanyi, Epitope-scan of Mycobacterium tuberculosis-specific carboxy-terminal residues of the hsp71 stress protein, *Mol. Immunol. 29*:1153 (1992).
25. F.I. Lamb, N.B. Singh, and M.J. Colston, The specific 18-kilodalton antigen of *Mycobacterium leprae* is present in *Mycobacterium habana* and functions as a heat-shock protein, *J. Immunol. 144*:1922 (1992).
26. R.J. Booth, D L. Williams, K.D. Moudgil, L.C. Noona, P M. Grandisonn, J.J. McKee, R.L. Prestidge, and J.D. Watson, Homologs of *Mycobacterium leprae* 18-kilodalton and *Mycobacterium tuberculosis* 19-kilodalton antigens in other mycobacteria, *Infect. Immun. 61*:1509 (1993).
27. D.P. Harris, B.T. Backstrom, R.J. Booth, S.G. Love, D R. Harding, and J D. Watson, The mapping of epitopes of the 18-kDa protein of *Mycobacterium leprae* recognised by murine T cells in a proliferation assay, *J. Immunol. 143*:2006 (1989).
28. A. Verbon, R.A. Hartskeerl, A. Schuitema, A.H.J. Kolk, D.B. Young, and R. Lathigra, The 14,000-molecular-weight antigen of *Mycobacterium tuberculosis* is related to the alpha-crystallin family of low-molecular-weight heat shock proteins, *J. Bacteriol. 174*:1352 (1992).
29. B-Y. Lee, S.A. Hefta, and P.J. Brennan, Characterisation of the major membrane protein of virulent *Mycobacteium tuberculosis, Infect. Immun. 60*:2066 (1992).
30. T.M. Doherty, B.T. Backstrom, R.L. Prestidge, S G. Love, D R.K. Harding, and J.D. Watson, Immune responses to the 18-kDa protein of *Mycobacterium leprae*. Similar B cell epitopes but different T cell epitopes seen by inbred strains of mice, *J. Immunol. 146*:1934 (1991).
31. A. Verbon, R.A. Hartskeerl, C. Moreno, and A.H.J. Kolk, Characterization of B cell epitopes on the 16K antigen of *Mycobacterium tuberculosis, Clin. Exp. Immunol. 89*:395 (1992).
32. K. Huygen, D. Abramowicz, P. Vandenbussche, F. Jacobs, J. de Bruyn, A. Kentos, A. Drowat, J.-P. van Vooren, and M. Goldman, Spleen cell cytokine secretion in *Mycobacterium bovis* BCG-infected mice, *Infect. Immun. 60*:2880 (1992).
33. I.M. Orme, E.S. Miller, A.D. Roberts, S.K. Furney, J.P. Griffin, K.M. Dobos, D. Chi, B. Rivoire, and P.J. Brennan, T lymphocytes mediating protection and cellular cytolysis during the course of *Mycobacterium tuberculosis* infection, *J. Immunol. 148*:186 (1992).
34. S.J. Brett, J.R. Lamb, J.H. Cox, J.B. Rothbard, A. Mehlert, and J. Ivanyi, Differential pattern of T cell recognition of the 65-kDa mycobacterial antigen following immunization with the whole protein or peptides, *Eur. J. Immunol. 19*:1303 (1989).
35. B.T. Backstrom, D.P. Harris, R.L. Prestidge, and J.D. Watson, Genetic control of immune responses to the 18-kDa protein of *Mycobacterium leprae, Cell Immunol. 142*:264 (1992).
36. T.M. Doherty, B.T. Backstrom, S.G. Love, D.R.K. Harding, and J.D. Watson, Identification of T cell stimulatory epitopes from the 18 kDa protein of *Mycobacterium leprae, Int. Immunol. 5*:673 (1993).

37. H.M. Vordermeier, D.P. Harris, R. Lathigra, E. Roman, C. Moreno, and J. Ivanyi, Recognition of peptide epitopes of the 16,000 MW antigen of *Mycobact erium tuberculosis* by murine T cells, *Immunology 80*:6 (1993).
38. E. Roman, C. Moreno, and D.B. Young, Mapping of Hsp70 binding sites on protein antigens, *Eur. J. Biochem.* (1994).
39. D.C. De Nagel, and S K. Pierce, A case for chaperones in antigen processing, *Immunol. Today 13*:86 (1992).
40. W.C.A. Van Shooten, D.G. Elferink, J. Van Embden, D.C. Anderson, and R.R. P. De Vries, DR3-restricted T cells from different HLA-DR3-positive individuals recognize the same peptide (amino acids 2–12) of the mycobacterial 65-kDa heat-shock protein, *Eur. J. Immunol. 19*:2075 (1989).
41. A.S. Mustafa, K.E.A. Lundin, and F. Oftung, Human T cells recognise mycobacterial heat shock proteins in the context of multiple HLA-DR molecules: studies with healthy subjects vaccinated with *Mycobacterium bovis* BCG and *Mycobacterium leprae, Infect. Immun. 61*:5294 (1993).
42. T.H.M. Ottenhoff, B. Kale, J.D.A. Embden, J.E.R. Thole, and R. Kiessling, The recombinant 65-kD heat shock protein of *Mycobacterium bovis* bacillus Calmette-Guerin/*M. tuberculosis* is a target molecule for $CD4^+$ cytotoxic T lymphocytes that lyse human monocytes, *J. Exp. Med. 168*:1947 (1988).
43. D.S. Kumararantne, A.S. Pithie, P. Drysdale, J.S.H. Gaston, R. Kiessling, P.B. Iles, C.J. Ellis, J. Innes, and R. Wise, Specific lysis of mycobacterial antigen-bearing macrophages by class II MHC-restricted polyclonal T cell lines in healthy donors or patients with tuberculosis, *Clin. Exp. Immunol. 80*:314 (1990).
44. W.H. Boom, R.S. Wallis, and K.A. Chervenak, Human *Mycobacterium tuberculosis*-reactive $CD4^+$ T-cell clones: heterogeneity in antigen recognition, cytokine production, and cytotoxicity for mononuclear phagocytes, *Infect. Immun. 59*:2737 (1991).
45. F. Oftung, A.S. Mustafa, T.M. Shinnick, R.A. Houghten, G. Kvalheim, M. Degre, K.E.A. Lundin, and T. Godal, Epitopes of the *Mycobacterium tuberculosis* 65-kilodalton protein antigen as recognized by human T cells, *J. Immunol. 141*:2749 (1988).
46. J. Henwood, J. Loveridge, J.I. Bell, and J.S.H. Gaston, Restricted T cell receptor expression by human T cell clones specific for mycobacterial 65-kDa heat-shock protein: selective *in vivo* expansion of T cells bearing defined receptors, *Eur. J. Immunol. 23*:1256 (1993).
47. J.R. Lamb, V. Bal, P. Mendez-Sampario, A. Mehlert, A. So, J.B. Rothbard, S. Jindal, R.A. Young, and D.B. Young, Stress proteins may provide a link between the immune response to infection and autoimmunity, *Int. Immunol. 1*:191 (1989).
48. M.E. Munk, B. Schoel, S. Modrow, R.W. Karr, R.A. Young, and S.H.E. Kaufmann, T lymphocytes from healthy individuals with specificity to self eptiopes shared by the mycobacterial and human 65KDa heat-shock protein, *J. Immunol. 143*:2844 (1989).
49. A.M. Anderton, R. van der Zee, and J.A. Goodacre, Inflammation activates self hsp 60-specific T cells, *Eur. J. Immunol. 23*:33 (1993).
50. M. Harada, G. Matsuzaki, Y. Yoshikai, N. Kobayashi, S. Kurosawa, H. Takimoto,

and K. Nomoto, Autoreactive and heat shock protein 60-recognizing CD4+ T-cells show antitumor activity against syngeneic fibrosarcoma, *Cancer Res. 53*:106 (1993).
51. F. Poccia, P. Piselli, S. Di Cesare, S. Bach, V. Colizzi, M. Mattei, A. Bolognesi, and F. Stirpe, Recognition and kiling of tumour cells expressing heat shock protein 65kD with immunotoxins containing saporin, *Br. J. Cancer 66*:427 (1992).
52. P. Minden, R.A. Houghten, J.R. Spear, and T.M. Shinnick, A chemically synthesized peptide which elicits humoral and cellular immune response to mycobacterial antigens, *Infect. Immun. 53*:560 (1986).
53. U. Sengupta, S. Sinha, G. Ramu, J. Lamb, and J. Ivanyi, Suppression of delayed hypersensitivity skin reactions to tuberculin by *M. leprae* antigens in patients with lepromatous and tuberculoid leprosy, *Clin. Exp. Immunol. 68*:58 (1987).
54. Z.H. Handzel, V. Buchner, A. Leviatan, Y. Bar-Khayim, Y. Burstein, H. Dockrell, and K.P.W.J. McAdam, Proliferative responses of leprosy patients, healthy contacts and BCG vaccinees to a major native 12-kDa protein of *Mycobacterium leprae, Immunol. Infect. Dis. 2*:237 (1992).
55. P.F. Barnes, V. Mehra, B. Rivoire, S-J. Fong, P.J. BrennanM. S. Voegtline, P. Minden, R.A. Houghten, B.R. Bloom, and R.L. Modlin, Immunoreactivity of a 10-kDa antigen of *Mycobacterium tuberculosis, J. Immunol. 148*:1835 (1992).
56. E. Adams, R.J. Garsia, L. Hellqvist, P. Holt, and A. Basten, T cell reactivity to the purified mycobacterial antigens p65 and p70 in leprosy patients and their household contacts, *Clin. Exp. Immunol. 80*:206 (1990).
57. E. Adams, W.J. Britton, A. Morgan, A. L. Goodsall, and A. Basten, Identification of human T cell epitopes in the *Mycobacterium leprae* heat shock protein 70-kD antigen, *Clin. Exp. Immunol. 94*:500 (1993).
58. H.M. Dockrell, N.G. Stoker, S.P. Lee, M. Jackson, K.A. Grant, N.F. Jouy, S.B. Lucas, R. Hasan, R. Hussain, and K.P.W.J. McAdam, T-cell recognition of the 18-kilodalton antigen of *Mycobacterium leprae, Infect. Immun. 57*:1979 (1989).
59. F. Oftung, T.M. Shinnick, A.S. Mustafa, K.E.A. Lundin, T. Godal, and A.H. Nerland, Heterogeneity among human T cell clones recognising an HLA-DR4, Dw4-restricted epitope from the 18-kDa antigen of *Mycobacterium leprae* defined by synthetic peptides, *J. Immunol. 144*:1478 (1990).
60. P. Launois, M. Niang N'Diaye, J.L. Sarthou, A. Drowart, J.P. van Vooren, J.L. Cartel, and K. Huygen, T cell reactivity against antigen 85 but not against the 18- and 65-kD heat shock proteins in the early stages of acquired immunity against *Mycobacterium leprae, Clin. Exp Immunol. 96*:86 (1994).
61. T. Koga, A. Wand-Württenberger, J. de Bruyn, M.E. Munk, B. Schoel, and S.H.E. Kaufman, T cells against a bacterial heat shock protein recognize stressed macrophages, *Science 245*:1112 (1989).
62. S. Khetan, K.B. Sainis, S. Rath, and R. Kamat, Murine CD8+ T suppressors against mycobacterial 65-kDa antigen compete for IL-2 and show lack of major histocompatibility complex–imposed restriction specificity in antigen recognition, *Eur. J. Immunol. 23*:2440 (1993).
63. E.M. Janis, S.H.E. Kaufman, R.H. Schwartz, and D.M. Pardoll, Activiation of γδ T cells in primary immune response to Mycobacterium tuberculosis, *Science 244*: 713 (1989).

64. Y. Yoshikai, G. Matsuzaki, T. Inoue, and K. Nomoto, An increase in number of T-cell receptor γ/δ-bearing T cells in athymic nude mice treated with complete Freund's adjuvant, *Immunology 70*:61 (1990).
65. A. Augustin, R.T. Kubo, and G.-K Sim, Resident pulmonary lymphocytes expressing the γ/δ T-cell receptor, *Nature 340*:239 (1989).
66. T. Inoue, Y. Yoshikai, G. Matsuzaki, and K. Nomoto, Early appearing γ/δ–bearing T cells during infection with clamette guerin bacillus, *J. Immunol. 146*:2754 (1991).
67. J.P. Griffin, K.V. Harshan, W.K. Born, and I.M. Orme, Kinetics of accumulation of γ/δ receptor-bearing T lymphocytes in mice infected with live mycobacteria, *Infect. Immun. 59*:4263 (1991).
68. D. Kabelitz, A. Bender, T. Prospero, S. Wesselborg, O. Janssen, and K. Pechhold, The primary response of human γ/δ^+ T cells to *Mycobacterium tuberculosis* is restricted to Vg9-bearing cells, *J. Exp. Med. 173*:1331 (1991).
69. R.L. O'Brien, Y-X. Fu, R. Cranfill, A. Dallas, C. Ellis, C. Reardon, J. Lang, S.R. Carding, R. Kubo, and W. Born, Heat shock protein Hsp60-reactive γδ cells: a large, diversified T-lymphocyte subset with highly focused specificity, *Proc. Natl. Acad. Sci. USA 89*:4348 (1992).
70. C.E. Roak, M. K. Vollmer, R.L. Cranfill, S.R. Carding, W.K. Born, and R.L. O'Brien, Liver γδ T cells: TCR junctions reveal differences in heat shock protein-60-reactive cells in liver and spleen, *J. Immunol. 150*:4867 (1993).
71. Y-X. Fu, M. Vollmer, H. Kalataradi, K. Heyborne, C. Reardon, C. Miles, R. O'Brien, and W. Born, Structural requirements for peptides that stimulate a subset of γδ T cells, *J. Immunol. 152*:1578 (1994).
72. N. Kobayashi, G. Matsuzaki, Y. Yoshikai, R. Seki, J. Ivanyi, and K. Nomoto, $\gamma\delta 5^+$ T cells from BALB/c mice recognize the murine heat shock protein 60 target cell specificity, *Immunology 81*:240 (1994).
73. I. Kaur, S.D. Voss, R.C. Gupta, K. Schell, P. Fish, and M.S. Sondel, Human peripheral γδ T cells recognize hsp60 molecules on Daudi Burkitt's lymphoma cells, *J. Immunol. 150*:2046 (1993).
74. J. Ivanyi and K. Sharp, Control by H-2 genes of murine antibody responses to protein antigens of *Mycobacterium tuberculosis, Immunology 59*:329 (1986).
75. S.J. Brett, and J. Ivanyi, Genetic influences on the immune repertoire following tuberculous infection in mice, *Immunology 71*:113 (1990).
76. T.A. Adeleye, M.J. Colston, R. Butler, and P.J. Jenner, The antibody repertoire to proteins of *Mycobacterium leprae*. Genetic influences at the antigen and epitope level, *J. Immunol. 147*:1947 (1991).
77. C. Barrios, A.R. Lussow, J. Van Embden, R. Van der Zee, R. Rappuoli, P. Constantino, J.A. Louis, P-H, Lambert, and G. Del Giudice, Mycobacterial heat-shock proteins as carrier molecules. II: The use of the 70-kDa mycobacterial heat-shock protein as carrier for conjugated vaccines can circumvent the need for adjuvants and Bacillus Calmette Guerin priming, *Eur. J. Immunol. 22*:1365 (1992).
78. P.S. Jackett, G.H. Bothamley, H.V. Batra, A. Mistry, D.B. Young, and J. Ivanyi, Specificity of antibodies to immunodominant mycobacterial antigens in pulmonary tuberculosis, *J. Clin. Microbiol. 26*:2313 (1988).
79. A. Elsaghier, C. Prantera, G. Bothamley, E. Wilkins, S. Jindal, and J. Ivanyi,

Disease association of antibodies to human and mycobacterial hsp70 and hsp60 stress proteins, *Clin. Exp. Immunol. 89*:305 (1992).
80. A.A.F. Elsaghier, E.G.L. Wilkins, P.K. Mehrotra, S. Jindal, and J Ivanyi, Elevated antibody levels to stress protein HSP70 in smear-negative tuberculosis, *Immunol. Infect. Dis. 1*:323 (1991).
81. J. Ivanyi, K. Sharp, P. Jackett and G. Bothamley, Immunological study of the defined constituents of mycobacteria, *Springer Semin. Immunopathol. 10*:279 (1988).
82. G. Bothamley, J. Swanson Beck, R.C. Potts, J.M. Grange, T. Kardjito, and J. Ivanyi, Specificity of antibodies and tuberculin response after occupational exposure to tuberculosis, *J. Infect. Dis. 166*:182 (1992).
83. G. Matsuzaki, Y. Yoshikai, M. Harada, and K. Nomoto, Autoreactive T cells from normal mice recognize mycobacterial 65 kd heat shock protein from *Mycobacterium bovis, Int. Immunol. 3*:215 (1990).
84. H. Nagasawa, M. Oka, K. Maeda, J-G. Chai, H. Hisaeda, Y. Ito, R. A. Good, and K. Himeno, Induction of heat shock protein closely correlates with protection against *Toxoplasma gondii* infection, *Proc. Natl. Acad. Sci. USA 89*:3155 (1992).
85. S. Kantengwa, Y.R.A. Donati, M. Clerget, I. Maridonneau-Parini, F. Sinclair, E. Mariethoz, M. Perin, A.D.M. Rees, D.O. Slosman, and B.S. Polla, Heat shock proteins: an autoprotective mechanism for inflammatory cells?, *Immunology 3*:49 (1991).
86. A.D.M. Rees, Y. Donati, G. Lombardi, J. Lamb, B. Polla, and R. Lechler, Stress-induced modulation of antigen-presenting cell function, *Immunology 74*:386 (1991).
87. Y. Mistry, D.B. Young, and R. Mukherjee, hsp 70 synthesis in Schwann cells in response to heat shock and infection with *Mycobacterium leprae, Infect. Immun. 60*:3105 (1992).
88. C. Moreno, J. Taverne, A. Mehlert, C.A. W. Bate, R. Brealey, A. Meager, G.A. W. Rook, and J.H.L. Playfair, Lipoarabinomannan from *Mycobacterium tuberculosis*induces the production of tumour necrosis factor from human and murine macrophages, *Clin. Exp. Immunol. 76*:240 (1989).
89. J.S. Friedland, R. Shattock, D.G. Remick, and G.E. Griffin, Mycobacterial 65-kD heat shock protein induces release of proinflammatory cytokines from human monocytic cells, *Clin. Exp. Immunol. 91*:58 (1993).
90. L.S. Schlesinger and M.A. Horwitz, Phagocytosis of *Mycobacterium leprae* by human monocyte-derived macrophages is mediated by complement receptors CR1 (CD35), CR3 (CD11b/CD18), and CR4 (CD11c/CD18) and IFN-g activation inhibits complement receptor function and phagocytosis of this bacterium, *J. Immunol. 147*:1983 (1991).
91. P.W. Nyberg, S.A.S. Nordman, and L. Linko, A synergistic interaction between Bacillus Calmette-Guerin (BCG) and NADPH oxidase stimulants on the production of reactive oxygen metabolites by human mononuclear phagocytes, *APMIS 102*:67 (1994).
92. C.J. Elson, R.N. Barker, S.J. Thompson, and N.A. Williams, Immunologically ignorant autoreactive T-cells, epitope spreading and repertoire limitation, *Immunol. Today* (1994).

93. L. Chouchane, F.S. Bowers, S. Sawasdikosol, R.M. Simpson, and T.J. Kindt, Heat-shock proteins expressed on the surface of human T cell leukemia virus type–infected cell lines induce autoantibodies in rabbits, *J. Infect. Dis. 169*:253 (1994).
94. N.A. Buchmeier and F. Heffron, Induction of *Salmonella* stress proteins upon infection of macrophages, *Science 248*:730 (1990).
95. V.L. Headley and S.M. Payne, Differential protein expression by *Shigella flexneri* in intracellular and extracellular environments, *Proc. Natl. Acad. Sci. USA 87*:4179 (1990).
96. B.K.R. Patel, D.K. Banerjee, and P.D. Butcher, Determination of *Mycobacterium leprae* viability by polymerase chain reaction amplification of 71-kDa heat-shock protein mRNA, *J. Infect. Dis. 168*:799 (1993).

17

Heat Shock Proteins in Fungal Infections

Bruno Maresca
International Institute of Genetics and Biophysics, Naples, Italy and Washington University School of Medicine, St. Louis, Missouri

George S. Kobayashi
Washington University School of Medicine, St. Louis, Missouri

I. INTRODUCTION

A general homeostatic mechanism that protects cells and entire organisms from the deleterious effects of environmental stresses is termed the heat shock response. Its product, heat shock proteins (hsps), play major roles in many cellular processes that affect host-parasite interactions. These proteins possess unique properties and have been implicated also in chronic degenerative diseases, infectious diseases, immunology, and cancer. The ability of pathogens to initiate infection, invade tissues, and survive in mammalian hosts is critically linked to the induction of specific gene products. All fungi that have been involved in diseases of humans exist as saprobes in nature except the normal symbionts of humans, *Malasezzia furfur* and opportunistic species of *Candida*. They gain no obvious benefits by parasitizing humans. At the onset of infection, fungi experience a major increase in environmental temperature. Since dimorphic pathogenic fungi exhibit distinct morphological phases during this process, developmentally regulated gene expression is of particular importance. As saprobes in nature at 25°C, they exist in a filamentous mycelial form but when they infect a suitable mammalian host, they must adapt to a temperature of 37°C. In this chapter, we deal with some of the unique characteristics of the heat shock response in pathogenic fungi in relation to disease and the role hsps play in the immune response.

The heat shock response and is characterized by a rapid induction of specific sets of genes due to transcriptional activation (1). The response is universal, having been observed in bacteria as well as in lower eukaryotes, plants, and humans. In addition to temperature increases, a wide variety of other stresses induce a similar response at the level of the cell. These include, for example, exposures to ethanol, uncouplers of oxidative phosphorylation, inhibitors of electron transport components, steroid hormones, prostaglandins, and heavy metals (2). In eukaryotes, temperature activation is mediated by a *cis*-element, or heat shock element (HSE), located 80–150 nucleotides upstream of the start site of RNA transcription of all heat-inducible genes (3). This element (nnTC-nnGAAnnTCnnnGAAnnTTCnnGAAn) consists of a highly conserved nucleotide sequence (4) that is specific for heat shock DNA-binding proteins, heat shock factor (hsf).

Hsps ranging between 6 and about 110 kDa are the products of the stress response (5). In some cases, hsps are constitutively produced at a substantial level; in others, closely related proteins encoded by separate genes (heat shock cognate genes, *hsc*), which have the same amino acid sequences but are not stress or heat inducible, are produced only at normal temperatures. The synthesis of hsps correlates with the acquisition of thermotolerance at otherwise lethal temperatures protecting organisms not only from lethality but also from heat-induced developmental defects. Furthermore, some, but not all, of the heat shock genes are also developmentally induced in many organisms (5).

Considerable advances have been made toward the understanding the possible functions of hsps. For example, members of the hsp 70 family have been shown to maintain mitochondrial proteins in an unfolded, translocational-competent state and function inside mitochondria to maintain proteins in the unfolded state as they are being transported (6). The induction of hsps is also involved in preventing denaturation of proteins in mammalian and bacterial cells (7). Furthermore, in *Saccharomyces cerevisiae*, hsps maintain mitochondrial ATPase in a functional state under severe heat shock condition (8), and hsp 104, together with other as yet unidentified factor(s), has been shown to be responsible for induced tolerance to extreme temperatures in yeast cells (9).

With respect to structure and function, hsps are among the most highly conserved proteins known. Heat shock genes are induced very early during infection in parasitic organisms and play a critical role during adaptation to the new environment. An understanding of the role of hsps both in adaptation to the host environment and as antigens requires a detailed analysis of their regulation. In most organisms, several genes encoding for hsp 70 exist, and only some of these are stress inducible. Therefore, it is crucial to have precise information from both structural and transcriptional analyses in order to assign to each particular protein a specific role in the host/parasite relationship. This is particularly important for hsp 70, since this protein is probably the most conserved protein in all organisms thus far analyzed.

II. HSPS AND FUNGAL DISEASES

It has been proposed that the heat shock response plays a fundamental role in infectious processes during host invasion. In addition to bacterial infections (e.g., *Mycobacterium, Shigella,* and *Salmonella*), parasites (e.g., *Leishmania, Trypanosoma, Plasmodium, Giardia,* and *Schistosoma*), hsps have been detected in many pathogenic fungi (e.g., *Aspergillus fumigatus, Fonsecae pedrosoi,* and *Histoplasma capsulatum*), when, at the time of infection, the organisms adapt to temperature increases. Although, heat shock response observed in these organisms is similar to that seen in eukaryotic cells that have been extensively studied, for example, *S. cerevisiae*, specific differences exist. For example, after experiencing heat shock yeast cells of *S. cerevisiae* stop growing, adapt, and then resume growth, whereas higher eukaryotic cells can only survive a transient increase in temperature.

Dimorphic fungi existing saprobically at 25°C must adjust to the constant temperature of a homeothermic mammalian host at 37°C. However, it is reasonable to assume that during host invasion the stress response is elicited independently of the temperature shock. In this regard, it would be interesting to analyze adaptation in parasites that do not experience heat shock when they invade their host; examples being parasites of fish and plants. Furthermore, in dimorphic pathogens, high temperatures result in both a heat shock response and a developmentally regulated phase transition. It is not clear thus far whether expression of the heat shock gene family is part of the process of differentiation itself or an epiphenomenon involved in adaptation to new temperatures and living conditions. These organisms must therefore adapt not only to higher temperatures but also to different environmental conditions such as redox potential, hormones, and the presence or absence of growth nutrients, and they must protect themselves from the oxidative products and enzymes of the host.

In addition to their role in host-pathogen interactions, stress proteins of pathogens assume an importance immunologically. Hsps of a large spectrum of infectious agents are among the dominant antigens recognized by the immune system (10). Members of the hsp 60 and hsp 70 families of bacterial and lower eukaryotic pathogens, often found at high levels, act as immunodominant antigens, and immunological reactions to these highly conserved proteins have been implicated in the pathology of neurodegenerative diseases (11), cancer (12), several autoimmune diseases (13,14,15), and inflammatory processes associated with pathogens (16–18). Although the host immune system is able to discriminate between its own hsp 70 and those of the invader, in a deranged immune system, hsps may ultimately cause autoimmune diseases in humans (19). It is conceivable that the induction of hsps by the pathogen and their concomitant recognition by the immune system is the result of an equilibrium established between host and pathogen during evolution. In fact, the immune system's recognition of a highly conserved protein such as hsp 70 could be the result of confusion with mammalian "self"–hsp 70 during the process of antigen presentation. On the other hand, these proteins may not be eliminated or modified by the pathogen as a self-defense mechanism against the host immune system unless its own heat shock response is greatly impaired. In yeast, for example, it has been shown that mutations in specific *hsp* 70 genes are lethal; this may be the case in most eukaryotes and pathogens.

III. THERMOTOLERANCE

The ability of an organism to withstand a lethal heat shock once it has been pretreated with moderate increases in heat is amply documented and correlates well with conditions that induce the optimal synthesis of hsps. However, these experiments have often been performed under nonphysiological conditions, such as exposure to a sudden very high temperature change followed by a rapid shift to low temperature in order to rescue the organism. Under natural conditions, temperature changes are gradual and thermotolerance results. Thus, in most eukaryotic organisms, thermotolerance plays a central physiological role in the maintenance of vital cell properties. For example, mitochondrial ATPase of *S. cerevisiae* becomes nonfunctional within minutes when cells are directly shifted to the upper temperature range (44°C), and it cannot survive at this temperature; however, inducing thermotolerance by a gradual temperature increase protects ATPase activity at high temperatures and cells survive. In contrast, thermotolerance probably plays a minor role during fungal infections, because these organisms face a sudden and drastic environmental temperature change when they colonize the host. In organisms such as these, the heat shock response is not an artificial phenomenon. At the onset of host invasion, there is no time to develop thermotolerance, and the entire cellular apparatus, including mRNA maturation machinery and membrane structures, must remain functional to allow the organism to survive and undergo proper morphogenesis. This could be achieved either by evolution of temperature-resistant structures or by the existence of a constitutive thermotolerant state before host invasion.

In *H. capsulatum*, a dimorphic fungus that exists in a multicellular filamentous state in nature and as unicellular budding yeasts in human tissue, the transition between the multicellular and unicellular phases can be reversibly induced in the laboratory by shifting the temperature from 25°C (mycelia) to 37°C (yeast). In virulent strains of this fungus, mitochondrial ATPase is coupled during phase transition at 37°C, and only under experimental conditions of

39–43°C (depending on the strain) is electron transport uncoupled (20). In the temperature-sensitive and nonvirulent strains, such as the Downs strain, ATPase uncouples at 37°C, causing profound changes in the pattern of the differentiation process, such that RNA and protein synthesis is blocked, and the time required to complete transition is prolonged (21). Only under experimental conditions, in which a thermotolerant state is elicited by a short incubation at 34°C or by osmotic stress, can ATPase remain coupled in these strains when cells are shifted to 37°C, and differentiation can proceed in a manner that occurs in virulent, thermotolerant strains. A similar pattern has also been described in strains of other pathogenic fungi such as *Blastomyces dermatitidis* and *Paracoccidioides brasiliensis* (22). Thus, in nonvirulent strains, the presence of temperature-sensitive, membrane-bound enzymes such as ATPase or structures such as spliceosome (see below), whose activities are dependent during phase transition at 37°C on high levels of hsps, may severely hamper the survival of these organisms during their normal phase transition; hence their lack of virulence.

IV. mRNA MATURATION

There are major differences in the manner *H. capsulatum* and other eukaryotic organisms retain functional RNA processing at their upper temperature ranges. In *Drosphila* (23), HeLa cells (24), *Caenorhabditis elegans* (25), and *S. cerevisiae* (8), mRNA splicing is interrupted at high temperatures. However, spliceosome activity can remain functional if thermotolerance is elicited (23). Yost and Lindquist hypothesized that the spliceosome is a temperature-sensitive structure whose activity is protected by the presence of hsps. Thus, in all these organisms, physiological fluctuations of temperatures that gradually induce hsps ensure that their spliceosomes remain functional at high temperatures (23). However, Minchiotti et al. (26) have shown that when phase transition is induced in *H. capsulatum* under nonphysiological temperatures up to 42°C, mRNA of *hsp 82* and β-tubulin are correctly processed, suggesting that temperature-resistant splicing is a characteristic feature of this organism.

Hsp 70 may be one of the proteins involved in the protection of the splicing events, since in the *S. cerevisiae ssa1 ssa2* double mutant (carrying mutations in two members of the hsp 70 family), unspliced mRNAs do not accumulate in cells shifted directly from 25 to 41°C (8). In this particular mutant, which is about 100-fold more thermotolerant than wild-type cells, the presence of high constitutive expression of hsp 70 at 25°C temperatures has been correlated with the ability to survive lethal temperatures and to protect mRNA processing. It is intriguing that mycelial cells of *H. capsulatum* have constitutively high levels of hsp 70 (27). Whatever the mechanism, it is reasonable to speculate that it is crucial that the spliceosome remains functional at high temperature, since only accurate mRNA processing will allow adaptation to the host environment and morphogenesis to occur. However, a more detailed analysis of mRNA maturation in other parasites is necessary to elucidate differences with other eukaryotic cells.

V. HEAT SHOCK GENE EXPRESSION

Constitutively high levels of hsp 70 have been found in several organisms. In germinating cells of the fungus *C. albicans* under non–heat shock conditions, three hsps have been described (28-, 58-, and 66-kDa proteins, with the latter possibly being a member of the hsp 70 family), whereas in a nongerminating variant, only hsp 18 and hsp 22 were detectable at 23°C (28). However, cells of *C. albicans* that have been starved synthesized seven proteins that represent nutritionally stressed proteins which have been shown to be different from those induced by heat shock (28).

Once mycelia of *H. capsulatum* have been fully transformed to the yeast phase at 37°C, the temperature necessary to induce a heat shock response is no longer 37 but 40°C. This holds true for other parasites: for example, the schistosomulum of *Schistosoma mansoni* predominantly synthesize hsp 70 at 37°C and cannot induce a heat shock response during the first 3 hr of differentiation. However, 15 hr later, fully differentiated schistosomula induce heat shock on exposure to 42°C (29). Similar data are available for a number of other organisms: For example, in fish, heat shock is induced at a lower temperature in cold-adapted animals than in warm-adapted animals (30). Hsps are also induced by other stresses, although it is difficult to imagine how a temperature-regulated hsf can also be directly activated by an osmotic shock, by an electron transport inhibitor, or by ethanol. In *P. brasiliensis*, mycelia and yeast cells when shifted to 40°C induce synthesis of four proteins of 134-, 82-, 28-kDa molecular weight (31). Mycelia incubated at 37°C did not induce the synthesis of any of these proteins, suggesting that hsps are not necessary at this temperature at which phase transition occurs. Furthermore, immunoprecipitation studies have also demonstrated the presence of a large amount of constitutively synthesized *hsc 70* in *S. mansoni* (32). Furthermore, in trypomastigotes, most of the hsps synthesized at 39°C are also present at 24°C. In agreement with the Yost and Lindquist hypothesis, it has been suggested that the induction of hsps in trypomastigotes could prepare *Trypanosoma* for infecting the mammalian host (33), as in other stages of parasites. Several heat shock genes from pathogens have been cloned and sequenced. However, most of the studies have been done at a structural level and not a transcriptional level, so that very little is known about the mechanisms of heat shock gene regulation in organisms.

Transcription and protein data obtained from *H. capsulatum* have shown that maximal heat shock response, which varies between 34 and 37°C, correlates to the level of virulence of the strain (26,27,34). Expression studies of cloned *hsp 70* (34) and *hsp 82* (26) have shown that 34°C is the temperature of maximal heat shock mRNA transcription in the nonvirulent Downs strain, whereas in the virulent G222B and G217B strains, maximal induction occurs at 37°C. Similar results have been found at the protein level (27). Under laboratory conditions, induction of thermotolerance resulted in higher levels of heat shock mRNA transcription (35).

VI. HSPS AS ANTIGENS

Stress proteins play important roles in host-parasite interactions and are of immunological importance, since some of them, hsp 60, hsp 70, and hsp 90, are among the most dominant antigens synthesized at the onset of infection by a large spectrum of pathogens (36–38) and recognized by the host's immune system (10,39). Recent studies provide a better understanding of antigen processing and presentation, suggesting that they are involved in the molecular processes that lead to assembly.

In murine studies, Gomez et al. (40) showed that HIS-62 (hsp 60) and HIS-80 (hsp 70) *H. capsulatum* induce cellular immune responses that are able to mediate protective immunity against subsequent challenge against a virulent strains (41). Although the 80-kDa antigen has sequence similarity with hsp 70, it differs antigenically, since murine T cells responded to 80-kDa antigen but not to a purified hsp 70. Presumably, the 80-kDa protein contains an epitope recognized by T cells that is not present in hsp 70. They also showed that mice immunized with viable yeast cells mounted a delayed-type hypersensitivity response to the 80-kDa antigen, indicating that it is indeed a target of the cellular immune response to *H. capsulatum* (42,43). In a recent report, Jevons et al. (44) described a murine monoclonal immunoglobulin G (IgG) antibody, MAb 69F, that also recognized an 80-kDa antigen that was identical to the protein isolated by Gomez et al. (45) based on molecular mass and N-terminal amino acid sequence. Enzymatic and chemical studies showed the molecule to be *O*-glycosylated and heat-inducible, and it was also found to be of cytoplasmic origin or membrane bound but not in the cell wall. Of interest, MAb 69F did not appreciably stain mycelia of *H. capsulatum* grown at 25°C, but when mycelial cells were incubated at 37°C for 24 hr, they exhibited the same general staining pattern as yeast cells. The MAb used by Gomez et al. (42–43) was directed primarily against *hsp 70* and recognized a band on Western blot at 80-kDa, whereas MAb 69F predominantly recognized the 80 kDa band and was reactive with cytoplasmic antigens prepared from *A. fumigatus, P. brasiliensis, B. dermatitidis* and the *duboisii* variant of *H. capsulatum,* suggesting that the two MAbs are directed against different epitopes (44).

A similar situation occurs in candidiasis. Although the pathogenesis of infections caused by *C. albicans* is unclear, an immunodominant 47-kDa antigen has been isolated from sera and urine of patients that is thought to play a role in systemic candidiasis (45). This antigen, which has 83% homology with hsp 90 of *S. cerevisiae*, was specific to all isolates of *C. albicans* tested and appears to be a breakdown product of hsp 90. Antibody directed against the 47-kDa antigen is protective in an animal model of infection. Related hsp 90s have also been identified in infections caused by *C. parapsilosis* and *A. fumigatus* (45).

In a related study, Costantino et al. (46) analyzed the humoral response of the CBA/H mice to systemic infection with *C. albicans* and isolated a 96-kDa protein that was distinct from the hsp 90 antigen. Radiolabeling experiments showed that this protein, which is not heat inducible, was associated with antibody reaction to the heat-inducible hsp 75 protein. Whereas passively administered antibodies to hsp 90 of *C. albicans* protected BALB/c mice against lethal challenge of yeast cells (47), Costantino et al. showed that in CBA/H mice, the antibody response is not directed against hsp 90 and concluded that hsp 90 of *C. albicans* cannot be involved in protective humoral immunity in CBA/H mice. At present, the role antibodies directed against the 96-kDa antigen play in protection of CBA/H mice is unknown (46).

In mucocutaneous candidiasis, Polonelli et al. (48) have studied in vitro conditions influencing the expression of antigenic determinants of *C. albicans* reactive with secretory IgG (sIgG) from patients with oral and vaginal disease. Using indirect immunofluorscence, they detected a surface antigen on yeast cells incubated at 37°C but not on those cultured at 25°C that reacted with sIgG from patients with mucocutaneous disease. They were able to show that alkaline extracts obtained under reducing conditions from cells grown at 25 and 37°C that cells grown at the elevated temperature for 20 min expressed greater quantities of material reactive to monoclonal antibody to the *hsp 70* gene family than control cells maintained at 25°C. Treatment of reactive material by periodate oxidation eliminated reactivity, which they attributed to cleavage of the polysaccharide (mannan) conjugated to protein. Although these observations have analogies to those seen under typical heat shock responses, there are differences which the authors ascribe to the presence of heat shock mannoproteins; however, further work will be required to establish this point.

VII. CONCLUSIONS

Mycotic infections have gained prominence as a result of advances in medicine, particularly in the clinical setting where the patient is immunosuppressed or otherwise debilitated. Of the recently described mycoses, candidiasis, cryptococcosis, invasive aspergillosis, and progressive disseminated histoplasmosis remain of particular interest because they are diagnostic signs of acquired immunodeficiency syndrome (AIDS). The clinical features of these infections and the genetic characteristics of the etiological agents present unique parasite-host interactions that make them valuable research models to study. As a consequence of infection, fungi induce high levels of heat shock proteins whose sequences are astonishingly conserved in evolution. A great deal of attention has focused on these proteins because of their potential role as immunogens. During the process of infection, hsps of fungi play a central role in the process of differentiation, adaptation, and protection from the host's killing mechanisms such as, for example, reactive oxygen metabolites and low pH. On the other hand, mammalian T cells proliferate in response to fungal hsps. It is now accepted that hsps are the targets of humoral and cell-mediated immune responses. However, the reasons why these proteins are immunodominant are still unclear. A possibility is that the abundance with which they are induced is the reason for their immunodominance (49). Other candidates include, for example, the presence of conserved epitopes, preferential processing, and an intrinsic role as virulence factors. Thus far, there is no experimental evidence to support any theory.

Although the host's immune system may be able to discriminate between its own hsp 70 and those of the parasite, an impaired immune system may ultimately cause autoimmune diseases in humans. It is reasonable to speculate that induction of hsps by the pathogen and their recognition by the host's immune system of these proteins is the result of an equilibrium established between mammals and pathogens during evolution. In fact, although the immune system recognizes a highly conserved protein such as hsp 70 that could be confused with the mammalian self–HSP 70 proteins during the process of antigen presentation, these proteins may not be eliminated or modified by the pathogen as a self-defense mechanism against the host unless its own heat shock response and its capacity to adapt is greatly impaired. In yeast, for example, it has been shown that certain *hsp 70* genes are essential, and this is probably the case in most eukaryotes and pathogens.

What is the primary sensor of the increase in temperature that leads to heat shock gene activation? How is heat shock activation regulated? Unless pathogenic fungi have different mechanisms of regulation, the answers to these questions must come from studies on the biology of these organisms and on their ability to adapt to survive in a complex environment. It has been suggested that specific binding of hsf could be mediated by the direct temperature sensitivity of hsf possibly by itself or that of additional transcription factors (50), or that hsf could be activated by a temperature-regulated kinase (51). Since simple eukaryotic organisms, pathogens, and numerous animals experience different thermal regimens, this model would then imply the existence of several hsfs that sense the diverse ΔT necessary for heat shock gene activation in different organisms or in the same organism depending whether it is growing in a cold or warm environment or even in different tissues of the same organism. Possibly, certain hsps act as a thermometer that transduces the signal to activate gene transcription through its temperature-dependent interaction with hsf (52). However, it has yet to be established how hsp 70, which is the most highly conserved hsp at the amino acid level, detects the different temperature shifts necessary for heat activation that exist in disparate organisms. Hsp 70 has been proposed negatively to regulate hsf at normal temperature, and heat shock, a depletion of the hsp 70 pool triggers a release of hsf which is then capable of binding to DNA (52). According to this hypothesis, the hsp 70/hsf structural conformation would be operative in a large range of temperatures and for seasonal acclimation; for example, a large number of temperature-sensitive sensors would be required.

Another mechanism that has been proposed as a sensor for heat shock gene activation is the dimerization of specific proteins during stress (7). Although this can occur for proteins such as p68 kinase, β-galactosidase, and luciferase in cells under heat stress, enzymatic activities are restored after heat shock. In parasites exposed to high temperature after infection and after heat shock, we must assume that the cytoplasmic milieu of insoluble proteins must rapidly change in order to sense the new higher temperature necessary to induce heat shock.

None of these models of heat shock gene regulation seem to be satisfactory, since they do not explain the biology of temperature adaptation of both parasites and stenothermal animals. Thus, the study of the heat shock response of parasites could bring about a better understanding of the basic mechanisms of heat shock gene regulation. Further studies are needed to determine whether in parasites the process of adaptation to higher temperatures is epiphenomenally related to virulence or whether it is strictly correlated to it. Should the latter be the case, then the capacity to interfere with the heat shock response may contribute to the development of new drugs and vaccines. It is obvious that a number of questions concerning the regulation of the heat shock response remain unanswered.

ACKNOWLEDGMENTS

Portions of the studies reported in this work were supported by a contract from Ministero della Sanità–Instituto Superiore di Sanità, VIII Progetto AIDS 1995 and NIH grant AI 29609.

REFERENCES

1. P.K. Sorger, M.J. Lewis, and H.R.B. Pelham, Heat shock factor is regulated differently in yeast and HeLa cells, *Nature 329*:81 (1987).
2. S. Lindquist, The heat-shock response, *Annu. Rev. Biochem. 55*:1151 (1986).
3. H.R.B. Pelham, Activation of heat shock genes in eukaryotes, *Trends Genet. 1*:31 (1985).

4. H.R.B. Pelham, A regulatory upstream promoter element in the *Drosphila* HSP70 gene, *Cell 30*:517 (1982).
5. P.K. Sorger and H.R.B. Pelham, Purification and characterization of a heat-shock element binding protein from yeast, *EMBO J. 6*:3005 (1987).
6. S. Lindquist, Heat-shock proteins and stresstolerance in microorganisms, *Curr. Opin. Genet. Dev. 2*:748 (1992).
7. M.F. Dubois, A.G. Hovanessian, and O. Bensaude, Heat shock induced denaturation of proteins, *J. Biol. Chem. 266*:9707 (1991).
8. E.J. Patriarca and B. Maresca, Acquired thermotolerance following heat shock protein synthesis prevents impairment of mitochondrial ATPase activity at elevated temperatures in *S. cerevisiae, Exp. Cell Res. 190*:57 (1990).
9. Y. Sanchez and S. Lindquist, HSP104 required for induced thermotolerance, *Science 248*:1112 (1990).
10. D. Young, T. Garbe, and R. Lathigra Heat shock protein in mycobacterial infection, *Heat Shock* (B. Maresca and S. Lindquist, eds.), Springer-Verlag, New York, 1991, p. 203.
11. M. Chopp, The roles of heat shock proteins and immediate early genes in central nervous system normal function and pathology, *Curr. Opin. Neurol. Neurosurg. 6*:6 (1993).
12. W.J., Welch, Mammalian stress response: cell physiology, structure/function of stress proteins, and implications for medicine and disease, *Physiol. Rev. 7*:1063 (1992).
13. E.R. de Graeff Meeder, W. van Eden, G.T. Rijkers, B.J. Prakken, B.J. Zegers, and W. Kuis, Heat-shock proteins and juvenile chronic arthritis, *Clin. Exp. Rheumatol. 11*:S25-8 (1993).
14. J.R. Lamb, V. Bal, P. Mendez-Samperio, A. Mehlert, A. So, J. Rothbard, S. Jindal, R.A. Young, and D.B. Young, Stress proteins may provide a link between the immune response to infection and autoimmunity, *Int. Immunol. 1*:191 (1989).
15. B. Twomey, S. McCallum, V. Dhillon, D. Isenberg, and D. Latchman, Transcription of the genes encoding the small heat shock protein ubiquitin is unchanged in patients with systemic lupus erythematosus, *Autoimmunity 13*:197 (1992).
16. G.S. Firestein, The immunopathogenesis of rheumatoid arthritis, *Curr. Opin. Rheumatol. 3*:398 (1991).
17. J.B. Winfield and W.N. Jarjour, Do stress proteins play a role in arthritis and autoimmunity?, *Curr. Topics Microbiol. Immunol. 167*:161 (1991).
18. D.R. Schultz and P.I. Arnold, Heat shock (stress) proteins and autoimmunity in rheumatic diseases, *Semin. Arthritis Rheum. 22*:357 (1993).
19. S.H.E. Kaufman, Heat shock proteins and the immune response, *Immunol. Today 11*:129 (1990).
20. E.J. Patriarca, G.S. Kobayashi, and B. Maresca, Mitochondrial activity and heat shock response during morphogenesis in the dimorphic pathogenic fungus *Histoplasma capsulatum, Biochem. Cell Biol. 70*:207 (1992).
21. B. Maresca, A.M. Lambowitz, B.V. Kumar, G.A. Grant, G.S. Kobayashi, and G. Medoff, Role of cysteine oxidase in regulating morphogenesis and mitochondrial activity in the dimorphic fungus *H. capsulatum, Proc. Natl. Acad. Sci. USA 78*:4596 (1981).

22. G. Medoff, A. Painter, and G.S. Kobayashi, Mycelial- to yeast-phase transitions of the dimorphic fungi *Blastomyces dermatitidis* and *Paracoccidioides brasiliensis, J. Bacteriol. 169*:4055 (1987).
23. H.J. Yost and S. Lindquist, RNA splicing is interrupted by heat shock and is rescue by heat shock protein synthesis, *Cell 45*:185 (1986).
24. U. Bond, Heat shock but not other stress inducers leads to the disruption of a subset of snRNPs and inhibition of in vitro splicing in HeLa cells, *EMBO J. 7*:3509 (1988).
25. R.J. Kay, R.H. Russnak, D. Jones, C. Mathias, and E.P. Candido, Expression of intron-containing *C. elegans* heat shock genes in mouse cells demonstrates divergence of 3′ splice site recognition sequences between nematodes and vertebrates, and an inhibitory effect of heat shock on the mammalian splicing apparatus, *Nucleic Acids Res. 15*:3723 (1987).
26. G. Minchiotti, S. Gargano, and B. Maresca, The intron-containing *hsp82* gene of the dimorphic pathogenic fungus *Histoplasma capsulatum* is properly spliced in severe heat shock conditions, *Mol. Cell Biol. 11*:5624 (1991).
27. G. Shearer, C. Birge, P.D. Yuckenberg, G.S. Kobayashi, and G. Medoff, Heat-shock proteins induced during the mycelial-to-yeast transitions of strains of *Histoplasma capsulatum, J. Gen. Microbiol. 133*:3375 (1987).
28. N. Dabrowa and D. Howard, Nutritional stress proteins in *Candida albicans, Infect. Immun. 44*:537 (1984).
29. R. Blanton, E.C. Loula, and J. Parker, Two heat-induced proteins are associated with transformation of *Schistosoma mansoni* cercariae to schistosomula, *Proc. Natl. Acad. Sci. USA 84*:9011 (1987).
30. T.J. Dietz and G.N. Somero, The threshold induction temperature of the 90-kDa heat shock protein is subject to acclimatization in eurythermal goby fishes (genus Gillichthys), *Proc. Natl. Acad. Sci. USA 89*:3389 (1992).
31. L.Z. Goldani, M.M. Picard, and A.M. Sugar, Synthesis of heat shock proteins in mycelia forms of *Paracoccidioides brasiliensis, J. Med. Microbiol. 40*:124 (1994).
32. M.L. Muhich, M.P. Hsu, and J.C. Boothroyd, Heat-shock disruption of trans-splicing in trypanosomes: effect on Hsp70, Hsp85 and tubulin mRNA synthesis, *Gene 264*:7107 (1989).
33. A., Alcina, A. Urzainqui, and L. Carrasco, The heat shock response in *Trypanosoma cruzi, Eur. J. Biochem. 172*:121 (1988).
34. M. Caruso, M. Sacco, G. Medoff, and B. Maresca, Heat shock 70 gene is differentially expressed in *Histoplasma capsulatum* strains with different levels of thermotolerance and pathogenicity, *Mol. Microbiol. 1*:151 (1987).
35. B.S. Polla, Heat shock proteins in host-parasite interactions, *Parasitol. Today* and *Immunol. Today* (combined March issues), A38-A41 (1991).
36. D.C. DeNagel and S.K. Pierce, Heat shock proteins in immune response, *Crit. Rev. Immunol. 13*:71 (1993).
37. R.C. Matthews and J. Bournie, The role of *hsp90* in fungal infection. *Immunol. Today 13*:345 (1992).
38. R.A. Young and T.J. Elliot, Stress proteins, infection and immune surveillance, *Cell 59*:5 (1989).

39. T.M. Shinnick, Heat shock proteins as antigens of bacterial and parasitic pathogens, *Curr. Top. Microbiol. Immunol. 167*:145 (1991).
40. F.J. Gomez, A.M. Gomez, and G.S. Deepe, Jr., A 80 kilodalton antigen from *Histoplasma capsulatum* that has homology to heat shock protein 70 induces cell-mediated immune response and protection in mice, *Infect. Immun. 6*:2565 (1992).
41. G. S. Deepe, Jr., Protective immunity in murie histoplasmosis: functional comparison of adoptively transferred T-cell clones and splenic T cells, *Infect. Immun. 56*:2350 (1988).
42. A.M. Gomez, J.C. Rhodes, and G.S. Deepe, Jr., Antigenicity and immunogenicity of an extract from the cell wall and cell membrane of *Histoplasma capsulatum* yeast cells, *Infect. Immun. 59*:330 (1991).
43. F.J. Gomez, A.M. Gomez, and G.S. Deepe, Jr., Protective efficacy of a 62 kilodalton antigen, HIS-62, from the cell wall and cell membrane of *Histoplasma capsulatum* yeast cells, *Infect. Immun. 59*:4459 (1991).
44. L. Jaevons, L. Hunt, and A. Hamilton, Immunochemical studies of heat shock protein 80 of *Histoplasma capsulatum, J. Med. Vet. Mycol. 32*:47 (1994).
45. R.C. Matthews and J. Bournie, The role or *hsp90* in fungal infection, *Immunol. Today 13*:345 (1992).
46. P.J. Costantino, K.M. Franklin, N.F. Gare, and J.R. Warmington, Production of antibodies to antigens of *Candida albicans* in CBA/H mice, *Infect. Immun. 62*:1400 (1994).
47. R.C. Matthews, J.P. Burnie, D. Howards, T. Rowland, and F. Walton, Autoantibody to heat shock protein 90 can mediate protection against systemic candidosis, *Immunology 74*:20 (1991).
48. L. Polonelli,M. Gerloni, S. Conti, P. Fiscaro, C. Cantelli, P. Portincasa, F. Almondo, P.L. Barea, F.L. Hernondo, and J. Ponton, Heat-shock mannoproteins as targets of secretory diseases in *Candida albicans, J. Infect. Dis.* (1994).
49. R.A. Young and T.J. Elliot, Stress proteins, infection and immune surveillance, *Cell 59*:5 (1989).
50. C. Wu, V. Zimarino, C. Tsai, B. Walker, and S. Wilson, Transcriptional regulation of heat shock genes et al., *Stress Proteins in Biology and Medicine* (R.I. Morimoto, A. Tissieres, and C. Georgopoulos, eds.), Cold Spring Harbor Laboratory Press, Cold Spring Harbor, NY, 1990, p. 429.
51. P.K. Sorger, Heat shock factor and the heat shock response, *Cell 65*:363 (1991).
52. E.A. Craig and C.A. Gross, Is *hsp70* the cellular thermometer?, *Trends Biochem. Sci. 16*:135 (1991).

18

Stress Proteins and Infertility

Steven S. Witkin
Cornell University Medical College, New York, New York

I. INTRODUCTION

The majority of women whose fertility is impaired owing to occlusion of the fallopian tubes have never been diagnosed as having a sexually transmitted disease (STD) and have never experienced symptoms consistent with a diagnosis of an upper genital tract infection (1). This absence of STD-associated history and symptoms is also prevalent in women who fail to become pregnant after the successful transfer of viable embryos into the uterus following in vitro fertilization (IVF). Nevertheless, there is recent evidence suggesting that infection by the obligate intracellular bacterium *Chlamydia trachomatis* is a major cause of both of these fertility problems. Furthermore, an immune response to a single *C. trachomatis* component, the 57-kDa heat shock protein (hsp), has been implicated in the pathogenesis of these conditions. This chapter reviews the evidence that immune sensitization to the chlamydial 57-kDa hsp is involved in infertility-related sequelae to a *C. trachomatis* infection of the female genital tract.

II. IMMUNE PATHOGENESIS OF *C. TRACHOMATIS* INFECTIONS

Although it is well known that *C. trachomatis* infections can cause infertility and adverse pregnancy outcome, the mechanism whereby this occurs has only recently begun to be elucidated. In an in vitro fallopian tube organ culture, addition of *C. trachomatis* induced little damage to the integrity of the fallopian tubes (2). Similarly in vivo, experimental inoculation of *C. trachomatis* into the fallopian tubes of nonhuman primates resulted in only a transient infection with no long-lasting damage (3). Repeated inoculations, however, led to permanent fallopian tube scarring (3). Analogous to the established mechanism of damage in trachoma, a chlamydial eye infection (4), it appeared that an immune response to a *C. trachomatis* infection played a prominent role in eliciting fallopian tube pathology.

In experimental trachoma in guinea pigs, a single chlamydial antigen, a 57-kDa protein loosely associated with the cell surface, was shown to be responsible for eliciting a delayed hypersensitivity response in animals whose eyes had previously been infected (i.e., were sensitized) with *Chlamydia* (5). Further studies established that the chlamydial 57-kDa protein was a member of the GroEL 60-kDa hsp family (6) and had extensive amino acid sequence homology to the hsps of other bacteria and humans (7).

III. FALLOPIAN TUBE OCCLUSION AND IMMUNE RESPONSES TO HSP

A. Humoral Immune Responses

Humoral immune responses to the *C. trachomatis* 57-kDa hsp have been associated with female infertility and upper genital tract infections. High-titer antibody to the chlamydial 57-kDa hsp was most prevalent in women who were infertile owing to fallopian tube occlusion (8,9) and in women with pelvic inflammatory disease who had an ectopic pregnancy (10). The relation between a humoral immune response to the chlamydial 57-kDa hsp and tubal infertility due to a *C. trachomatis* infection has been demonstrated (9).

B. Cell-Mediated Immune Responses

An association between a cell-mediated immune response to the *C. trachomatis* 57-kDa hsp and fallopian tube infection (salpingitis) has also been demonstrated. In the presence of purified recombinant *C. trachomatis* 57-kDa hsp (11), peripheral blood lymphocytes (PBLs) from 9 of 18 (50%) women with salpingitis, none of 10 women with a lower genital tract cervicitis, and 3 of 42 (7.1%) healthy women proliferated (12). Furthermore, it appeared that among the patients with salpingitis, those women with a previous history of salpingitis or ectopic pregnancy had the highest prevalence of sensitization to hsp. This suggested that a prolonged or repeated exposure to *C. trachomatis* might be necessary to induce immune sensitization to hsp in most women. Additional testing utilizing the 57-kDa protein that had been freed of possible endotoxin contamination demonstrated that PBLs from 6 of 14 (42.9%) women with salpingitis, but none of the PBLs from women without evidence of an upper genital tract infection, still exhibited a significant lymphoproliferative response (12). In a subsequent study, PBL responses to the recombinant chlamydial 57-kDa hsp were identified in 4 of 10 (40%) women with recurrent salpingitis, in 1 of 9 (11.1%) women with an initial episode of salpingitis, and in 1 of 32 (3.1%) women with no history of salpingitis (13).

The ability of purified recombinant chlamydial 57-kDa hsp to elicit a cell-mediated immune response in the fallopian tubes of monkeys who had previously been infected with *C. trachomatis* has recently been demonstrated (14). Furthermore, sites within the fallopian tube that had not previously been infected with *C. trachomatis* also responded immunologically, indicating that a systemic delayed hypersensitivity response to hsp could be elicited following infection at a distant site.

C. Immune Responses to Synthetic Peptides Corresponding to Conserved Hsp Epitopes

The conservation of protein epitopes between an infectious microorganism and its host provides an opportunity for the induction of infection-induced autoimmune responses. To test whether chlamydial salpingitis induced sensitization to conserved epitopes, patients' PBLs were tested for their ability to respond to five synthetic peptides corresponding to regions of identity or near identity between the *C. trachomatis* and human hsp (13). Since hsp functions as a molecular chaperone and readily binds other proteins, the use of synthetic peptides in these investigations assured that no false-positive responses occurred as a result of stimulation by a contaminating native or recombinant peptide. PBLs from none of the patients or controls responded to three of the peptides, whereas one women each with a first or recurrent episode of salpingitis responded to a fourth peptide (amino acid sequence DIAILTGG) corresponding to amino acids 291–298 in the chlamydial hsp. PBLs from 4 of 10 (40%) women with recurrent salpingitis, but none of 9 women with a first episode of salpingitis and none of 32 women with no evidence of salpingitis, responded to the fifth 9 amino acid peptide (amino acid sequence AVKAPGFGD) corresponding to amino acids 275–283 in the chlamydial hsp. Interestingly, this peptide is also expressed in the *Escherichia coli* and *Mycobacteria* 60-kDa hsp, and the homologous mycobacterial peptide has been shown to stimulate T lymphocytes obtained from some healthy individuals (15).

IV. IN VITRO FERTILIZATION OUTCOME AND IMMUNE RESPONSES TO HSP

Women who are infertile owing to blockage of the fallopian tubes are prime candidates for IVF. The possible effect on IVF outcome of being sensitized to the chlamydial 57-kDa hsp was, therefore, investigated in a prospective study (16). Endocervical samples were obtained from 216 women at the time of oocyte retrieval and tested by ELISA for immunoglobulin A (IgA) antibodies reactive with the recombinant *C. trachomatis* 57-kDa hsp. Among the 68 women whose IVF cycle resulted in a term birth, only 5 (7.3%) were hsp antibody positive. In marked contrast, this antibody was detected in 41 (27.7%) of the women whose IVF cycle did not result in a live birth. In the majority of the positive women, there was no evidence of pregnancy after embryo transfer, indicating that immune sensitization to hsp in the genital tract may result in interference with embryo implantation in the uterus and/or other early pregnancy events.

The means by which sensitization to hsp leads to very early stage pregnancy failure remains to be elucidated. Initial studies in my laboratory have implicated an autoimmune mechanism in this process. Cervical IgA antibodies in women whose embryos failed to implant after IVF, but not cervical IgA from women with successful pregnancies after embryo transfer, reacted with synthetic peptides corresponding to hsp epitopes expressed in both the chlamydial and human hsp. The epitopes tested were those shown previously by other investigators to elicit a B-lymphocyte response in women infected with *C. trachomatis* (17). Since there is evidence that the human trophoblast (18), as well as the endometrium during the early stages of pregnancy (19), expresses the 60-kDa hsp, immune rejection of the embryo might result from the following scenario: In those women undergoing a cycle of IVF who were previously sensitized to conserved epitopes of the 60-kDa hsp, perhaps as a consequence of a chronic asymptomatic chlamydial infection, expression of hsp by the trophoblast and/or endometrium would lead to reactivation of hsp-sensitized lymphocytes. This would induce a local inflammatory response and cytokine release and interfere with the immune regulatory mechanisms which normally prevent rejection of the semiallogeneic embryo.

V. CONCLUSIONS

Repeated or long-term chronic infection of the female upper genital tract with *C. trachomatis* can induce immune sensitization to the chlamydial 57-kDa hsp as well as to conserved epitopes also present in the homologous human hsp. Since the extent of the immune response to the chlamydial hsp appears to be genetically determined (20), susceptibility to hsp-mediated infertility and/or early-stage pregnancy failure might be limited only to those women in which a potent immune response to conserved epitopes present in both the human and chlamydial hsp could be elicited. Whether analysis of female partners of infertile couples and prospective IVF patients for the presence of immune sensitivity to conserved hsp epitopes would have predictive value in terms of the likelihood for a successful pregnancy outcome remains to be evaluated. If the utility of this testing could be proven, there would then be a strong need to develop protocols to overcome this immune-mediated block to fertility.

REFERENCES

1. W. Cates, Jr., M. R. Joesoef, and M. B. Goldman, Atypical pelvic inflammatory disease: can we identify clinical predictors?, *Am. J. Obstet. Gynecol. 169*:341 (1993).
2. G. R. Hutchinson, D. Taylor-Robinson, and R.R. Dourmashkin, Growth and effect of chlamydiae in human and bovine oviduct organ cultures, *Br. J. Venereol. 55*:194 (1979).
3. D.L. Patton, P. Wolner-Hanssen, S.J. Cosgrove, and K.K. Holmes, The effects of *Chlamydia trachomatis* on the female reproductive tract of the Macaca nemestrina after a single tubal challenge and following repeated cervical inoculations, *Obstet. Gynecol 76*:643 (1990).
4. H. R. Taylor and E. Young, Immune mechanisms in chlamydial eye infection: cellular immune responses in chronic and acute disease, *J. Infect. Dis. 150*:745 (1984).
5. R.P. Morrison, K. Lyng, and H.D. Caldwell, Chlamydial disease pathogenesis. Ocular hypersensitivity elicited by a genus-specific 57-kD protein, *J. Exp. Med. 169*:663 (1989).
6. R.P. Morrison, R.J. Belland, K. Lyng, and H.D. Caldwell, Chlamydial disease pathogenesis. The 57-kD chlamydial hypersensitivity antigen is a stress response protein, *J. Exp. Med. 170*:1271 (1989).
7. R.P. Morrison, D.S. Manning, and H.D. Caldwell, Immunology of *Chlamydia trachomatis* infections. Immunoprotective and immunopathogenic responses, *Sexually Transmitted Diseases* (T.C. Quinn, ed.), Raven Press, New York, 1992, p. 57.
8. R.C. Brunham, I.W. Maclean, B. Binns, and R.W. Peeling, Chlamydia trachomatis: its role in tubal infertility, *J. Infect. Dis. 152*:1275 (1985).
9. B. Toye, C. Laferriere, P. Claman, P. Jessamine, and R. Peeling, Association between antibody to the chlamydial heat-shock protein and tubal infertility, *J. Infect. Dis. 168*:12136 (1993).
10. E.A. Wager, J. Schachter, P. Bavoil, and R.S. Stephens, Differential human serologic response to two 60,000 molecular weight *Chlamydia trachomatis* antigens, *J. Infect. Dis. 162*:922 (1990).
11. Y. Yuan, K. Lyng, Y.X. Zhang, D.D. Rockey, and R.P. Morrison, Monoclonal antibodies define genus-specific, and cross-reactive epitopes of the chlamydial 60-kilodalton heat shock protein (hsp60): specific immunodetection and purification of chlamydial hsp60, *Infect. Immun. 60*:2288 (1992).
12. S.S. Witkin, J. Jeremias, M. Toth, and W.J. Ledger, cell-mediated immune response to the recombinant 57-kDa heat-shock protein of *Chlamydia trachomatis* in women with salpingitis, *J. Infect. Dis. 167*:1379 (1993).
13. S.S. Witkin, J. Jeremias, M. Toth, and W.J. Ledger, Proliferative response to conserved epitopes of the *Chlamydia trachomatis* and human 60 kilodalton heat-

shock proteins by lymphocytes from women with salpingitis, *Am. J. Obstet. Gynecol. 171*: 455 (1994).
14. D.L. Patton, Y.T. Sweeney, and C.C. Kuo, Demonstration of delayed hypersensitivity in Chlamydial trachomatis salpingitis in monkeys: a pathogenic mechanism of tubal damage, *J. Infect. Dis. 169*:680, (1994).
15. M.E. Munk, B. Schoel, S. Modrow, R.W. Karr, R.A. Young, and H.E. Kaufman, T lymphocytes from healthy individuals with specificity to self-epitopes shared by the mycobacterial and human 65-kilodalton heat shock protein, *J. Immunol. 143*: 2844 (1989).
16. S.S. Witkin, K.M. Sultan, G.S. Neal, J. Jeremias, J.A. Grifo, and Z. Rosenwaks, Unsuspected Chlamydia trachomatis infection and in vitro fertilization outcome. *Am. J. Obstet. Gynecol. 171*: 1208 (1994).
17. Y. Yi, G. Zhong, and R.C. Brunham, Continuous B-cell epitopes in Chlamydia trachomatis heat shock protein 60, *Infect. Immun. 61*:1117 (1993).
18. K. Heybourne, Y. Fu, A. Nelson, A. Farr, R. O'Brien, and W. Born, Recognition of trophoblasts by $\gamma\delta$ T cells, *J. Immunol. 153*: 2918 (1994).
19. L. Mincheva-Nilsson, V. Baranov, M. M. Yeung, S. Hammarstrom, and M.L. Hammarstrom, Immunomorphologic studies of human decidua-associated lymphoid cells in normal early pregnancy, *J. Immunol. 152*:2020 (1994).
20. G. Zhong, and R.C. Brunham, Antibody responses to the chlamydial heat shock proteins hsp60 and hsp70 are H-2 linked, *Infect. Immun. 60*:3143 (1992).

19

Heat Shock Proteins in Visceral Leishmaniasis

Paulo Paes de Andrade and Cynthia Rayol de Andrade
Federal University of Pernambuco, Recife, Pernambuco, Brazil

I. INTRODUCTION

Heat shock proteins (hsps) are among the most abundant and conserved proteins in nature. They can be found in simple prokaryotic organisms as well as in higher eukaryotes and show an amazingly high degree of conservation of their amino acid sequences. As expected, their functions in different organisms are usually similar, as are their structural characteristics and molecular weights. It came therefore as a surprise the observation that hsps could be antigenic in malaria and schistosomiasis (1,2). The *Plasmodium* protein had about 70% amino acid homology to hsp 70 of *Drosophila melanogaster*, was cytoplasmatic and could be found in all stages of asexual development in the blood and in gametocytes but not in sporozoites. The helminth protein was roughly homologue to the low molecular weight hsps also seen in *Drosophila*. By this time, the heat shock responses of some other parasites were already known, but it was not promptly realized that many of these inducible proteins could be antigens in natural and experimental infections. Indeed, the differential expression of some genes of *Leishmania mexicana* had already been described 2 years earlier by Hunter et al. (3), who in 1984 observed seven actively synthesized proteins during temperature-induced in vitro differentiation from the promastigote to the amastigote form. These proteins corresponded in molecular weight to the well-known heat shock proteins seen in *Drosophila*. Also for *Trypanosoma brucei* and *Leishmania major*, it was then suggested that heat shock genes could be responsible for differentiation from the insect to the vertebrate forms (4). Other stress stimuli, such as exposure to sodium arsenite, as well as heat shock, were soon reported to induce the increased synthesis of some proteins in *L. tropica, L. enrietti*, and *L. donovani* (5). The heat shock response was then reported in *Schistosoma mansoni*, which differentially expresses a group of proteins under stress, one being soon identified as the parasite hsp 70 homologue (6). This protein was also found to be a major immunogen that always elicited an antibody response in infected humans and in experimental animals immunized against potentially protective irradiated cercariae (7). The same hsp homologue was also found to be a major antigen of three other helminths, *Brugia malayi, B. pahangi,* and *Onchocerca volvulus* (8,9), whereas another important antigen of *S. mansoni* was reported to be homologous to hsp 90 (10).

Hsps were seen in other *Leishmania* species (11), in *Trypanosoma cruzi* (12,13), and in its mitochondrion (14), but the formal proof of hsp antigenicity among trypanosomatids had to wait until the year 1990. MacFarlane et al. (15) and Andrade et al. (16) reported that patients with kala-azar had antibodies against an antigen related to the hsp 70 family. Although patients with Chagas' disease did not have cross-reactive antibodies to the leishmanial hsp 70 homologue, they also had specific antibodies to a similar protein from *Trypanosoma cruzi* (17). It was also shown that patients with mixed *T. cruzi/L. braziliensis* infections also developed a humoral response against hsps from both parasites (18). Table 1 summarizes the first reports on parasite hsps either in cell biology (as regulatory proteins or homologues of already described hsps from yeast or *Drosophila*) or in immunology (as antigens recognized by the humoral response of naturally or experimentally infected mammals). A total of 35 different parasites were seen to express either heat shock protein cognates (*hsc,* constitutively expressed homologous proteins found in most eukaryotes) or hsps as a response to environmental stresses, and in about half of them these proteins were antigens recognized by antibodies from infected mammalian hosts. We review in the next sections the humoral and cellular responses against hsps in leishmaniasis, with special focus on kala-azar, and briefly discuss the current hypotheses on the role of these proteins in pathogenesis and protection.

II. ANTI-HSP ANTIBODIES IN VISCERAL LEISHMANIASIS

Leishmania parasites cause a panoply of clinical diseases that range from self-limiting granulomatous lesions of the skin to severe mucosal destruction and from asymptomatic infections to fatal visceral involvement. Visceral leishmaniasis is an endemic disease in tropical and subtropical countries and constitutes an important public health problem in Brazil's northeast region (43,44). There is no reliable estimate on the prevalence of the disease worldwide, but it is known that its mortality and morbidity are very high in many countries. In Brazil, the mortality rate is estimated to be about 10%. New outbreaks in In-

Table 1 Heat Shock Proteins and Heat Shock Protein Cognates Found in Parasites

	In cell biology			As antigen		
Parasite	PY	(ref.)	Class	PY	(ref.)	Class
Babesia divergens				1991	(19)	Hsp 70
Brugia malayi				1987	(8)	Hsp 70
B. pahangi	1992	(20)	Small Hsps	1987	(8)	Hsp 70
Crithidia fasciculata	1992	(21)	Hsp 90			
C. fasciculata (mitochondrion)	1993	(22)	Hsp 70			
Eimeria bovis				1988	(23)	Hsp 70
Entamoeba histolytica				1992	(24)	Hsp 70
Giardia lamblia	1990	(25)	ND, non–hsp 70	1992	(26)	Hsp 60, hsp 70
Leishmania aethiopica	1991	(27)	Hsp 70			
L. braziliensis	1991	(27)	Hsp 70	1992	(18)	Hsp 70
L. chagasi				1992	(28)	Hsp 70, hsp 90
L. donovani	1985	(5)	Many	1990	(15)	Hsp 70
L. enrietti	1985	(5)	Many			
L. guyanensis	1991	(27)	Hsp 70			
L. major	1985	(4)	Hsp 70, hsp 90			
L. major (mitochondrion)	1993	(29)	Hsp 70			
L. mexicana	1984	(3)	Many			
	1990	(30)	Hsp 90			
L. mexicana (mitochondrion)	1990	(31)	Hsp 22			
L. panamensis	1991	(27)	Hsp 70			
L. peruviana	1992	(32)	Hsp 70			
L. tropica	1985	(5)	Many			
Leishmania sp. (from lizards)	1988	(11)	Many			

(continued)

Table 1 Heat Shock Proteins and Heat Shock Protein Cognates Found in Parasites

	In cell biology			As antigen		
Parasite	PY	(ref.)	Class	PY	(ref.)	Class
Mesocestoides corti				1991	(33)	Hsp 60, hsp 70
Onchocerca volvulus				1989	(9)	Hsp 70
Plasmodium berghei	1993	(34)	Hsp 70			
P. chabaudi	1993	(35)	Hsp 70			
P. cynomolgi	1992	(36)	Hsp 70			
P. falciparum	1986	(1)	Hsp 70	1986	(1)	Hsp 70
				1988	(37)	Hsp 90
Schistosoma bovis	1986	(2)	Hsp 22			
S. hematobium	1986	(2)	Hsp 22			
S. japonicum				1987	(38)	Hsp 70
S. mansoni	1986	(2)	Hsp 22	1986	(2)	Hsp 22
	1987	(6)	Hsp 70	1987	(7)	Hsp 70
				1989	(10)	Hsp 90
Theileria annulata	1992	(39)	ND			
Toxoplasma gondi				1992	(40)	ND
Trichomonas vaginalis	1993	(41)	Many			
Trypanosoma brucei	1985	(4)	Hsp 70, hsp 90			
Trypanosoma cruzi	1987	(12)	Hsp 70	1990	(17)	Hsp 70
	1988	(13)	Many			
	1993	(42)	Hsp 60			
T. cruzi (mitochondrion)	1989	(14)	Hsp 70			

PY, publication year; ND = not done.

dia and Sudan point toward the possibility of an explosive increase of the disease in this decade (45,46). Canids are thought to be the main reservoir for the parasite, *Leishmania chagasi*, in South America, but only dogs are important in the maintenance of the disease (47). Since no human-to-human transmission was ever proved in American visceral leishmaniasis, the Brazilian Ministry of Health decided to fight the disease by controlling the canine reservoir in endemic areas. The serodiagnosis of canine kala-azar is presently based on indirect immunofluorescence results (48). About 2 million serological tests are done each year for canine kala-azar (M. M. Lacerda, personal communication, National Workshop on Leishmaniasis, September 1993, Recife, Brazil), and more than 100,000 dogs are subsequently eliminated. In spite of the continued surveillance, new foci of visceral leishmaniasis were described in the last years, predominantly in previously uninhabited areas in the outskirts of cities.

The immunological diagnosis of kala-azar is of major importance in the establishment of the diagnosis and in determining the beginning of therapy for visceral leishmaniasis, (49) because a definitive diagnosis of the disease based only on clinical data or on clinical data supported by laboratory findings may be difficult or even impossible owing to the multiplicity of clinically similar diseases (50). The efficiency of serodiagnosis for visceral leishmaniasis, however, is frequently impaired by antibody cross reactivity to some of the diseases which are clinically similar (51) to kala-azar. This cross reactivity is caused by common epitopes found in proteins, carbohydrates, and other antigenic molecules from *Leishmania* and in those from other parasites (specially *Trypanosoma cruzi*) or even from the host. As in human visceral leishmaniasis, anti-*Leishmania*-specific antibodies are also commonly found in dogs naturally infected with *L. chagasi*, although titers in different serological tests are usually lower than those observed in the course of human infection. Immunofluorescence, however, suffers from the lack of specificity and results are plagued by false positives, especially when sera from stray dogs are examined. The use of crude *Leishmania* antigens for serodiagnosis therefore precludes a high specificity for immunological tests, particularly in regions where Chagas' disease is endemic. The defined and limited number of epitopes offered by the use of purified recombinant *Leishmania* antigens potentially increases the specificity of a serodiagnostic test for kala-azar.

A. Human Visceral Leishmaniasis and the Humoral Response to Hsp

In the search for fusion proteins that could be used for diagnosis, MacFarlane et al. (15) used a λgt11 library constructed from *L. donovani* promastigote mRNA (52) and found one clone that was recognized by serum antibodies from an African patient with kala-azar. Sequence analysis of this clone revealed that its 600-bp insert coded for the carboxyterminal region of an antigen related to the 70-kDa heat shock protein family. The full-length sequence of the corresponding gene (1959 nucleotides) was determined and found to have a high degree of similarity with the hsps of other organisms: 70% similarity with human hsp 70, 85% with *Trypanosoma brucei* hsp 70, and more than 95% with *Leishmania major* hsp 70. Genes encoding the antigen were present on a single chromosome as a series of approximately 12 3.7-kb direct tandem repeats. The antigen was identified as a 70-kDa heat shock cognate protein (*hsc 70*) because of its constitutive expression during heat shock. Moreover, it was expressed at all stages of the parasite life cycle. The *hsc 70* fusion protein was recognized by 9 of 17 patients with kala-azar, but not by any of cutaneous leishmaniasis sera tested or by a pool of 20 Chagas' disease sera.

It was somewhat surprising that sera from patients with infections due to closely related parasites as *L. braziliensis* or *T. cruzi* did not recognize a rather conserved leishmanial protein, hsp 70. Interestingly, although constitutively expressed schistosome homologues of hsp 70 elicit a dominant antibody response in humans infected with either *Schistosoma japonicum* or *S. mansoni*, in each case, the parasite antigens are immunologically distinct and non–cross reactive (53). Similarly, Kumar et al. (54) observed specific antibodies against *Plasmodium falciparum* hsp 70 homologues in the sera of *P. falciparum*–infected individuals but not in sera from people exposed to a different human malarial parasite, *P. vivax*. On the other hand, Engman et al. (17) observed that although hsp 70 is a major polypeptide antigen in Chagas' disease, antibodies to *T. cruzi* hsp 70 do not react with the human hsp 70 homologue even though the proteins display 73% amino acid sequence identity.

The antigenicity of leishmanial hsps in human kala-azar was further confirmed and extended by Andrade et al. (28). The quest for a diagnostic antigen again motivated the investigators. They screened two λZAP cDNA libraries from *L. donovani* and *L. chagasi* with the serum of a Brazilian patient with kala-azar. From the >200 phages that were positive, 17 were further studied by in vivo excision of the pBluescript plasmid and transformation of DH5α *Escherichia coli* strain. DNA sequence analysis of six clones revealed that three of them encoded the highly conserved hsp 90 and two coded for hsp 70. One clone had no homology to any sequence in the Genebank database. Southern blot analysis of the 17 clones with leishmanial hsp 70 and hsp 90 probes showed that 5 clones belonged to the hsp 90 family (29%), whereas 7 were hsp 70 homologues (41%). Four kala-azar sera were tested in ELISA using electroeluted, partially purified peptides representing about 250 amino acids of the hsp 70 carboxyterminal or the final 453 carboxyterminal amino acids of hsp 90. All sera strongly reacted against hsp 70 and three with hsp 90. Seven Chagas' disease sera were selected on the basis of their reactivity against *Leishmania chagasi* antigens in Western blot and proved to be negative when tested against both hsp 70 and hsp 90.

Curiously, none of them were significantly reactive toward antigens above 60 kDa, which is where the native hsp 70 and hsp 90 are located.

Using deletion constructs prepared from recombinant *L. donovani* hsp 70 that extended from the amino terminal into the carboxyterminal region, Wallace et al. (55) proved that the antigenic part is represented by the carboxyterminal moiety. More than 50% of the sera tested did not recognize an hsp 70 construct lacking the last 160 carboxyterminal amino acids. When a shorter construct lacking the last 227 amino acids was tested, two of three of the sera failed to recognize the polypeptide. The deletion of the 331 carboxyterminal amino acids completely abolished the antigenic properties of hsp 70. The same investigators used affinity-purified recombinant hsp 70 in Western blots to evaluate the presence of cross-reactive antibodies in sera from patients with cutaneous leishmaniasis (21 samples), lepromatous leprosy (8 samples), malaria (8 samples), schistosomiasis (8 samples), and Chagas' disease (8 samples). None of the 53 sera cross reacted with *L. donovani* recombinant hsp 70. Finally, they mapped an immunodominant B-cell epitope to the hexapeptide EADDRA extending from amino acids 529–534. This epitope was not recognized by Chagas' disease, schistosomiasis, or cutaneous leishmaniasis sera. The corresponding epitope by *L. major* hsp 70, EEDDKA, was not recognized by kala-azar sera.

To further study the potential of leishmanial hsps for the diagnosis of human kala-azar, we prepared SDS lysates of the 17 recombinants described earlier (28) and of the nonrecombinant strain, which were then run in SDS-polyacrylamide gels and blotted onto nitrocellulose membranes. The Western blots were developed with sera from 17 kala-azar, 6 Chagas' disease, and 6 control sera. Individual differences in antigen recognition were observed among patients with kala-azar, although most antigens were identified by all patients tested. Chagas' disease and control sera were not reactive. We also partially purified some of the recombinant polypeptides representing different parts of the carboxyterminal moiety of hsp 70 and hsp 90 and five fusion proteins with no DNA sequence homology to hsp and coated ELISA plates. Sera to be tested were previously immunoadsorbed against a wild-type *E. coli* sonicate. Figure 1 represents the results obtained with a panel of sera from Brazilian patients with kala-azar, from patients with Chagas' disease, and from healthy controls. The clones C4 and C7, belonging to the hsp 70 family, were specifically recognized by kala-azar sera, as well as the clones D2 and D9, which were later seen to share common epitopes with hsp 70 (see below), although they did not hybridize with a hsp 70 DNA probe. Fusion proteins C1 and C8 were shown to be hsp 90 homologues and were recognized by some of the kala-azar sera, although no clear cut-off point could be established for the ELISA. Again some kala-azar sera recognized the non-hsp fusion proteins, but no clear cut-off value could be defined for the test.

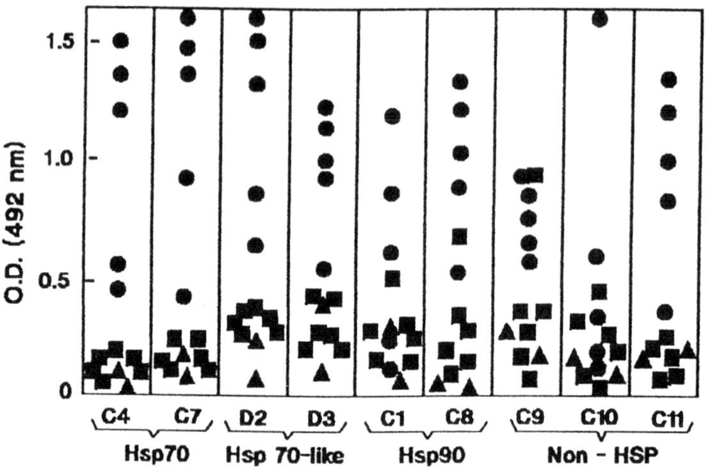

Figure 1 Absorbance readings obtained in ELISA for kala-azar (closed circles), Chagas' disease (squares), and control (triangles) sera (dilution 1:200). Microplates were coated with recombinant proteins of the hsp 70 (C4 and C7) and hsp 90 (C1 and C8) groups, with hsp 70-like recombinant proteins (D2 and D3), and with the three non-hsp recombinant proteins (C9, C10, and C11).

The immunological relationship of the supposedly non-hsp fusion polypeptides was investigated by the use of selective immunoadsorptions of a pool of four kala-azar sera against each fusion protein and subsequent use of the adsorbed sera in ELISAs against every single fusion protein. The recombinant hsps were immunologically grouped in the same way as they were by sequence or Southern hybridization data, as well as three of the non-hsp fusion proteins. The last two, D2 and D3, could be grouped as hsp 70 homologues on the basis of immunoadsorption data.

In an experiment complementary to the use of deletion constructs presented by Wallace et al. (55), we studied the recognition of fusion polypeptides representing different sizes of the carboxyterminal part of leishmanial hsp 70 by a panel of randomly selected Brazilian kala-azar sera. Our results are represented in Table 2. A fragment (C2) extending only about 200 amino acids from the carboxyterminus failed to be recognized by 4 of 10 sera tested. The inclusion of about 30 amino acids to this polypeptide was enough to render it antigenic for all 10 sera tested. These results suggest that the B-cell epitopes recognized by Brazilian patients with kala-azar are not concentrated near the proposed immunodominant hexapeptide but are possibly spread over the carboxyterminal half of hsp 70.

Table 2 Reactivity of a Panel of Brazilian Kala-azar Sera Against Different Fragments of Fusion Hsp 70.

		Hsp 70 carboxy-terminal fragment						
Sera		C6	D4	D5	C7	C4	D1	C2
BR1		+	+	+	+	+	+	+
BR2		+	+	+	+	+	+	–
BR3		+	+	+	+	+	+	–
BR4		+	+	+	+	+	+	+
BR5		+	+	+	+	+	+	+
BR6		+	+	+	+	+	+	–
BR7		+	+	+	+	+	+	+
BR8		+	+	+	+	+	+	+
BR9		+	+	+	+	+	+	+
BR10		+	+	+	+	+	+	–
	MW (kDa)	65	57	56	40	34	33	29
	Leishmanial moiety (aa)	500	425	433	291	241	233	200

Uneducated estimates of fragment sizes are expressed in kDa (for the whole fragments) and in amino acid length (for the leishmanial moiety.

The surprisingly high percentage (82%) of clones encoding hsps or immunologically similar polypeptides among our 17 clones prompted us further to study the human humoral response to hsp in kala-azar. We developed Western blots from *L. chagasi* proteins with a pool of four kala-azar sera adsorbed against either nonrecombinant DH5 alpha, transformed DH5 alpha expressing hsp 70 or hsp 90, or against a lysate containing equal amounts of both hsp 70 and hsp 90. Adsorptions against hsp 70- and hsp 90-containing lysates significantly reduced or completely abolished serum reaction against the two most prominent bands in the blot, at 70 and 90 kDa, respectively. Simultaneous adsorption against both hsps abolished serum reaction against both bands (data not shown). In a broader study using immunoblots for the differential diagnosis of visceral leishmaniasis, we also showed (56) that hsp 70 is one of the most intensively recognized antigens in visceral leishmaniasis, although rarely in other clinically similar, nonparasitic diseases.

B. Canine Visceral Leishmaniasis and Anti-Hsp Antibodies

To study the humoral response anti-*Leishmania* hsp 70 in canine kala-azar, we used the clone C7 isolated previously and coding for about 290 amino acids of the carboxyterminal part of *L. chagasi* hsp 70. ELISA plates were coated with

250 ng of partially purified C7 and used to study a panel of dog sera comprising 27 presumably positive samples (based on immunofluorescence results from the National Health Foundation (NHF), Brazil), 14 samples from healthy stray dogs from an endemic area, and 7 samples from control kennel beagles. Figure 2 shows the results obtained by ELISA. The canine sera from a nonendemic area were used to establish the cut-off optical density (OD) reading: mean + 3 SD = 0.073. None of the sera from apparently healthy animals from the endemic areas were positive, whereas only 17 out of 29 sera from NHF-immunofluorescence–positive dogs were positive by the test.

Since the canine infections were not parasitologically proven, we decided to establish the Western blot as the gold standard to compare all other tests. Each serum from the NHF-immunofluorescence–positive group was analyzed. Sera from animals bred in nonendemic areas were used as negative controls. Four bands at 20, 40, 72, and 87 kDa were recognized by most dogs. Other secondary bands were also less frequently recognized. A clear pattern for positive dogs was therefore established. From the 29 dogs of the positive group, only 17

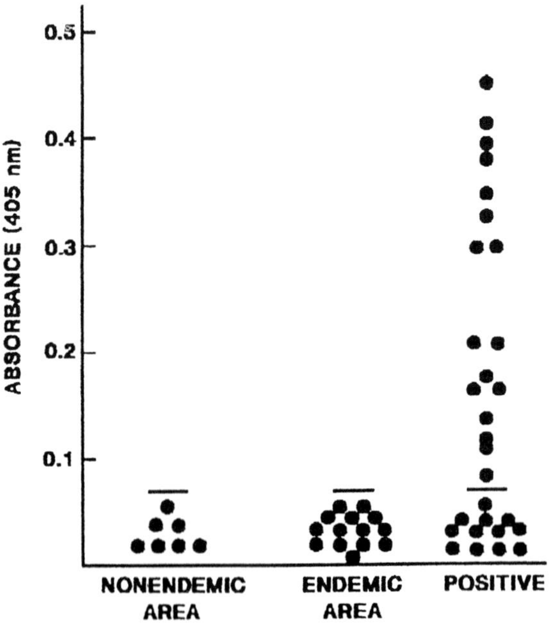

Figure 2 ELISA results using C7 recombinant polypeptide as antigen. Sera from animals bred outside the endemic areas were not reactive as well as those looking healthy in endemic areas. Only 17 (63%) of the sera previously characterized as positive were reactive in the assay.

Table 3 Sensitivity and Specificity of Indirect Immunofluorescence, ELISA Using Trypsin Treated, Formalin-Fixed Promastigotes of *L. chagasi* and ELISA Using the C7 Polypeptide Representing a Carboxyterminal Part of *Leishmania chagasi* Hsp 70

Test	Specificity (%)	Sensitivity (%)
Immunofluorescence	66.6	100
ELISA with promastigotes	95.7	88.2
ELISA with C7 polypeptide	96.9	94.0

(59%) were also positive in Western blots (data not shown). From the group of 17 dogs diagnosed as positive for visceral leishmaniasis by the recombinant ELISA, 16 were confirmed by Western blot. On the other hand, none of the apparently healthy animals from endemic areas were positive either by Western blot or by the recombinant ELISA.

Taking the Western blot as a standard, the results from immunofluorescence (according to our own protocol, which makes the test more specific than the technique used at the NHF), conventional ELISA (using whole *L. chagasi* promastigotes as antigens) (57) and recombinant ELISA were compared (58). From these data, the sensitivity and specificity of the three tests were calculated and are depicted in Table 3.

The recombinant ELISA was equivalent to the conventional ELISA, although a little more specific and sensitive. The high sensitivity of the indirect immunofluorescence is due to the use of a low titer as the cut-off value for the test, which was a technical decision to avoid missing seropositive dogs in control campaigns.

In conclusion, the results above demonstrate that a strong anti–heat shock protein humoral response is a hallmark in visceral leishmaniasis, both human and canine. They also strongly support the use of recombinant *Leishmania* hsp 70 as antigen for the serodiagnosis of kala-azar.

III. ANTI-HSP CELLULAR IMMUNE RESPONSE IN VISCERAL LEISHMANIASIS

Both recovery from and resistance to *Leishmania* infections seem to depend on the triggering of a T-cell response that controls the parasite, leading to granuloma formation in the cutaneous form or to resolution of the visceral disease. T cells may also be involved in the pathogenesis in both mucosal and in visceral leishmaniasis. In contrast to the humoral response, anti-hsp cellular responses in kala-azar were virtually unknown. Russo et al. (59) recently described experiments showing that γδ T cells from patients with cutaneous leishmaniasis or from individuals recovered from a kala-azar proliferate in response to *L. donovani* hsp 70. γδ T cells were enriched directly from peripheral blood mononuclear cells of a patient with mucosal leishmaniasis. Highly enriched populations of αβ and γδ T cells were cultured with *Leishmania* lysates or with recombinant hsp 70. Although αβ T cells strongly proliferate in response to *Leishmania* antigen, they we poor responders to hsp 70 when compared with the γδ T cells, which conversely were poor responders to the total parasite antigen.

That γδ T cells were involved in both cutaneous and visceral leishmaniasis was already known (60–64), but there was no defined antigen associated with their proliferation. Pfeffer et al. (65) had previously shown that γδ T cells from normal individuals failed to respond to total *Leishmania* antigens or leishmanial components. These cells accumulate in the lesions of patients with localized American cutaneous leishmaniasis (60,61) and, based on the limited diversity of γδ T-cell receptors observed, it was suggested that they could recognize a small set of antigens perhaps common to distinct pathogens and/or those expressed by the host (61). γδ T cells would then be clonally selected in lesions by these antigens and later undergo oligoclonal expansion. Increased levels of γδ T cells among PBMC were also observed in visceral leishmaniasis (59,62). South American patients with Kala-azar patients were seen to have γδ T cells coexpressing only CD8 (59), whereas a large proportion of γδ T cells from African patients had $CD4^{+}CD8^{-}$ phenotypes. In both cases, γδ T-cell receptors were associated with CD3.

The cellular immune response anti–hsp 70 in visceral leishmaniasis was also recently reported by Carvalho (66) and Costa et al. (67). Lymphocytes from *L. chagasi*-infected but asymptomatic patients were stimulated in vitro with crude leishmanial antigens and recombinant *L. chagasi* antigens, including hsp 70. Lipophosphoglycan (LPG-AP), a 42-kDa phosphatase and hsp 70 induced the proliferation of T cells and interferon gamma (IFN-γ) production, but not interleukin-4 (IL-4) production. Significant levels of IL-2 were also detected in 7 of 10 subjects' PBMC when stimulated with hsp 70. Their results strengthen the concept that these antigens (and particularly hsp 70, which was the strongest stimulator of T-cell proliferation) are highly immunogenic during natural infections and may be associated with the development of a protective immune response.

We recently evaluated the in vivo cellular response of dogs to hsp 70. In a pilot experiment, preparative for a canine vaccine trial, delayed-type hypersensitivity (DTH) responses were measured 48 hr after intradermal injection of 200 μl of a buffered solution containing C7, the recombinant polypeptide representing the last 290 amino acids of *L. chagasi* hsp 70. As a control, we used Leishvacin, a mixture of five different *Leishmania* species used as antigen for intradermal reactions in Brazil, according to the protocol proposed by Genaro et al. (68).

Most seropositive but asymptomatic dogs were DTH positive. Seronegative and diseased dogs were negative. Only three dogs reacted against Leishvacin, which were seropositive and asymptomatic. This result suggests that hsp is also a potent antigen in vivo.

IV. CONCLUSIONS

In many parasitic diseases, hsps are antigens recognized both by antibodies and T cells from the infected host. There are different, sometimes conflicting, explanations for these observations (see ref. 69 for a review). *Leishmania* hsps are abundant proteins and that alone could increase their immunogenicity in natural infections. Immunological priming can be possibly excluded on the basis of the specificity of the humoral reaction against hsp 70 and hsp 90 in kala-azar. Perhaps because of their special structure and biological function, hsps can be efficiently processed and presented on the surface of antigen-presenting cells. Again, hsps could be virulence factors in some forms of leishmaniasis (70). These hypotheses deserve further studies.

The molecular basis for the specificity observed in the humoral response against many hsps is far from being understood. How does the host avoid the production of antibodies to the conserved regions in hsps? Indeed, when immunized against a parasite's hsp, experimental animals produce antibodies that recognize homologues from other parasites (2,6,23,55), presumably binding to conserved as well as nonconserved sequences. In a natural infection, however, antibodies are usually parasite specific and are thought to be directed against nonconserved epitopes. But are there enough divergent segments in the carboxyterminal region of hsps? It is known that this is the less conserved region, but a closer inspection of *L. donovani* hsp 70 amino acid sequence shows that only three blocks of four or more consecutive amino acids are different between this hsp and that of *Trypanosoma cruzi*. The most divergent region of hsp 70, extending from amino acids 527–582 and encompassing the putative immunodominant epitope EADDRA, is represented in Figure 3 below. The sequences for *T. cruzi* hsp 70 and for *E. coli* DnaK are compared. In boldface are the few amino acids that are different between *L donovani* and *T. cruzi*. Amino acids substitutions from one group to another (there are four groups: neutral-nonpolar, neutral-polar, acidic, and basic), which are prone to produce important changes in the protein 3-dimensional structure, are indicated with asterisks. Downstream from the proposed immunodominant epitope is a small divergent region of four amino acids. Other divergent segments do not contain many group substitutions. This points toward the extreme paucity of divergent regions even in the less conserved carboxyterminal moiety. Hedstrom et al. (53) came to the same conclusion after inspection of *Schistosoma* hsp 70 sequence. Thus, strik-

Aminoacid		527																												
T. cruzi	...	K	Y	E	A	E	D	K	D	Q	V	R	Q	I	D	A	K	N	G	L	E	N	Y	A	F	S	M	K	N	...
L. donovani	...	K	Y	E	A	D	D	R	A	Q	R	D	R	V	E	A	K	N	G	L	E	N	Y	A	Y	S	M	K	N	...
E. coli	...	A	N	A	E	A	D	R	K	F	E	E	L	V	Q	T	R	N	Q	G	D	H	L	L	H	S	T	R	A	...
									*		*	*	*																	

T. cruzi	...	A	V	N	D	P	N	V	A	G	K	I	E	E	A	D	K	K	T	I	T	S	A	V	E	E	A	L	E	...
L. donovani	...	T	L	G	D	S	N	V	S	G	K	L	D	D	S	D	K	A	T	L	N	K	E	I	D	V	T	L	E	...
E. coli	...	Q	V	E	E	A	-	-	G	D	K	L	P	A	D	D	K	T	A	I	E	S	A	L	T	-	A	L	E	...
		*		*		*									*			*					*			*	*			
Aminoacid																												582		

Figure 3 Comparison of the inferred aminoacid sequence of *T. cruzi* hsp 70, *L. donovani* hsp 70, and *E. coli* DnaK near the putative immunodominant hexapeptide EADDRA. Asterisks mark amino acid substitutions from different groups (there are four amino acid groups: neutral-nonpolar, neutral-polar, acidic, and basic). In boldface are all amino acids which are not the same between the *L. donovani* and the *T. cruzi* sequences.

ingly limited diversity seems to be sufficient to elicit a discriminatory antibody response to these parasite host-like antigens. It is possible that most epitopes found in the carboxyterminal region are indeed conformational epitopes. Indeed, results of molecular modeling suggest that most of the epitopes recognized by the immune system of the monkey in an immunogenic and antigenic part of *Plasmodium* hsp 70 are created by folding of the molecule (71)

It is not presently known if anti-hsp cellular responses in visceral leishmaniasis have a role in protection or in the pathogenesis of the disease. A role in protection is suggested both by experiments showing that lymphocytes from asymptomatic patients with kala-azar strongly proliferate in vitro in response to hsp 70 and produce INF-γ and IL-2 (66,67) and by the presence of positive DTH against hsp 70 in seropositive asymptomatic dogs. On the other hand, γδ T cells isolated from patients with kala-azar secreted high levels of IL-4 and IL-6 (62). Since γδ T cells proliferate against *Leishmania* hsp, these results together would suggest a role of hsps in the pathogenesis of kala-azar. Nevertheless, a potential role for γδ T cells in the control of infection with *L. major* in the murine model was suggested by Rosat et al. (63): In both susceptible BALB/c and resistant CBA/J mice, blocking of the γδ T-cell receptor by a monoclonal antibody anti–γδ TCR resulted in the development of larger lesions that contained increased numbers of parasites and in delayed healing of cutaneous lesions in otherwise resistant CBA/J mice. These investigators suggest that γδ T cells could be involved in host defense against this parasite. A final position on the role of hsps in the protection/pathogenesis of visceral leishmaniasis awaits more results, which are bound to come in the near future in this fast-moving research field.

ACKNOWLEDGMENTS

We thank the Brazilian National Research Council (CNPq), the Ministry of Education (CAPES), the Ministry of Science and Technology (RHAE), the Brazilian Agency for Research and Development (FINEP), the Alexander von Humboldt Foundation (Germany), the German Bureau of Academic Interchange (DAAD), and the Japan International Cooperation Agency (JICA) for fellowships and support.

REFERENCES

1. A.E. Bianco, J.M. Favaloro, T.R. Burkot, J.G. Culvenor, P.E. Crewther, G.V. Brown, R.F. Anders, R.L. Coppel, and D.J. Kemp, A repetitive antigen of *Plasmodium falciparum* that is homologous to heat shock protein 70 of *Drosophila melanogaster*, *Proc. Natl. Acad. Sci. USA 83*(22):8713 (1986).
2. V. Nene, D.W. Dunne, K.S. Johnson, D.W. Taylor, and J.S. Cordingley, Sequence and expression of a major egg antigen from *Schistosoma mansoni*. Homologies to heat shock proteins and alpha-crystallins, *Mol. Biochem. Parasitol. 21*:179 (1986).
3. K.W. Hunter, C.L. Cook, and E.G. Hayunga, Leishmanial differentiation in vitro: induction of heat shock proteins, *Biochem. Biophys. Res. Commun. 125*:755 (1984).
4. L.H. Van-der-Ploeg, S.H. Giannini, and C.R. Cantor, Heat shock genes: regulatory role for differentiation in parasitic protozoa, *Science 228*:1443 (1985).
5. F. Lawrence and M. Robert-Gero, Induction of heat shock and stress proteins in promastigotes of three *Leishmania* species, *Proc. Natl. Acad. Sci. USA 82*:4414 (1985).
6. P.D. Yuckenberg, F. Poupin, and T.E. Mansour, *Schistosoma mansoni*: protein composition and synthesis during early development; evidence for early synthesis of heat shock proteins, *Exp. Parasitol. 63*:301 (1987).
7. R. Hedstrom, J. Culpepper, R.A. Harrison, N. Agabian, and G. Newport, A major immunogen in *Schistosoma mansoni* infections is homologous to the heat-shock protein Hsp70, *J. Exp. Med. 165*:1430 (1987).
8. M.E. Selkirk, P.J. Rutherford, D.A. Denham, F. Partono, and R.M. Maizels, Cloned antigen genes of *Brugia* filarial parasites, *Biochem. Soc. Symp. 53*:91 (1987).
9. N.M. Rothstein, G. Higashi, J. Yates, and T. V. Rajan, *Onchocerca volvulus* heat shock protein 70 is a major immunogen in amicrofilaremic individuals from a filariasis-endemic area, *Mol. Biochem. Parasitol. 33*:229 (1989).
10. K.S. Johnson, K. Wells, J.V. Bock, V. Nene, D.W. Taylor, and J.S. Cordingley, The 86-kilodalton antigen from *Schistosoma mansoni* is a heat-shock protein homologous to yeast HSP-90, *Mol. Biochem. Parasitol. 36*:19 (1989).
11. K.A. Ul'masov, A. Ovezumkhammedov, K.K. Karaev, and M.B. Evgen'ev, [Molecular mechanisms of adaptation to hyperthermia in higher organisms. III. Induction of heat-shock proteins in two *Leishmania* species] (original in Russian), *Mol. Biochem. Mosk. 22*:1583 (1988).

12. D.M. Engman, L.V. Reddy, J.E. Donelson, and L.V. Kirchhoff, *Trypanosoma cruzi* exhibits inter- and intra-strain heterogeneity in molecular karyotype and chromosomal gene location, *Mol. Biochem. Parasitol. 22*:115 (1987)
13. A. Alcina, A. Urzainqui, and L. Carrasco, The heat-shock response in *Trypanosoma cruzi, Eur. J. Biochem. 172*:121 (1988).
14. D.M. Engman, L.V. Kirchhoff, and J.E. Donelson, Molecular cloning of mtp70, a mitochondrial member of the hsp70 family, *Mol. Cell Biol. 9*:5163 (1989).
15. J. MacFarlane, M.L. Blaxter, R.P. Bishop, M.A. Miles, and J.M. Kelly, Identification and characterisation of a *Leishmania donovani* antigen belonging to the 70-kDa heat-shock protein family, *Eur. J. Biochem. 190*:377 (1990).
16. C.R. Andrade, P.P. Andrade, L.V. Kirchhoff, and J.E. Donelson, Use of recombinant HSP-like polypeptides in the diagnosis of visceral leishmaniasis, *Mem. Inst. Oswaldo Cruz 85*(Suppl.):92 (1990).
17. D.M. Engman, E.A. Dragon and J.E. Donelson, Human humoral immunity to hsp70 during *Trypanosoma cruzi* infection, *J. Immunol. 144*:3987 (1990).
18. P. Levy-Yeyati, S. Bonnefoy, G. Mirkin, A. Debrabant, S. Lafon, A. Panebra, E. Gonzalez-Cappa, J.P. Dedet, M. Hontebeyrie-Joskowicz, and M.J. Levin, The 70-kDa heat-shock protein is a major antigen determinant in human *Trypanosoma cruzi/ Leishmania braziliensis braziliensis* mixed infection, *Immunol. Lett. 31*:27 (1992).
19. B. Carcy, E. Precigout, A. Valentin, A. Gorenflot, R.T. Reese, and J. Schrevel, Heat shock response of *Babesia divergens* and identification of the hsp70 as an immunodominant early antigen during ox, gerbil and human babesiosis, *Biol. Cell. 72*:93 (1991).
20. E. Devaney, A. Egan, E. Lewis, E.V. Warbrick, and R.M. Jecock, The expression of small heat shock proteins in the microfilaria of *Brugia pahangi* and their possible role in development, *Mol. Biochem. Parasitol. 56*:209 (1992).
21. K. Nadeau, M.A. Sullivan, M. Bradley, D.M. Engman, and C.T. Walsh, 83-kilodalton heat shock proteins of trypanosomes are potent peptide-stimulated ATPases, *Protein Sci. 1*:970 (1992).
22. P.N. Effron, A.F. Torri, D.M. Engman, J.E. Donelson and P.T. Englund, A mitochondrial heat shock protein from *Crithidia fasciculata, Mol. Biochem. Parasitol. 59*:191 (1993).
23. N.P. Robertson, R.T. Reese, J.M. Henson, and C.A. Speer, Heat shock-like polypeptides of the sporozoites and merozoites of *Eimeria bovis, J. Parasitol. 74*:1004 (1988).
24. S. Ortner, B. Plaimauer, M. Binder, G. Wiedermann, O. Scheiner, and M. Duchene, Humoral immune response against a 70-kilodalton heat shock protein of *Entamoeba histolytica* in a group of patients with invasive amoebiasis, *Mol. Biochem. Parasitol. 54*:175 (1992).
25. A. Aggarwal, V.F. de-la-Cruz, and T.E. Nash, A heat shock protein gene in *Giardia lamblia* unrelated to HSP70, *Nucleic Acids Res. 18*:3409 (1990).
26. D.S. Reiner, T. M. Shinnick, F. Ardeshir, and F.D. Gillin, Encystation of *Giardia lamblia* leads to expression of antigens recognized by antibodies against conserved heat shock proteins, *Infect. Immun. 60*:5312 (1992).
27. T. Hanekamp and P.J. Langer, Molecular karyotype and chromosomal localization

of genes encoding two major surface glycoproteins, gp63 and gp46/M2, hsp70, and beta-tubulin in cloned strains of several *Leishmania* species, *Mol. Biochem. Parasitol. 48*:27 (1991).

28. C.R. de Andrade, L.V. Kirchhoff, J.E. Donelson, and K. Otsu, Recombinant *Leishmania* Hsp90 and Hsp70 are recognized by sera from visceral leishmaniasis patients but not Chagas' disease patients, *J. Clin. Microbiol. 30*:330 (1992).
29. S. Searle and D.F. Smith, *Leishmania major*: characterisation and expression of a cytoplasmic stress-related protein, *Exp. Parasitol. 77*:43 (1993).
30. M. Shapira and G.Pedraza, Sequence analysis and transcriptional activation of heat shock protein 83 of Leishmania mexicana amazonensis, *Mol. Biochem. Parasitol. 42*:247 (1990).
31. E. Pinelli and M. Shapira, Temperature-induced expression of proteins in *Leishmania mexicana amazonensis*. A 22-kDa protein is possibly localized in the mitochondrion, *Eur. J. Biochem. 194*:685 (1990).
32. J.M. Blackwell, Leishmaniasis epidemiology: all down to the DNA, *Parasitology 104*(Suppl):S19 (1992).
33. D.M.Estes and J.M. Teale, Biochemical and functional analysis of extracellular stress proteins of *Mesocestoides corti, J. Immunol. 147*:3926 (1991).
34. N. Kumar, H. Nagasawa, J.B. Sacci Jr, B.J. Sina, M. Aikawa, C. Atkinson, P. Uparanukraw, L.B. Kubiak, A.F. Azad, and M.R. Hollingdale, Expression of members of the heat-shock protein 70 family of the exoerythrocytic stages of *Plasmodium berghei* and *Plasmodium falciparum, Parasitol. Res. 79*:109 (1993).
35. G. Langsley, J.C. Barale, and D. Mattei, Isolation from a *Plasmodium chabaudi* chromosome 7 specific library of a novel gene encoding from a protein with multiple GGMP repeats homologous to hsp70, *Mol. Biochem. Parasitol. 59*:331 (1993).
36. V. Eckert, L. Sanchez, A. Cochrane, and V. Enea, *Plasmodium cynomolgi*: the hsp 70 gene, *Exp. Parasitol. 75*:323 (1992).
37. M. Jendoubi and S. Bonnefoy, Identification of a heat shock-like antigen in *P. falciparum*, related to the heat shock protein 90 family, *Nucleic Acids Res. 16*:10928 (1988).
38. B.J. Scallon, B.J. Bogitsh, and C.E. Carter, Cloning of a *Schistosoma japonicum* gene encoding a major immunogen recognized by hyperimmune rabbits, *Mol. Biochem. Parasitol. 24*:237 (1987).
39. L. Heine, D.E. Rebeski, W. Leibold, K.T. Friedhoff, and E. Gunther, Induction of the 68 kDa major heat-shock protein in different *Theileria annulata*- and virus-transformed bovine lymphoblastoid cell lines, *Vet. Immunol. Immunopathol. 33*:271 (1992).
40. H. Nagasawa, M. Oka, K. Maeda, C. Jian-Guo, H. Hisaeda, Y. Ito, R.A. Good, and K. Himeno, Induction of heat shock protein closely correlates with protection against *Toxoplasma gondii* infection, *Proc. Natl. Acad. Sci. USA 89*:3155 (1992).
41. S.R. Davis and W.B. Lushbaugh, Oxidative stress and *Trichomonas vaginalis*: the effect of hydrogen peroxide in vitro, *Am. J. Trop. Med. Hyg. 48*:480 (1993).
42. M. Giambiagi-de-Marval, K. Gottesdiener, E. Rondinelli, and L.H. Van-der-Ploeg, Predicted amino acid sequence and genomic organization of *Trypanosoma cruzi* hsp 60 genes, *Mol. Biochem. Parasitol. 58*:25 (1993).

43. J.B. Vieira, M.M. Lacerda, and P.D. Marsden, National reporting on leishmaniasis: the Brazilian experience. *Parasitol. Today 6*:339 (1990).
44. G. Grimaldi, Jr., and D. McMahon-Pratt, Leishmaniasis and its etiological agents in the New World: an overview, *Progress in Clinical Parasitology* (T. Sun, ed.), Field and Wood Medical Publishers, Philadelphia, 1991, pp. 73–118.
45. P. DeBeer, A. el-Harith, L.L. Deng, S.J. Semiao- Santos, B. Chantal, and M. van Grootheest, A killing disease epidemic among displaced Sudanese population identified as visceral leishmaniasis, *Am. J. Trop. Med. Hyg. 44*:283 (1991).
46. TDR/WHO, Kala-azar surges in two fronts, *TDR News 37*:1 (1991).
47. World Health Organization, The Leishmaniasis. WHO Technical Report Series 701, WHO, Geneva, 1984.
48. M.C.A. Marzochi, S.G. Coutinho, P.C. Sabroza, P.P. Soza, M.A. Souza, L.M. Toledo, and F.B. Rangel Fo., Leishmaniose visceral canina no Rio de Janeiro - Brasil. *Cad. Saude Publ. (RJ) 1*:432 (1985).
49. M. Reincke, B. Allolio, E. Hackenberg, S. Cohen, W. Winkelman, and W.T.O. Kaufmann, In Europa erworbene viszerale Leishmaniose (Kala-Azar), *Med. Klin. 85*:220 (1990).
50. P.D. Marsden and T.C. Jones, Clinical manifestations, diagnosis and treatment of leishmaniasis, *Human Parasitic Infections, Vol. 1. Leishmaniasis* (K.-P. Chang, and R.S. Bray, eds.), Elsevier, Amsterdam, 1985.
51. M.E. Camargo, American trypanosomiasis (Chagas' disease), *Laboratory Diagnosis of Infectious Diseases: Principles and Practice* (A. Balows, W.J. Hausler, Jr., and E.H. Lennette, eds.), Springer-Verlag, New York, 1988, pp. 744–753.
52. M.L. Blaxter, M.A. Miles, and J.M. Kelly, Specific serodiagnosis of visceral leishmaniasis using a *Leishmania donovani* antigen identified by expression cloning, *Mol. Biochem. Parasitol. 30*:259 (1988).
53. R. Hedstrom, J. Culpepper, V. Schinski, N. Agabian, and G. Newport, Schistosome heat-shock proteins are immunologically distinct host-like antigens, *Mol. Biochem. Parasitol. 29*:275 (1988).
54. N. Kumar, Y. Zhao, P. Graves, J. Perez-Folgar, L. Maloy, H. Zheng, Human immune response directed against *Plasmodium falciparum* heat shock-related proteins, *Infect. Immun. 58*:1408 (1990).
55. G.R. Wallace, A.E. Ball, J. MacFarlane, S.H. el-Safi, M.A. Miles, and J.M. Kelly, Mapping of a visceral leishmaniasis-specific immunodominant B-cell epitope of *Leishmania donovani* Hsp70, *Infect. Immun. 60*:2688 (1992).
56. A. Hoerauf, P.P. Andrade, C.R. Andrade, W. Solbach, and M. Roellinghoff, Immunoblotting as a valuable tool to differentiate human visceral leishmaniasis from lympho-proliferative disorders and other clinically similar diseases, *Res. Immunol. 143*:375 (1992).
57. E.A.E.R. Mohammed, E.P. Wright, P.A. Kager, J.J. Laarman, and K.W. Pondman, ELISA using intact promastigotes for immunodiagnosis of kala-azar, *Trans. R. Soc. Trop. Med. Hyg. 79*:344 (1985).
58. P.M.M.F. Moura, Proteínas recombinantes de Leishmania no diagnóstico do calazar canino, M.Sc. Thesis, Federal University of Rio de Janeiro, Brazil, 1994.
59. D.M. Russo, R.J. Armitage, M. Barral-Netto, A. Barral, K. H. Grabstein, and

S.G. Reed, Antigen-reactive gamma delta T cells in human leishmaniasis, *J. Immunol. 151*:3712 (1993).

60. P. Esterre, J.P. Dedet, C. Frenay, M. Chevallier, and J.A. Grimaud, Cell populations in the lesion of human cutaneous leishmaniasis: a light microscopical, immunohistochemical and ultrastructural study, *Virchows Arch. A. Pathol. Anat. Histopathol. 421*:239 (1992).
61. K. Uyemura, J. Klotz, C. Pirmez, J. Ohmen, X.H. Wang, C. Ho, W.L. Hoffman, and R. L. Modlin, Microanatomic clonality of gamma delta T cells in human leishmaniasis lesions, *J. Immunol. 148*:1205 (1992).
62. S. Raziuddin, A.W. Telmasani, M. el-Hag-el-Awad, O. al-Amari, and M. al-Janadi, Gamma delta T cells and the immune response in visceral leishmaniasis, *Eur. J. Immunol. 22*:1143 (1992).
63. J.P. Rosat, H.R. MacDonald, and J.A. Louis, A role for gamma delta+ T cells during experimental infection of mice with *Leishmania major, J. Immunol. 150*:550 (1993).
64. M. Alaibac, G. Harms, K. Zwingenberger, J. Morris, R. Yu, and A.C. Chu, Gamma delta T lymphocytes in oriental cutaneous leishmaniasis: occurrence and variable delta gene expression, *Br. J. Dermatol. 128*:388 (1993).
65. K. Pfeffer, B. Schoel, H. Gulle, H. Moll, S. Kromer, S.H. Kaufmann, and H. Wagner, Analysis of primary T cell responses to intact and fractionated microbial pathogens, *Curr. Top. Microbiol. Immunol. 173*:173 (1991).
66. E.M. Carvalho, Restoration of T cell responsiveness in human visceral leishmaniasis by cytokine and cytokine antagonists, *Mem. Inst. Oswaldo Cruz 88* (Suppl.):40 (1993).
67. S.R. de Costa, Y. Skeiky, M. Freire, A. D'Oliveira Jr., S. Reed, and E. Carvalho, Análise da resposta imune a antígenos recombinantes de L. chagasi em indivíduos curados de leishmania visceral e em portadores de infecções assintomáticas, *Rev. Soc. Bras. Med. Trop. 27*(Suppl. 1):181 (1994).
68. O. Genaro, W. Mayrink, M.S.M. Michalik, M. Dias, C.A. Costa, and M.N. Melo, Naturally occurring visceral leishmaniasis in dogs: clinical aspects, *Mem. Inst. Oswaldo Cruz 83*:43 (1988).
69. T.M. Shinnick, Heat shock proteins as antigens of bacterial and parasite pathogens, *Current Topics in Microbiology and Immunology* (S.H.E. Kaufman, ed.), Springer-Verlag, 1991, pp. 145–160.
70. R.M. Smejkal, R. Wolff, and J.G. Olenick, *Leishmania braziliensis panamensis:* increased infectivity resulting from heat shock. *Exp. Parasitol. 65*:1 (1988).
71. S.J. Richman, T.S. Vedvick, and R.T. Reese, Peptide mapping of conformational epitopes in a human malarial parasite heat shock protein, *J. Immunol. 143*:285 (1989).

20

Role of Heat Shock Protein Immunity in Transplantation

Rene J. Duquesnoy and Ricardo Moliterno
University of Pittsburgh Medical Center, Pittsburgh, Pennsylvania

I. INTRODUCTION

It is now well established that acute cellular rejection of organ transplants is mediated by donor-specific alloactivated T lymphocytes that infiltrate the allograft. Such lymphocytes can have a direct cytotoxic effect on allograft target cells, and more likely they recruit through lymphokine release other inflammatory cells which cause graft injury. However, it has become apparent that donor-specific alloreactive T cells comprise a rather small proportion of graft-infiltrating lymphocytes involved with rejection. This can be demonstrated by limiting dilution analysis assays which permit the determination of the frequencies of different subsets of donor-specific lymphocytes in infiltrating cells isolated directly from the graft (1–4) or indirectly after interleukin-2 (IL-2)–induced propagation (5,6). The importance of this concept is readily visualized from the histological hallmark of acute cellular rejection: a vast number of graft-infiltrating lymphocytes, most of which are apparently not donor specific. Immunostaining generally shows the presence of varying proportions of CD4 and CD8 subsets of CD3 lymphocytes expressing the $\alpha\beta$ T-cell receptor (TCR), suggesting that such cells carry the surface markers necessary for antigen recognition. Before we address the question what other antigens might be recognized by graft-infiltrating lymphocytes, we refer to another study which further illustrates our concept.

This study dealt with the IL-2–induced lymphocyte propagation from coronary vessel walls isolated from transplanted human hearts removed because of severe accelerated arteriopathy due to chronic rejection (7). These lymphocyte cultures contained high proportions of TCR γδ cells which failed to show donor HLA-specific alloreactivity. Others have also identified TCR γδ cells in IL-2–propagated lymphocyte cultures from long-term surviving heart transplants undergoing chronic rejection (8) and rejecting kidney transplants (9). Since antigen recognition by TCR γδ cells may not be major histocompatibility complex (MHC) restricted (10–11), we conclude that graft-infiltrating TCR γδ cells must recognize other antigens.

Many studies have been conducted to test the postulate that graft-infiltrating lymphocytes recognize "tissue-specific" or "organ-specific" antigens, whereas others consider the so-called minor histocompatibility antigens, including the male sex-linked antigen H-Y (12–15). Although these antigens may play a role in graft rejection, we have considered an alternative strategy toward identifying the "other" antigens recognized by graft-infiltrating cells. This strategy is based on the concept that during the initiation of cellular rejection, infiltrating donor-specific lymphocytes induce a stress response within the allograft which then triggers the recruitment and activation of heat shock protein–reactive lymphocytes.

Many reports have indicated that heat shock proteins (hsps) can be recognized by T cells during various immunologically mediated inflammatory processes (16–20). Injurious stimuli to cells induce an increased production of hsps which could lead to their cell surface expression and subsequent recognition by the immune system. Most relevant are findings with experimental models and in clinical situations that hsp-specific T cells are involved in the pathogenesis of several autoimmune diseases, including rheumatoid arthritis, insulin-dependent diabetes, and multiple sclerosis (21–24). Moreover, hsps have been shown to function as tumor-rejection antigens in several experimental models (25–27).

Recently, we have conducted three studies that provide experimental support of the concept about the role of hsp immunity in transplant rejection.

II. HSP-INDUCED T-LYMPHOCYTE PROPAGATION FROM HEART TRANSPLANT BIOPSIES CORRELATES WITH CELLULAR REJECTION

The first study (28) dealt with the incubation of heart transplant biopsies with soluble *Mycobacterium tuberculosis* extracts (MTEs), a source of hsp recognized by human T cells. Studies on 299 biopsies from 89 heart transplant patients demonstrated that MTE can induce lymphocyte propagation from heart transplant biopsies, especially during rejection. A highly significant correlation was seen between MTE and Il-2–induced lymphocyte growth, and an accelerated growth was seen for cultures incubated with MTE + Il-2. A second series of experiments with 47 biopsy cultures showed the propagation of lymphocytes induced by recombinant mycobacterial hsp65. These data suggest that hsp-reactive T lymphocytes are recruited into the allograft during rejection.

T-cell phenotype analysis of biopsy-derived lymphocytes showed higher frequencies of $CD8^+$ cells in MTE- and hsp 65–propagated lymphocytes from early posttransplant biopsies, whereas later on, most cultures had a predominance of $CD4^+$ cells. Of special interest is the presence of TCR $\gamma\delta$ cells in biopsy lymphocyte cultures. During the first year posttransplant, approximately 20–30% of biopsy-derived lymphocytes contained significant numbers of TCR $\gamma\delta$ T cells ($>$ 10%) regardless of propagation stimulus. On the other hand, more than 50% of long-term transplant biopsies yielded $\gamma\delta$ T cells, especially when cultured with MTE and hsp 65. No associations were found between the TCR $\gamma\delta$ cell frequencies and histological rejection grade.

MTE has been reported to stimulate preferentially the proliferation of the γ^{9+},δ^{2+} subset circulating TCR $\gamma\delta$ T lymphocytes (29). However, our data on MTE and hsp-propagated graft-infiltrating lymphocytes showed almost exclusively the δ^{1+} rather than the γ^{9+} phenotype. They are consistent with the experience of Vaessen et al. (9), who reported that 30% of IL-2–propagated lymphocyte cultures from long-term biopsies contain TCR $\gamma\delta$ cells primarily of the $\delta 1$ subset.

Several reports have indicated that TCR $\gamma\delta$ cells can recognize hsp expressed by mycobacteria and stressed mammalian cells and that they may participate in autoimmune reactions (10,30–32). It is possible that hsp might be recognized by $CD4^-CD8^-$TCR $\gamma\delta$ cells we can propagate from coronary disease, a manifestation of long-term heart transplant survivors with graft coronary disease due to chronic rejection (8). Since a major function of TCR $\gamma\delta$ cells deals with the recognition of stressed cells during chronic disease (10), we postulate the involvement of hsp-reactive TCR $\gamma\delta$ cells in graft coronary disease. Relevant to this concept are the recent findings by Wick's group on the upregulated hsp expression and the presence of hsp-reactive TCR $\gamma\delta$ cells in arteriosclerotic lesions in an experimentally induced rabbit model (33,34).

III. HSP REACTIVITY OF LYMPHOCYTES ISOLATED FROM HETEROTOPIC RAT CARDIA ALLOGRAFTS

This study (35) deals with a rat model of heterotopic MHC-incompatible cardiac allografts (ACI into Lewis) whereby graft-infiltrating lymphocytes and spleen cells were tested in vitro with different recombinant mycobacterial hsp preparations (kindly provided by Dr. J.D.A. van Embden, National Institute of Public Health and Environmental Protection, Bilthoven, The Netherlands). As expected, allograft lymphocytes showed primed proliferative responses to irradiated spleen cells from the donor. This proliferation was markedly augmented by hsp 65 (threefold) and hsp 71 (fivefold), whereas hsp 10 and the protein control ovalbumin had no effect. Since allograft lymphocytes did not respond to hsp alone or in the presence of autologous spleen cells, the MLR augmentation by hsp seemed related to donor cell stimulation. One possibility is that hsps interact with donor cells, perhaps via a mechanism of antigen processing and presentation. This could mean that hsp antigens might be presented in context with allo-MHC.

Another possibility is that on donor stimulator–induced activation, alloreactive T cells in the graft would release IL-2 and other lymphokines which would then be used to augment the responses of hsp-reactive cells. Indeed, proliferation of allograft lymphocytes to hsp 65 and hsp 71 in context with syngeneic spleen cells was seen if small quantities of IL-2 had been added to the cultures. In contrast, hsp-specific proliferation was never observed with syngraft lymphocytes even after addition of IL-2. Spleen cells from allograft and syngraft recipients showed hsp augmentation of alloproliferation, but the magnitude was less than that with allograft lymphocytes. Culture conditions have been established to generate hsp 65– and hsp 71–specific T-lymphocyte lines and clones from allograft-infiltrating cells. Certain clones exhibit hsp reactivity in context with autologous antigen-presenting cells and exogenous IL-2 (35).

All hsp preparations from the World Health Organization (WHO) contained small quantities of endotoxin, which might influence lymphocyte proliferation. Removal of endotoxin by polymyxin B treatment (36) did not significantly affect the hsp 65–induced proliferation of allograft cells. In contrast, it caused a marked reduction in the hsp 71–specific response in context with either donor spleen cells or autologous spleen cells + IL-2. Addition of lipopolysaccharide did not restore the hsp 71 responsiveness, suggesting that the stimulatory activity of the hsp 71 preparation was not due to endotoxin. Since polymyxin B did not deplete hsp 71 from the preparation, it is possible that it altered the stimulatory activity of hsp 71 through a conformational change. Hsp 70 molecules have two distinct conformations associated with high- and low-affinity binding of peptide (38–40). Polymyxin B, a cyclic peptide antibiotic, is a rather selective antagonist of protein kinase C and shows inhibitory binding with the chaperone calmodulin (37). It is also possible that polymyxin B might function as an antagonist through inhibitory binding with *M. tuberculosis* hsp 71.

Thus far, our studies on the hsp reactivity of allograft lymphocytes have been done with recombinant protein preparations of mycobacterial rather than mammalian origin. In various autoimmune disease models, the reactivity of mycobacterial hsp-specific lymphocytes may reflect recognition of antigenic determinants shared with host self-proteins (16–19). Throughout evolution, the molecular structures of the different types of hsps have been highly conserved between the species (41). Sequence homologies have been reported between hsp

60 and a wide range of autoantigens (42). T cells can recognize shared epitopes of mycobacterial and human hsp 65 (16–19).

At present, it is not known whether the reactivity of allograft lymphocytes toward mycobacterial hsp reflects a specific response against epitopes shared with host hsp. Since the transplanted rats were maintained under specific pathogen-free conditions, it seems unlikely that the hsp reactivity of allograft lymphocytes was due to any microbial infection of the graft. Recent studies have indicated that graft-infiltrating lymphocytes react with murine grp 78, which also showed cross-reactivity with mycobacterial hsp 71 (43).

IV. HSP EXPRESSION IN RAT CARDIAC ALLOGRAFTS

The proliferation data with the human and rat cardiac transplant models are consistent with the concept that heart allografts are infiltrated by hsp-reactive T cells recruited as a component of the lymphocyte-mediated immune response causing cellular rejection. The third series of studies deals with hsp expression in rat heart grafts (43). This was done by immunoblotting of proteins extracted from allograft and syngraft stromal tissues. A kinetic analysis during the first 5 days posttransplant suggested that three types of stressful stimuli appeared to increase hsp expression in the allograft. The first was a physiological stress secondary to the trauma of the transplant procedure and ischemia/reperfusion injury and this would occur in allogeneic and syngeneic grafts. During the first day after transplantation, both types of grafts showed higher expression of hsp 72 and grp 78, and to a lesser extent, hsp 60 and grp 75. On the second and third days, the expression of grp 78 and a 96-kDa band was markedly higher in allografts than in syngrafts, and this may reflect an immunologically mediated stress response in the allograft when infiltrating hsp-reactive lymphocytes become first detectable in the allograft. The third type of stress appears related to the inflammatory process associated with rejection. On the fourth and fifth days posttransplant, the allografts showed strong expression of at several lower molecular weight proteins reacting, including a 40-kDa band detected by anti-hsp 60 and a 50-kDa band reacting with anti–grp 78. The appearance of lower molecular weight hsp cross-reactive proteins might reflect a proteolytic degradation of hsp which had increased expression earlier during the posttransplant period.This process may generate large quantities of hsp-derived peptides which may be presented by MHC molecules to graft-infiltrating T cells. Other interpretations of the lower molecular bands in later allografts are that they resulted from alternate splicing of mRNA, or that they represent other stress proteins that cross react with antibodies against hsp 60 and hsp 70 family members. These findings extend the data of Currie et al. (44) on the increased synthesis of a 71-kDa stress protein in transplanted rat hearts and demonstrate a relationship between altered hsp expression and cardiac allograft rejection. Immunostaining of transplant tissues may yield additional information about cell surface expression of hsps. Our preliminary results suggest increased staining for hsp 70 during rejection. (J. Demetris, personal communication). Two recent abstracts indicate that the upregulation of hsp expression in kidney and liver transplants may serve as a diagnostic marker of graft injury due to ischemia and rejection (45,46). Similarly, Clancy et al. (47) described an increased production of hsp 70 in lymphoid tissues during acute graft-versus-host (GVH) disease in an F1 hybrid rat model and preimmunization with bovine hsp 70 increases GVH disease susceptibility. Rose and coworkers have shown by Western blotting that pretransplant serum levels of cardiac-specific antibodies correlate with heart transplant rejection (48,49). Much of the antibody activity was directed toward the 60- and 70-kDa hsps in cardiac extracts.

V. DIFFERENT ASPECTS OF THE ROLE OF HSPS IN TRANSPLANT IMMUNITY

Although these data must be considered preliminary, they seem consistent with the concept about the role of hsps in autoimmune disease and tumor immunity. Several investigators have presented evidence that hsps can function as tumor rejection antigens. Srivastava and his group have shown that hsp 70 and two other hsps (96 kDa and 84–86 kDa) isolated from methylcholanthrene-induced murine sarcomas elicit specific tumor immunity (26,50,51). Sato and coworkers have obtained evidence with several experimental models that hsps may play a vital role in antitumor resistance (27,52). As an example, activated H-*ras* oncogene transformed rat fibrosarcoma cells have surface expression of *hsc 70* which can be recognized by BCG-primed $CD4^-$, $CD8^-$ T cells. Hsp 60–specific CD^{4+} T cells show antitumor activity against a murine sarcoma (53). Lukacs et al. (54) have shown that injection of mice with tumor cells transfected with mycobacterial hsp 65 induces immune protection against the original tumor.

Numerous reports demonstrate the role of hsps in autoimmune disease (16–20). Hsp 60–reactive TCR $\alpha\beta$ and $\gamma\delta$ cells appear in synovial fluids from patients with rheumatoid arthritis and reactive arthritis (55), and in spinal fluids and chronically demyelinated lesions in patients with multiple sclerosis (56). Hsps have also clearly been implicated in the pathogenesis of autoimmune diseases. Two examples illustrate this concept. In insulin-dependent diabetes of NOD mice, Elias et al. (22) have demonstrated hsp 60–reactive T cells that are specific for an epitope shared between mycobacterial and mammalian hsp 60 and corresponding to an amino acid sequence in positions 437–460. Administration of this peptide prevented diabetes in this model (22). Van Eden's group has shown in the adjuvant arthritis model in Lewis rats that T cells specific for the 180–188 sequence of mycobacterial hsp 60 confer disease (21), and that immunization with that epitope is protective (57,58).

Thus, the role of hsps in autoimmune diseases might be through a mechanisms whereby certain bacterial infections elicit immune responses to microbial hsps which then would react against stressed host cells that expressed cross-reacting hsp antigens (23–25). However, since many healthy individuals have also bacterial hsp-specific T lymphocytes and antibodies which can recognize self-hep (59), another concept has been proposed that the expression of self-hsp is intended to permit an immune surveillance of stressed autologous cells (23–26). Accordingly, the immune system contains a repertoire of autoreactive cells directed against a relatively limited spectrum of dominant self-antigens which include several hsps (26). Mor and Cohen (60) have shown the enrichement of hsp 65–reactive T cells in lesions of experimental autoimmune encephalomyelitis induced by transfer of myelin basic protein–specific T lymphocytes, suggesting that the inflammatory lesion itself attracted anti–self-hsp T cells. The activity of hsp-reactive T cells is regulated by a tightly controlled immune network, and their function would be in the selective removal of aberrant cells from host tissues, perhaps through a guarded and transient inflammatory process; autoimmune disease may occur if this autoreactivity goes out of control (26).

We propose that transplant immunity is a regulated system of alloreactive and self–(hsp) reactive lymphocytes. A balanced immune system would promote transplant acceptance, whereas dysregulation would induce allograft rejection. Self-surveillance must be tightly regulated to avoid unwanted activation of alloreactive and/or hsp-reactive lymphocytes. This concept may explain the findings published in 1978 by Russell et al. (61) that administration of spleen cells and BCG (a source of hsp) caused an irreversible rejection of long-term surviving kidney allografts in dogs. These agents alone induced only a transient increase in serum creatinine levels, but their combination seemed to cause an imbalanced system of alloreactive and hsp-reactive immunity, thereby inducing rejection rather than graft acceptance.

Our data on the IL-2 augmentation of hsp 65– and hsp 70–reactive lymphocytes from cardiac allografts (35) seem to indicate that IL-2 might be an important lymphokine involved in the regulation of hsp-reactive lymphocytes. During cellular rejection, infiltrating donor-specific alloreactive lymphocytes produce IL-2 which would be utilized for the activation and expansion of hsp-reactive lymphocytes recruited into the allograft as a mechanism of augmenting the immunologically mediated inflammatory process.

In organ transplantation, we must also consider the influence of preservation/reperfusion injury which may lead to graft infiltration by inflammatory cells and which might increase the risk of rejection. Several investigators have reported an upregulation of hsp expression following ischemia/reperfusion injury of cardiac tissue (62–64). Increased hsp expression conveys a cytoprotective effect (63–66), but it may also stimulate the recruitment of hsp-reactive lymphocytes. In cases of short-term ischemia, a substantial intragraft production of IL-

2 and other cytokines seems unlikely and, therefore, infiltrating hsp-reactive lymphocytes would eventually migrate away. In contrast, following more severe ischemic injury and, as we propose, during graft rejection, the infiltrating hsp-reactive lymphocytes would be activated by various cytokines like IL-2, and this would further expand the inflammatory process.

Heat treatment may also have a profound effect on immune cells. This can be readily demonstrated in the mixed leukocyte reaction, whereby mild heat treatment of cells abolishes their in vitro stimulatory activity (67–69) and also seems to promote anergy of responders and perhaps even the generation of suppressor cells (70). Exogenous IL-2 restores the stimulatory activity (71). Pretransplant administration of heated donor cells can also induce in vivo donor-specific unresponsiveness and prolongation of graft survival (72,73). The mechanism of heat-induced loss of the stimulatory of allogeneic cells is largely unknown. Mild heat treatment has a minimal effect on cell viability and it does not seem to decrease surface expression of HLA-DR antigens (67). Since hyperthermia causes such a marked upregulation of hsps, we believe that hsps must play a significant role in the heat-induced changes in the stimulatory ability of allogeneic cells. In other words, heat stress may also generate an hsp-related negative signal in T-cell alloactivation. This concept can readily be incorporated into our model that a regulated balance between alloreactive and hsp-reactive lymphocytes is required for transplant survival.

VI. DEVELOPMENT OF A WORKING MODEL FOR THE ROLE OF HSPS IN TRANSPLANT IMMUNITY

This working model considers four phases in the process of T-cell–induced stress response and the recruitment and activation of hsp-reactive lymphocytes during transplant rejection. Transplant rejection is inititiated by graft-infiltrating alloreactive T cells which through direct cytotoxic effects and indirectly through cytokine release and inflammatory cell recruitment and activation induce allograft injury and tissue stress (phase 1).

Stressed target cells in the transplanted tissue respond by upregulation of the production of various hsps, especially the hsp 60 and hsp 70 families (phase 2). Although hsp 60 operates as a molecular chaperone primarily within the mitochondrial compartment, there has been convincing evidence that hsp 60 can be recognized by T lymphocytes, and that mycobacterial and/or mammalian hsp 60–specific immune responses play an important role in several autoimmune diseases (16–20).

The hsp 70 family has several distinctive features which predict a greater complexity of hsp 70–specific immune responsiveness. At least eight members have been identified, including two or three controlled by genes linked to the MHC complex (74,75). Most information about mammalian hsp 70 pertains to the inducible 72-kDa (hsp 72), the constitutively expressed 73-kDa (*hsc 73*), the PBP 72/74 molecules involved with antigen processing (76–78), and the 78-kDa (grp 78 or BiP) chaperone in the endoplasmic reticulum which can bind proteins and peptides (79,80). Our Western blotting and immunostaining data have suggested an upregulation of hsp 72, *hsc 73*, and grp 78 in rat allografts undergoing rejection. Certain members of the hsp 70 family may be expressed on the cell surface of antigen-presenting cells and they may play a role in antigen presentation to T cells (76–78). Hsp 60 can also be expressed at the cell surface (81–83) and, therefore, might be recognized by T cells, particularly γδ T cells (10).

Increased hsp production by stressed cells might be associated with positive and negative influences on transplant immunity (phase 3). The positive signals deal with the recognition and activation of lymphocytes, especially hsp-reactive T cells, which now have been amply demonstrated in various immunologically mediated inflammatory lesions in autoimmune disease, in tumor immunity, and as our data show, in transplant rejection. T cells might react directly with hsp molecules expressed on the cell surface or they might recognize hsp peptides presented by MHC molecules at the cell surface. Lymphokines such as IL-2 will play an important role in the hsp reactivity of graft-infiltrating lymphocytes. It is well known that rejection is associated with increased expression of MHC antigens on allograft tissue and graft-infiltrating lymphocytes produce interferon gamma (IFN-γ), a potent stimulator of MHC antigen expression. Interestingly, IFN-γ and other cytokines such as tumor necrosis factor α (TNF-α) also increase hsp synthesis in mammalian cells (84–87).

Conversily, increased hsp production might be expected to exert a cytoprotective effect as indicated by reports describing increased resistance of stressed cells to subsequent stress stimuli (88,89) and, as shown recently, with hsp gene–transfected cells (90–93). This could mean that under certain conditions, stressed cells might develop an increased resistance to immunologically mediated injury. Moreover, stressed cells may generate negative signals for lymphocyte activation, as has been amply demonstrated by the loss of stimulatory activity of heated cells (67–73).

Most of these hsps will eventually be broken down by various proteases, and this will generate vast quantities of hsp peptides which can be expected to be taken up by the newly synthesized MHC molecules and transported to the cell surface (phase 4). This concept predicts that much of the increased MHC expression on allograft tissue might reflect enhanced presentation of self-hsp peptides. We expect, therefore, that graft-infiltrating cells also contain lymphocytes which recognize hsp peptides presented by allo-MHC. It is also possible that stressed cells release hsps and their break-down products which could be taken up and processed by macrophages which would then present hsp peptides to T cells in context with self-MHC.

VII. CONCLUSIONS

This model on the role of hsps in transplant immunity may also increase our understanding of the various mechanisms affecting transplant outcome, including infection and chronic rejection. During infection, host immune responses have been shown to be directed primarily to microbial hsps (16–18). The link between autoimmune disease and infection has been interpreted with a mechanism based on cross reactivity between microbial and mammalian hsps. However, another and probably better interpretation is that immune responses to microbial hsps cause a dysregulation of self-hsp immunity which then might lead to autoimmune disease (18,19). In transplantation, the relation between infection and rejection was recognized long ago (93). Although several mechanisms have been proposed, including bacterial superantigens, this relation may have a similar basis as in autoimmune disease, whereby infection-induced dysregulation of self-hsp immunity would perturb its balance with alloreactivity, thereby promoting graft rejection.

Further considerations about the role of hsps might contribute to a better understanding of chronic rejection which is targeted primarily against the vascular endothelium and epithelial cells in allografts. The immune responses can be cellular- and antibody-mediated, and they can be directed against various antigens, including MHC alloantigens. Although some studies have indicated that HLA-mismatching increases chronic rejection, we and others have found that matching for HLA-DR is associated with a more progressive vanishing bile duct syndrome, a manifestation of chronic liver transplant rejection (94,95). Cytomegalovirus (CMV) hepatitis, which occurs more frequently in DR-matched liver allografts, presents also an increased risk for chronic rejection. CMV infection is also associated with more chronic rejection of heart and lung transplants (96,97). Although many postulates have been forwarded, our understanding of the mechanism(s) of chronic rejection remains unclear. It must be apparent, however, that chronic rejection is a slowly progressing, immunologically mediated inflammatory process which causes a chronic state of tissue stress. This may lead to a stress-induced upregulation of hsp expression and subsequent activation of hsp immunity, which would further be augmented by certain infections, like CMV, thereby creating a dysregulated balance between alloimmunity and hsp immunity. The pathogenesis of chronic rejection and many autoimmune diseases may involve similar immunological mechanisms. Since hsp immunity plays a significant role in autoimmune disease, it seems likely that hsp-reactive lymphocytes are involved in chronic rejection.

In summary, our experience supports the concept about the involvement of hsp-reactive lymphocytes in transplant rejection. Since recent data have shown that such cells react toward murine grp 78 (43), we believe that the immune recognition of self-hsp antigens expressed by stressed tissues in the allograft might be an important mechanism affecting transplant outcome.

ACKNOWLEDGMENTS

These studies were supported by NIAID grant AI-23567 and by the Pathology Education and Research Foundation. R.M. is a recipient of a CAPES scholarship from the Brazilian government.

REFERENCES

1. C.G. Orosz, N.E. Zinn, L. Sirined, and R.M. Ferguson, In vivo mechanisms of alloreactivity. I Frequency of donor-reactive T lymphocytes in sponge matrix allografts, *Transplantation 41*:75 (1986).
2. C.G. Orosz, D.R. Bishop, and R.M. Ferguson, In vivo mechanisms of alloreactivity. VI. Evidence that alloantigen deposition initiates both local and systemic mechanisms that influence CTL accumulation at a graft site, *Transplantation 48*:818 (1989).
3. A.J. Suitters, M.L. Rose, M.J. Dominguez MJ, and M.H. Yacoub, Selection for donor-specific cytotoxic T lymphocytes within the allografted human heart, *Transplantation 49*:1105 (19990).
4. S.M. Stepkowski, and T. Ito, Frequency of alloantigen-specific T cytotoxic cells in high- an low-responder recipients of class I MHC-disparate heart allografts, *Transplantation 50*:112 (1990).
5. C. Kaufman, Propagation and Characterization of Allograft Infiltrating and Circulating Lymphocyte Populations in Human Cardiac Transplant Recipients, Ph.D. Dissertation, University of Pittsburgh, Pittsburgh, PA, 1991
6. A.J. Ouwehand, C.C. Baan, D.L. Roelen, L.M.B. Vaessen, A.H.M.M. Balk, N.H.P.H. Jutte, E. Bos, F.H.J. Claas, W. Weimar, Detection of cytotoxic T cells with high affinity for donor antigens in the transplanted heart as prognostic factor for graft rejection, *Transplanttion 56*:1223 (1993).
7. R.J. Duquesnoy, C. Kaufman, T.R. Zerbe, M.C. Woan, and A. Zeevi, Presence of CD4, CD8 double negative and T cell receptor-gamma delta positive T cells in lymphocyte cultures propagated from coronary arteries from heart transplant patients with graft coronary disease, *J. Heart Lung Transpl. 11*:583 (1992).
8. L.M.B. Vaessen, A.J. Ouwehand, C.L. Baan, N.H.P.M., Jutte, A.H.M.M. Balk, F.J.H. Claas, and W. Weimar, Phenotypic and functional analysis of T cell receptor $\gamma\delta$-bearing cells isolated from human hearts, *Transplantation 147*:846 (1991).
9. A.D. Kirk, S. Ibrahim, D.V. Dawon, F. Sanfilippo, and O.J. Finn, Characterization of T cells expressing the γ/δ antigen receptor in human renal allografts. *Hum. Immunol. 36*:11 (1993).
10. W. Born, R. Harbeck, and R.L. O'Brien, Possible links between autoimmune system and stress response: the role of $\gamma\delta$ T lymphocytes, *Semin. Immunol. 3*:43 (1991).
11. M. Kronenberg, Antigens recognized by $\gamma\delta$ T cells. *Curr. Opin. Immunol. 6*:64 (1994).
12. R.J. Duquesnoy and D.V. Cramer, Immunological mechanisms of cardiac transplant rejection, *Cardiovasc. Clin. 20*:87 (1990).

13. J.F. Burdick and Clow, Rejection of murine cardiac allografts. I. Relative roles of major and minor antigens, *Transplantation 42*:67 (1986).
14. G.J. Cerilli and L. Brasile, Tissue specific antigens - a role in organ transplantation: a theory for the existence of tissue specific antigens, *Organ Transplantation and Replacement* (G.J. Cerilli, ed.), Lippincott, Philadelphia, PA, 1987, pp. 208–222.
15. E. Goulmy, Minor histocompatibility antigens in man and their role in transplantation, *Transplant Rev. 2*:29 (1988).
16. R.A. Young, Stress proteins and immunology, *Ann. Rev. Immunol. 8*:401 (1990).
17. S.H.E. Kaufmann, Heat shock proteins as antigens in immunity against infection and self, *The Biology of Heat Shock Proteins and Molecular Chaperones* (R.I. Morimot, A. Tissieres, and C. Georgopoulos, ed.), Cold Spring Harbor Laboratory Press, Plain View, NY, 1994, pp. 495–531.
18. W. Van Eden, Heat shock proteins as immunogenic bacterial antigens with the potential of inducing and regulating autoimmune arthritis, *Immunol. Rev. 121*:5 (1991).
19. I.R. Cohen and D.B. Young, Autoimmunity, microbial immunity and the imnunological homunculus, *Immunol. Today 12*:105 (1991).
20. R.D. Schultz, and P.I. Arnold, Heat shock (stress) proteins and autoimmunity in rheumatic diseases, *Semin. Arthritis Rheum. 22*:357 (1993).
21. J.M. Hogervorst, C.J.P. Boog, J.P.A. Wagenaar, M.H.M. Wauben, R. Van der Zee R, and W. van Eden, T cell reactivity to an epitope of the mycobacterial 65-heat shock protein (hsp65) corresponds with arthritis susceptibility in rats and is regulated by hsp65-specific cellular responses, *Eur. J. Immunol. 21*:1289 (1991).
22. D. Elias, D. Markovits, T. Reshef T, R. van der Zee, and I.R. Cohen, Induction and therapy of autoimmune diabetes in non-obese diabetic (NOD/Lt) mouse by a 65-kDa heat shock protein, *Proc. Natl. Acad. Sci. USA 87*:1576 (1990).
23. S.J. Thompson, G.A. Rook, R.J. Brealy, R. Van der Zee, and C.J. Elson, Autoimmune reactions to heat-shock proteins in pristane-induced arthritis, *Eur. J. Immunol. 20*:2479 (1990).
24. M. Salvetti, G. Buttinelli, M. Ristori, M. Carbonari, M. Cherchi, M. Fiorelli, M.G. Grasso, L. Toma, C. Pozili, T-lymphocyte reactivity to the recombinant mycobacterial 65 and 70 kDa heat shock proteins in multiple sclerosis, *J. Autoimmunol. 5*:691 (1992).
25. S.J. Ullrich, E.A. Robison, L.W. Law, M. Willingham, and E. Appella, A mouse tumor-specific transplantation antigen is a heat shock protein, *Proc. Natl. Acad. Sci. USA 83*:3121 (1986).
26. P.K. Srivastava, Peptide-binding heat shock proteins in the endoplasmic reticulum: role in immune response to cancer and in antigen presentation, *Adv. Cancer Res. 62*:153 (1993).
27. Y. Tamura, N. Tsuboi, N. Sato, and K. Kikuchi, 70 kDa heat shock cognate protein is a transformation-associated antigen and a possible target for the host's anti-tumor immunity. *J. Immunol. 151*:5516 (1993).
28. R. Moliterno, M. Woan, C. Bentlejewski, A. Zeevi, S. Pham, B.P. Griffith, and R.J. Duquesnoy, Heat shock protein-induced T lymphocyte propagation for endomyocardial biopsies in heart transplantation, *J. Heart Lung Transplant. 14*:329 (1995).

29. G. Panchamoorthy, J. McLean, R.L. Modlin, C.T. Morita, S. Ishikawa, M.B. Brener MB, and H. Band, A predominance of the T cell receptor Vγ2/Vδ2 subset in human mycobacteria-responsive T cells suggests germiline gene encoded recognition, *J. Immunol. 147*:3360 (1991).
30. A. Haregewoin, G. Soman, R.C. Horn, and R.W. Finberg, Human γδ+ T cells respond to mycobacterial heat-shock protein, *Nature 340*:309 (1989).
31. J. Holoshitz, F. Konig, J.E. Coligan, J. deBruyn, and S. Strober, Isolation of CD4CD8- mycobacteria-reactive T lymphocyte clones from rheumatoid arthritis synovial fluid, *Nature 339*:226 (1989).
32. E. Hermann, A.W. Lohse, M.J. Mayet, R. Van Der Zee, W. Van Eden, P. Probst, T. Poralla T, K.-H. Meyer zum Buschenfelde and B. Fleischer, Stimulation of synovial fluid mononuclear cells with the human 65-kD heat shock protein or with live enterobacteia leads to preferential expansion of TCR-γ/δ+ lymphocytes, *Clin. Exp. Immunol. 89*:427 (1992).
33. Q. Xu, R. Kleindienst, W. Waitz, H. Dietrich, and G. Wick, Increased expression of heat shock protein 65 coincides with a population of infiltrating T lymphocytes in artherosclerotic lesions of rabbits specifically responding to heat shock protein 65, *J. Clin. Invest. 91*:2693 (1993).
34. Q. Xu, H. Dietrich, J.H. Steiner, A.M. Gown, B. Schoel, G. Mikuz, S.H.E. Kaufmann, and G. Wick, Induction of arteriosclerosis in normocholesterolemic rabbits by immunization with heat shock protein 65, *Arteriosclerosis Thrombosis 12*:789 (1992).
35. R. Moliterno, L. Valdivia, F. Pan, and R.J. Duquesnoy, Heat shock protein-reactivity of lymphocytes isolated from heterotopic rat cardiac allografts, *Transplantation 59*:594 (1995).
36. A. Issekutz, Removal of gram negative bacteria from solution by affinity chromarography, *J. Immunol. Methods. 61*:275 (1993).
37. L. Hegeman, L.A.A. Van Rooijen, J. Traber, and B.H. Schmidt, Polymyxin B is a selective and potent antagonist of calmodulin, *Eur. J. Pharmacol. 207*:17 (1991).
38. K. Liberek, D. Skowyra, M. Zylicz, C. Johnson, and C. Georgopoulos, The Escherichia coli DnaK chaperone, the 70 kDa heat shock protein eukaruotic equivalent, changes conformation upon ATP hydrolysis, thus triggering its dissociation from a bound target protein, *J. Biol. Chem. 266*:14491 (1991).
39. D.R. Palleros, K.L. Reid, J.S. McCarty, G.C. Walker, and A.L. Fink, DnaK, hsp73, and their molten globules. Two different ways heat shock proteins respond to heat, *J. Biol. Chem. 267*:5279 (1992).
40. K. Park, G.C. Flynn, J.E. Rothman, and G.D. Fasman, Conformational change of chaperone Hsc70 upon binding to a decepeptide: a circular dichroism study, *Protein Sci. 2*:325 (1993).
41. S. Jindal, A.K. Dudani, B. Singh, C.B. Harley, and R.S. Gupta, Primary structure of human mitochondrial protein homologous to the bacterial and plant chaperonins and to the 65-kilodalton mycobacterial antigen, *Mol. Cell. Biol. 9*:2279 (1989).
42. D.B. Jones, A.F.W. Coulson, and G.W. Duff, Sequence homologies between hsp60 and autoantigens, *Immunol. Today 14*:115 (1993).
43. J. Qian, R. Moliterno, M. Donovan-Peluso, K. Liu, J. Suzow, L. Valdivia, F. Pan,

and R.J. Duquesnoy, Expression of stress proteins and lymphocyte reactivity in heterotopic cardiac allografts undergoing cellular rejection, *Transp. Immunol.* (in press).

44. R.W. Currie, V.K. Sharma, S.M.Stepkowski, and R.F. Payce, Protein synthesis in heterotopically transplanted rat hearts, *Exp. Cell Biol. 55*:46 (1987).
45. J. Hart and V. Kwasiborski, Heat shock protein expression in liver allografts, *Lab. Invest. 70*:132A (abstract 739) (1994).
46. F. Mohtasham and J. Pullman, Hsp70 expression as a marker of tubular injury in renal allograft biopsies, *Lab. Invest. 70*:159A (abstract 932) (1994)
47. J. Clancy, Role of HSP70 in the pathogenesis of acute graft-versus host disease in the adult F1 Hybrid rat, IBC Conference on Heat Shock Proteins, Boston, MA, Sept. 29–30, 1994.
48. N. Latif, C.S. Baker, M.J. Dunn, M.L. Rose, P. Brady, and M.H. Yacoub, Frequency and specificity of antiheart antibodies in patients with dilated cardiomyopathy detected using SDS-PAGE and Western blotting, *J. Am. Coll. Cardiol. 22*:1378 (1993).
49. N. Latif, M.L. Rose, M.H. Yacoub, and M.J. Dunn, Association of pre-transplant antibodies with clinical course following cardiac transplantation, *J. Heart Lung Transplant.* (1994).
50. H. Udono and P.K. Srivastava, Heat shock protein 70-associated peptides elicit specific cancer immunity, *Exp. Med. 178*:1391 (1993).
51. H. Udono and P.K. Srivastava, Comparison of tumor-specific immunogenicities of stress-induced proteins gp96, hsp90 and hsp70, *J. Immunol. 152*:5398 (1994).
52. Y. Tamura, N. Tsuboi, N. Sato, and K. Kikuchi, 70 kDa heat shock cognate protein is a transformation-associated antigen and a possible target for the host's antitumor immunity, *J. Immunol. 151*:5516 (1993).
53. M. Harada, G. Matsuzaki, Y. Yoshikai, N. Kobayashi, S. Kurosawa, H. Takimoto, and K. Nomoto, Autoreactive and hsp60-recognizing CD4+ T cells show anti-tumor activity against syngeneic sarcoma, *Cancer Res. 53*:106 (1993).
54. K.V. Lukacs, D.B. Lowrie, R.W. Stokes, and M.J. Colston, Tumor cells transfected with a bacterial heat-shock protein gene lose tumorigenicity and induce protection against tumors, *J Exp. Med. 178*:343 (1993).
56. G. Birnbaum, L. Kotilinek, and L. Albrecht, Spinal fluid lymphocytes from a subgroup of multiple sclerosis patients respond to mycobacterial antigens, *Ann. Neurol. 34*:294 (1993).
57. M.H. Wauben, C.J. Boog, R. Van der Zee, and W. Van Eden, Towards peptide immunotherapy in rheumatoid arthritis:competitor-modulator concept, *J. Autoimmune 5*:205 (1992).
58. X.D.Yang, J.Gasser, and U. Feige, Prevention of adjuvant arthritis in rats by a nonapeptide from the 65-kD mycobacterial heat shock protein: specificity and mechanism, *J. Clin. Exp. Immunol. 87*:99 (1992).
59. M.E. Munk, B. Schoel, S. Modrow, R.W. Karr, R.A. Young, S.H.E. Kaufmann, T lymphocytes from healthy individuals with specificity to self epitopes shared by the mycobacterial and human 65kDa heat shock protein, *J. Immunol. 143*:2844 (1989).

60. F. Mor and I.R. Cohen, T cells in the lesion of experimental autoimmune encephalomyelitis—enrichment for reactivities to myelin basic protein and to heat shock proteins, *J. Clin. Invest. 90*:2447 (1992).
61. P.S. Russell, C.M. Chase, R.B. Colvin, and J.M.D. Plate, Immune destruction of long-surviving, H-2 incompatible kidney transplants in mice, *J. Exp. Med. 147*:1469 (1978).
62. R.W. Currie, V.K. Sharma, S.M. Stepkowski, and R.F. Payce, Effectts of ischemia and perfusion temperature on the synthesis of stress induced (heat shock) proteins in isolated and perfused hearts, *J. Mol. Cell. Cardiol. 19*:795 (1987).
63. M.S. Marber, D.S. Latchman, J.M. Walker, and D.M. Yellon, Cardiac stress protein elevation 24 hours after brief ischemia or heat stress is associated with resistance to myocardial infarction, *Circulation 88*:1264 (1993).
64. R.W. Currie, R.M. Tanguay, and J.G. Kingma, Heat-shock response and limitation of tissue necrosis during occlusion/reperfusion in rabbit hearts, *Circulation 87*:963 (1993).
65. M.M. Hutter, R.E. Sievers, V. Barbosa, and C.L. Wolfe, Heat shock protein induction in rat hearts. A direct correlation between the amount of heat-shock protein induced and the degree of myocardial protection, *Circulation 89*:355 (1994).
66. R.S. Williams, J.A. Thomas, M. Fina, Z. German, and I. Benjamin, Human heat shock protein 70 (hsp70) protects murine cells from injury during metabolic stress, *J. Clin. Invest. 92*:503 (1993).
67. J.A. Brewer, J.A. Hank, T. Wendel, and P.M. Sondel, Heated lymphocytes express HLA-DR antigens despite their inability to stimulate MLC, *Tissue Antigens 22*:246 (1983).
68. C.M. Boyer, D.D. Kostyu, C.S. Brissette, and D.B. Amos, Functional defect of heat-inactiated human lymphocytes in mixed lymphocyte culture, *Cell Immunol. 101*:440 (1986).
69. R. Loertscher, M. Abbud-Filo, and A.B. Leichtman, Differential effct of gamma-irradiated and heat-treated lymphocytes on T cell activation and interleukin-2 and interleukin-3 release in the mixed lymphocyte reaction, *Transplantation 44*:673 (1987).
70. N.E. Goeken, Human suppressor cell induction in vitro: preferential activation by class I MHC antigen, *Immunology 132*:2291 (1984).
71. C.S. Brissette-Storkus, D.D. Kostyu, C. Boyer, and J.R. Dawson, Induction of HLA-specific CTL to non-immunogenic, heat-inactivatedlymphocytes by interleukin-2. *Transplantation 50*:862 (1990).
72. A.J. Freeman, and Baird, Immunosuppression mediated by heat-treated cells, *Transplantation 55*:1439 (1993).
73. G.P. Martinelli, C. Horowitz, K. Chiang, D. Racelis, and H. Schanzer, Pretransplant conditioning with donor-speific transfusions using heated blood and cyclosporine, *Transplantation 43*:140 (1987).
74. C.A. Sargent, I. Dunham, J. Trowsdale, and D. Campbell, Human major histocompatibility complex contains genes for the major heat shock protein hsp70, *Proc. Natl. Acad. Sci. USA 86*:1968 (1989).
75. W. Wurst, C. Benesch, B. Drabant, E. Rothermel, B. Beneck, and E. Gunther,

Localization of heat shock protein 70 genes inside the major histocompatibility complex close to class III genes, *Immunogenet. 30*:46 (1989).
76. A. Van Buskirk, D.C. DeNagel, L.E. Guagliardi, F.M. Brodsky, and S.K. Pierce, Cellular and subcellular distribution of PBP72/74, a peptide-binding protein that plays a role in antigen processing, *J. Immunol. 146*:500 (1991).
77. D.C. DeNagel, and S.K. Pierce, Heat shock proteins in immune responses, *Crit. Rev. Immunol. 13*:71 (1993).
78. G.C. Manara, P. Sansoni, L. Badiali-De Giorgi, L. Gallinella, C. Ferrari, V. Branti, F.F. Fagnoni, C.L. Ruegg, G. De Panfilis, and G. Pasquinelli, New insights suggesting a possible role of a heat shock protein 70-kD family-related protein in antigen processing/presentation phenomenon in humans, *Blood 82*:2865 (1993).
79. I.G. Haas and M.Wabl, Immunoglobulin heavy chain binding protein, *Nature 306*: 387 (1983).
80. J.R. Gething and J.F. Sambrook, Protein folding in the cell, *Nature 355*:33 (1992).
81. I. Kaur, S.D. Voss, R.S. Gupta, K. Schell, P. Fisch, and P.M. Sondel, Human peripheral gd T cells recognize hsp60 molecules on Daudi Burkitt's lymphoma cells, *J. Immunol. 150*:2046 (1993).
82. M.T. Ferm, K. Soderstrom, S. Jindal, A. Gronberg, I. Ivanyi, R. Young, and R. Kiessling, Induction of human hsp60 expression in monocytic cell lines, *Int. Immunol. 4*:305 (1992).
83. B.J. Soltys and R.S. Gupta, Immunoelectron microscopy suppotrs hsp60 presence at nonmitochondrial sites, Cold Spring Harbor Meeting on Hsp, Cold Spring Harbor, NY, Abstract 285 (1994).
84. B.Y. Rubin, S.L. Anderson, R.M. Lunn, N.K. Richardson, G.R. Hellerman, L.J. Smith, and L.J. Old, Tumor necrosis factor and IFN induce a common set of proteins, *J. Immunol. 141*:1180 (1988).
85. T. Koga, A. Wand-Wurttenberger, J. deBruyn, M.E. Munk, B. Schoel, and S.H.E. Kaufman, T cells against a bacterial heat shock protein recognize stressed macrophages, *Science 245*:1112 (1989).
86. U. Steinhoff, B. Schoell, and S.H.E. Kaufmann, Lysis of interferon-g activated Schwann cells by crossreactive CD8 a/b T cells with specificity to the mycobacterial 65 kDa heat shock protein, *Int. Immunol. 2*:279 (1990).
87. A. Sztankay, K. Trieb, P. Lucciarini, E. Steiner, and B. Grubeck-Loebenstein, Interferon gamma and iodide increase the inducibility of the 72kd heat shock protein in cultured human thyroid epithelial cells, *J. Autoimmune 7*:-219 (1994).
88. M.J. Welch, J.P. Suhan, Cellular and biochemical events in mammaian cells during and after recovery from physiological stress, *J. Cell Biol. 103*:2035 (1986).
89. G.C. Li, and W.J. Werb, Correlation between synthesis of heat shock proteins and development of thermotolerance in Chinese hamster fibroblasts, *Proc. Natl. Acad. Sci. USA 79*:3218 (1982).
90. G.C. Li, L. Li, Y.-K. Liu, J.Y. Mak, L. Chen, and W.M.F. Lee, Thermal response of rat fibroblasts stably transfected with the human 70-kDa heat shock protein-encoding gene, *Proc. Natl. Acad. Sci. USA 88*:1681 (1991).
91. C.E. Angelidis, L. Lazarides, and G.N. Pagoulatos, Constitutive expression of heat shock protein 70 in mammalian cells confers thermoresistance, *Eur. J. Biochem. 199*:35 (1991).

92. R. Mestril. S.-H. Chi, R. Sayen, K. O'Reilly, and W.H. Dillman, Expression of inducible stress protein 70 in rat heart myogenic cells confers protection against simulated ischemia-induced injury. *J. Clin. Invest. 93*:759 (1994).
93. R.L. Simmons, R. Weil, M. Tallent, G.M. Kjellstrand, and J.S. Najarian, Do mild infections trigger the rejection of renal allografts?, *Transpl. Proc. 2*:419 (1970).
94. R. Manez, L.T. White, S. Kusne, M. Martin, A.J. Demetris, T.E. Starz, and R.J. Duquesnoy, Association between donor-recipient HLA-DR compatibility and cytomegalovirus hepatitis and chronic rejection in lver transplantation, *Transpl. Proc. 25*:908 (1993).
95. P.T. Donaldson, J. O'Grady, and R.D. Portman, Evidence for an immune response to HLA class I antigens in the vanishing bile syndrome after liver transplantation, *Lancet 1*:945 (1987).
96. M.T. Grattan, G.E. Noren-Cabal, V.A. Starnes, P.E. Oyer, E.B. Stinson, and N.E. Shumway, Cytomegalovirus infection is associated with cardic allograft rejection and atherosclerosis, *J.A.M.A. 261*:3561 (1988).
97. A. Zeevi, M.E. Uknis, K.J. Spichty, M. Tector, R.J. Keenan, C. Rinaldo, S. Yousem, S. Duncan, I. Paradis, J.H. Dauber, B.P. Griffith, and R.J. Duquesnoy, Proliferation of cytomegalovirus primed lymphocytes in bronchoalveolar lavages from lung transplant patients, *Transplantation 54*:635 (1992).

21

Lupus and Heat Shock Proteins

Breda N. Twomey, Veena B. Dhillon, David S. Latchman, and David Isenberg
University College London, London, England

I. INTRODUCTION

Systemic lupus erythematosus (SLE), or simply lupus, is an autoimmune rheumatic disease affecting women of child-bearing years. The clinical manifestations of SLE are diverse and involve many different organ systems. The main features of lupus include joint pains, skin rashes, fever, chest pains, headaches, and kidney inflammation. The clinical diversity is matched by an apparent diversity of autoantibodies detectable in the serum of patients and a broad range of T-cell abnormalities.

Heat shock proteins (hsps) were originally identified on the basis of their increased synthesis following exposure to elevated temperatures. They have since been shown to be induced following exposure to a number of stressful stimuli, including certain disease states. Hsps are also expressed in almost every cell type constitutively and perform vital cellular housekeeping functions. They function as molecular chaperones, mediating the correct folding and association of multimeric complexes of newly synthesized proteins, the translocation of proteins within the cell, and the degradation of denatured proteins. They provide a pivotal role in the regulation of cell growth, proliferation, and differentiation, and thus it is not surprising that they have been structurally conserved throughout evolution.

Links between hsps and autoimmune disease have been explored during this decade, and there is much supporting evidence. Both the hsps and their response to stressful stimuli are remarkably similar from bacteria to humans. Hsps have been shown to be immunodominant antigens during bacterial and mycobacterial infections. It is interesting that peripheral T cells carrying the γ/δ T-cell receptor (TCR) can recognize the mycobacterial 65-kDa hsp (1,2). The reactivity of these somewhat primitive γ/δ T cells to mycobacteria was major histocompatibility complex (MHC)–unrestricted and displayed limited V-region usage, reminiscent of bacterial "superantigen" recognition by αβ cells. Given that mycobacterial and self-hsps are both immunodominant and structurally similar, that circulating T- and B- cell clones can recognize these antigens, and that infection may lead to increased expression of both mycobacterial and self-hsps, the concept that hsps may be involved in the etiopathogenesis of autoimmunity has developed. In particular, "molecular mimicry" may be involved in mediating autoimmune disease (3) and more specifically SLE.

However, the presence of autoantibodies and/or T cells directed against human hsps alone is insufficient to generate autoimmune disease, since such antibodies are also found in some normal individuals (4,5) and in individuals with tuberculosis (6). Therefore, abnormalities in the expression or localization of the endogenous human hsps may be required to trigger autoimmune reactivity from a response initially mounted against bacterial hsps (7). Against this background, a number of genetic susceptibility factors are also implicated in determining whether autoimmunity will develop or if the invading pathogen will be eliminated and the immune "balance" restored.

Hsps have been implicated as having a role in antigen processing and presentation (8). In agreement with this role, the hsp 70 gene has been mapped within the MHC region in a number of species, including humans (9).

This chapter outlines the differential expression patterns of hsps and of autoantibodies to the hsps that have been observed in patients with SLE and in the lupus-prone MRL/lpr mouse strain and discusses the clinical relevance of these findings.

II. EXPRESSION OF HSPS IN PERIPHERAL MONONUCLEAR CELLS IN SLE

The overexpression of both hsp 70 and hsp 90 in peripheral blood mononuclear cells (PBMCs) pooled from a number of patients with SLE was first detected by Deguchi et al. (10). Subsequently, in our laboratories, it was shown that hsp 90 protein levels in PBMCs were elevated in 15% of patients with SLE ($n = 57$) compared with normal controls and compared with patients with other autoimmune diseases such as rheumatoid arthritis and osteoarthritis (11). These patients had high hsp 90 levels had active disease. In a second more comprehensive study involving 102 patients with SLE and 59 normal control subjects, both hsp 90 and hsp 72 (the stress-inducible form of hsp 70) were significantly elevated in the PBMCs of patients with SLE whereas hsp 60 and hsp 73 (the constitutive isoform of hsp 70) remained unchanged (12). Hsp 90 was elevated in 32% of the patients with SLE, but there was no correlation between elevated hsp 90 and hsp 72 in these patients. Thus, overexpression of one hsp does not necessarily imply overexpression of another, which suggests that some quite specific signals in these patients are responsible for the expression patterns and they are not merely determined by the ongoing stress of the disease.

The expression of hsp 90 on the surface of both lymphocytes and monocytes from a group of 62 patients with SLE was measured by flow cytometry and showed some interesting results. Twenty percent of the patients with SLE displayed a significant elevation of surface expression of hsp 90 on lymphocytes compared with normal controls; the monocytes in these patients also showed higher levels of surface expression compared with the levels in normal controls, although it did not reach statistical significance (13). The lymphocyte populations that displayed elevated surface expression of hsp 90 were identified as B cells and $CD3^+$ and $CD4^+$ T cells. There appeared to be two groups of patients with SLE; one group showed surface expression of hsp 90 on lymphocytes, whereas the other group did not. In the former group, there was a correlation with total hsp 90 levels. No surface expression of hsp 60 was detected in these patients with SLE using the procedures employed in the hsp 90 studies. These are significant findings in view of the fact that hsp 90 is a cytoplasmic protein in normal cells and hence unavailable to immune detection. Owing to either increased intracellular expression or to an aberrant transport mechanism in SLE lymphocytes, hsp 90 is detectable on the cell surface where it could cause the activation of B or T cells and therefore play an instrumental role in the breakdown of immune tolerance. There is already some evidence for the presence of autoantibodies to hsp 90 in SLE which will be discussed later.

Alternatively, hsp 90 may play a more indirect role in antigen presentation, and thus explain the cell surface localization of hsp 90. It has been shown that peptide products of hsp 90 are among several endogenous proteins whose peptides are bound to HLA-B27 in a lymphoblastoid cell line (14). In another study, several tumor cell lines were found to express 2–10 times higher levels of hsp 90 when compared with normal cells, and the hsp 90 was localized to the cell surface (15). Interestingly, in immunoprecipitation experiments, this hsp 90 coprecipitated with MHC 1 molecules, again forging a link with antigen presentation. Fibroblasts transformed with herpes simplex virus accumulate high levels of hsp 90 during lytic infection, some of which becomes located on the cell surface (16). Thus, it appears that in a number of different situations of hsp 90 overexpression, the excess protein becomes localized to the cell surface either by an active mechanism or simply reflecting increased whole cell levels, where it may inadvertently become involved in an autoimmune process.

III. CLINICAL RELEVANCE OF HSP OVEREXPRESSION

As discussed above, hsp 90 was significantly elevated in those patients with SLE who had severely active disease compared with those patients who had inactive or mildly active disease using the UCH/Middlesex scoring system (12). Using a more sophisticated system, the British Isles Lupus Assessment Group's (BILAG) system for analyzing disease activity, disease activity was assessed in eight different disease categories or organ systems. Hsp levels were correlated with active or inactive disease in these eight individual disease categories in each patient at the time of sampling. Patients with SLE with active neuropsychiatric (NP) and cardiorespiratory (CR) lupus had significantly elevated hsp 90 compared with those who were inactive for disease in these categories (Fig. 1) (12). There were no differences in hsp 72 or hsp 60 levels in active compared with inactive BILAG disease categories, but hsp 73 was higher in inactive compared with active categories, and significantly so in NP and renal BILAG categories. There was no relationship between hsp 72 or hsp 60 levels and activity grading of the disease, but mean hsp 73 levels were higher in inactive and mild lupus compared with severe disease.

Elevated surface expression of hsp 90 was also higher in those patients with a high global BILAG score, indicating a correlation with more severe disease (13). Surface expression of hsp 90 correlated significantly with active disease in the musculoskeletal BILAG system.

In view of the well-documented association of hsp 90 with steroid hormone receptors (17,18), thus maintaining their inactive non-DNA–binding conformation, the possibility that the elevation of hsp 90 might be a consequence of the steroid treatment received by many of the patients with SLE was investigated. However, there was no differences in the levels of hsp 90 or any of the hsps in patients on no, low, or high doses of oral prednisolone (11). There was no significant correlation between hsp levels and anti–double-stranded DNA (dsDNA) antibody titer or increased erythrocyte sedimentation rate (ESR). There were no significant differences in hsp levels in patients who possessed some of the other classic hallmarks of active SLE; namely, serum levels of antinuclear antibodies (ANA), anti-Sm, anti-Ro, anti-La, or anti-RNP antibodies. The lack of correlation between hsp levels and markers of general disease activity is paralleled by the findings from the FACS studies, where surface hsp 90 expression did not show any correlation with ESR, anti-ds DNA titers, C3 levels, or with markers of T-cell activation such as HLA-DR and interleukin (IL-2) (13). This lack of correlation suggests that other factors are involved in determining hsp overexpression, which might include underlying genetic influences (discussed later). Alternatively, ESR and anti-dsDNA titers may be rather poor indicators of global disease activity; and the correlation between ESR and anti-dsDNA titers was only modest ($r = 0.25$).

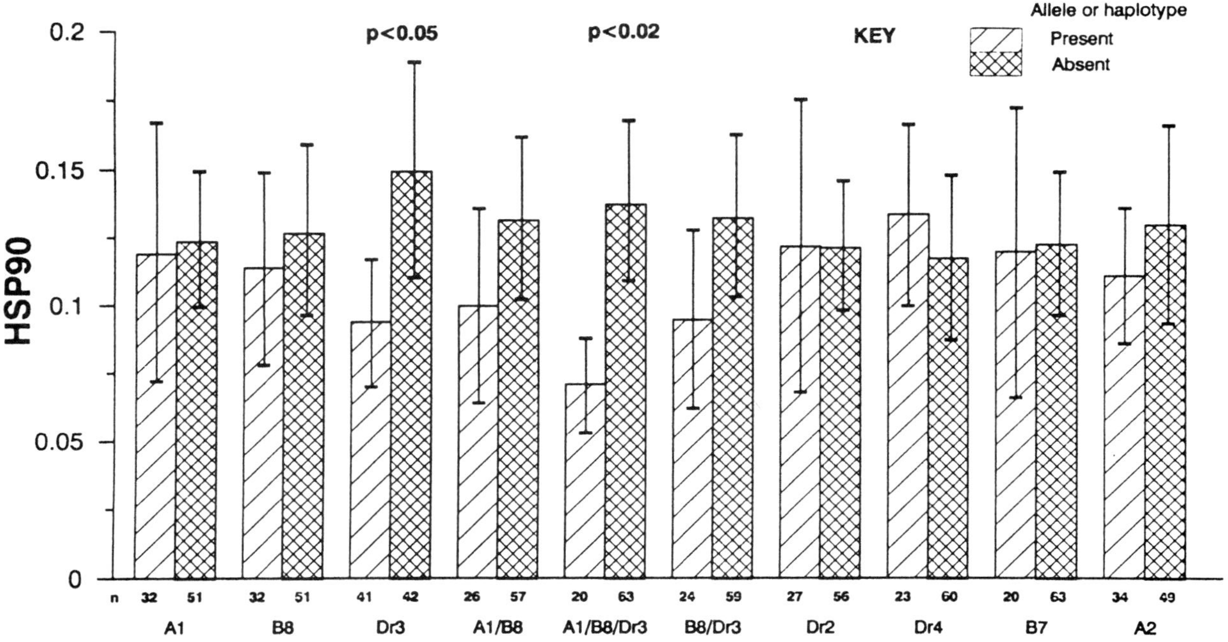

Figure 1 Relationship between HLA haplotype and hsp 90 expression in peripheral blood mononuclear cells in SLE.

There was significant elevation of hsp 90 in patients with antiphospholipid antibodies (APAs), determined by the presence of lupus anticoagulant (LAC) measured from coagulation times, and/or anticardiolipin antibody (ACL) assayed by ELISA (19,20). There was also significant elevation of hsp 90 in patients with thrombocytopaenia. There was a high frequency of APAs in patients with active NP SLE compared with the frequency of thrombocytopaenia in these patients, and compared with the frequency of APA or thrombocytopaenia in patients with active CR SLE. Conversely, no such correlations were observed in patients with elevated hsp 72 levels. These findings suggest that hsp 90 is particularly elevated in patients with the antiphospholipid syndrome (APS). A role for some APAs, via their reactivity with the cofactor β_2 glycoprotein I (β_2GPI), in generating disease by promoting thrombosis is broadly accepted (21–23). However, thrombosis in association with ACL in this SLE population was rare (7%) (23). Furthermore, there is a wide range of reactivity of APAs in SLE, and binding the β_2GPI is not an absolute requirement (25–27). Thus, if hsp 90 is related to the pathogenesis of APS in SLE, it is likely that its role involves factors other than, or additional to, thrombosis.

One mechanism for disease in APS may occur by virtue of a shared antigenicity of the phoshodiester groups which form a component part of both DNA and phospholipids and with which anti-dsDNA antibodies may react (28). In addition, there is considerable evidence for commonality of antiendothelial cell antibodies (AECAs) and ACL, both in the genetic background driving their production, and in their reactivities (29–31). Thus, there is increasing evidence to suggest that APAs may also contribute to disease by a mechanism other than thrombosis. If there is overexpression of hsp 90, then it may be involved in this mechanism by perhaps sharing an epitope with antigens with which ACL and/or AECAs react.

The relationship between HLA status and hsp levels was also examined. Hsp 90 levels were significantly higher in patients who were DR3 negative and A1/B8/DR3–negative compared with those who were positive for these allo/haplotypes (19). Hsp 72 levels did not differ significantly between patients who were positive or negative for any of the haplotypes tested. Although negativity for HLA A1, B8, or DR3 alone or in combination was significantly associated with nonwhite ethnic origin, there was no significant differences in hsp 90 or hsp 72 levels between white and nonwhite patients. It is interesting to note that Bachelor et al. (32) have shown that the association between SLE and DR3 is secondary to a primary association with C4 null alleles, so hsp 90 may be overexpressed in SLE patients who suffer from complement deficiency. Alternatively, overexpression of the hsp 90 gene may be linked to the positive or negative presence of other, as yet undefined, alleles leaving the association of hsp 90 with genetic factors still an open question.

IV. MECHANISMS OF HSP OVEREXPRESSION IN SLE

It is important to determine whether the altered hsp expression that has been observed in SLE is paralleled by changes at the level of gene transcription as a first step in understanding their role in mediating the disease process. The rate of transcription of hsp 90α and hsp 90β genes, the inducible hsp 70 gene, and the three genes encoding the small hsp ubiquitin were determined by nuclear run-on assays. The mean level of transcription of both hsp 90α and hsp 90β genes was higher in the SLE group compared with normal controls, but only the elevation in transcription of the hsp 90β gene reached statistical significance (33). There was no correlation between transcription of the hsp 90α and hsp 9)β genes in individual patients with SLE. A clear correlation was observed between the level of hsp 90 protein and the level of hsp 90β transcription; in contrast, the level of hsp 90α transcription showed no correlation with the level of hsp 90 protein. Thus, the enhanced levels of hsp 90 protein are dependent on the enhanced transcription of the β rather than the α gene (Fig. 2).

The mean level of transcription of the inducible hsp 70 gene was marginally higher in the SLE group than in the control group, but this difference was not statistically significant (34). However, a small proportion of patients with SLE did show greatly elevated hsp 70 transcription. Deguchi and Kishimoto (35) also detected elevated transcription of this gene in a small group of patients with SLE. When the levels of the inducible hsp 72 protein were compared with the levels of hsp 70 gene transcription no relationship was found. Similarly, there was no

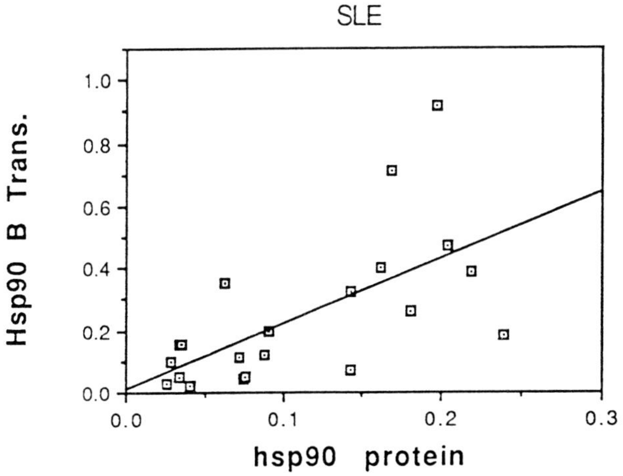

Figure 2 Relationship between Hsp90β gene transcription and hsp 90 protein levels ($r = 0.636$, $p = 0.002$). (From Ref. 33.)

correlation between hsp 72 protein levels and hsp 70 mRNA levels in individual patients with SLE (34). These findings suggest that posttranscriptional processes such as increased translatability of hsp 70 mRNA or increased stability of the hsp 70 protein must be responsible for determining the levels of hsp 70 protein in patients with SLE. Such posttranscriptional processes have been shown to be of particular importance in the control of hsp expression (36).

In common with the situation at the protein level, there also appears to be differential overexpression of hsp genes at the level of transcription. Elevated hsp 90β gene transcription did not occur in those patients that showed elevation of hsp 70 transcription; hence, there was no correlation between transcription of the two genes. Transcription of the three genes encoding the small hsp ubiquitin were found to be unchanged in patients with SLE compared with normal controls (37). Thus, the elevated rates of transcription observed for hsp 70 and hsp 90 do not reflect a generalized induction of the heat shock genes owing to the stress of the disease or the presence of fever. The mechanisms responsible for the enhanced rates of transcription are unlikely to be mediated by the heat shock consensus element (HSE), which is located in the upstream region of all heat shock genes and mediates the induction of these genes in response to heat shock and other stresses (38,39). Activation mediated by the HSE would be expected to produce increased transcription of all these genes rather than the specific elevation of one or the other in different patients with no change in ubiquitin gene transcription. However, it is becoming clear that the transcription regulatory regions of heat shock genes are far from simple and as has been described for the human hsp 70 gene include, along with multiple copies of HSE, the *cis*-acting promoter elements CCAAT, Sp1, ATF, and TFIID (40), whose regulation has been shown to be complex (41).

In conclusion, elevation of hsp 90 levels of PBMCs from patients with SLE is dependent on overexpression of the corresponding hsp 90β gene, whereas elevation of hsp 72 levels relies more on posttranscriptional processes. This suggests that distinct gene regulatory mechanisms acting at different levels control the observed elevation of the individual hsps in different groups of patients with SLE.

V. HSP EXPRESSION PATTERN IN THE LUPUS-PRONE MRL/lpr MOUSE

The MRL/MP-lpr/lpr (MRL/lpr) mouse is homozygous deficient for the gene encoding the Apo-1/Fas antigen and develops a lupus-like disease that is characterized by lymphoproliferation, the production of a high titer of autoantibodies to nuclear antigens, and the appearance of abnormal lymphocyte subsets (28,42). Expression of hsps in these mice showed that hsp 90 was elevated in the spleen and not in any other tissue compared with MRL/Mp-+/+ (MRL/+) mice, which do not show signs of disease, and BALB/c control mice (43). Temporal studies of this expression detected highest levels of hsp 90 protein in 2-month-old mice which precede the onset of disease as measured by the levels of antibodies to single- and double-stranded DNA; antibody was not detected in 3-month-old mice but increased thereafter. Elevation of hsp 70 expression, shown by RNA and transcription analysis, increased progressively with age in the kidneys of MRL/lpr mice (44). Levels of the inducible hsp 72 protein were also observed to display a progressive age-dependent elevation in both spleen and liver compared with BALB/c controls (43). In parallel experiments, no increases were observed in the levels of hsp 65 or of the constitutive hsp 73 when a variety of MRL/lpr tissues were compared with their BALB/c equivalents.

In summary, hsp 90 is specifically elevated in the splenic tissue of MRL/1pr mice and precedes disease onset, whereas hsp 72 is elevated in several different tissues and is age dependent, increasing as the disease worsens. The elevation of hsp 72 may represent a stress response due to the ongoing disease process which normally results in the death of the mice at 5–7 months of age. On the otherhand, overexpression of hsp 90 may have a crucial role in mediating the onset of disease. As mentioned above, MRL/lpr mice have a primary genetic defect in the Apo-1/Fas gene, which regulates programmed cell death or apoptosis in lymphoid cells leading to the observed lymphoproliferation in these mice. It has recently been shown that heat stress which results in hsp overexpression can protect cells from apoptosis (45). Thus, the inactivation of the Apo-1 antigen may partially mediate its inhibitory effect on apoptosis via increased hsp 90 levels which subsequently play a role in the pathogenesis of the disease by enhancing the survival of lymphoid cells.

Whatever the precise nature of the roles of elevated hsp 90 and hsp 72 in MRL/lpr mice, it is interesting to note the similarities with expression of hsps in the PBMCs of patients with SLE with regard to specificity of particular hsps in different subsets of patients and correlations with active disease.

VI. CIRCULATING AUTOANTIBODIES IN SLE

As indicated earlier in this chapter, a major feature of lupus is the presence of a variety of antinuclear antibodies. In addition, given the surface expression of hsp 90 that we have demonatrated (13), it is important to identify antibodies to this and other heat shock proteins in patients with lupus.

The initial descriptions of antibodies to hsp 90 by Winfield and colleagues produced conflicting data. In their first report (46), they found that 6 of 15 patients with lupus had raised levels of anti–hsp 90 antibodies. Subsequently, they reported that less than 10% of patients with lupus had these antibodies and could not be distinguished from a variety of other rheumatic diseases (47).

We have recently developed an ELISA for detecting anti–hsp 90 antibodies and in addition our system has incorporated the native protein rather than the reduced form of hsp 90 used by Winfield and colleagues. We have found that approximately 35% of patients with lupus have raised immunoglobulin M (IgM) anti–hsp 90 antibodies and 26% have IgG antibodies (48). As with expression of the protein, the antibody response appears to be restricted in several ways. Notably, those individuals with raised hsp 90 antibodies were statistically significant more likely to be HLA A1 B8 DR3 negative, to be correlated with a low serum C3 level, and to have active renal disease.

In contrast to these observations, we have not found antibodies to another heat shock protein, hsp 70, to be elevated. However, it may be that antibodies to other heat shock proteins are linked to different diseases. For example, a recent report has documented a link between antibodies to the constitutive bovine and human 73-kDa (but not the inducible 72-kDa isoform) in patients with the so-called mixed connective tissue disease (49). Another report suggests that there is an association of IgG antibodies to the human 70-kDa hsp with rheumatoid arthritis (50), although this was not confirmed in another study, albeit using a different methodology (47).

VII. CONCLUSIONS

Taken together, the observations described in this chapter suggest that an important subset of patients with lupus (notably those who are A1 B8 DR3 negative) show a striking predisposition to overexpress a particular heat shock protein, hsp 90. Hsp 90 is thus overexpressed in the peripheral blood mononuclear cells (especially the B cells and $CD4^+$ T cells), particularly in patients with active central nervous system and cardiorespiratory diseases. Approximately 20% of patients with lupus demonstrate surface expression of this heat shock protein and approximately one-third of patients with lupus (again those more likely to be A1 B8 $DR3^-$) have raised levels of antibodies to hsp 90.

Overexpression of the hsp 90 protein is linked to increased transcription of the hsp 90β gene and interesting parallels of protein expression have been demonstrated in the MRL/lpr mouse, a lupus model. In contrast, similar studies with other heat shock proteins, notably hsp 70, have failed to demonstrate these features. These observations lead us to suggest that there is indeed something particular about the expression of hsp 90 and the formation of antibodies to it, and this dysregulation may have a role in the aetiopathogenesis of disease in a subset of patients with lupus.

REFERENCES

1. J. Holoshitz, F. Koning, J.E. Colligan, J. De Bruyn, and S. Strober, Isolation of CD4⁻ CD8⁻ mycobacteria-reactive T lymphocyte clones from rheumatoid arthritis synovial fluid, *Nature 339*:226 (1989).
2. R.L. O'Brien, M.P. Happ, A. Dallas, E. Palmer, R. Kubo, and W.K. Born, Stimulation of a major subset of lymphocytes expressing T cell receptor $\gamma\delta$ by an antigen derived from Mycobacterium tuberculosis, *Cell 57*:667 (1989).
3. B.A. Oldstone, Molecular mimicry and autoimmune disease, *Cell 50*:819 (1987).
4. J. Kindas-Mügge, G. Steiner, and J.S. Smolen, Similar frequency of autoantibodies against 70-kD class heat shock proteins in healthy subjects and systemic lupus erythematosus patients, *Clin. Exp. Immunol. 92*:46 (1993).
5. J. Field and W. Jarjour, Do stress proteins play a role in arthritis and autoimmunity?, *Immunol. Rev. 121*:193 (1991).
6. A. Rees, A. Scoging, A. Mehlert, D.B. Young, and J. Ivanyi, Specificity of the proliferative response of human CD8 clones to mycobacterial antigens, *Eur. J. Immunol. 18*:1881 (1988).
7. W. Van Eden, Heat shock proteins as immunogenic bacterial antigens with the potential to induce and regulate auto-immune disease, *Immunol. Rev. 121*:1 (1991).
8. S.K. Pierce, D. C. De Nagel, and A.M. Van Buskirk, A role for heat shock proteins in antigen processing and presentation, *Curr. Top. Microbiol. Immunol. 167*:83 (1991).
9. E. Günther, Heat shock protein genes and the major histocompatibility complex, *Curr. Top. Microbiol. Immunol. 167*:57 (1991).
10. Y. Deguchi, S. Negoro, and S. Kishimoto, Heat shock protein synthesis by human peripheral mononuclear cells from SLE patients, *Biochem. Biophys. Res. Commun. 148*:1063 (1987).
11. P.M. Norton, D.A. Isenberg, and D.S. Latchman, Elevated levels of the 90 kd heat shock protein in a proportion of SLE patients with active disease, *J. Autoimmun. 2*:187 (1989).
12. V. Dhillon, S. McCallum, P. Norton, B. Twomey, F. Erkellor-Yüksel, P. Lydyard, D. Isenberg, and D. Latchman, Differential heat shock protein overexpression and its clinical relevance in systemic lupus erythematosus, *Ann. Rheum. Dis. 52*:436 (1993).
13. F. Erkeller-Yüksel, D.A. Isenberg, V.B. Dhillon, D.S. Latchman, and P.M. Lydyard, Surface expression of heat shock protein 90 by blood mononuclear cells from patients with systemic lupus erythematosus, *J. Autoimmun. 5*:803 (1992).
14. T.S. Jardetzky, W.S. Lane, R.A. Robinson, D.R. Madden, and D.C. Wiley, Identification of self peptides bound to purified HLA-B27, *Nature 353*:326 (1991).
15. M. Ferrarini, S. Heltai, M.F. Zocchi, and C. Rugarli, Unusual expression and localization of heat shock proteins in human tumour cells, *Int. J. Cancer 51*:613 (1992).
16. N. B. La Thangue and D. S. Latchman, A cellular protein related to heat shock protein 90 accumulates during herpes simplex virus infection and is overexpressed in transformed cells, *Exp. Cell Res. 178*:169 (1988).

17. S.L. Kost, D.F. Smith, W.P. Sullivan, W.J. Welch, and D.O. Toft, Binding of heat shock proteins to the avian progesterone receptor, *Mol. Cell. Biol. 9*:3829 (1989).
18. K.A. Hutchison, M.J. Czar, L.C. Scherrer, and W.P. Pratt, Monovalent cation selectivity for ATP-dependent association of the glucocorticoid receptor with hsp70 and hsp90, *J. Biol. Chem. 267*:14047 (1992).
19. V.B. Dhillon, S. McCallum, D.S. Latchman, and D.A. Isenberg, Elevation of the 90kD heat shock protein in specific subsets of systemic lupus erythematosus patients, *Q. J. Med.* 37:215 (1994).
20. S.A. Krilis and J. Hunt, Antiphospholipid syndrome: new aspects and mechanisms, *Lupus 1 (Suppl.1)*:12 (1992).
21. I. Schousboe, Effect of beta-2 glycoprotein I on the activity of adenylate cyclase in platelet membranes, *Thromb. Res. 32*:291 (1983).
22. J. Nimbf, E.M. Bevers, and P.H.H. Bomans, Prothrombinase activity of human platelets is inhibited by beta-2 glycoprotein I, *Biochem. Biophys. Acta 884*:142 (1986).
23. J.G. Worrall, M.L. Snaith, J.R. Bachelor, and D.A. Isenberg, SLE: a rheumatological analysis of clinical features, serology and immunogenetics of 100 SLE patients during longterm followup, *J. Med. 74*:319 (1990).
24. M.V. Cuesta, J.J. Vazquez, A. Monereo, et al., Endothelial cell participation in thrombosis of patients with lupus anticoagulant, *Lupus 1(Suppl. 1)*:46 (1992).
25. S. Loizou, C.G. Mackworth-Young, C. Cofiner, and M.J. Walport, Heterogeneity of binding reactivity to different phospholipids of antibodies from patients with systemic lupus erythematosus (SLE) and syphilis, *Clin. Exp. Immunol. 80*:171 (1990).
26. S.S. Pierangli, S.A. Davis, and E.N. Harris, Beta2-glycoprotein enhances binding of antiphospholipid antibodies to cardiolipin but is not the target antigen, *Lupus 1 (Suppl.1)*:44 (1992)
27. I. Domenech, M.A. Khamashta, P. Harrison, and G.R.V. Hughes, Antiphospholipid antibodies differ in cofactor requirement for binding to different phospholipids, *Lupus 1(Suppl. 1)*:47 (1992).
28. W.J.W. Morrow and D. Isenberg, Systemic Lupus Erythematosus, *Autoimmune Rheumatic Diseases* (W.J.W. Morrow and D.A. Isenberg, eds.), Blackwell, Oxford, UK, 1987, p. 48.
29. M. Galezzi, G.D. Sebastiani, G. Passiu, et al, Anticardiolipin antibodies and anti-endothelial cell antibodies may share a common genetic background in SLE, *Lupus 1 (Suppl. 1)*:33 (1992).
30. C.G. Mackworth-Young, F. Andreottti, and I. Harmer, Endothelium-derived haemostatic factors and the antiphospholipid syndrome, *Lupus 1 (Suppl. 1)*:108 (1992).
31. N.J. Lindsey, R.A. Dawson, F.I. Henderson. M. Greaves, and P. Hughes, Stimulation of von Willebrand Factorantigen release by immunoglobulin from thrombosis prone patients with systemic lupus erythematosus and the anti-phospholipid syndrom, *Br. J. Rheumatol. 32*:123 (1993).
32. J.R. Bachelor, A.H.L. Fielder, and M.J. Walport, Family study of the major histocompatibility complex in HLA DR3 negative patients with systemic lupus erythematosus, *Clin. Exp. Immunol. 70*:364 (1987).

33. B.M. Twomey, V.B. Dhillon, S. McCallum, D.A. Isenberg, and D.S. Latchman, Elevated levels of the 90kD heat shock protein in patients with systemic lupus erythematosus are dependent upon enhanced transcription of the hsp90β gene, *J. Autoimmun. 6*:495 (1993).
34. B.M. Twomey, V. Amin, D.A. Isenberg, and D.S. Latchman, Elevated levels of the 70kD heat shock protein in patients with systemic lupus erythematosus are not dependent on enhanced transcription of the hsp70 gene, *Lupus 2*:297 (1993).
35. Y. Deguchi and S. Kishimoto, Enhanced expression of the heat shock protein gene in peripheral blood mononuclear cells of patients with active systemic lupus erythematosus, *Ann. Rheum. Dis. 49*:893 (1990).
36. H.J. Yost, R.B. Petersen, and S. Lindquist, RNA metabolism: strategies for regulation in the heat shock response, *TIG 6*:223 (1990).
37. B. Twomey, S. McCallum, V. Dhillon, D. Isenberg, and D. Latchman, Transcription of the genes encoding the small heat shock protein ubiquitin is unchanged in patients with systemic lupus erythematosus, *Autoimmunity 13*:197 (1992).
38. P.K. Sorger, Heat shock factor and the heat shock response, *Cell 65*:363 (1991).
39. K. Abravaya, M.P. Myers, S.P. Murphy, and R.I. Morimoto, The human heat shock protein hsp70 interacts with HSF, the transcription factor that regulates heat shock gene expression, *Genes Dev. 6*:1153 (1992).
40. K.D. Sarge and R.I. Morimoto, Surprising features of transcriptional regulation of heat shock genes, *Gene Express. 1*:169 (1991).
41. Watowich and R.I. Morimoto, Complex regulation of heat shock- and glucose-responsive genes in human cells, *Mol. Cell. Biol. 8*:393 (1988).
42. B.S. Andrew, R.A. Eisenberg, A.N. Theofilopoulos, S. Izui, C.B. Wilson, P.J. McLonahey, E.D. Murphy, J.B. Roths, and F.J. Dixon, Spontaneous murine-lupus like syndromes, *J. Exp. Med. 148*:1198 (1978).
43. G.B. Faulds, D.A. Isenberg, and D.S. Latchman, The tissue-specific elevation in synthesis of the 90kd heat shock protein precedes the onset of disease in lupus-prone MRL/lpr mice, *J. Rheumatol. 21*:214 (1994).
44. Y. Deguchi, Enhanced expression of heat shock protein gene in kidney lymphoid cells of lupus-prone mice during growing process, *Autoimmunity 10*:1 (1991).
45. C. Mailhos, M.K. Howard, and D.S. Latchman, Heat shock protects neuronal cells from programmed cell death by apoptosis, *Neuroscience 55*:621 (1993).
46. S. Minota, S. Koyasu, I. Yahara, and J.B. Winfield, Autoantibodies to the heat shock protein hsp90 in SLE, *J. Clin. Invest. 81*:106 (1988).
47. W. Jarjour, B. Jefferies, J. Davis, W.Welch, T. Nomura, and J.B. Winfield, Antibodies to human stress proteins, *Arthritis Rheum. 34*:1133 (1991).
48. S.E. Conroy, G.B. Faulds, W.Williams, D.S.Latchman, and D.A. Isenberg, Detection of autoantibodies to the 90 kD heat shock protein in SLE and other autoimmune diseases. *Q. J. Rheum. 35*:428 (1994).
49. N. Mairesse, M.F. Kahn, and T. Applebloom, Antibodies to the constitutive 73 kD heat shock protein: a new marker for mixed connective tissue disease, *Am. J. Med. 95*:595 (1993).
50. G. Tsoulfa, G.A.W. Rook, G.M. Bahl, et al., Elevated IgG antibody levels to the mycobacterial 65 kD heat shock protein are characteristic of patients with rheumatoid arthritis, *Scand J. Immunol. 30*:519 (1989).

22

Stress Protein Expression in Sarcoidosis

H. Bielefeldt-Ohmann
Queensland University of Technology, Brisbane, Queensland, Australia

Janelle M. Staton
Curtin University of Technology Perth, Western Australia, Australia

Joanne E. Dench
Royal Perth Hospital, Perth, Western Australia, Australia

I. INTRODUCTION

Sarcoidosis is a multiorgan disorder characterized, in affected tissues, by a mononuclear inflammatory process, noncaseating granulomas, and distortion of normal tissue architecture (1,2). Although all tissue types and organs can be affected, the tissues most commonly involved, at least as determined by clinical manifestations, are those of the lower respiratory tract (1,2). The disease manifests itself variably, taking either a subclinical course, or individuals may present with acute or subacute to chronic symptoms that eventually resolve spontaneously. In a number of cases, sarcoidosis is insidious, and it may result in progressive fibrosis of affected tissues. In the case of lung involvement, about 7% of patients will progress to respiratory failure, right ventricular failure, and death within 10–15 years (3). A more detailed description of the clinical features, diagnosis, and treatment of sarcoidosis is beyond the scope of this chapter, and readers are referred to the abundant literature on the subject (e.g., refs. 1–4).

All available data suggest that sarcoidosis is a relatively common disease of worldwide distribution (reviewed in ref. 1). The highest incidence occurs in the 20- to 40-year-old age group, and it is seen slightly more frequently in females.

Currently, there is no conclusive evidence that occupational, environmental, or genetic factors are important in determining susceptibility to the disease. However, certain HLA phenotypes appear to be linked to specific clinical manifestations of the disease. Patients of the HLA-A1, HLA-Cw7, or HLA-DR3 phenotypes are likely to present with acute arthritis, anterior uveitis, and erythema nodosum (5), and patients who express HLA-B13 or HLA-DR5j are likely to develop chronic persistent disease (6,7) (see below).

II. SARCOIDOSIS ETIOLOGY AND HISTOPATHOLOGY

A. Etiology

The etiology of sarcoidosis has still not been unambiguously established, although the evidence for a mycobacterial infection is gaining momentum (8). The similarities in clinical and histopathological (see Section II.B) features between sarcoidosis and tuberculosis, tuberculous leprosy, and certain types of *Mycobacterium avium* spp. infections (8) have suggested a common cause. This contention has been supported by immunological (see below) and epidemiological (9,10) findings. Although a study by Bocart et al. (11), using the powerful technique of polymerase chain reaction (PCR), seemed to present strong evidence against a link between mycobacteria and sarcoidosis (12), more recent studies by Saboor et al. (13) and Fidler et al. (14), using a similar technique, found a strong link between *M. tuberculosis* and sarcoidosis. In support of the latter two studies, recent research have shown mycobacterial rRNA to be raised in spleens from patients with sarcoidosis (15), and mycobacteria with genomic sequence-similarity to *M. avium* spp. have been cultured from sarcoid skin lesions (16).

Although other agents, both infectious and noninfectious, have been more easily dismissed as etiological culprits (reviewed in refs. 1 and 17), the idea that sarcoidosis might be, at least partially, an autoimmune disease has been more pervasive (2,8,17). With the discovery of bacterial stress proteins (SPs) as major antigens for the T- and B-cell response to many pathogens, and the significant sequence homology between prokaryotic and eukaryotic SPs, a link between infectious diseases and autoimmunity was widely hypothesized (18–20). This was particularly true in those autoimmune diseases in which the critical target antigen (21) at the site of disease had not been clearly identified. Included in this category was sarcoidosis, where several possible mechanisms for a SP-driven autoimmune response could be considered. Thus, the initial event might involve (1) “molecular mimicry” or cross reactivity between SPs of the invading pathogen and the host (22,23), or (2) true autoreactivity following upregulation, and perhaps modified expression/presentation (20), of host SPs in the course of an infection (for reviews on these aspects, see refs. 19,20,24,25 and other chapters in this volume).

Considering the concurrence of multiorgan involvement and discrete inflammatory processes (granulomas) in sarcoidosis-affected tissues, it seemed more likely that the role of SPs, if any, was one of "perpetuators" of the disease rather than "triggers." In other words, a SP-driven inflammatory reaction only occurs after an initial infection has resulted in expression/upregulation of both microbial and host SP, with a subsequent clonal expansion of cross-reactive T cells and development of a self-sustaining hypersensitivity-type inflammation (26,27).

B. Histopathology and Immunohistology of Granulomas

The characteristic pathological feature of sarcoidosis, and the cause for much dispute regarding the etiology (see Section II.A), is the noncaseating granulomatous inflammation in affected tissues. The granuloma consists of a compact aggregate of mature macrophages and epithelioid cells, interspersed by lymphocytes, and surrounded by a mantle of monocytes and lymphocytes. In situ phenotyping of cells in the granulomas has demonstrated that, like in tuberculoid leprosy, $CD4^+$ cells distribute evenly throughout the granulomas, whereas $CD8^+$ cells show a predilection for the surrounding mantle (26,28–30) and occur in much lower numbers than $CD4^+$ lymphocytes. Moreover, the $\gamma\delta^+$ T-cell subset appears to be extremely rare in granulomatous lesions from patients with sarcoidosis irrespective of their stage of development (31).

In addition to high level expression of major histocompatibility complex (MHC) class II antigen on the mononuclear phagocytic cells in granulomas (32,33), the $CD4^+$ cells present within the granulomas may also express HLA-DR (34–36), suggesting that they are in a state of activation.

A sequential immunohistological study of biopsies from patients with pulmonary sarcoidosis has demonstrated that alveolitis is likely to represent the first step leading to the granuloma formation (35). In the acute phase of the alveolitis, T lymphocytes predominate in the alveolar lumen and septa, caused by a cellular redistribution from the peripheral blood to the lung as well as in situ lymphocyte proliferation (2,37,38). Similar mechanisms govern the granuloma formation in other tissues (28) as well as account for a substantial increase in monocyte-macrophages (39–41), which subsequently come to dominate the alveolar inflammatory response in chronic pulmonary disease (42).

III. IMMUNOLOGY OF SARCOIDOSIS

A. T-Lymphocyte Activities in Sarcoidosis

Despite the systemic nature of the disease, activated T cells are observed only in the affected organs, whereas T cells of the peripheral blood remain quiescent or may even exhibit decreased reactivity when compared with peripheral blood mononuclear cells (PBMCs) of healthy individuals. The latter phenomenon has given rise to the designation "peripheral anergy" (43,44), but it can be ascribed to a compartmentalization of lymphocytes and monocytes to affected tissues (28,35,41,44), and it thus should not be construed as suggesting that individuals with sarcoidosis are immunosuppressed. Notably, patients with sarcoidosis are no more susceptible to infections than healthy individuals without sarcoidosis, unless they are on a high-dose/long-term corticosteroid therapy, as is commonly used.

Most of what is so far known about immune functions in sarcoidosis have been obtained in studies of lymphocytes and macrophages procured by bronchoalveolar lavage (BAL). It could reasonably be questioned whether BAL cells are representative of cells in the alveolar interstitia, including granulomas. A number of studies have attempted to address this question, and it has been found that phenotypically there is good correlation between subset ratios in BAL and lung parenchyma (45,46) as well as in, at least some, functional aspects (47). Another caveat to bear in mind when assessing immune functions in sarcoidosis is the grouping of patients. In most studies, patients are grouped according to whether they present with an "active" or "inactive" inflammatory disease—a classification which relies on the lymphocyte to macrophage ratio in BAL. However, from a clinical point of view, the patients are more appropriately categorized as having acute or chronic disease (48,49). Some of the discrepancies noted in the literature on sarcoidosis immunology may arise from these different approaches to grouping of the study subjects.

There appears to be a clear association between the activity of the disease and the numbers, phenotypes, and functional characteristics of T lymphocytes that are found in the alveolitis. In active disease, the alveolitis is dominated by T cells, predominantly of the $CD4^+$ phenotype, which also exhibit low surface expression of CD3/T-cell receptor (TCR) and enhanced expression of CD29, HLA-DR, VLA-1, and CD2. This is accompanied by increased levels of mRNA transcripts for the β chain of the TCR (30,36,50–52). These findings suggest that BAL T cells from patients with sarcoidosis have undergone recent stimulation in vivo via the $\alpha\beta$ TCR. Furthermore, this stimulation is likely to be restricted to the site of inflammation, as the markedly enhanced "spontaneous" release of interleukin-2 (IL-2) and IL-2R by BAL T cells retrieved from patients with pulmonary sarcoidosis were shown not to be constitutive but would cease when the cells were cultured in vitro (53-55). It seems likely that this locally produced IL-2 contributes to autocrine and/or paracrine T-cell proliferation in the affected tissues (56) as well as plays a role in local macrophage activation by stimulating tumor necrosis factor α (TNF-α) production (57,58). Spontaneous release of high levels of IL-2 has also been implicated in the pathogenesis of autoimmune disease (59), although its role remains controversial (60–62). $CD3^+$ T lymphocytes in granulomas and BAL are also the source of significant amounts of "spontaneously" released interferon gamma (IFN-γ) (35,47,63).

A number of studies have demonstrated restricted Vβ8 (64), Cβ1 (65), and Vα2.3 (52) gene usage by $CD4^+$ T lymphocytes in BAL from patients with sarcoidosis. These findings might, despite the small study populations in all three studies, be taken to suggest presence of a specific antigen or set of antigens in the lungs, and possibly other affected tissues, of patients with sarcoidosis. Furthermore, the compartmentalization in the lung of $V\alpha 2.3CD4^+$ cells was correlated with expression of the HLA-DR3(w17),DQw2 haplotype (52). As previously mentioned, the HLA-DR3 haplotype has been linked to certain clinical manifestations in sarcoidosis (5–7) and a good prognosis (66), but it is also associated with such "proven" autoimmune diseases as SLE, juvenile diabetes, and Gravel's disease (67).

Despite a single report of increased numbers of $\gamma\delta^+$ T cells in BAL in a subgroup of patients with pulmonary sarcoidosis (68), this finding could not be substantiated when a larger group of patients was examined (30). The latter result agrees with the previously mentioned study of granulomatous lesions (31), and it thus seems to preclude any speculations about a connection between SPs and γδ T cells in sarcoidosis, such as has been suggested for multiple sclerosis (69) and other pathological conditions, including autoimmune reactions (20, 25, 70,71).

B. Macrophage Activities in Sarcoidosis

In addition to local activation of T lymphocytes, the macrophages in sarcoid granulomas as well as alveolar macrophages retrieved from patients with pulmonary sarcoidosis carry several markers of activation. The cells, independent of differentiation stage (monocyte–macrophage–epithelioid cell), express high levels of HLA-DRα (32,33) and "spontaneously" produce significant amounts of TNF-α, IL-1α, and IL-1-β, and colony-stimulating factors (CSFs), as shown by both in situ immunolabeling for the proteins (47), measurement of activity in extracts from granulomatous tissue (47), gene expression in freshly retrieved cells (58,72–74), and assessment of in vitro production by alveolar macrophages (42,75–77). An additional indication that these cells are "primed" in vivo, for example, by IL-2 and IFN-γ (see Section III.A), is their heightened responsiveness to further in vitro stimulation by, for example, lipopolysaccharide (LPS), when compared with alveolar macrophages from normal control subjects (72). Although no correlation between signs of T-cell activation and alveolar macrophage activity was found in chronic sarcoidosis (42), it may be of importance in the early phases of granuloma formation (see Section IV). Later, in the chronic phase, the macrophage activities may be maintained by an autocrine/paracrine mechanism(s) (42), including a self-sustaining recruitment (30) and proliferation (39).

IV. PATHOGENESIS OF THE GRANULOMATOUS REACTION

Based on what is currently known about cellular activities in sarcoidosis, the cellular composition of the sarcoid granulomas (26,31), and antigen-specific granuloma formation in general (78), the following sequence of events can be proposed: On entry of an infectious agent (e.g. *Mycobacterium* spp.), an acute inflammatory response is induced (27,79,80) which includes early production of T-cell and monocyte chemotactic factors such as IL-8, TNF-α, and the chemokine molecules (79,81–84) and upregulation of adhesion molecules on endothelia and epithelia. A T-cell–mediated antigen-specific response would soon lead to IL-2 and IFN-γ production, which would contribute to local macrophage activation (85–87). These cells in turn would produce IL-1 and TNF-α, probably in sequential manner (88). Sustained IL-2, IFN-γ, and TNF-α production, induced by persistent microbes, microbial antigens, or host antigens, would lead to granuloma formation (78,85–87,89,90) and eventually fibrosis (17,91). Other factors, such as transforming growth factor β (TGF-β) may also contribute to this outcome (92,93).

This sequence of events appears to be very similar for a whole range of antigen-specific hypersensitivity reactions, including tuberculosis, leprosy, berylliosis, and schistosome egg reactions (27,85–88). The reaction in sarcoidosis may differ from a number of other granulomatous diseases by being generally more insidious with less prominent systemic effects (27). It remains to be determined whether this is due to a characteristic of the etiological agent, for example, microbial virulence factors (8,94), or it is because the pathogenesis involves a "pathogen free" phase driven by endogenous autoantigens such as SPs.

In order to assess the potential role of SPs in sarcoidosis, we have analyzed the expression of SP 60/SP 65 and SP 70 in sarcoid granulomatous tissues as well as T-cell and serological reactivity to mycobacterial SP 65 and SP 70.

V. SP EXPRESSION IN TISSUES, MONOCYTES, AND MACROPHAGES IN SARCOIDOSIS

Tissue biopsies, collected for diagnostic purposes and prior to any treatment, from patients with various interstitial lung diseases (ILDs), including sarcoidosis, were immunolabeled using SP-specific murine monoclonal antibodies (MAbs). The ML30 MAb was produced against *M. leprae* (95) and recognizes *M. lepra* SP 65 as well as a cross-reactive sequence in human SP 60 (reviewed in ref. 20). Two other SP 60–specific MAb, LK-1 and LK-2, were produced against human SP 60 (96). LK-2 cross reacts with bacterial SP 65, whereas LK-1 is monospecific for human SP 60 (96). The C92F3A-5 MAb recognizes the inducible form of the human SP 70 complex, SP 72; the N27F3-4 MAb recognizes both SP 72 and SP 73 (98).

Both SP 65–specific MAb-labeled specifically and with high intensity, monocyte-macrophages, and epithelioid cells in sarcoid granulomas (Fig. 1). The LK-2 MAb gave similar results, although with much less staining intensity (Table 1) (33). Except for the occasional alveolar macrophage, no other cells or structures bound the SP 60/SP 65–specific MAb in any of the other ILD biopsies studied (Table 1). Expression of the SP 70 complex has not as yet been studied extensively, but preliminary immunocytochemistry results suggest low or no detectable expression in lung biopsies from patients with ILDs, including sarcoidosis.

To confirm and extend these results in vitro alveolar macrophages, retrieved by BAL and enriched for by Percoll gradient centrifugation, were examined for SP expression by metabolic labeling and autoradiography or by immunoblotting. They were found to express high levels of SP 90, SP 70/SP 72/SP 73, and SP 60 irrespective of the health status of the donor (Fig. 2). Furthermore, when the cells were stimulated, under adherence-free conditions in vitro, with cyto-

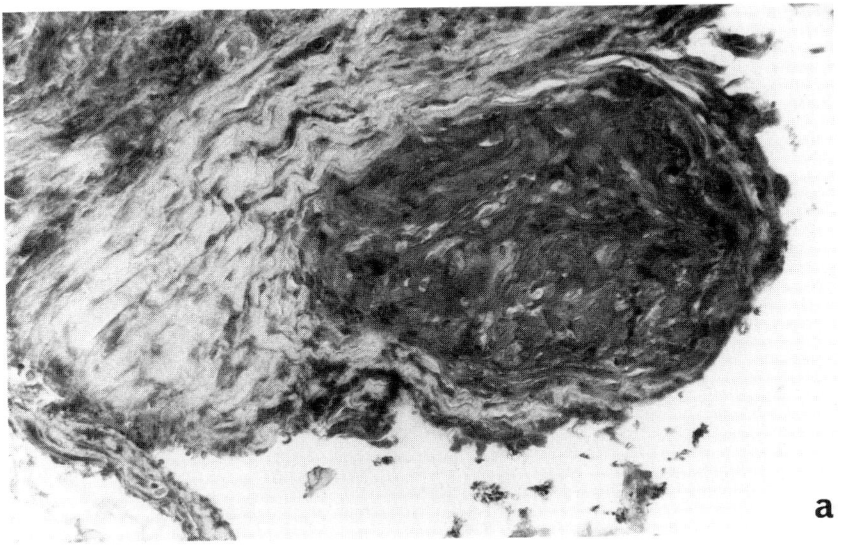

Figure 1 SP 60/SP 65 expression in granulomas in lung biopsies from patients with sarcoidosis detected by immunocytochemistry (33). (a,b) Tissue sections labeled with the ML30 MAb. The positively labeled granulomas appear darker (encircled by dashed line; red in the original preparation) than the surrounding tissue. (c) Lung biopsy from patient with fibrosing alveolitis (not prominent in this section). A few scattered alveolar macrophages are seen to label positively for SP 65/SP 60 (arrows), whereas the surrounding lung parenchyma does not express the antigen. Sections immunolabeled by a triple-layer APAAP protocol (22,33).

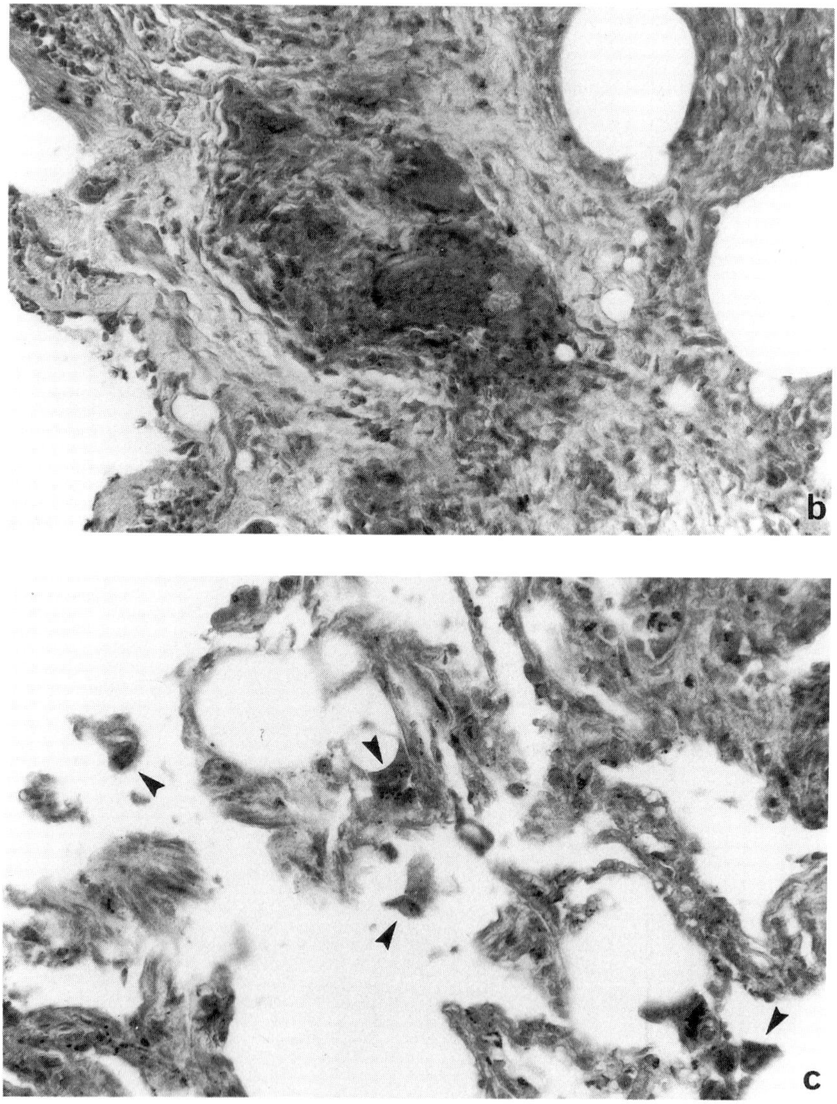

Figure 1 Continued.

kines known to be produced and present in sarcoid granulomas, including IFN-γ, IL-1β, and TNF-α, no further upregulation of synthesis and expression was apparent (Fig. 2) (33,99). Thus, it would appear that human alveolar macrophages express SPs maximally in vivo, perhaps as an autoprotective mechanism

Table 1 SP 65 Expression in Tissues from Patients with Sarcoidosis and other ILDs

Diagnosis	Tissue	n	SP expression[a]			HLA-DR Expression
			ML30	LK-1	LK-2	
Sarcoidosis	TBB[b]	6	–	ND	ND	+ to ++
Sarcoidosis	TBB	3	+ to++	ND	ND	++ to +++
Sarcoidosis	Lnmed	1	–	–	–	+++
Sarcoidosis	Lnmed	2	+	– to ++	+	++++
Sarcoidosis	Lnmed	4	++	+ to +++	+ to +++	4+ to 5+
Sarcoidosis	Lnmed	8	+++	+ to +++	+ to +++	3+ to 5+
Sarcoidosis	Lnmed	2	++++	++ to 4+	ND	4+ to 5+
CFA	Lung	6	–	ND	ND	+
EAA	Lung	3	–	ND	ND	+
Bronchitis obliterans[b]	Lung	3	– to +	ND	ND	+
Necrotizing granuloma[c]	TBB	1	–	ND	ND	++

Abbreviations: TBB, transbronchial biopsy; Ln Med, Mediastinal lymph node; CFA, crytogenic fibrosing alveolitis; EAA, extrinsic allergic alveolitis.
[a]Using the three MAb listed SP 65/SP 60–labeling intensity and degree of tissue involvement was judged on a scale from (–) to indicate no detectable SP, increasing to (5+) for maximal intensity, and/or tissue involvement.
[b]Or nonspecific fibrosis.
[c]Of unknown ethiology/diagnosis.
Source: modified from Ref. 33.

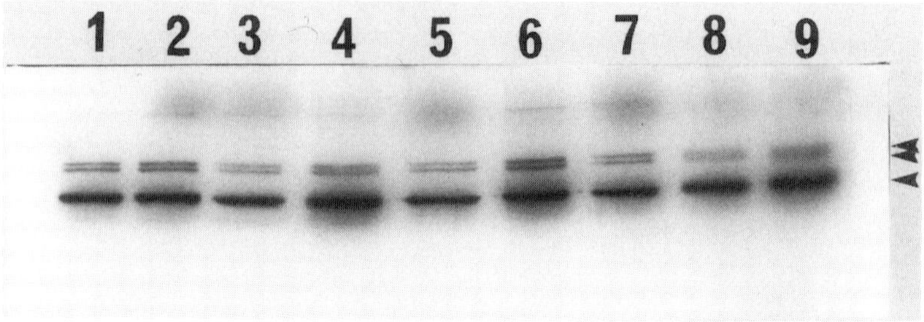

Figure 2 SP expression in isolated human alveolar macrophages following short-term exposure to heat or proinflammatory cytokines: lane 1, PGE_2; lane 2, 37°C control; lane 3, PGD_3; lane 4, IFN-γ; lane 5, 45°C heat shock (20 min); lane 6, TNF-α; lane 7, IL-1β; lane 8, PGF_2; lane 9, IL-6. SP (indicated by arrowheads) were visualized by Western blotting using a cocktail of SP60- and SP 72/SP 73–specific MAbs and ECL detection (33).

in a milieu where they are constantly bombarded with exogenous stimuli as well as exposed to their own potentially cytotoxic products.

In contrast to the alveolar macrophages, peripheral blood monocytes from patients with sarcoidosis as well as clinically normal donors expressed only low or undetectable levels of SP 60, SP 72/SP 73, and SP 90. However, their expression was readily upregulated on short-term exposure to IFN-γ (33) (Fig. 3).

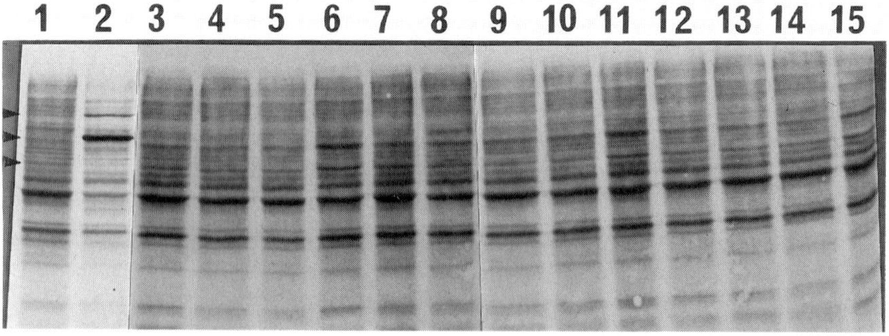

Figure 3 SP expression in isolated human peripheral blood monocytes following short-term (6 hr) exposure to proinflammatory cytokines or heat: lane 1, 37°C control; lane 2, 45°C for 20 min; lane 3, IL-1; lane 4, TNF-α; lane 5, IL-6, IFN-γ, lane 7, PGE_2; lane 8, PGD_3, lane 9, PGF_2a; lane 10, IL-1 + TNF-α; lane 11, IFN-γ + TNF-α; lane 12, IL-6 + TNF-α; lane 13, IL-1 + IL-6 + TNF-α; lane 14, IL-1 + PGE_2; lane 15, IFN-γ + $PGF_{2\alpha}$. Changes in protein expression was detected by metabolic labeling and autoradiography of gel-electrophoresis–separated proteins (33,99). Arrowheads indicate MW of 90, 70–73, and 60 kDa.

This result is in good agreement with the immunocytochemistry findings, where the smaller, more monocyte-like macrophages in the granulomas were the more intensely labeled, and it suggest that they may be recent "arrivals" either from the blood or as a result of in situ replication (30,39).

VI. IMMUNE RECOGNITION OF SPs IN SARCOIDOSIS

A. T-Lymphocyte Responses

T lymphocytes isolated from either peripheral blood or BAL of normal volunteers or patients with sarcoidosis and other ILDs were analyzed with respect to recognition of mycobacterial SP 65 either in soluble form (unpublished data) or bound to nitrocellulose particles (33). Responders and nonresponders distributed evenly between disease groups (Fig. 4a). A similar finding has been made by others both in the context of other diseases (18,100) and in studies involving patients with sarcoidosis (101). However, if the responses were categorized according to the activity state of the ILD or acuteness of inflammation, it became apparent that SP reactivity might reflect active inflammation in general irrespective of etiology (Fig. 4b). This conclusion is corroborated by a recent report by Anderton et al. (102), who found that SP 65–reactive T cells are activated during an acute inflammation induced in the absence of bacterial SP 65.

B. Serum Reactivity

Using a sensitive immunoblotting protocol, the serum antibody reactivities to mycobacterial SP 70 and SP 65 in patients with sarcoidosis, other ILDs and infections were analyzed (33,103). Although the majority of patients with sarcoidosis showed antibody reactivity to SP 65, they appeared not to differ significantly from other disease groups (Fig. 5) (33,103). As suggested by Kiessling et al. (20), the SP 65 serum reactivity might represent a response to protracted or repeated exposure to nonmycobacterial stimuli such as commensal organisms or clinically silent, superficial mucosal infections with *Candida albicans* (104) or *Chlamydia* spp. (105,106), or it might follow vaccination with nonmycobacterial vaccines (107).

In contrast to results recently reported by Kindås-Mügge et al. (108) and Del Guidice et al. (107), SP 70–specific serum reactivity was not detected in clinically healthy individuals but also not in the sarcoidosis group of our study (Fig. 5). Reactivity to SP 70 was seen, however, in individuals with acute infectious or noninfectious active inflammation such as caused by asbestosis. Thus, as for the T-cell proliferative responses to Sp 65, it appeared that SP 70 reactivity might reflect inflammation activity rather than a specific etiology (33,109).

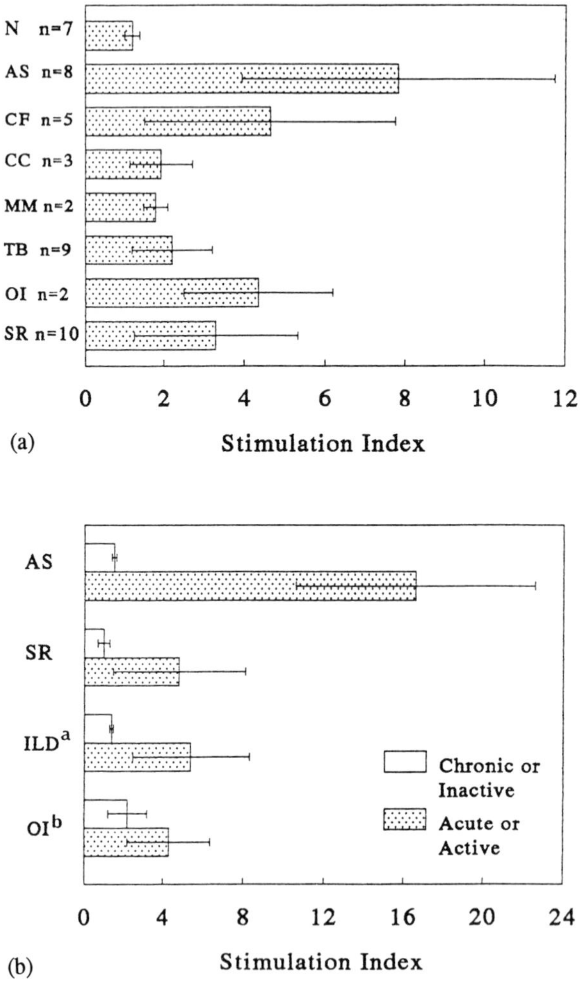

Figure 4 Proliferative response of peripheral blood T lymphocytes from healthy subjects and patients with various ILDs or infections to nitrocellulose-bound mycobacterial SP 65. The results shown represent mean stimulation index (SI) of n individuals ± SEM (From Ref. 33). (Top) Patients grouped according to diagnosis. N, normal; AS, asbestosis; CF, cryptogenic fibrosing alveolitis; CC, "chronic cough"; MM, malignant mesothelioma; TB, pulmonary tuberculosis; OI, other pulmonary infections; SR, sarcoidosis. (Bottom) Patients grouped according to diagnosis and acuteness/activity of disease. [a]Includes cryptogenic fibrosing alveolitis, silicosis, mesothelioma, and "chronic cough." [b]Includes mycobacterial, streptococcal, and staphylococcal lung infections.

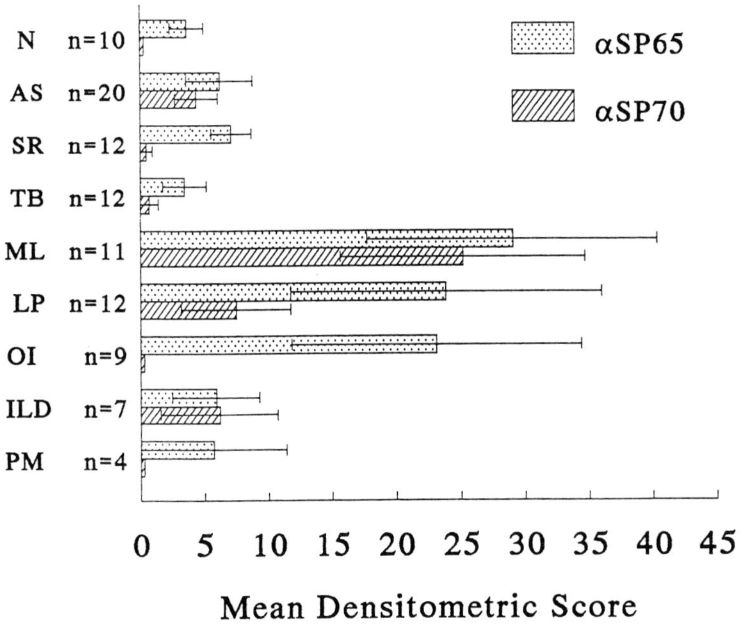

Figure 5 SP 65-and SP 70-serum antibody reactivity in healthy subjects and patients with various ILDs or infections. Sera were assayed against *M. tuberculosis* SP 65 and *M. bovis* SP 70 by immunoblotting with ECL visualization and densitometric quantitation (33). All densitometric values were standardized against a reference serum from a patient with melioidosis. Data shown are mean scores of n individuals ± SEM for sera tested at a 1/1000 dilution. N, normal; AS, asbestosis; SR, sarcoidosis; TB, pulmonary tuberculosis; ML, melioidosis; LP, leprosy; OI, other pulmonary infections (including infections with *M. avium*, *M. bovis, S. aureus*, and *Streptococcus* spp. [all HIV negative]); ILD, interstitial lung diseases (including CFA, "chronic cough," and postradiation fibrosis); PM, pulmonary malignancies.

VII. CONCLUSIONS: SP IN SARCOIDOSIS—A REFLECTION OF INFLAMMATORY ACTIVITY, NOT ETIOLOGY?

So far, our studies have failed to substantiate a causative role for SPs in the granulomatous inflammation characteristic of sarcoidosis. Although SP 60 protein is expressed at high levels in the granulomas of, at least some, patients with sarcoidosis, this is not reflected in the peripheral T- and B-cell responses to SP 65 and SP 70. It might be argued that the T- and B-cell reactivities are directed toward epitopes specific for the human SP homologues and therefore not detected using mycobacteria-derived SP, as used in our studies (19,33,110,111).

However, the lack of a proliferative response in BAL T cells from patients with sarcoidosis to autologous alveolar macrophages, which constitutively express high levels of SPs (see Fig. 2), could be taken to suggest that a cognate Sp 60/SP 70 response has not been missed (33,112). Future investigations should, nevertheless, be aimed at clarifying this point, by for example, employing nested peptides of the human SP60 and SP 70 sequences (113,114).

Thus, the hypothesis currently favored is schematically depicted in Figure 6. It is suggested that on respiratory infection with a slow-growing mycobacterium (8), an alveolitis is induced, initially dominated by T lymphocytes, followed by a substantial monocyte influx. The T cells are stimulated either by antigen-specific mechanisms or via bacteria-derived superantigens (115), resulting in proliferation and cytokine production, notably IFN-γ. The locally produced IFN-γ will induce an immediate (<6 hr) stress protein response in the immigrating monocytes, thereby conferring autoprotection on the cells which are likely to be further stimulated by the infectious agent(s) and its cellular products. Similar processes occur in other organs following a bacteriemia (30).

Although it certainly cannot be excluded that the process could become self-sustaining, the recent findings by Fidler et al. (14) and others (13,15,16) suggest that the agent may persist and thereby contribute to a chronic granulomatous

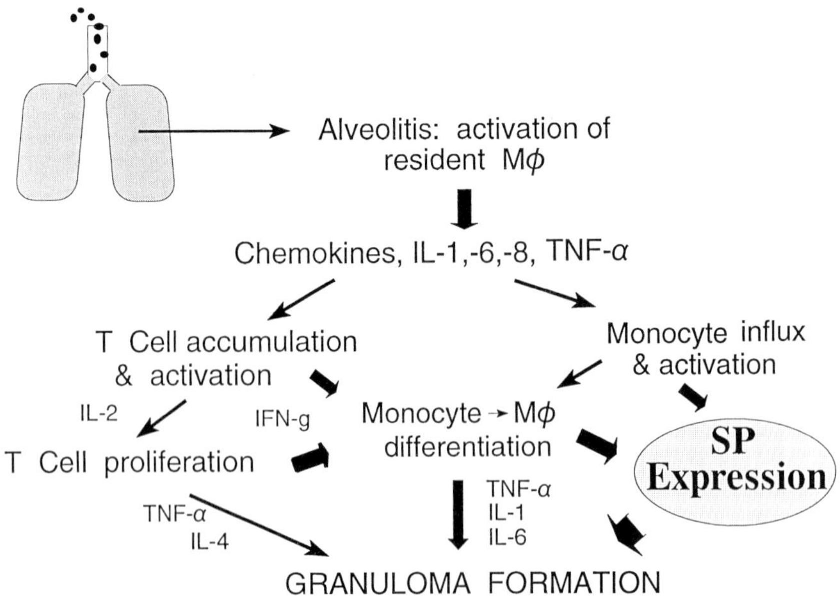

Figure 6 Schematic depiction of events leading to granuloma formation and SP expression in sarcoidosis.

host response (92). The occasionally observed SP 65–specific B- and T-cell responses in patients with sarcoidosis may be unrelated to the disease process (20,33). Alternatively, it may reflect slight changes in some individuals in the processing/presentation of bacterial or host SP, whereby normal SP tolerance, perhaps maintained via the constitutive SP expression in alveolar macrophages (see Fig. 2), is broken (20,116,117), or other normal immune regulatory mechanisms are abolished (118). In contrast, the SP 70–specific antibody response may be an indicator of recent bacterial infection (109) or, more generally, of active inflammation as suggested by the frequently high responsiveness in patients with asbestosis (see Fig. 5) (33).

ACKNOWLEDGMENTS

We thank Dr. D.R. Fitzpatrick for critical review of this manuscript.

REFERENCES

1. J.G. Scadding, and D.N. Mitchell, *Sarcoidosis*. 2nd ed. Chapman and Hall, London, 1985, pp. 72–100.
2. R.G. Crystal, P.B. Bitterman, S.I. Rennard, and B.A. Keogh, Interstitial lung disease of unknown cause. Disorders characterized by chronic inflammation of the lower respiratory tract, *N. Engl. J. Med. 310*:154, 235 (1984).
3. J.G. Scadding, The prognosis of intrathoracic sarcoidosis in England, *Br. Med. J. 2*:1165 (1961).
4. G. Semenzato, M. Chilosi, E. Ossi, L. Trentin, G. Pizzolo, A. Cipriani, C. Agostini, R. Zanibello, G. Marzeer, and G. Gasparotto, Bronchoalveolar lavage and lung histology. Comparative analysis of inflammatory and immunocompetent cells in patients with sarcoidosis and hypersensitivity pneumonitis, *Am. Rev. Respir. Dis. 132*:400 (1985).
5. E. Neville, D.G. James, D.A. Brewerton, D.C.O. James, C. Cockburn, and B. Fenichal, HLA antigens and clinical features of sarcoidosis, *Proceedings of the 8th International Conference on Sarcoidosis*. (W. James-William and B.H. Davies, eds.). Alpha Omega, Cardiff, 1980, pp. 201–205.
6. D.J. James, and W. James-William, Immunology of sarcoidosis, *Am. J. Med. 72*:5 (1982).
7. S. Abe, E. Yamaguchi, S. Makimura, N. Okazaki, H. Kunikane, and Y. Kawakami, Association of HLA-DR with sarcoidosis. Correlation with clinical course, *Chest 92*:488 (1987).
8. G.A.W. Rook and J.L. Stanford, Slow bacterial infections or autoimmunity?, *Immunol. Today 13*:160 (1992).
9. G. Brett, Epidemiological trends in TB and sarcoidosis in the district of London between 1958 and 1963, *Tubercle 46*:412 (1965).
10. S.A. Parkes, S. Baker, R.E. Bourdillon, C.R.H. Murray, and M. Rakshit, Epidemiology of sarcoidosis in the Isle of Man. I. A case controlled study, *Thorax 42*:420 (1987).

11. D. Bocart, D. Lecossier, A. de Lassence, D. Valeyre, J.-P. Battesti, and A.J. Hance, A search for mycobacterial DNA in granulomatous tissues from patients with sarcoidosis using the polymerase chain reaction, *Am. Rev. Respir. Dis. 145*: 1142 (1992).
12. M. Joyce-Brady, "Tastes great, less filling". The debate about mycobacteria and sarcoidosis, *Am. Rev. Respir. Dis. 145*:986 (1992).
13. S.A. Saboor, N. McI. Johnson, and J.J. McFadden, The use of the polymerase chain reaction to detect mycobacterial DNA in tuberculosis and sarcoidosis, *Lancet 339*:1012 (1992).
14. H.M. Fidler, G.A. Rook, N.McI. Johnson, and J. McFadden, *Mycobacterium tuberculosis* DNA in tissue affected by sarcoidosis, *Br. Med. J. 306*:546 (1993).
15. I.C. Mitchell, J.L. Turk, and D.N. Mitchell, Detection of mycobacterial rRNA in sarcoidosis with liquid-phase hybridisation, *Lancet 339*:1015 (1992).
16. D.Y. Graham, D.C. Markesich, D.C. Kalter, M.T. Moss, J. Hermon-Tailor, and F.A.K. El-Zaatari, Mycobacterial aetiology of sarcoidosis, *Lancet 340*:52 (1992).
17. P.D. Thomas, and G.W. Hunninghake, Current concepts of the pathogenesis of sarcoidosis, *Am. Rev. Respir. Dis.135*: 747 (1987).
18. J.R. Lamb, V. Bal, P. Mendez-Samperio, A. Mehlert, A. So, J. Rothbard, S. Jindal, R.A. Young, and D.B. Young, Stress proteins may provide a link between the immune response to infection and autoimmunity, *Intern. Immunol. 1*:191 (1989).
19. J.S.H. Gaston, Heat shock proteins and autoimmunity, *Semin. Immunol. 3*:35 (1991).
20. R. Kiessling, A. Grönberg, J. Ivanyi, K. Söderström, M. Ferm, S. Kleinau, E. Nilsson, and L. Klareskog, Role of hsp60 during autoimmune and bacterial inflammation, *Immunol. Rev. 121*:91 (1991).
21. N.R. Rose and C. Bona, Defining criteria for autoimmune diseases (Witebsky's postulates revisited), *Immunol. Today 14*:426 (1993).
22. D.R. Fitzpatrick, M. Snider, L. McDougall, T. Beskorwayne, L.A. Babiuk, T.J. Zamb, and H. Bielefeldt-Ohmann, Molecular mimicry: a herpesvirus glycoprotein antigenically related to a cell surface glycoprotein expressed by macrophages, polymorphonuclear leucocytes, and platelets. *Immunology 70*:504 (1990).
23. W. van Eden, J. Holoshitz, Z. Nevo, A. Frenkel, A. Klajman, and I.R. Cohen, Arthritis induced by a T cell clone that responds to Mycobacterium tuberculosis and to proteoglycans, *Proc. Natl. Acad. Sci. USA 82*:5117 (1985).
24. W. van Eden, Heat-shock proteins as immunogenic bacterial antigens with the potential to induce and regulate autoimmune arthritis, *Immunol. Rev. 121*:5 (1991).
25. D.B. Young, Heat-shock proteins: immunity and autoimmunity, *Curr. Opin. Immunol. 4*:396 (1992).
26. R.L. Modlin, F.M. Hofman, P.R. Meyer, O.P. Sharma, C.R. Taylor, and T.H. Rea, In situ demonstration of T lymphocyte subsets in granulomatous inflammation: leprosy, rhinoscleroma and sarcoidosis, *Clin. Exp. Immunol. 51*:430 (1983).
27. G.A.W. Rook, J. Taverne, C. Leveton, and J. Steele, The role of gamma-interferon, vitamin D3 metabolites and tumour necrosis factor in the pathogenesis of tuberculosis, *Immunology 62*:229 (1987).

28. G. Semenzato, A. Pezzuto, M. Chilosi, and G. Pizzolo, Redistribution of T lymphocytes in the lymph nodes of patients with sarcoidosis, *N. Engl. J. Med. 306*:48 (1982)
29. A.C.M.Th. van Maarsseven, H. Mullink, C.L. Alons, and J. Stam, Distribution of T-lymphocyte subsets in different portions of sarcoid granulomas: immunohistologic analysis with monoclonal antibodies, *Hum. Pathol. 17*:493 (1986).
30. G. Semenzato, R. Zambello, L. Trentin, and C. Agostini, Cellular Immunity in sarcoidosis and hypersensitivity pneumonitis, *Chest 103*:139S (1993).
31. A. Tazi, I. Fajac, P. Soler, D. Valeyre, J.P. Battesti, and A.J. Hance, Gamma/delta T-lymphocytes are not increased in number in granulomatous lesions of patients with tuberculosis or sarcoidosis, *Am. Rev. Respir. Dis. 144*:1373 (1991).
32. A.C.M.Th. van Maarsseven, H. Mullink, R. Rensen, and J. Stam, HLA-DR antigens in epithelioid cell granulomas of sarcoidosis using semithin frozen sections, *Sarcoidosis 2:*148 (1985).
33. J.M. Staton, J.E. Dench, B. Currie, D.R. Fitzpatrick, R. Himbeck, R. Allen, J. Bruce, B.W.S. Robinson, and H. Bielefeldt-Ohmann, Stress protein expression and immune recognition in sarcoidosis and other chronic interstitial lung diseases, *Immunol. Cell Biol. 73*:23 (1995).
34. U. Costabel, K.J. Bross, K.H. Rühle, G.W. Løhr, and H. Matthys, Ia-like antigens on T-cells and their subpopulations in pulmonary sarcoidosis and in hypersensitivity pneumonitis. Analysis of bronchoalveolar and blood lymphocytes, *Am. Rev. Respir. Dis. 131*:337 (1985).
35. G. Semenzato, C. Agostino, and R. Zambello, Activated T cells with immunoregulatory functions at different sites of involvement in sarcoidosis. Phenotypoic and functional evaluations, *Ann. N.Y. Acad. Sci. 465*:56 (1986).
36. R. Gerli, S. Darwish, L. Broccucci, V. Minotti, F. Spinozzi, C. Cernetti, A. Bertotto, and P. Rambotti, Analysis of CD4-positive T cell subpopulation in sarcoidosis, *Clin. Exp. Immunol. 73*:226 (1988).
37. R.G. Crystal, W.C. Roberts, G.W. Hunninghake, J.E. Gadek, J.D. Fulmer, and B.R. Line, Pulmonary sarcoidosis: a disease characterized and perpetuated by activated T-lymphocytes, *Ann. Intern. Med. 94*:73 (1981).
38. G. Semenzato, Immunology of interstitial lung diseases: cellular events taking place in the lung of sarcoidosis, hypersensitivity pneumonitis and HIV infection, *Eur. Respir. J. 4*:94 (1991).
39. P.B. Bitterman, L.E. Salzman, S. Adelberg, V.J. Ferrans, and R.G. Crystal, Alveolar macrophage replication. One mechanism for the expansion of the mononuclear phagocyte population in the chronically inflamed lung, *J. Clin. Invest. 74*:460 (1984).
40. F. Malavasi, A. Funaro, G. Bellone, F. Caligaris-Cappio, G. Semenzato, A.P.M. Cappa, E. Ferrero, F. Novelli, M. Alessio, S. Demaraia, P. Dellabona, Definition of CB12 monoclonal antibody as a differentiation marker specific for human monocytes and their bone marrow precursors, *Cell. Immunol. 97*:276 (1986).
41. C. Agostini, L. Trentin, R. Zambello, M. Luca, M. Masciarelli, A. Capriani, G. Marcer, and G. Semenzato, Pulmonary alveolar macrophages in patients with sarcoidosis and hypersensitivity pneumonitis: characterization by monoclonal antibodies, *J. Clin. Immunol. 7*:64 (1987).

42. J. Strausz, D.N. Männel, S. Pfeifer, A. Borkowski, R. Ferlinz, and J. Müller-Quernheim, Spontaneous monokine release by alveolar macrophages in chronic sarcoidosis, *Int. Arch. Allergy Appl. Immunol. 96*:68 (1991).
43. M. Horsmanheimo, Correlation of tuberculin-induced lymphocyte transformation with skin test reactivity and with clinical manifestations of sarcoidosis, *Cell. Immunol. 10*:329 (1974).
44. B.N. Hudspith, K.C. Flint, D. Geraint-James, J. Brostoff, and N. McI. Johnson, Lack of immune defiency in sarcoidosis: compartmentalization of the immune response, *Thorax 42*:250 (1987).
45. I.L. Paradis, J.H. Dauber, and B.S. Rabin, Lymphocyte phenotypes in bronchoalveolar lavage and lung tissue in sarcoidosis and idiopathic pulmonary fibrosis, *Am. Rev. Respir. Dis. 133*:855 (1986).
46. P.G. Holt, U.R. Kees, M.A. Schon-Hegrad, A.H. Rose, J. Ford, N. Bilyk, R. Bowman B.W.S. Robinson, Limiting dilution analysis of T cells extracted from solid lung tissues, *Immunology 64*:649 (1988).
47. M. Asano, T. Minagawa, M. Ohmichi, and Y. Hiraga, Detection of endogenous cytokines in sera or in lymph nodes obtained from patients with sarcoidosis, *Clin. Exp. Immunol. 84*:92 (1991).
48. A. Cantin, and R.G. Crystal, Interstitial pathology: an overview of the chronic interstitial lung disorders, *Int. Arch. Allergy Appl. Immunol. 76*:83 (1985).
49. Y.H. Lin, P.L. Haslam, and M. Turner-Warwick, Chronic pulmonary sarcoidosis: relationship between lung lavage cell counts, chest radiograph, and results of standard lung function tests, *Thorax 40*:501 (1985).
50. E. Yamaguchi, N. Okazaki, A. Itoh, K. Furuya, S. Abe, and Y. Kawakami, Enhanced expression of CD2 antigen on lung T cells, *Am. Rev. Respir. Dis. 143*:829 (1991).
51. R.M. duBois, M. Kirby, B. Balbi, C. Saltini, G. Crystal, T-lymphocytes that accumulate in the lung in sarcoidosis have evidence of recent stimulation of T-cell antigen receptor, *Am. Rev. Respir. Dis. 145*:1205 (1992).
52. J. Grunewald, C.H. Janson, A. Eklund, M. Öhrn, O. Ollerup, U. Persson, and H. Wigzell, Restricted Vα2.3 gene usage by CD4$^+$ T lymphocytes in bronchoalveolar lavage fluid from sarcoidosis patients correlates with HLA-DR3, *Eur. J. Immunol. 22*:129 (1992).
53. P. Pinkston, P.B. Bitterman, and R.G. Crystal, Spontaneous release of interleukin-2 by lung T-lymphocytes in active pulmonary sarcoidosis, *N. Engl. J. Med. 308*:793 (1983).
54. J. Müller-Quernheim, C. Saltini, P. Sondermeyer, and R.G. Crystal, Compartmentalized activation of the interleukin 2 gene by lung T lymphocytes in active pulmonary sarcoidosis, *J. Immunol. 137*:3475 (1986).
55. E.C. Lawrence, K.P. Brousseau, M.B. Berger, C.C. Kurman, L. Marcon, and D.L. Nelson, Elevated concentrations of soluble interleukin-2 receptors in serum samples and broncoalveolar lavage fluids in active sarcoidosis, *Am. Rev. Respir. Dis. 137*:759 (1988).
56. M.L. Toribio, J.C. Gutierrez-Ramos, L. Pezzi, M.A.R. Marcos, and C. Martinez-A, Interleukin-2-dependent autocrine proliferation in T-cell development, *Nature 342*:82 (1989).

57. W.W. Hancock, L. Kobzik, A.J. Colby, C.J. O'Hara, A.G. Cooper, and J.J. Godleski, Detection of lymphokines and lymphokine receptors in pulmonary sarcoidosis: immunohistologic evidence that inflammatory macrophages express IL-2 receptors, *Am. J. Pathol. 123*:1 (1986).
58. S.M. Dubinett, M. Huang, S. Dhanani, D. Kelley, A. Lichtenstein, W.W. Grody, and L.E. Mintz, *In situ* regulation of pulmonary macrophage TNF-α mRNA expression by IL2, *Chest 103*:91S (1993).
59. A. Altman, A.N. Theofilopoulos, R. Weiner, D.H. Katz, and F.J. Dixon, Analysis of T cell function in autoimmune murine strains. Defects in production of and responsiveness to interleukin 2, *J. Exp. Med. 154*:791 (1981).
60. M. Linker-Israeli, A.C. Bakke, R.C. Kitridou, S. Gendler, S. Gillis, and D.A. Horwitz, Defective production of interleukin 1 and interleukin 2 in patients with systemic lupus erythematosus (SLE), *J. Immunol. 130*:2651 (1983).
61. J.C. Gutierrez-Ramos, J.L. Andreu, Y. Revilla, E. Vinuela, and C. Martinez-A, Recovery from autoimmunity of MLR/lpr mice after infection with an interleukin-2/vaccinia recombinant virus, *Nature 346*:271 (1990).
62. J.C. Gutierrez-Ramos, I.M. de Alboran, and C. Martinez-A, In vivo administration of interleukin-2 turns on anergic self-reactive T cells and leads to autoimmune disease, *Eur. J. Immunol. 22*:2867 (1992).
63. P.L. Mosley, C. Hemken, M. Monick, K. Nugent, and G.W. Hunninghake, Interferon and growth factor activity for human lung fibroblasts. Release from bronchoalveolar cells from patients with active sarcoidosis, *Chest 89*:657 (1986).
64. D.R. Moller, K. Konishi, M. Kirby, B. Balbi, and R.G. Crystal, Bias toward use of a specific T cell receptor β-chain variable region in a subgroup of individuals with sarcoidosis, *J. Clin. Invest. 82*:1183 (1988).
65. N. Tamura, D.R. Moller, B. Balbi, and R.G. Crystal, Preferential usage of T-cell antigen receptor β-chain constant region Cβ1 in T-lymphocytes in the sarcoid lung, *Am. Rev. Respir. Dis. 139*:A60 (1989).
66. E. Hedfors, and F. Lindström, HLA-B8/DR3 in sarcoidosis. Correlation to acute onset disease with arthritis, *Tissue Antigens 22*:200 (1983).
67. J. Tiwari, and T. Terasaki (eds.), *HLA and Disease Association*, Springer-Verlag, New York (1985).
68. B. Balbi, D.R. Moller, M. Kirby, K.J. Holroyd, and R.G. Crystal, Increased numbers of T lymphocytes with γδ-positive antigen receptors in a subgroup of individuals with pulmonary sarcoidosis, *J. Clin. Invest. 85*:1353 (1990).
69. K. Selmaj, C.F. Brosnan, and C.S. Raine, Colocalization of lymphocytes bearing γδ T-cell receptor and heat shock protein hsp65$^+$ oligodendrocytes in multiple sclerosis, *Proc. Natl. Acad. Sci. USA 88*:6452 (1991).
70. R.A. Young, and T.J. Elliott, Stress proteins, infection and immune surveillance, *Cell 59*:5 (1989).
71. W. Born, L. Hall, J. Boymel, T. Shinnick, D. Young, P. Brennan, and P. O'Brien, Recognition of a peptide antigen by heat shock reactive γδ T lymphocytes, *Science 249*:67 (1990).
72. M. Spatafora, A. Merendino, G. Chiappara, M. Gjomarkaj, M. Melis, V. Bellia, and G. Bonsignore, Lung compartmentalization of increased TNF releasing ability by mononuclear phagocytes in pulmonary sarcoidosis, *Chest 96*:542 (1989).

73. H. Kreipe, J.R. Radzun, K. Heidorn, J. Barth, J.K. Kallee, W. Petermann, J. Gerdes, and M.R. Parwaresh, Proliferation, macrophage colony-stimulating factor, and macrophage colony-stimulating factor-receptor expression of alveolar macrophages in active sarcoidosis, *Lab. Invest. 62*:697 (1990).
74. M.D. Wewers, C. Saitini, S. Sellers, M.J. Tocci, E.K. Bayne, J.A. Schmidt, and R.G. Crystal, Evaluation of alveolar macrophages in normals and in individuals with active pulmonary sarcoidosis for the spontaneous expression of the interleukin-1β gene, *Cell. Immunol. 107*:479 (1987).
75. G.W. Hunninghake, Release of interleukin-1 by alveolar macrophages of patients with active pulmonary sarcoidosis, *Am. Rev. Respir. Dis. 129*:569 (1984).
76. P.R. Bachwich, J.P. Lynch, J.W. Larrick, M. Spengler, and S.L. Kunkel, Tumor necrosis factor production by human sarcoid alveolar macrophages, *Am. J. Pathol. 125*:421 (1986).
77. J. Müller-Quernheim, S. Pfeifer, D. Männel, J. Strausz, and R. Ferlinz, Lung-restricted activation of the alveolar macrophage/monocyte system in pulmonary sarcoidosis, *Am. Rev. Respir. Dis. 145*:187 (1992).
78. S.L. Kunkel, S.W. Chensue, R.M. Strieter, J.P. Lynch, and D.G. Remick, Cellular and molecular aspects of granulomatous inflammation, *Am. J. Respir. Cell. Mol. Biol. 1*:439 (1989).
79. H. Bielefeldt-Ohmann, L.A. Babiuk, and R. Harland, Cytokine synergy with viral cytopathic effects and bacterial products during the pathogenesis of respiratory tract infection, *Clin. Immunol. Immunopathol. 60*:153 (1991).
80. Y. Rosen, T.J. Thanassiades, S. Moon, and H.A. Lyons, Nongranulomatous interstitial pneumonitis in sarcoidosis: relationship to development of epithelioid granulomas, *Chest 74*:122 (1978).
81. J.S. Friedland, R. Shattock, D.G. Remick, and G.E. Griffin, Mycobacterial 65-kD heat shock protein induces release of proinflammatory cytokines form human monocytic cells, *Clin. Exp. Immunol. 91*:58 (1993).
82. J.S. Friedland, D.G. Remick, R. Shattock, and G.E. Griffin, Secretion of interleukin-8 following phagocytosis of *Mycobacterium tuberculosis* by human monocyte cell lines, *Eur. J. Immunol. 22*:1373 (1992).
83. H. Bielefeldt Ohmann, M. Campos, L. McDougall, M.J.P. Lawman, and L.A. Babiuk, Expression of tumor necrosis factor-α receptors on bovine macrophages, lymphocytes and polymorphonuclear leukocytes, internalization of receptor bound ligand and some functional effects, *Lymphokine Res. 9*:43 (1990).
84. J.J. Oppenheim, C.O.C. Zachariae, N. Mukaida, K. Matsushima, Properties of the novel proinflammatory supergene "intercrine" cytokine family, *Annu. Rev. Immunol 9*:617 (1991).
85. S.M. Wahl, J.B. Allen, S. Dougherty, V. Evequoz, D.H. Pluznik, R.L. Wilder, A.R. Hand, and L.M. Wahl, T lymphocyte-dependent evolution of bacterial cell wall-induced hepatic granulomas, *J. Immunol. 137*:2199 (1986).
86. M.E.A. Mielke, H. Rosen, S. Brocke, C. Peters, and H. Hahn, Protective immunity and granuloma formation are mediated by two distinct tumor necrosis factor alpha- and gamma interferon–dependent T cell-phagocyte interactions in murine Listeriosis: dissociation on the basis of phagocyte adhesion mechanisms, *Infect. Immun. 60*:1875 (1992).

87. P.J. Haley, Mechanisms of granulomatous lung disease from inhaled beryllium: the role of antigenicity in granuloma formation, *Toxicol. Pathol. 19*:514 (1991).
88. S.W. Chensue, I.G. Oterness, G.I. Higashi, C.S. Forsch, and S.L. Kunkel, Monokine production by hypersensitivity (*Schistosoma mansoni* egg) and foreign body (Sephadex bead)-type granuloma macrophages. Evidence for sequential production of IL-1 and tumor necrosis factor, *J. Immunol. 142*:1281 (1989).
89. P. Amiri, R.M. Locksley, T.G. Parslow, M. Sadick, E. Rector, D. Ritter, and J.H. McKerrow, Tumour necrosis factor α restores granulomas and induces parasite egg-laying in schistosome-infected SCID mice, *Nature 356*:604 (1992).
90. V. Kindler, A.-P. Sappino, G.E. Grau, P.-F. Piguet, and P. Vassalli, The inducing role of tumor necrosis factor in the development of bactericidal granulomas during BCG infection, *Cell 56*:731 (1989).
91. P.F. Piguet, M.A. Collart, G.E. Grau, A.-P. Sappino, and P. Vassalli, Requirement of tumour necrosis factor for development of silica-induced pulmonary fibrosis, *Nature 344*:245 (1990).
92. T.J. Broekelmann, A.H. Limper, T.V. Colby, and J.A. McDonald, Transforming growth factor β_1 is present at sites of extracellular matrix gene expression in human pulmonary fibrosis, *Proc. Natl. Acad. Sci. USA 88*:6642 (1991).
93. J. Maeda, N. Ueki, T. Ohkawa, N. Iwahashi, T. Nakano, T. Hada, and K. Higashino, Local production and localization of transforming growth factor-beta in tuberculous pleurisy, *Clin. Exp. Immunol. 92*:32 (1993).
94. H. Shiratsuchi, Z. Toossi, M.A. Mettler, and J.J. Ellner, Colonial morphotype as a determinant of cytokine expression by human monocytes infected with *Mycobacterium avium*, *J. Immunol. 150*:2945 (1993).
95. J. Ivanyi, S. Sinha, R. Aston, D. Cussel, M. Keen, and U. Sengupta, Definition of species specific and cross-reactive antigenic determinants of *Mycobacterium leprae* using monoclonal antibodies, *Clin. Exp. Immunol. 52*:528 (1983).
96. C.J.P. Boog, E.R. de Graeff-Meeder, M.A. Lucasses, R. van der Zee, M.M. Voorhorst-Ogink, P.J.S. van Kooten, H.J. Geuze, and W. van Eden, Two monoclonal antibodies generated against human hsp60 show reactivity with synovial membranes of patients with juvenile chronic arthritis, *J. Exp. Med. 175*:1805 (1992).
97. W.J. Welch and J.P. Suhan, Cellular and biochemical events in mammalian cells during and after recovery from physiological stress, *J. Cell. Biol. 103*:2035 (1986).
98. S. Minota, B. Cameron, W.J. Welch, and J.B. Winfield, Autoantibodies to the constitutive 73kD member of the HSP70 family of heat shock proteins in systemic lupus erythematosus, *J. Exp. Med. 168*:1475 (1988).
99. J.E. Dench, A.H. Rose, B.W.S. Robinson, and H. Bielefeldt-Ohmann, Mononuclear phagocytic cells from patients with sarcoidosis express heat shock proteins, *Proceedings 22nd Annual Scientific Meeting Australasian Society for Immunology*, (1992), p. 57.
100. H.P. Fisher, C.E.M. Sharrock, M.J. Colston, and G.S. Panayi, Limiting dilution analysis of proliferative T cell responses to mycobacterial 65-kDa heat-shock protein fails to show significant frequency differences between synovial fluid and peripheral blood of patients with rheumatoid arthritis, *Eur. J. Immunol. 21*:2937 (1991).

101. A. Faith, C. Moreno, R. Lathigra, E. Roman, M. Fernandez, S. Brett, D.M. Mitchell, J. Ivanyi, and A.D.M. Rees, Analysis of human T-cell epitopes in the 19,000 MW antigen of *Mycobacterium tuberculosis*: influence of HLA-DR, *Immunology 74*:1 (1991).
102. S.M. Anderton, R. van der Zee, and J.A. Goodacre, Inflammation activates self hsp60-specific T cells, *Eur. J. Immunol. 23*:33 (1993).
103. J.E. Dench, H. Bielefeldt-Ohmann, A.H. Rose, and B.W.S. Robinson, Patients with sarcoidosis express reactivity to a 65-kilodalton heat shock protein, *J. Leuk. Biol., Suppl. 3*:21 (1992) (Abstract).
104. L. Ivanyi and J. Ivanyi, Elevated antibody levels to mycobacterial 65-kDa stress protein in patients with superficial candidiasis, *J. Infect. Dis. 162*:519 (1990).
105. E.A. Wagar, J. Schachter, P. Bavoil, and R.S. Stephens, Differential human serological response to two 60,000 molecular weight *Chlamydia trachomatis* antigens, *J. Infect. Dis. 162*:922 (1990).
106. H. Bielefeldt-Ohmann, Unpublished data (1992).
107. G. Del Giudice, A. Gervaix, P. Costantino, C.-A. Wyler, C. Tougne, E.R. de Graeff-Meeder, J. van Embden, R. van der Zee, L. Nencioni, R. Rappuoli, S. Suter, and P.-H. Lambert, Priming to heat shock proteins in infants vaccinated against pertussis, *J. Immunol. 150*:2025 (1993).
108. I. Kindås-Mügge, G. Steiner, and J.S. Smolen, Similar frequency of autoantibodies against 70-kD class heat shock proteins in healthy subjects and systemic lupus erythematosus patients, *Clin. Exp. Immunol. 92*:46 (1993).
109. H. Bielefeldt-Ohmann, J. M. Staton, and B. Currie, Stress-protein responses in acute melioidosis, Unpublished data (1995).
110. M.P. Davenport, K.R. McKenzie, A. Basten, and W.J. Britton, The variable C-terminal region of the *Mycobacterium leprae* 70-kilodalton heat shock protein is the target for humoral immune responses, *Infect. Immun. 60*:1170 (1992).
111. E.R. de Graeff-Meeder, R. van der Zee, G.T. Rijkes, H.-J. Schuurman, W. Kuis, J.M.J. Bijlsma, B.J.M. Zegers, and W. van Eden, Recognition of human 60kD heat shock protein by mononuclear cells from patients with juvenile chronic arthritis, *Lancet 337*:1368 (1991).
112. J.M. Staton and H. Bielefeldt-Ohmann, Unpublished data (1993).
113. H.M. Geysen, R.H., Meloen, and S.J. Barteling, Use of peptide synthesis to probe viral antigens for epitopes to a resolution of a single aminoacid, *Proc. Natl. Acad. Sci. USA 81*:3998 (1984).
114. R. van der Zee, W. van Eden, R.H. Meloen, A. Noordzij, and J.D.A. van Embden, Efficient mapping and characterization of a T cell epitope by the simultaneous synthesis of multiple peptides, *Eur. J. Immunol. 19*:43 (1989).
115. A. Herman, J.W. Kappler, P. Marrack, and A.M. Pullen, Superantigens: mechanism of T-cell stimulation and role in immune responses, *Annu. Rev. Immunol. 9*:745 (1991).
116. J.W. Kappler, U. Staerz, J. White, and P.C. Marrack, Self-tolerance eliminates T cells specific for Mls-modified products of the major histocompatibility complex, *Nature 332*:35 (1988).
117. H-M. Vordermeier, D. P. Harris, G. Friscia, E. Roman, H-M. Surcel, C. Moremo, G. Pasvol, and J. Ivanyi, T cell repertoire in tuberculosis: selective anergy

to immunodominant epitope of the 38-kDa antigen in patients with active disease, *Eur. J. Immunol. 22*:2631 (1992).

118. J.B.A.G. Haanen, T.H.M. Ottenhoff, R.F.M. Lai a Fat, H. Soebono, H. Spits, and R.R.P. de Vries, *Mycobacterium leprae*-specific T cells from a tuberculoid leprosy patient suppress HLA-DR3-restricted T cell responses to an immunodominant epitope on 65-kDa hsp of mycobacteria, *J. Immunol. 145*:3898 (1990).

23

Stress Proteins in Inflammatory Liver Disease

Ansgar W. Lohse and Hans Peter Dienes
Johannes Gutenberg-University, Mainz, Germany

I. INTRODUCTION

Only very few studies have thus far dealt with the expression of heat shock proteins (hsps) in normal and inflamed liver, and little is known to date on specific disease associations and immune responses to hsps in liver diseases. These studies have almost exclusively concentrated on the 60 kDa heat shock protein, hsp 60.

The liver is not only involved as a bystander organ in many systemic diseases, it is also, more than most other organs, the focus of many chronic inflammatory diseases. Hepatitis B virus infection is the most prevalent infectious disease in the world, and chronic hepatitis C appears to be similarly common worldwide, although with a different geographical distribution. Alcoholic hepatitis as the inflammatory lesion in many patients with alcoholic liver disease is thought to be largely immune mediated. Although systemic autoimmune diseases like lupus erythematosus rarely affect the liver, the liver itself is the target organ of autoimmune attack in autoimmune hepatitis (which in itself is probably a heterogeneous group of diseases), primary biliary cirrhosis, amd primary sclerosing cholangitis.

The acute phase response in systemic inflammatory conditions is centered on the liver, where protein synthesis is rapidly adjusted to the need of the endangered organism. It is likely that this massive response is associated with a marked increase in the expression of stress proteins, but data on this are not yet available.

II. HSP 60 EXPRESSION IN NORMAL LIVER

The normal liver seems to express only very low levels of hsp 60. Immunohistology using the ML30 murine monoclonal antibody was used in two studies. In the report by Koskinas et al., very low level hsp 60 expression was found in less than 5% of hepatocytes, if at all, in healthy individuals or in those with fatty liver disease (1). These investigators did not say if staining in nonparenchymal liver cells could be observed in the normal liver. Broomé et al., using the same antibody, observed some granular cytoplasmic staining in all normal hepatocytes (2). We have studied normal liver sections using the monoclonal antibodies LK-1 and LK-2, and we also observed only minimal staining. Immunoelectron microscopy showed that this staining was mitochondrial (3).

Using freshly isolated and cultured guinea pig hepatocytes, Kupffer cells, and sinusoidal endothelial cells, we were able to attain more information on the distribution of hsp 60 in the normal liver. Western blot analysis of solubilized proteins from these cells showed marked expression in normal hepatocytes, significant expression in Kupffer cells, and only minimal expression in normal sinusoidal endothelial cells (Fig. 1) (3). Culture of Kupffer cells at 42°C for 1 hr prior to solubilization lead to a marked increase in hsp 60 expression (Fig. 2), thus confirming the nature of it as a stress protein also in Kupffer cells. Preparation of subcellular fragments of human liver homogenate confirmed that the maximum expression was in mitochondria followed by nuclei and microsomes (3). Expression in the cytoplasm was only minimal. It thus appears that the cytoplasmic staining observed by Broomé et al. (2) may either have been some nonspecific binding or, more likely, mitochondrial binding not identified as such owing to the insensitivity of light microscopy in distinguishing intracellular compartments. From our results, it can be summarized that hsp 60 is expressed in the normal liver; in particular in hepatocytes and less so in Kupffer cells. Like elsewhere in the body, it is a mitochondrial, heat stress inducible protein.

III. HSP 60 EXPRESSION IN AUTOIMMUNE LIVER DISEASES

Expression of and immunity to hsp 60, as explained elsewhere in this book, has been thought to play a critical role in the immunopathogenesis and prepetuation of autoimmune diseases. In liver disease, therefore, attention has focused on autoimmune liver disease. We have found markedly increased expression in biopsy specimens from patients with active autoimmune hepatitis (3). Very strong staining of hepatocytes and moderately strong staining of Kupffer cells was observed. In addition, sinusoidal lining cells showed some staining as well. Comparison with conventional histology showed increased expression to be

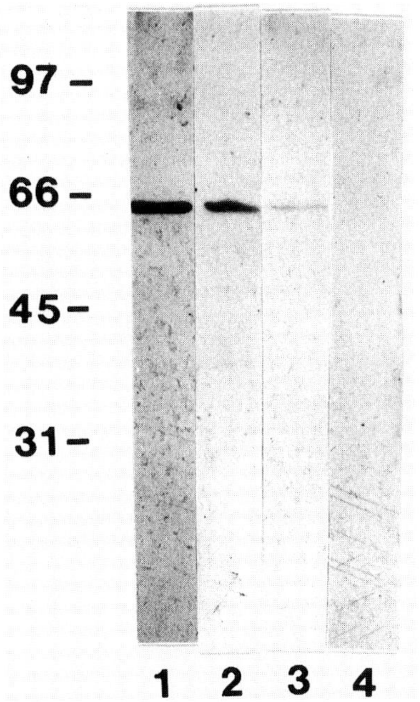

Figure 1 Immunoblotting analysis of solubilised proteins from guinea pig hepatocytes (lane 1), Kupffer cells (lane 2), and endothelial cells (lane 3) using monoclonal anti-hsp 60 antibody LK-1. Lane 4 shows the negative control without secondary antibody on hepatocyte membrane. (From Ref. 19.)

centred in areas of active inflammation. It was remarkable that increased hsp 60 expression was irregular even within inflammatory zones. Although some hepatocytes showed pronounced granular staining, directly neighboring hepatocytes could be found to express the normal minimal amounts of hsp 60. This patchy staining was confirmed by immunoelectron microscopy, with neighbouring cells expressing markedly different levels of hsp 60 (3). Immunoelectron microscopy showed that the increased hsp 60 expression in inflammation was also almost exclusively confined to the mitochondria. In addition, staining within the mitochondria was found to be patchy (Fig. 3).

Broomé et al. examined biopsies from the two other kinds of autoimmune liver disease, primary biliary cirrhosis (PBC) and primary sclerosing cholangitis (PSC) (2). Marked granular staining of hepatocytes was observed in both conditions, but Kupffer cells and endothelial cells were found not to stain with the ML30 anti–hsp 60 antibody. In both PBC and PSC, the bile ducts are the focus of the presumably autoimmune attack, and it was thus interesting to observe

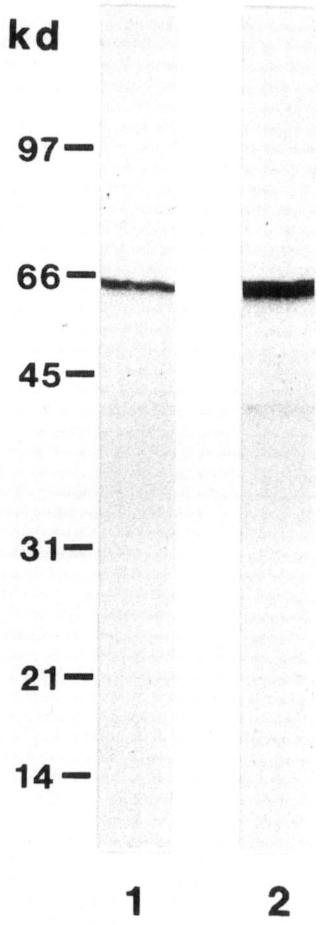

Figure 2 Immunoblotting analysis of soublised guinea pig Kupffer cells incubated for 1 hr at 37°C (left lane) or 42°C (right lane) with monoclonal anti–hsp 60 antibody LK-1 showing increased expression induced by heat stress. (From Ref. 19.)

that almost all bile ducts (but not all bile duct cells) showed marked hsp 60 expression in these two conditions (2). Bile duct staining was not observed in viral hepatitis (see below) or autoimmune hepatitis (see above). We have since had very similar results in liver biopsy specimens from patients with PBC (unpublished observation).

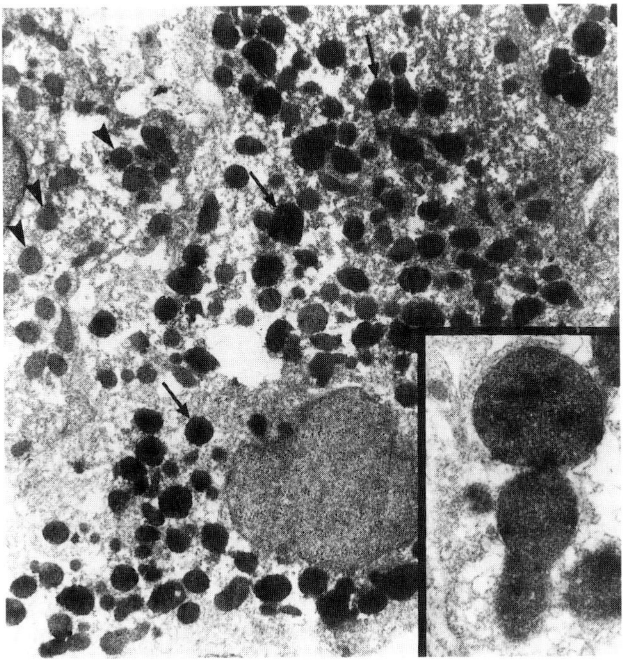

Figure 3 Immunoelectron microscopy specimen of a biopsy from a patient with chronic active autoimmune hepatitis showing intense mitochondrial staining in some hepatocytes (arrows) but no significant staining in mitochondria of a directly neighboring hepatocyte (arrowheads) (× 6500). Insert: higher magnification showing speckled staining pattern within the mitochondria (× 22 000). (From Ref. 19.)

Mycobacterial infection has repeatedly been hypothesized to be linked to the pathogenesis of PBC (4). Unfortunately, the study of Broomé et al. (2) does not use other hsp 60–specific antibodies to distinguish between (myco-) bacterial hsp 60 and human hsp 60. We are presently investigating this aspect using the LK-1 and LK-2 antibodies that can distinguish between mammalian and bacterial hsp 60. From the data thus far, it appears that all hsp 60 expression in PBC is human; thus arguing strongly against a mycobacterial infection.

Jorge et al. have examined expression of hsp 90 in primary biliary cirrhosis and found an increased expression both in hepatocytes and bile duct cells (5). Intriguingly, this increased hsp 90 expression was paralleled by an increase in estrogen receptor expression, which is remarkable in view of the marked female preponderance of primary biliary cirrhosis. The pattern of expression of hsp 60 and hsp 90 is very similar in these studies, and the question arises if hsp overexpression in PBC is simply an expression of inflammation.

IV. HSP 60 IN ALCOHOLIC HEPATITIS

Except for its induction by alcohol, the pathogenesis of alcoholic hepatitis is poorly understood. It seems likely that it is an immune-mediated liver disease, presumably due to an immune response against acetylated self-proteins (6). Genetic linkage (to HLA-B8 [7]) and autoantibodies support such a hypothesis (8). It is therefore of interest to study the expression of and response to stress proteins in alcoholic hepatitis. It could be shown that both the expression of hsp 60 (1) and of hsp 70 (9) is markedly increased in alcoholic hepatitis, and, like in the other liver diseases, this increased expression is most marked in hepatocytes, which also constitute the target cell of the inflammatory process.

Alcoholic hepatitis is associated with raised immunoglobulin A (IgA) levels, and IgA deposits can be found in the liver, especially in the perisinusoidal region. The King's College group has also looked at antibodies to hsp 60 and found raised IgA titers in comparison to normal controls and other liver diseases in patients with alcoholic liver disease (10). Interestingly, they not only found raised IgA anti–hsp 60 titers in alcoholic hepatitis, but they also found that these raised IgA anti–hsp 60 titers occurred only in alcoholic hepatitis, but not in alcoholics without liver damage (1,10). Nonetheless, some IgA anti–hsp 60 could also be found in liver disease controls, and the question remains to what extent the raised titers were simply an expression of raised overall IgA levels or due to a specific immune response. The results were unfortunately not related to serum immunoglobulin levels.

V. HSP 60 IN VIRAL HEPATITIS

Chronic viral hepatitis types B and C are very common diseases. The reasons why a viral infection can cause persistent mild to moderate organ-specific inflammation without inducing elimination of the virus are incompletely understood. Two of the reasons are viral escape due to mutations and impaired host immune response either due to neonatal infection inducing tolerance or due to genetic factors. It appears that the tissue damage in these conditions is mostly immune mediated rather than due to a directly cytotoxic effect of the virus. What role do stress proteins play in this chronic inflammatory condition?

The answer is not at all clear yet. From the study by Brommé et al. (2) and our work, it is clear that chronic viral hepatitis is also associated with a marked overexpression of hsp 60 in sites of active inflammation. Indeed, we could not find a discernible difference in hsp 60 expression between autoimmune and viral hepatitis (3). Both the degree of overexpression and its pattern seemed identical. It thus appears that hsp 60 expression is not related to the specific immunopathogenesis of either of these conditions.

Nonetheless, it is conceivable that the overexpression of this important self-antigen may be of immunoregulatory importance in the disease process. It is not unlikely that overexpression of hsp 60 and an hsp 60–specific immune response may be perpetuating the chronic hepatitis. On the other hand, response to the viral antigens may be somehow dampened by the presence of this intrinsically regulated self-antigen, which seems to be central in the immune systems regulatory network, and which Cohen named the immunological homunculus.

Increased hsp expression thus seems to be a feature of all types of inflammatory liver diseases, and is colocalized with the inflammatory infiltrates. The question thus arises, if increased expression on HSP60 in inflammatory liver diseases must be simply considered an epiphenomenon or does it have a role in pathogenesis? Future studies will in particular have to assess humoral and cellular immune responses to the stress proteins with various liver diseases, and to relate these findings to histological and clinical disease activity. It will be particularly important to find out if expression of hsp 60 and immunity to hsp 60 is associated with prognosis. Although initially it was thought that immune responses to heat shock proteins are pathological and at least partly responsible for the pathogenesis of autoimmune process, studies in experimental and clinical arthritis, as discussed elsewhere in this book, make it appear that it may be beneficial to respond to self-hsps. Two alternative hypotheses therefore need to be tested in immune-mediated liver disease: Is hsp expression and immunity to hsp associated with a good prognosis and thus a protective mechanism, or is it a marker of progressive disease, presumably by perpetuating the chronic inflammatory process in the liver?

REFERENCES

1. J. Koskinas, V.R. Winrow, G.L.A. Bird, et al., Hepatic 60-kD heat-shock protein responses in alcoholic hepatitis, *Hepatology 17*:1047 (1993).
2. U. Broomé, A. Scheynius, and R. Hultcrantz, Induced expression of heat-shock protein on biliary epithelium in patients with primary sclerosing cholangitis and primary biliary cirrhosis, *Hepatology 18*:298 (1993).
3. A.W. Lohse, H.P. Dienes, J. Herkel, E. Hermann, W. van Eden, and K.H. Meyer zum Büschenfelde, Expression of the 60 kDa heat shock protein in normal and inflamed liver, *J. Hepatol. 19*:159 (1993).
4. L. Vilagut, J. Vila, O. Vinas, et al., Mycobacterium gordonae in primary biliary cirrhosis: possible etiology for the disease, *J. Hepatol. 18*(Suppl. 1):S7 (1993).
5. A.D. Jorge, A.O. Stati, L.V. Roig, G. Ponce, O.A. Jorge, and R.D. Ciocca, Steroid receptors and heat shock proteins in patients with primary biliary cirrhosis, *Hepatology 18*:1108 (1993).
6. R.N.M. MacSween, R.S. Anthony, and M. Farquharson, Antibodies to alcohol-altered hepatocytes in patients with alcoholic liver disease, *Lancet 2*:803 (1981).

7. J.B. Saunders, A.D. Wodak, A. Haines, et al., Accelerated development of alcoholic cirrhosis with HLA B8, *Lancet 2:*1381 (1982).
8. C.A. Laskin, E. Vidins, L.M. Blendis, and C.A. Soloninka, Autoantibodies in alcoholic liver disease, *Am. J. Med. 89*:129 (1990).
9. R. Omar, M. Pappolla, and B. Saran, Immunocytochemical detection of the 70 kD heat shock protein in alcoholic liver disease , *Arch. Pathol. Lab. Med. 114*:589 (1990).
10. V. Winrow, G.L. Bird, J. Koskinas, D.R. Blake, R. Williams, and G.J.M. Alexander, Circulating IgA antibody againts a 65 kDa heat shock protein in acute alcoholic hepatitis, *J. Hepatol. 20*:359 (1993).

24

Helicobacter pylori–Associated Chronic Gastritis

Lars Engstrand
University Hospital, Uppsala, Sweden

I. INTRODUCTION

The recognition that *Helicobacter pylori* is a major cause of chronic inflammation of the human gastric antral mucosa has revolutionized our perception of gastroduodenal disease. Once acquired, the bacteria persists for decades in the mucous layer overlaying the gastric epithelium. A high priority is being placed on studies of this new bacterial pathogen that focus on understanding the mechanisms used by the organism to induce disease (1). The importance of microbial and immunological factors has drawn attention to studies of gut-associated immunity. The inflammatory events in *H. pylori*-associated gastritis are deep in the tissue removed from the area of colonization, indicating that extracellular products from the bacteria may diffuse into the mucosa (2). The continuous presence of inflammatory response and immunological damage could be a consequence of these products. The connection between autoimmunity, slow bacterial infections, and pathogenesis has been suggested in disease such as sarcoidosis, Crohn's disease, and rheumatoid arthritis (3), but the possibility of cross-reactive autoimmunity has not been clarified yet. Expression of stress proteins in *H. pylori*-diseased gastric mucosa and the presence of anti-*H. pylori* stress protein antibodies in infected patients suggest that a role of stress proteins in gastroduodenal diseases is possible.

II. THE HSP 60 PROTEIN OF *H. PYLORI*

A stress protein produced by *H. pylori* has been identified by several groups (4–6). Its molecular weight is reported to be within a range of 54–62 kDa, probably due to technical differences. This protein is present in all *H. pylori* strains, seems to be conserved in the species of *H. pylori*, and belongs to the heat shock protein (hsp) 60 family (6). The N-terminal sequence shows similarity with several members of the hsp 60 chaperonin family, including the human mitochondrial chaperonin (4,5). Western immunoblots following polyacrylamide gel electrophoresis in SDS demonstrate that a monoclonal antibody, ML30, prepared against the 65 kDa heat shock protein of *Mycobacterium lepræ*, recognizes the *H. pylori* hsp 60 (Fig. 1). Cloning and molecular characterization of a gene coding for the hsp 60 protein of *H. pylori* has recently been reported (6). The ultrastructure of *H. pylori* hsp 60 is composed of seven subunits and appears as a disk-shaped molecule with a diameter similar to that of *H. pylori* urease (7). Observed heterogenicity in earlier preparation of *H. pylori* urease was probably caused by contaminating hsp 60 particles; that is, copurification of these two macromolecules (8). The exact function of this urease-associated stress protein is not known. We have speculated that *H. pylori*

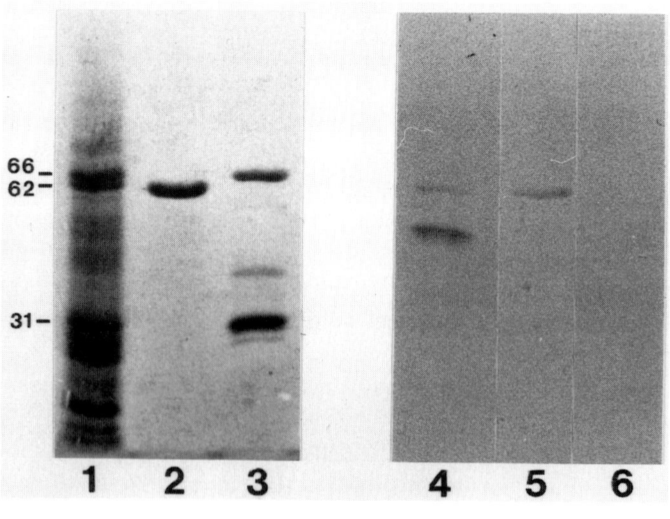

Figure 1 SDS-PAGE and Western blot of (lanes 1 and 4) the high molecular weight cell-associated proteins of *H. pylori* (HM-CAP), (lanes 2 and 5) purified HSP62 and (3 and 6) purified urease. Reactions in lanes 4–6 were with a monoclonal antibody, ML30, diluted 1:50.

hsp 60 functions in the transmembrane export of the urease subunit proteins or stabilizes the urease complex in the face of gastric acidity (4).

III. ANTI-HSP 60 ANTIBODIES IN *H. PYLORI*-INFECTED PATIENTS

The antigenicity of *H. pylori* hsp 60 has been investigated by means of Western blot using sera from patients with *H. pylori*-associated gastritis (6,9). The immunogenic urease enzyme exposed on the surface of *H. pylori* appears to be specific for this organism with little antigenic cross-reactivity to other bacteria present in the gut (10). The high molecular weight cell-associated proteins of *H. pylori* (HM-CAP) contain both urease and *H. pylori* hsp 60 (see Fig. 1) and is reported to be highly effective in diagnostic ELISAs (11). Serum immunoglobulin G (IgG) antibody responses detectable with HM-CAP demonstrate that both *H. pylori* urease and hsp 60 are immunogenic and give a strong response (9). This response decreases after successful treatment with no predominance for either antigen. Thus, the hypothesis that the serum immune response should be less dependent on the continued presence of *H. pylori* (due to a high degree of homology with other bacterial hsp 60 homologues) could not be confirmed in our study (9). However, cross-reacting anti-*H. pylori* hsp 60 antibodies in an ongoing infection may be involved in tissue damage (see below).

IV. EXPRESSION OF *H. PYLORI* HSP 60 IN GASTRIC EPITHELIAL CELLS

We have previously shown that a monoclonal antibody (MAb) raised against *Mycobacteria leprae* (ML30) reacts with *H. pylori* hsp 60 and with epithelial cells at the site of *H. pylori* infection (9,12). Frozen sections were stained by the peroxidase-antiperoxidase method (13). The MAb ML30 (kindly provided by Prof. J. Ivanyi, Tuberculosis & Related Infections Unit, Hammersmith Hospital, London, England) shows a widespread staining of human tissues (14). The binding of the MAb was visualized as described earlier (15). We found that the staining was present in epithelial cells in two different patterns: (1) intense staining in the area of Golgi apparatus, and (2) widespread staining in the cytoplasm (Fig. 2). *H. pylori* located above the epithelium were also stained by ML30 (Fig. 3). Induced expression of major histocompatibility complex (MHC) class II antigens was also observed in the gastric epithelium. Neither HLA-DR nor *H. pylori* hsp 60 expression were observed in *H. pylori*-negative biopsies. Long-term follow-up of patients after successful treatment showed the continuing presence of intraepithelial *H. pylori* hsp 60 and HLA-DR, which could reflect activation of the local immune response even though the bacteria are eradicated

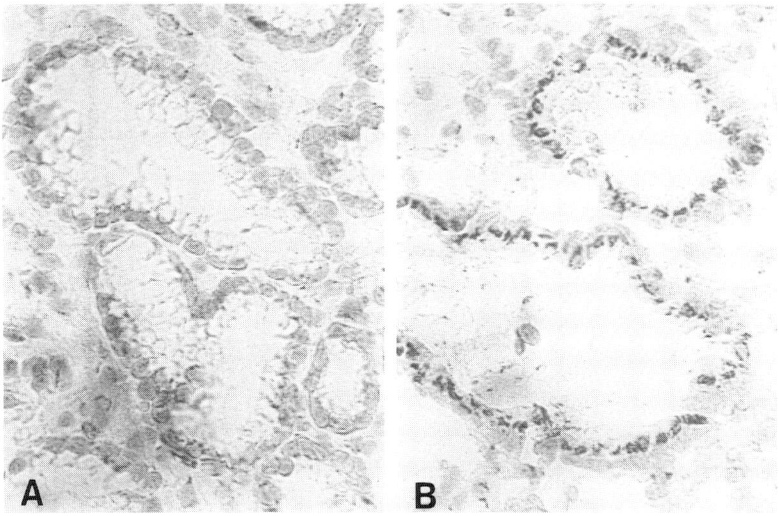

Figure 2 Immunoperoxidase staining of frozen sections of gastric biopsy specimens from (A) a patient with normal gastric mucosa and (B) a patient with *H. pylori*–associated chronic gastritis of the antrum. Note the presence of *H. pylori* hsp 60 in the epithelial cells (B). The monoclonal antibody, ML30, was used as primary antibody. The sections are counterstained with Mayer hematoxylin.

from the gastric mucosa (16). The patients with a normal histopathology after treatment (and no visible *H. pylori*) also stained negative for HLA-DR and *H. pylori* hsp 60.

We do not know if the epithelial cells express autologous hsp 60 because of stress imposed by the bacteria, or if *H. pylori* itself expresses hsp 60 inside the epithelial cells owing to stress imposed by the host. Evidence is now accumulating that *H. pylori* invades gastric epithelial cells and survives intracellularly (17,18). Thus, *H. pylori* hsp 60 may be implicated as a survival factor, similar to stress proteins of the hsp 60 family in Salmonellal (19), chlamydial (20), and mycobacterial (21) infections.

The presence of intraepithelial *H. pylori* hsp 60 after successful treatment indicates that continuous infection by the organism may occur in patients who are eradicated judging from ordinary standard methods; that is, culture, breath test, and histology. Coexpression of *H. pylori* hsp 60 and HLA-DR by the epithelial cells indicates an ongoing activation of the local immune response in this situation. Thus, epithelial cell expression of the antigen may act as a triggering factor in the ongoing chronic inflammation. Furthermore, treatment failure may be explained by intraepithelial survival in spite of the fact that extracellular killing of the organism is efficient (22). Our findings support the hypoth-

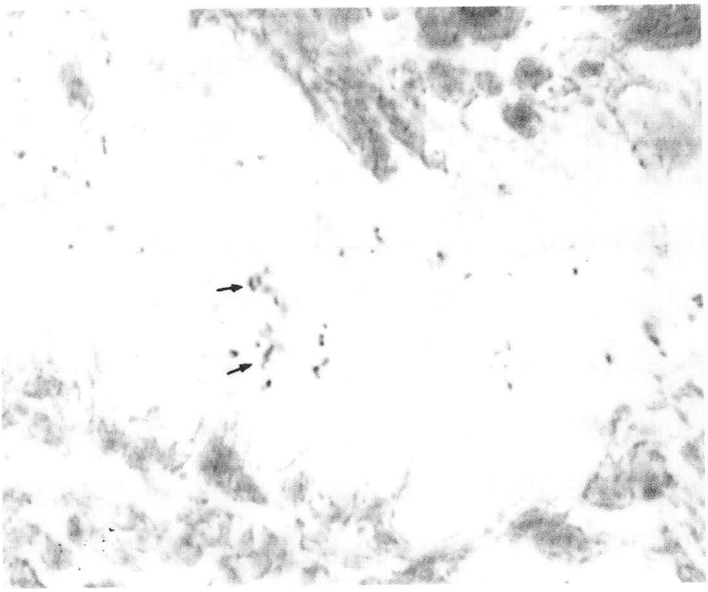

Figure 3 Immunoperoxidase-stained frozen section from a patient with *H. pylori*-associated chronic gastritis. *H. pylori* are shown stained with MAb ML30 (arrows).

esis that *H. pylori* has a capacity of intracellular survival and is by this protected from extracellular antibacterial treatment.

V. ROLE OF *H. PYLORI* HSP 60 IN THE PATHOGENESIS OF GASTRITIS

Given the high conservation between *H. pylori* hsp 60 and its human homologue, it might be possible that humoral anti–hsp 60 antibodies act as autoantibodies and/or that complexes of these antibodies with autoantigen could play a role in the tissue damage that occurs in *H. pylori*-associated gastritis. It has been shown that *H. pylori* infection induces antibodies cross reacting with human gastric mucosa (23). This induction of autoantibodies could be explained by the presence of anti *H. pylori* hsp 60 antibodies. However, why do high levels of antibodies against *H. pylori* hsp 60 appear when this antigen should be well tolerated by the host immune system owing to the conserved status? This could be explained in two different ways as suggested by Macchia et al. (6): (1) specific epitopes in *H. pylori* hsp 60 are the targets for the immune response, or (2) failure in the host tolerance is due to antibodies reacting with epitopes that are common to *H. pylori* hsp 60 and its human homologue. Future research will hopefully give us more information about the immunogenicity of *H. pylori* hsp 60 and if cross-reacting autoantibodies play a role in the pathogenesis of gastritis.

The continuous expression of *H. pylori* hsp 60 may act as a triggering factor for activation of T cells, especially intraepithelial T cells which are increased in *H. pylori*-associated gastritis. We lack information about the significance of these observations, but it could be that such T cells play a role in the local host defense against *H. pylori*. The possibility of autoreactivity through mimicry may explain the immunopathological effects of *H. pylori*. *H. pylori* hsp 60 is 75% homologous to the human homologue (4,5), and shared antigenic epitopes from bacteria and tissue could fit the concept of molecular mimicry and autoimmune pathology (3,24). We have previously demonstrated an increased number of γ/δ T cells and CD8-reacting T lymphocytes (suppressor-cytotoxic T cells), located mainly within the epithelium, in patients with *H. pylori*-associated chronic gastritis of the antrum (12,15). Could it be that cytotoxic T cells in the epithelium, triggered by immunogenic peptides (i.e., hsp 60 from stressed *H. pylori* and/or autologous hsp 60) present in the gastric epithelial cells, initiate a sequence of autoreactive events due to cross reactivity through "antigenic mimicry"? *H. pylori* hsp 60 may be the target for intraepithelial T cells in a "first-line" defense against *H. pylori* and thereafter act as a maintenance factor in the chronic inflammation that will persist for years. The role of *H. pylori* hsp 60 as a triggering factor for T-cell activation in the gastric mucosa calls for further investigation.

Another hypothesis that has been raised by Blaser is that induction of inflammation is adaptive (2): *H. pylori* release extracellular products which diffuse into the mucosa and activate inflammatory cells. The inflammation by itself causes tissue damage that could result in the release of nutrients into the mucous layer that should be beneficial to *H. pylori*. In this case, the chronic antigenic stimulation leads to downregulation of the inflammation by T cells in order to minimize the clinical outcome of the infection. Thus, insufficient downregulation leading to uncontrolled inflammation and in the long-term atrophic gastritis would not be beneficial for the bacteria, as nutrient supply from the host will diminish. A key molecule in the interaction between the host and the bacteria could be *H. pylori* hsp 60.

The diversity in clinical and pathological outcome of *H. pylori* infection are probably explained by both bacterial and host factors. Differences in host response is ripe for exploration, and studies of genetic differences could help to explain familiar clustering of gastroduodenal ulcer and gastric cancer. We have recently found that genetic factors are important in *H. pylori*-associated chronic gastritis (25).

Future studies on T-cell response, cross-reacting antibodies, and genetic factors will hopefully be the subject of a great deal of interest and help us to understand the role of *H. pylori* hsp 60 in the pathogenesis of *H. pylori*.

ACKNOWLEDGMENTS

The author wishes to thank the Swedish Medical Research Council (projects 10617 and 10848), the Swedish Society of Medicine, the Professor Nanna Svartz Foundation, the Tore Nilssons Foundation, and the Åke Wibergs Foundation for grant support.

REFERENCES

1. A. Lee, J. Fox, and S. Hazell. Pathogenicity of *Helicobacter pylori*: a perspective, *Infect. Immun. 61*:1601 (1993).
2. M.J. Blaser, Helicobacter pylori: microbiology of a "slow" bacterial infection, *Trends Microbiol. 1*:255 (1993).
3. G.A.W. Rook, and J.L. Stanford, Slow bacterial infections or autoimmunity? *Immunol. Today 13*:160 (1992).
4. D.J. Evans, Jr., D.G. Evans, L. Engstrand, and D.Y. Graham, Urease-associated heat shock protein of *Helicobacter pylori*, *Infect. Immun. 60*:2125 (1992).
5. B.E. Dunn, R.M. Roop, II, C. Sung, S.A. Sharma, G.I. Perez-Perez, and M.J. Blaser, Identification and purification of a cpn60 heat shock protein homolog from *Helicobacter pylori*. *Infect. Immun. 60*:1946 (1992).
6. G. Macchia, A. Massone, D. Burroni, C. Covacci, S. Censini, and R. Rappuoli, The Hsp60 protein of *Helicobacter pylori*: structure and immune response in patients with gastroduodenal diseases, *Mol. Microbiol. 9*:645 (1993).
7. J.W. Austin, P. Doig, M. Stewart, and T. Trust, Structural comparison of urease and GroEL analog from *Helicobacter pylori*, *J. Bacteriol. 174*:7470 (1992).
8. J.W. Austin, P. Doig, M. Stewart, and T. Trust, Macromolecular structure and aggregation states of *Helicobacter pylori* urease, *J. Bacteriol. 173*:5663 (1991).
9. L. Engstrand, D.J. Evans, Jr., D.G. Evans, F.A.K. El-Zaatari, and D.Y. Graham, Serum IgG antibodies to the 62-KDa heat shock protein of Helicobacter pylori: a pathogenic role for stress proteins in patients with gastritis?, *Immunol. Infect. Dis. 3*:227 (1993).
10. D.G. Newel, A. Lee, P.R. Hawtin, M.J. Hudson, A.R. Stacey, and J. Fox, Antigenic conservation of the ureases of spiral- and helical-shaped bacteria colonising the stomachs of man and animals, *E.E.M.S. Microbiol. Lett. 65*:183 (1989).
11. D.J. Evans, Jr., D.G. Evans, D.Y. Graham, and P.D. Klein, A sensitive and specific serologic test for detection of *Campylobacter pylori* infection, *Gastroenterology 96*:1004 (1989).
12. L. Engstrand, A. Scheynius, and C. Påhlson, An increased number of γ/δ T cells and gastric epithelial cell expression of the groEL stress-protein homologue in *Helicobacter pylori*-associated chronic gastritis of the antrum, *Am. J. Gastroenterol. 86*:976 (1991).
13. L.A. Sternberger, *Immunocytochemistry*. John Wiley & Sons, New York, 1979.
14. D.J. Evans, P. Norton, and J. Ivanyi, Distribution of tissue sections of the human groEL stress protein homologue, *Acta Pathol. Microbiol. Immunol. Scand. 98*:437 (1990).

15. L. Engstrand, A. Scheynius, C. Påhlson, L. Grimelius, A. Schwan, and S. Gustavsson, Association of *Campylobacter pylori* with induced expression of class II transplantation antigens on gastric epithelial cells, *Infect. Immun. 57*:827 (1989).
16. L. Engstrand, A. Genta, A. Scheynius, M.F. Go, and D.Y. Graham, Presence of intraepithelial H. pylori heat shock protein (Hsp62) after successful treatment—a reflection of continuing chronic inflammation? Abstract. The VIIth Meeting on Gastroduodenal Pathology and Helicobacter pylori, Houston, TX, 1994.
17. D.G. Evans, D.J. Evans, Jr., and D.Y. Graham, Adherence and internalization of *Helicobacter pylori* by HEp-2 cells, *Gastroenterology 102*:1557 (1992).
18. G. Bode, P. Malfertheiner, and H. Ditschuneit, Pathogenic implications of ultrastructural findings in Campylobacter pylori related gastroduodenal disease, *Scand. J. Gastroenterol. 23*:(Suppl. 142):25 (1988).
19. N.A. Buchmeier and F. Heffron, Induction of Salmonella stress proteins upon infection of macrophages, *Science 248*:730 (1990).
20. M.C. Cerrone, J.J. Ma, and R.S. Stephens, Cloning and sequence of the gene for heat shock protein 60 from *Chlamydia trachomatis* and immunological reactivity of the protein, *Infect. Immun. 59*:79 (1991).
21. D. Young, R. Lathigra, D. Hendrix, D. Sweetser, and R.A. Young, Stress proteins are immune targets in leprosy and tuberculosis, *Proc. Natl. Acad. Sci. USA 85*:4267 (1988).
22. K. Hultén, O. Cars, E. Hjelm, and L. Engstrand, In vitro activity of azithromycin against intracellular *Helicobacter pylori*. *J. Antimicrob. Chemother.* 1995 (in press).
23. R. Negrini, L. Lisato, I. Zanella, L. Cavazzini, S. Gullini, V. Villanacci, C. Poiesi, A. Albertini, and S. Ghielmi, Helicobacter pylori infection induces antibodies cross-reacting with human gastric mucosa, *Gastroenterology 101*:437 (1991).
24. S.H.E. Kaufmann, Heat shock proteins and the immune response, *Immunol. Today 11*:129 (1990).
25. H. Malaty, L. Engstrand, N. Pedersen, and D.Y. Graham, Helicobacter pylori infection: genetic and environmental influences—a twin study, *Ann. Intern. Med. 120*:982 (1994).

25

Jejunal Epithelial Cell Stress Protein Expression in Gluten-Induced Enteropathy (Celiac Disease)

Markku Mäki
University of Tampere, Tampere, Finland

Immo Rantala
University Hospital of Tampere, Tampere, Finland

I. INTRODUCTION

Stress proteins, or heat shock proteins (hsps) are phylogenetically conserved proteins synthesized during stress and thus providing the cells with a protective mechanism against environmental insult (1). Stress proteins share sequence homology with a wide range of autoantigens (2,3). Especially in rheumatoid arthritis, stress proteins have been thought to function as autoantigens (4,5) and to be the target antigen, for example, for γ/δ T-cell–receptor-bearing lymphocytes (6,7).

Gluten-induced enteropathy, or celiac disease, is a permanent condition in which the ingested cereal proteins, especially wheat gluten or gliadin, trigger an immunologically mediated small intestinal enteropathy (villous atrophy with crypt hyperplasia) in genetically susceptible individuals carrying the human leukocyte antigen—DQ2 α/β heterodimer encoded by the gene alleles DQA1*0501 and DQB1*0201 (8–10). Certain tissue autoantibodies, called reticulin or endomysium antibodies, typically occur in celiac disease (11). The disease is self-perpetuating if the causative agent, the trigger, is continuously present. On a gluten-free diet, the lesion heals because of the renewal and proliferative capacity of the intestinal mucosa. Gluten challenge again will lead to mucosal deterioration (12). In fact, autoimmune mechanisms have been hypothesized to be operative in generating the jejunal damage in gluten-sensitive enteropathy (13).

The human GroEL stress protein homologue, 65-kDa heat shock protein (hsp 65), has been suggested to serve as the prime target in autoimmune diseases. The monoclonal antibody (MAb) ML30, which is made against mycobacterial hsp 65 (14), reacts with human hsps, and stains many human tissues, including the epithelial cells in the small intestine (15). In celiac disease, the density of jejunal intraepithelial γ/δ T cells is highly increased, and this finding is also disease specific (16). As hsp 65 and γ/δ T cells are often linked together when discussing infection and autoimmunity, the question arises whether the expression of hsp 65 is enhanced in the jejunal celiac lesion.

II. PATIENTS AND METHODS

The study group consisted of 77 consecutive children referred to the outpatient pediatric clinic at the University Hospital of Tampere, Finland, for symptoms and signs suggesting celiac disease. They all underwent jejunal biopsy using either a pediatric- or adult-size Watson suction capsule. Part of the biopsy specimen was freshly frozen in Tissue-Tek (Miles Inc., Elkhart, IN) embedding mounting medium and stored at −70°C for later use. Celiac disease diagnosis was confirmed in 20 children both histologically and on later follow-up according to the acknowledged criteria of the European Society for Pediatric Gastroenterology and Nutrition (12,17). Children with normal villous architecture (n = 57) were considered not to suffer from celiac disease. The median age of the children was 6.5 years, with a range of 0.8–15.8 years. The median age of those 20 children found to suffer from celiac disease was 8.5 years, with a range of 1.3–15.8 years. Further, frozen biopsies were available from two children with celiac disease when on a long-term gluten-free diet and from one of them at the time of mucosal relapse after subsequent gluten challenge.

The expression of hsp 65 was recognized in 5-μm thick cryostat tissue sections by the MAb ML30. This MAb (kindly provided by Prof. J. Ivanyi, Tuberculosis and Related Infections Unit, Hammersmith Hospital, London, UK) recognizes hsps of mycobacteria and the GroEL homologue in different human tissues (15).

The frozen jejunal biopsy sections were freshly cut, acetone fixed, and primary antibodies were revealed using the Vectastain Elite ABC kit (Vector Laboratories, Burlingame, CA) according to the manufacturer's instructions. The sections were incubated with ML30 (1:4000) overnight and endogenous peroxidase was blocked using 0.3% H_2O_2 in methanol for 30 min. Peroxidase was developed using diaminobenzidine (DAB, BDH Laboratory Supplies, Poole,

UK) giving the brown staining. Nuclei were counterstained with Harris' hematoxylin.

Epithelial cell apical areas were blind tested separately in the villi and the crypts for expression of hsp 65, and the general impression of the whole biopsy specimen intensity was recorded and graded ± (faint), + (moderate), or ++/+++ (strong).

III. RESULTS

The hsp 65 expression in the jejunal mucosa was confined to epithelial cells, and no significant staining was observed in the the basement membrane area or in the lamina propria. Within epithelial cells, the hsp 65 expression was detected in the apical parts above the nuclei. However, in biopsy specimens with faint or strong expression, the staining was not always uniform. There were also areas with no visible staining. Only faint staining spots were sometimes observed in the biopsy specimens, but more often the expression was moderate to strong. Figure 1 (top) (see color plate) shows moderate crypt epithelial staining in a child with normal villous architecture on routine histology in contrast to Figure 1 (bottom), where the expression of hsp 65 is extremely strong and fills the whole apical parts of the epithelial cells (total villous atrophy with crypt hyperplasia on routine histology). In the two adolescents with celiac disease but on a long-term strict gluten-free diet, only tiny spots were seen in one patient, and moderate expression in the other (Fig. 2, top) (see color plate). In the latter case, a second biopsy specimen after gluten challenge and with histological mucosal relapse, an enhanced hsp 65 expression was observed with grading +++ (Fig. 2, bottom).

When the children were divided according to their final diagnoses, clearly a strong staining pattern was observed in patients suffering from gluten-induced enteropathy and celiac disease (Table 1). However, there were also cases with strong expression of epithelial hsp 65 among individuals ingesting normal amounts of gluten and having symptoms or signs suggesting celiac disease but with normal villous architecture; thus being excluded for the diagnosis of celiac disease.

IV. DISCUSSION

The present study shows that ingested gluten induces enhanced upregulation of jejunal epithelial hsp 65 in individuals shown to have celiac disease. In the present study, we used the MAb ML30 on freshly cut frozen tissue specimens, as the staining in our formalin-fixed paraffin-embedded sections was not successful. The staining was not for some reasons always uniform throughout the

Table 1 Hsp 65 Expression in Jejunal Epithelial Cells in Children Biopsied for Suspected Celiac Disease

Hsp 65 expression	Celiac disease (N)	Excluded for celiac disease (N)
Faint/moderate	4	43
Strong	16	14

The children were divided into two groups: confirmed celiac disease and excluded for celiac disease (normal jejunal mucosal morphology). χ^2 16.87, $P < 0.001$.

biopsy specimen, but we observed that a blind-tested subjective arbitrary grading worked where the intensity and amount of hsp 65 in individual epithelial cell mitochondrial areas was assessed. We also want to stress that the age (median age 8.5 years) of the children with celiac disease in this study is not the classic one (malabsorption syndrome in young infancy).

It is intriguing to speculate that the gluten-induced enhanced expression of epithelial cell hsp 65 in genetically susceptible individuals is of pathogenetic importance. Hsps play a major role under stress conditions by protecting the host against environmental insult. Gluten could be such an insult for certain genetically susceptible individuals. The relation of stress proteins to γ/δ T-cell recognition may be of relevance in the pathogenesis even if the role of intraepithelial γ/δ T cells in the jejunal mucosa of patients with celiac disease is not well understood. Hsp 65 could be the target antigen for autoreactive γ/δ T cells. However, one can speculate that γ/δ T cells are not effector cells, as they are abundantly present intraepithelially and also during the gluten-free diet period when the mucosa has healed (16). On the other hand, T-cell release from an anergic state could result in a destructive autoimmune response in the intestinal epithelium (18), and in celiac disease, these events could be locally triggered by gluten.

In the present series, we noticed in many patients with a normal jejunal mucosal morphology a strong expression of epithelial cell cytoplasmic hsp 65. These children were suspected to have a gluten-induced enteropathy and therefore biopsied. Gluten might indeed be an environmental insult even for these patients. They may belong to the entity known as "latent celiac disease," in which the mucosa later deteriorates to that typical of celiac disease (19,20). Also, increased density of intraepithelial γ/δ T cells has been demonstrated in the normal jejunal mucosa at the latent stage of the disease before the onset of overt celiac disease and mucosal deterioration (21). Gluten could also cause epithelial cell stress in first-degree relatives of patients with celiac disease. Thirty percent of the healthy family members with normal small bowel villous architecture have an increased density of γ/δ T cells in their mucosa (22). Another gluten-induced disease, in which hsp 65 could be overexpressed in epithelial cells in normal or flat mucosa, is dermatitis herpetiformis (23). This gluten-induced skin disease may develop in genetically susceptible subjects without enteropathy but with a high density of γ/δ T cells in normal mucosa (24).

V. CONCLUSIONS

Ingested prolamins and wheat gluten could be an environmental insult in genetically susceptible individuals leading to enhanced expression of jejunal epithelial cell stress proteins. Immunohistochemical staining of frozen jejunal biopsy specimens of patients with celiac disease (median age 8.5 years) showed overexpression of the human GroEL stress protein homologue, 65-kDa hsp 65, using the monoclonal antibody ML30. Strong staining of hsp 65 was also seen in many children suspected but excluded for celiac disease. As intraepithelial γ/δ T cells also are abundantly present in the celiac disease lesion, it is intriguing to speculate hsp 65 to be involved in the pathogenesis and to be the target antigens for autoreactive T cells.

ACKNOWLEDGMENTS

The Celiac Disease Study Project is supported by the Medical Research Council, Academy of Finland, the Sigrid Jusélius Foundation, and the Research Fund of the University Hospital of Tampere.

REFERENCES

1. S.H.E. Kaufmann, Heat shock proteins in health and disease, *Int. J. Clin. Lab. Res. 21*:221 (1992).
2. D.B. Young, Heat-shock proteins—immunity and autoimmunity, *Curr. Opin. Immunol. 4*:396 (1992).
3. D.B. Jones, A.F.W. Coulson, and G.W. Duff, Sequence homologies between hsp60 and autoantigens, *Immunol. Today 14*:115 (1993).
4. E.R. De Graeff-Meeder, R. van der Zee, G.T. Rijkers, et al., Recognition of human 60 kD heat shock protein by mononuclear cells from patients with juvenile chronic arthritis, *Lancet 337*:1368 (1991).
5. A.J. Quayle, K.B. Wilson, S.G. Li, et al., Peptide recognition, T-cell receptor usage and HLA restriction elements of human heat-shock protein (hsp)-60 and

mycobacterial 65-kDa hsp-reactive T-cell clones from rheumatoid synovial fluid, *Eur. J. Immunol. 22*:1315 (1992).
6. W. Born, L. Hall, A. Dallas, et al., Recognition of a peptide antigen by heat shock-reactive gamma/delta T lymphocytes, *Science 249*:67 (1990).
7. A. Haregewoin, B. Singh, R.S. Gupta, and R.W. Finberg, A mycobacterial heat-shock protein-responsive gamma/delta T cell clone also responds to the homologous human heat-shock protein: a possible link between infection and autoimmunity, *J. Infect. Dis. 163*:156 (1991).
8. J.S. Trier, Celiac disease, *N. Engl. J. Med. 325*:1709 (1991).
9. M.N. Marsh, Gluten, major histocompatibility compelx, and the small intestine. A molecular and immunobiologic approach to the spectrum of gluten sensitivity ('celiac sprue'), *Gastroenterology 102*:330 (1992).
10. L.M. Sollid and E. Thorsby, HLA susceptibility genes in celiac disease: genetic mapping and role in pathogenesis, *Gastroenterology 105*:910 (1993).
11. M. Mäki, K. Holm, V. Lipsanen, et al., Serological markers and HLA genes among healthy first-degree relatives of patients with coeliac disease, *Lancet 338*:1350 (1991).
12. M. Mäki, M.L. Lähdeaho, O. Hällström, M. Viander, and J.K. Visakorpi, Post-pubertal gluten challenge in coeliac disease, *Arch. Dis. Child. 64*:1604 (1989).
13. M. Mäki, O. Hällström, and A. Marttinen, Reaction of human non-collagenous polypeptides with coeliac disease autoantibodies, *Lancet 338*:724 (1991).
14. J. Ivanyi, S. Sinha, R. Aston, D. Cussell, M. Keen, and K. Sengupta, Definition of species specific and cross-reactive antigenic determinants of Mycobacterium leprae using monoclonal antibodies, *Clin. Exp. Immunol. 52*:528 (1983).
15. D.J. Evans, P. Norton, and J. Ivanyi, Distribution in tissue sections of the human groEL stress-protein homologue, *A.P.M.I.S. 98*:437 (1990).
16. E. Savilahti, A. Arato, and M. Verkasalo, Intestinal gamma/delta bearing T lymphocytes in celiac disease and inflammatory bowel disease in children. Constant increase in celiac disease, *Pediatr. Res. 28*:579 (1990).
17. J.A. Walker-Smith, S. Guandalini, J. Schmitz, D.H. Schmerling, and J.K. Visakorpi, Revised criteria for diagnosis of coeliac disease, *Arch. Dis. Child. 65*:909 (1990).
18. Nagler-Anderson C, McNair LA, Cradock A. Self-reactive, T cell receptor-gamma/delta-positive, lymphocytes from the intestinal epithelium of weanling mice, *J. Immunol. 149*:2315 (1992).
19. M. Mäki, K. Holm, S. Koskimies, et al., Normal small bowel biopsy followed by coeliac disease, *Arch. Dis. Child. 65*:1137 (1990).
20. A. Ferguson, E. Arranz, and O. O'Mahoney, Clinical and pathological spectrum of coeliac disease—active, silent, latent, potential, *Gut 34*:150 (1993).
21. M. Mäki, K. Holm, P. Collin, and E. Savilahti, Increase in gamma/delta T cell receptor bearing lymphocytes in normal small bowel mucosa in latent coeliac disease, *Gut 32*:1412 (1991).
22. K. Holm, M. Mäki, E. Savilahti, V. Lipsanen, P. Laippala, and S. Koskimies, Intraepithelial gamma/delta T-cell-receptor lymphocytes and genetic susceptibility to coeliac disease, *Lancet 339*:1500 (1992).

23. T. Reunala and M. Mäki, Dermatitis herpetiformis: a genetic disease, *Eur. J. Dermatol. 3*:519 (1993).
24. E. Savilahti, T. Reunala, and M. Mäki, Increase of lymphocytes bearing the gamma/delta T cell receptor in the jejunum of patients with dermatitis herpetiformis, *Gut 33*:206 (1992).

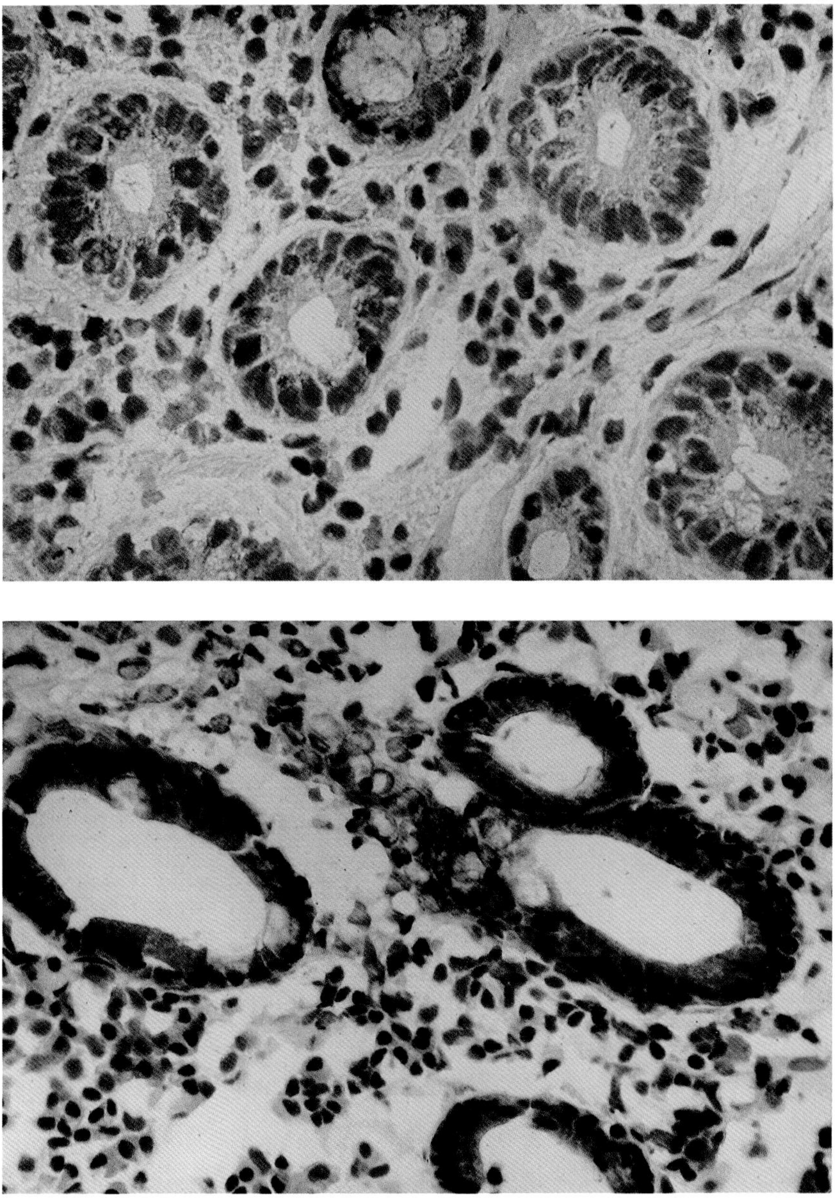

Figure 25.1 (Top) Moderate expression of human groEL stress protein homologue, 65-kDa heat shock protein (hsp 65) expression in jejunal epithelial crypts in a 9-year-old boy suspected to have celiac disease but with normal jejunal mucosal morphology. Brown staining of hsp 65 is seen in the cytoplasm apical to the epithelial cell nuclei. (Bottom) Strong staining of hsp 65 in epithelial cell cytoplasm in flat jejunal mucosa of a 12-year-old girl with untreated celiac disease.

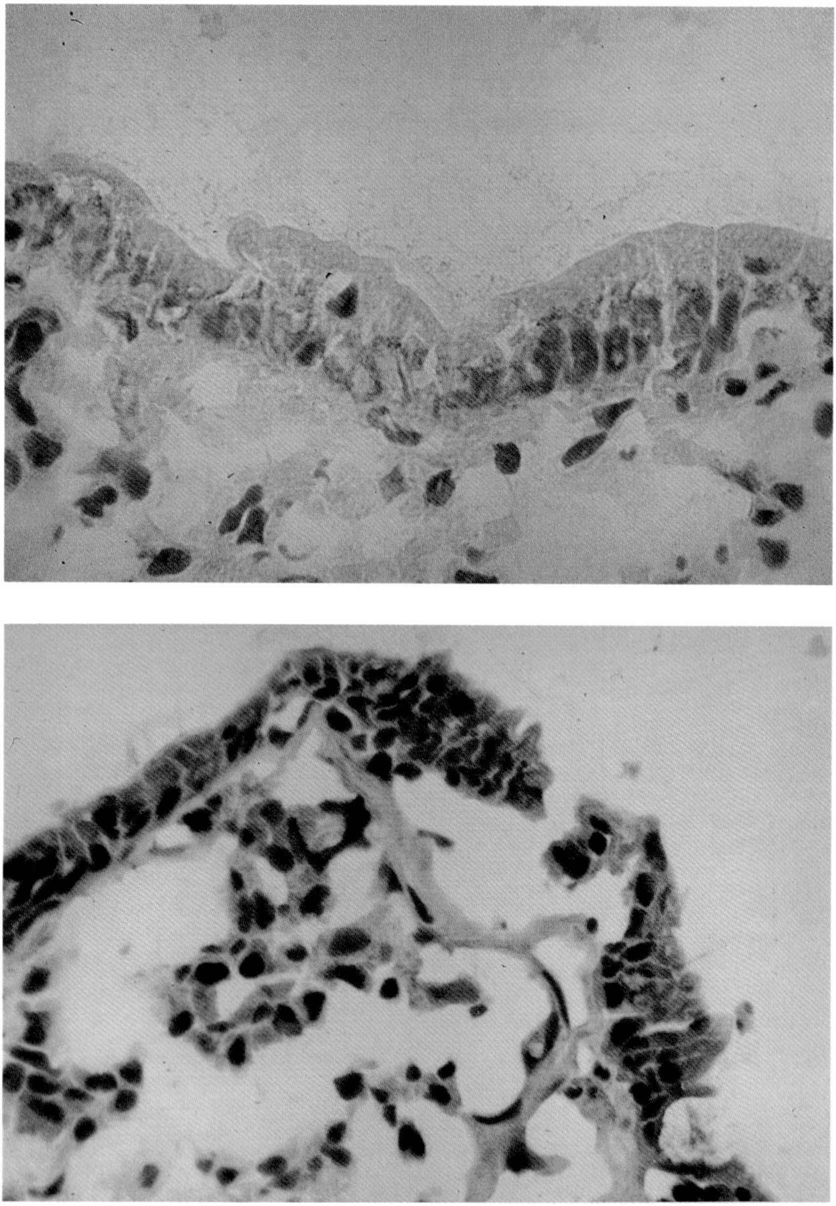

Figure 25.2 (Top) Moderate hsp 65 expression in normal jejunal mucosa in an adolescent with celiac disease but on gluten-free diet (brown apical staining of epithelial cell cytoplasm). (Bottom) Strong hsp 65 expression in epithelial cell cytoplasms in the same adolescent after gluten challenge. The jejunal mucosa had deteriorated to that typical for celiac disease.

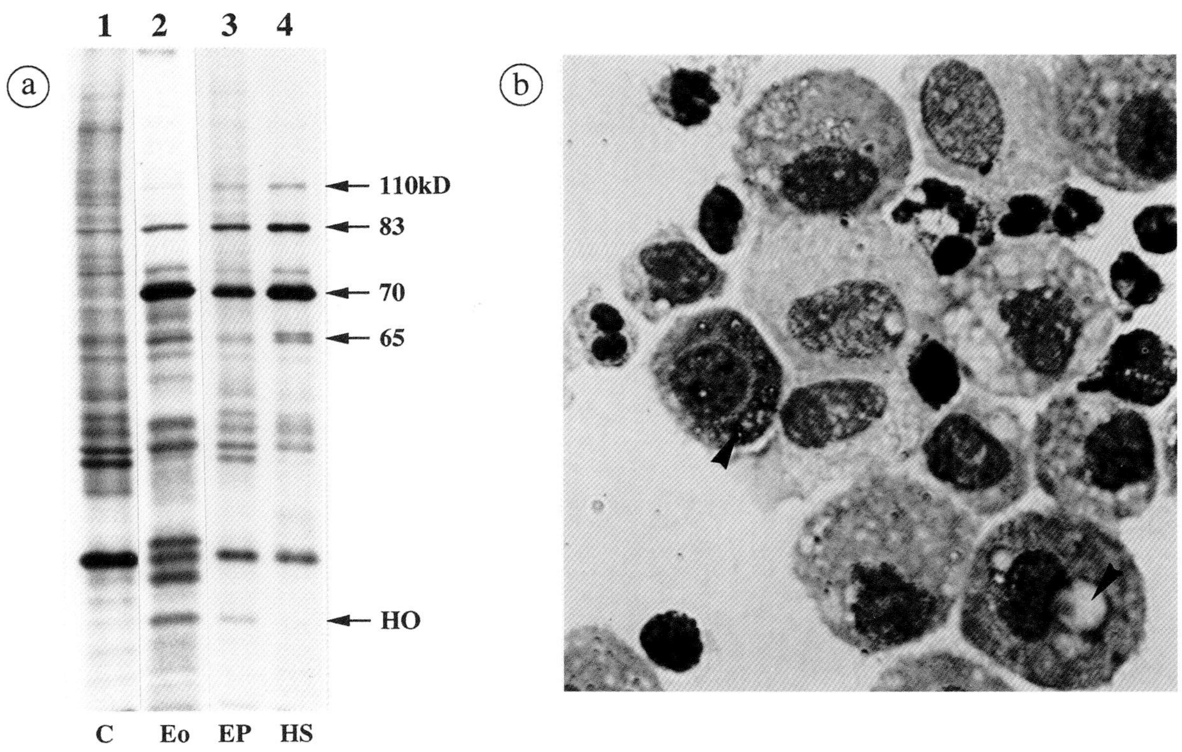

Figure 30.1 Alveolar eosinophilia and phagocytosis of eosinophils by AMs lead to stress protein synthesis by the latter cells. a, Protein synthesis by AMs. Autoradiographs of SDS-PAGE (10% polyacrylamide) of AM lysates after labeling of the cells with [^{35}S]- methionine (6 μCi/ml for 90 min) under control conditions (lane 1), after heat shock (44°C, 20 min) (lane 4), erythrophagocytosis (lane 3), and ex vivo labeling of AM from a patient with alveolar eosinophilia (lane 2). Arrows point to the major hsp. HO is induced during phagocytosis of eosinophils along with the classical hsp. b, Phagocytosis by the AM of eosinophilic material or whole eosinophils (arrows). (Adapted from Ref. 5.)

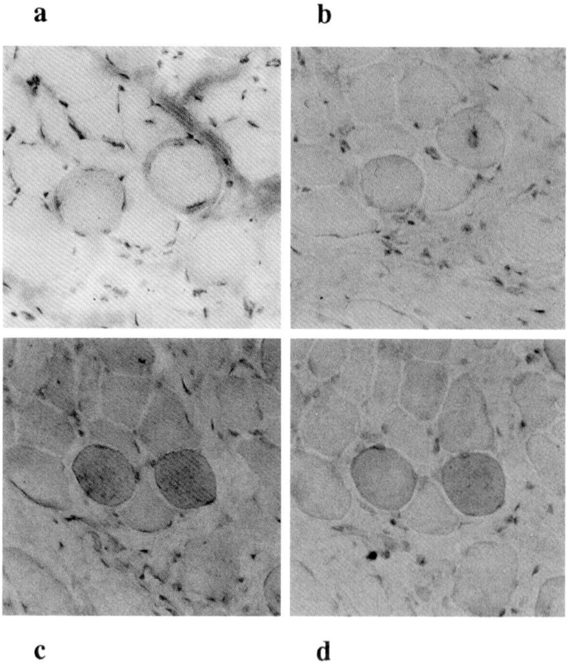

Figure 31.1 Hsp 70 and ubiquitin expression in muscle of patients with DMD. a, Hematoxylin-eosin; b, control (IgGl); c, hsp 70; d, ubiquitin. (For detailed methods, see ref. 18.) (Adapted from Ref. 18.).

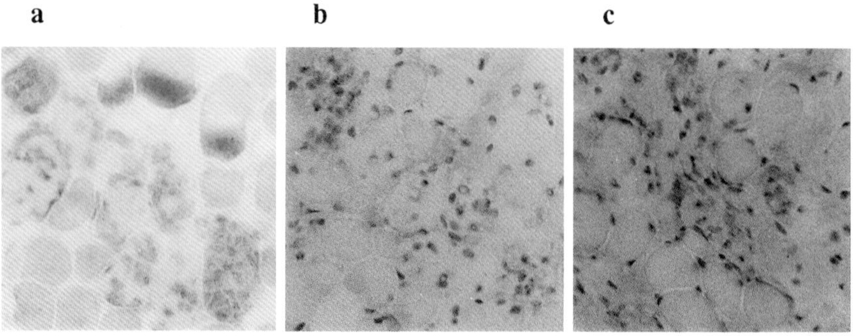

Figure 31.2 Expression of hsp 90 in necrotic fibers of patients with DMD. a, Acid phosphatase; b, control (IgGl); c, hsp 90. (For detailed methods, see Ref. 18.) (Adapted from Ref. 18.)

26

Identification of Endogenous Heat Shock Protein 60 as an Autoantigen in Autoimmune NOD Mouse Diabetes

Katrina Brudzynski
Robarts Research Institute, University of Western Ontario, London, Ontario, Canada

I. INTRODUCTION

Research into heat shock proteins (hsps) as cell molecular chaperons has taken on a new intensity and rapidly expanded into the field of immunology. The recent event which attracted the attention of immunologists was the finding that hsp 60 molecular chaperons are potent immunogens implicated in many inflammatory and autoimmune diseases (1–5). The observation that anti–hsp 60 immune responses are a common denominator in such diverse diseases as insulin-dependent diabetes (6–8), rat adjuvant arthritis, and rheumatoid arthritis (9–13) has taken many in the immunological field by surprise. After initial progress in the recognition of pathological conditions in which hsp 60 immunity has occurred, research is now focusing on elucidating the molecular mechanism underlying hsp 60 autoimmunity and the significance of hsp 60 immunity in the pathogenesis of a particular disease. In this chapter, the progress in defining the role of endogenous hsp 60 in type I diabetes in NOD mice is described.

II. HSP 60 PROTEIN AS A MOLECULAR CHAPERON

A 60-kDa heat shock protein, hsp 60, has been described as a plastid-specific chaperonin which, through interaction with proteins, guides them to assume the correct three-dimensional structure (14,15). This class of molecular chaperonins is represented by *Escherichia coli* GroEl (16,17), the ribulose-1,5-biphosphate carboxylase (RuBisCO) subunit binding protein (14,18), and mitochondrial hsp 60 (19,20). The folding reaction involving the hsp 60 molecular chaperonin is a two-step process: the formation of a binary complex between target protein and the chaperonin in an ATP-independent manner and the release of properly folded protein using energy from ATP hydrolysis (21). Formation of the complex is conferred by the polypeptide-binding domain, whereas subsequent cooperative ATP hydrolysis is regulated by a smaller cochaperonin, GroES (22–24). The structural features of the polypeptide-binding domain that enable recognition of target polypeptide are not yet known, nor is the structure of the target polypeptides that support polypeptide-chaperonin interactions. Since most chaperons interact with a wide variety of proteins which differ in their primary structure, it is unlikely that recognition depends on a specific amino acid sequence of target proteins. Rather the hsp 60 chaperons recognize conformational features such as hydrophobic surfaces of amphiphilic α-helixes (25), judging by the binding preferences to folding intermediates of the molten-globule structure (26).

This ability to distinguish proteins in an extended but not fully denatured form may be critical for cell viability under physiological conditions and for the cell survival after stress. In normal cell physiology, the binding of mitochondrial and chloroplast chaperons facilitates folding and assembly of newly imported polypeptides, which are transported in a "translocation-competent" extended form, as well as newly translated proteins synthesized within these organelles (21, 27,28). In stressed cells, the binding of GroEl as well as mitochondrial hsp 60 protects against aggregation of insoluble complexes and prevents complete protein denaturation/inactivation (26).

A folding complex includes one or two molecules of substrate protein and usually a 14-subunit hsp 60 oligomer. As seen in electron microscopy, hsp 60 oligomer consists of two disks stacked together. Each disk consists of seven 60-kDa subunits arranged in a ring. The inside pore created by the "double-donut" structure is the primary site of interaction with polypeptides (29,30). The cooperative action of GroES heptamer is usually required for full function of hsp 60 (22–24,30).

Until recently, this approximately 840-kDa hsp 60 folding machinery has been observed only in mitochondria and chloroplasts. By contrast, the distinct isoforms of the hsp 70 molecular chaperon, encoded for by different genes, were found to be present in various cellular compartments. In the past 2 years, long-expected cytosolic version of hsp 60 chaperonin has been identified as a TCP-1 complex (31). A TCP-1 complex was originally observed in mouse tissues (31,32), but it has also been found in the cytosol of plant and yeast cells (33,34). TCP-1 exists as a homo-oligomer or constitutes a part of a multisubunit heteromeric toroidal complex (TRiC) (35,36). Although the primary structures of TCP-1/TRiCs bears only 20% structure similarity to hsp 60 (35), they both have architecture characteristic of the molecular chaperonins consisting of two stacked rings each containing seven to nine subunits (35,36). TRiCs are thought to be associated with the microtubule-network and to be involved in folding of tubulin and actin (36–39).

The molecular chaperon function is essential for basic cellular functions in cells of prokaryotic as well as eukaryotic organisms. Preservation of the hsp 60 function is reflected by the considerable sequence conservation of the hsp 60 molecule in evolution. Sequence comparison of bacterial and eukaryotic hsp 60 showed 45–50% sequence identity (40), whereas the human and rodent hsp 60 molecules differ in only 13 amino acids of 573 residues (41).

From an immunological perspective, the sequence conservation of hsp 60s, their cellular abundance, and their induction by environmental stress are features that may underlie the mechanism of hsp 60 immunity, including type I diabetes in NOD mice.

III. HSP 60 AS AN AUTOANTIGEN IN TYPE I DIABETES

A. NOD Mice as an Animal Model of Autoimmune Diabetes

The NOD mouse is an animal model of insulin-dependent diabetes mellitus with striking similarities to human type I diabetes (42,43). Similarly to the human disease, the induction of diabetes in these mice is under polygenic control and involves at least three autosomal recessive genes located outside of chromosome 17 [location of the major histocompatibility complex (MHC)] and MHC-linked gene(s) responsible for the generation and activation of autoreactive T cells inherited in dominant fashion [44-46]). Although genetic predisposition appears to be an important prerequisite in the development of type I diabetes, it seems that certain environmental factors (viral or bacterial infections, chemicals, diet) are mainly responsible for inducing the initial lesions and/or precipitating the disease.

Of primary significance in the pathogenesis of type I diabetes is the progressive infiltration of islet cells of the pancreas by mononuclear cells (insulitis). The infiltration of islets encompasses several stages. First (at about 5 weeks of age), the immune cells surround and invade islet vasculature, arteries, and venules (47). At approximately 6–7 weeks of age, lymphocytes begin to accumulate around the islet cells (nondestructive peri-insulitis) and then invade the islets, targeting insulin-secreting β cells (destructive intrainsulitis) at 8–13 week of age. Insulitis is observed in both sexes; however, in NOD mice, the incidence is higher in females (75% in our colony) than in males (25%).

The gradual destruction of insulin-secreting β cells and the consequent insulin deficiency gives rise to overt diabetes (at 20–35 weeks of age).

Several lines of evidence indicate that T cells play a critical role in the development of insulitis and diabetes. Characterization of their autoantigen targets has become essential in understanding specific β-cell destruction. Over the past years, several laboratories have isolated T-cell clones responding to islet cells (48,49). Islet-specific $CD4^+$ T cells isolated from spleen and lymph nodes of diabetic NOD mice have been shown to recognize and destroy islet transplants when incorporated to the graft sites (48). In adoptive transfer studies, $CD4^+$ T-cell clones derived from islet-infiltrating T cells were able to induce insulitis in diabetes-resistant NOD mice (50). However, both $CD4^+$ and $CD8^+$ T cells were required for transfer of the diseases (50). The studies on biopsies from human diabetic pancreas revealed that the specific destruction of β cells is carried out by $CD8^+$ effector cells (51) which recognize islet cell antigens in the context of MHC class I molecules.

Some islet cell antigens, to which the T-cell and antibody immune responses are directed, have been identified recently (52–56). It appears (5–7) that mycobacterial hsp 60 is one of the putative antigens stimulating proliferation of diabetogenic $CD4^+$ $CD8^+$ T cells.

B. Immune Responses to Mycobacterial Hsp 60 in Diabetes

The potential role of the heat shock protein hsp 60 in the development of diabetes in NOD mice was first postulated by Elias et al. (6,7). This hypothesis was based on the detection of T-cell and antibody responses against a 62-kDa molecule cross reactive with a mycobacterial 65-kDa heat shock protein. These hsp 65 cross-reactive T-cell and antibody responses developed coincidently with the progression of insulitis and were followed by induction of anti-insulin antibodies and anti-idiotypic antibodies (6). An isolated diabetogenic T-cell clone was able to transfer disease to naive mice. The importance of the hsp 60 cross-reactive immunity in the pathogenesis of NOD mice diabetes was supported further by the finding that administration of hsp 60 antigen or synthetic peptide epitope to young NOD mice protected them against diabetes (7). These results provided strong evidence that one of the antigens in diabetes is related to hsp 60.

Since the epitope on mycobacterial hsp 60 recognized by diabetogenic T helper cells was similar to human hsp 60 (7), it was assumed that the endogenous protein bearing this epitope might serve as an autoantigen. The natural candidate then for hsp 65 cross-reactive immunity was mitochondrial hsp 60.

The concept of hsp 60 as an antigenic target in type I diabetes evoked much controversy (57–59). It was hard to envision how a ubiquitous protein such as hsp 60, present in every cell in the body, could play a role in an organ-specific disease, and why a mitochondrial antigen, an intracellularly located protein physically separated from access to the immune system by mitochondrial membranes, could elicit immune responses.

Therefore, in order to determine the significance of the hsp 60 proteins in the pathogenesis of type I diabetes, we have undertaken systematic analysis of the localization and distribution of endogenous antigens recognized by hsp 60 antibodies in pancreatic β cells.

C. Intracellular localization of Hsp 60 in Islet β Cells

The cellular distribution and localization of hsp 60 proteins in pancreatic cells was analyzed by various experimental approaches but mainly by immunological criteria such as immunoelectron microscopy and immunoblotting. A panel of antibodies used in these studies contained a polyclonal (P1) (60) and two monoclonal antibodies (MAbs) raised against human hsp 60, MAb P1 (61) and MAb 13.2 (provided by Dr. S. Jindal), and a monoclonal antibody raised against yeast hsp 60 (MAb 149) (62). All the above antibodies specifically recognized human recombinant hsp 60 (huhsp 60) as well as a 62-kDa protein in islet cell lysates and in lysates of secretory granules on Western blots.

Immunoelectron microscopy with these antibodies revealed the presence of a number of hsp 60 cross-reactive proteins distributed among different cellular compartments. Interestingly, the distribution of hsp 60 cross-reactive proteins differed in stressed and nonstressed cells. In addition to mitochondria, the hsp 60 cross-reactive proteins have been detected in secretory granules of β cells in healthy control mice (Fig. 1) (65). In β cells of islets stressed by an inflammatory process, hsp 60 immunoreactivity has been found in the cytoplasm (8,65). The antibody binding to a secretory granule hsp 60–like protein was not accidental or attributable to the specificity of a particular antibody. First, recognition of secretory granule and mitochondrial hsp 60 was abolished after pretreatment of a polyclonal rabbit antibody raised against huhsp 60 with recombinant huhsp 60 prior to immunostaining. Second, the same pattern of secretory granule and mitochondrial staining was observed with another two monoclonal antibodies; MAb 149, raised against yeast hsp 60, and MAb 13.22, raised against human hsp 60 (Fig. 2a, b).

Identification of hsp 60 immunoreactivity outside mitochondria revealed that epitopes of mitochondrial chaperonins are preserved in proteins located in different cellular compartments. Preservation of these epitopes among hsp 60–like proteins suggests that they might be functionally important. Such structural and functional similarity to hsp 60 has been found recently in peroxisomal F_1-ATPase (63). The enzyme is involved in protein folding and assembly of imported precursors into mitochondria and peroxisomes. A sequence comparison of the α subunit and mitochondrial hsp 60 revealed two highly conserved amino acids stretches (64).

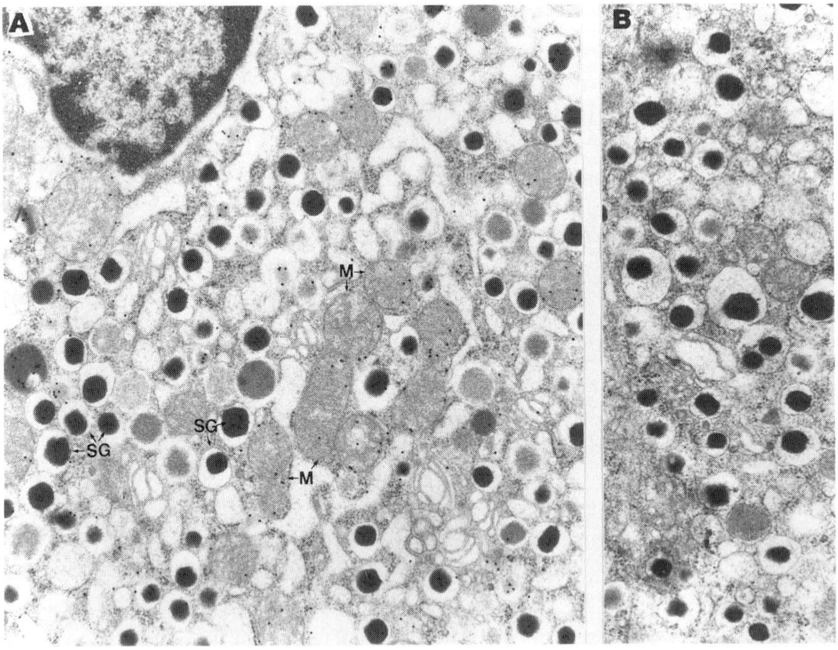

Figure 1 Immunoelectron microscopic localization of hsp 60 antibody binding sites in pancreatic β cells of control, healthy mouse. (A) Immunogold labeling with rabbit polyclonal hsp 60 antibody (1:400) indicates preferential binding to secretory granules (SGs) and mitochondria (M). (B) In control experiments, no staining of SGs and mitochondria was observed with normal rabbit serum used as a primary antibody.

On the other hand, identification of hsp 60 immunoreactivity outside mitochondria may also indicate that there is more than one candidate for hsp 60 cross-reactive immune responses in pancreatic β cells.

D. Expression of Hsp 60 in Islet Cells of Prediabetic and Diabetic NOD Mice

A distinctively different distribution of hsp 60 proteins has been observed in β cells of islets infiltrated by mononuclear cells. The preferential association of hsp 60 antibodies with secretory granules was lost in stressed β cells, with a concomitant marked increase of hsp 60 immunoreactivity in the cytosol (65,8). Electron microscopy revealed that the cytosolic hsp 60 immunoreactivity was associated with β cell microvesicles and a microtubule-like network (Fig. 3 and 4). Polyclonal and monoclonal hsp 60 antibodies (anti-P1 and MAb P1) used in these studies were raised against hsp 60 protein P1 (40,41,61). Interestingly, the P1 protein has initially been discovered as a microtubule-related protein in mutants of Chinese hamster ovary cells (CHO) resistant to the microtubule inhibitor (60,66). In CHO cells, it appeared in roughly equimolar amounts with tubulin.

The redistribution of hsp 60 in β cells of prediabetic NOD mice appeared to be consequence of insulitis. Progression of insulitis from nondestructive, peri-islet to destructive, intraislet insulitis was associated with a gradual reduction of hsp 60 antibody binding to secretory granules and increased binding to the cytosol (65).

The redistribution was accompanied by pathological changes in cell morphology. The changes which occurred in inflamed islets included an increased number of mitochondria, swelling of mitochondria and secretory granules, an increased number of microvesicles, fragmentation of the Golgi apparatus, and the collapse of the intermediate filament cytoskeleton. The latter was judged indirectly, being based on the alteration in secretory function of β cells and secretory granule biogenesis (65).

Based on these data, we have postulated that the hsp 60 redistribution is a part of the general β cell response to stress caused by islet inflammation. In this context, the reason for an appearance of hsp 60 immunoreactivity associated with microvesicles and β cell microtubules could be a practical one directed to restore cellular trafficking and the collapsed cytoskeleton. On the other hand, elevated cytoplasmic levels of hsp 60 might be used as an indicator of cellular stress.

In view of the strong immunogenicity of hsp 60 proteins, the emerging question was whether altered distribution of hsp 60 would change the antigenicity of β cells and evoke autoimmune responses in diabetes-prone mice.

E. Identification of Endogenous Hsp 60 as the β Cell Autoantigen

Screening prediabetic and diabetic NOD mice sera demonstrated the presence of autoantibodies that specifically bound huhsp 60 and 62-kDa islet cells antigen on Western blots (8). Autoantisera obtained form prediabetic and diabetic NOD mice recognized secretory granules and mitochondria on thin sections of pancreata of healthy, control SJL mice in immunoelectron microscopy experiments, giving a staining pattern similar to that of hsp 60 antibodies. A double labeling with the hsp 60 antibody and diabetic NOD mice sera confirmed colocalization of antigenic targets for these antibodies. Moreover, preincubation of thin sections of pancreas with hsp 60 antibody blocked the binding of the diabetic mice sera and vice versa (8).

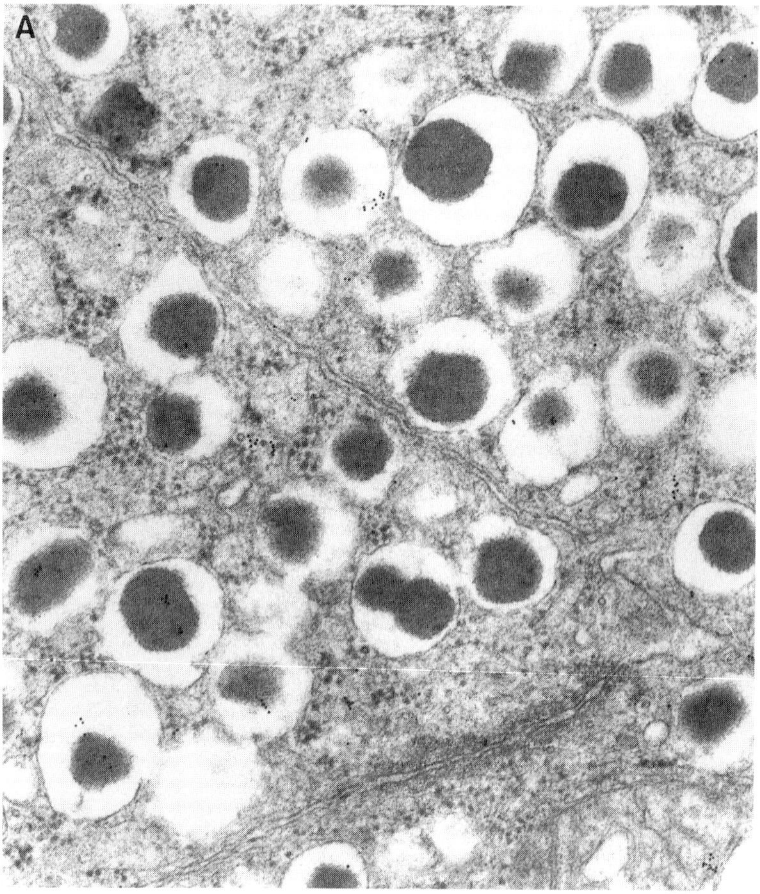

Figure 2 Immunolocalization of binding sites of different hsp 60 antibodies in thin section of islet β cells. Immunoelectron microscopic detection and localization of hsp 60 proteins in mouse β cells by monoclonal antibody 149 raised against yeast hsp 60 (A) and by monoclonal antibody 13.2 raised against human hsp 60 (B). In both cases, immunogold labeling is concentrated over SGs and mitochondria.

The dramatic changes in distribution of hsp 60 antibody binding sites within the β cells of NOD mice affected by insultits have been shown to overlap the relocalization of binding sites of diabetic mice sera. In double-labeling experiments, carried out on inflamed islets of NOD mice, the pattern of staining by these two antisera was identical and was characterized by reduced binding to secretory granules and increased binding to the cytoplasm (8). Taken together,

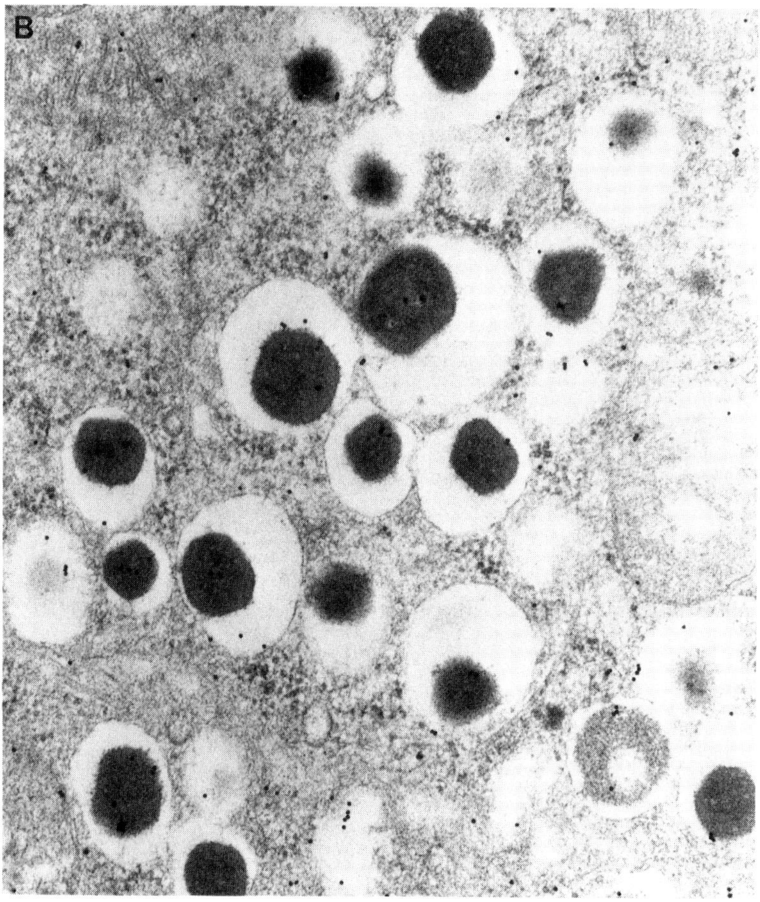

Figure 2 Continued.

these results indicate that both hsp 60 antibody and diabetic mice sera recognized the same hsp 60 antigen, or protein, which is structurally closely related to hsp 60.

Evidence for implicating endogenous hsp 60–related protein in type I diabetes is also supported by the study of Jones et al. (67). These investigators showed that a rat insulinoma cell line contains endogenous protein recognized by monoclonal antibody against mycobacterial hsp 65 as well as by sera from diabetic patients. The relative levels of this protein have been shown to increase under stress such as heat shock or treatment with interleukin-1β.

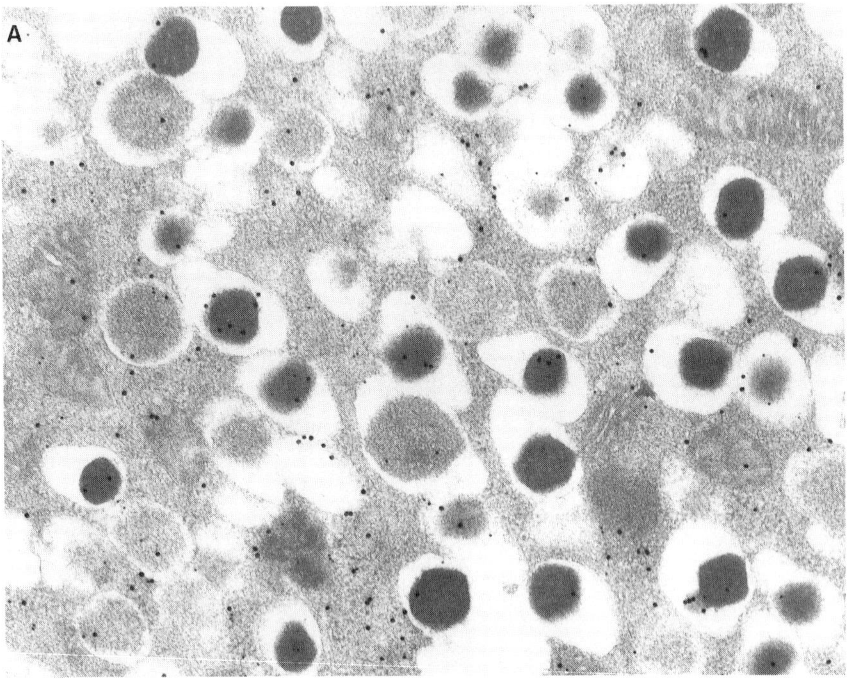

Figure 3 Recognition of hsp 60–related protein(s) associated with microtubule-like structure visualized by immunoelectron microscopy and computerized image analysis. (A) The original EM picture showing the distribution of hsp 60–related proteins in thin sections of NOD mouse pancreatic β cells detected by double-immunogold technique with hsp 60 monoclonal antibody (anti-P1, 20 nm gold) and human diabetic sera containing anti–hsp 60 antibodies (10 nm). (B and C) Digitized and computer enhanced image of the above EM picture, presented in a reversed pseudocolor (B) and in a reversed gray scale (C). The remaining sequence (D–G) shows the images resulting from increasing the black-level threshold, eliminating the brightes whites (cellular structures).

F. Link Between Insulitis and Hsp 60 Autoimmunity

The altered distribution and increased levels of cellular hsp 60 that occur following insulitis raised the possibility that these changes may underlie the anti–hsp 60 immune responses in NOD mice diabetes. A combined methodological approach including immunocytochemistry and immunoblotting allowed a systematic analysis of the relationship between the progression of insulitis, redistribution of hsp 60, and the induction of hsp 60 autoantibodies. It appears that there is a temporal correlation between insulitis-caused redistribution of hsp 60

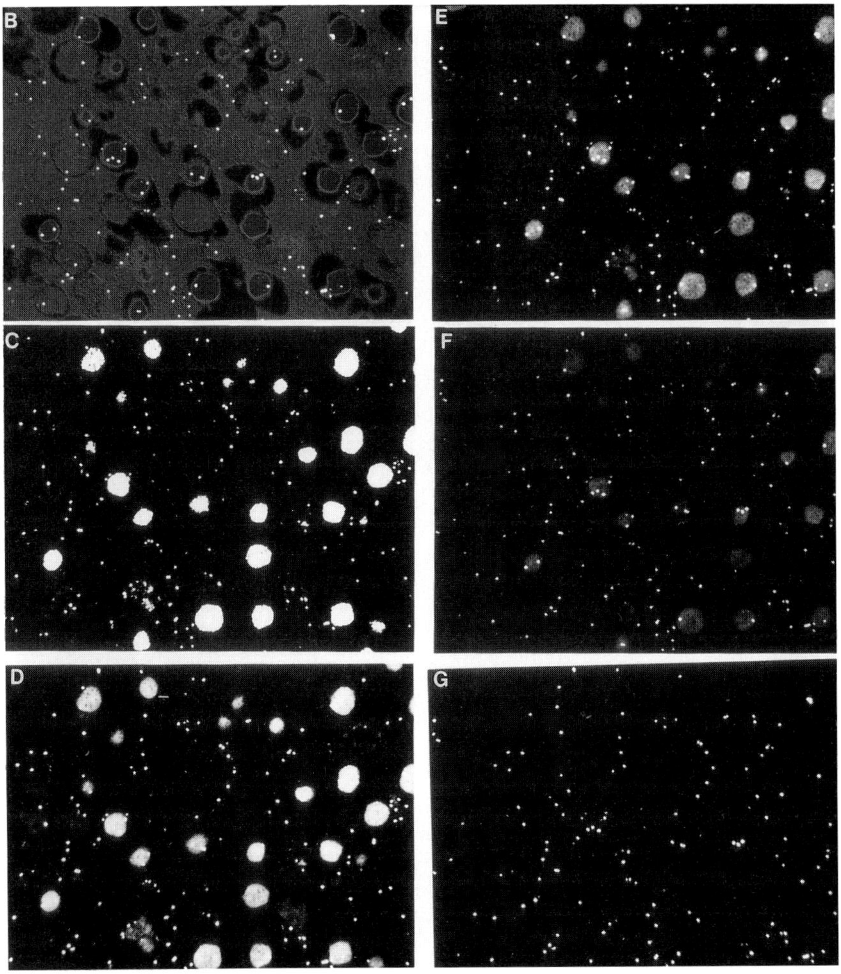

Figure 3 Continued.

inside the β cells and induction of hsp 60 autoantibodies (68). The induction and subsequent development of humoral hsp 60 immunity paralleled progression of insulitis. A maximal hsp 60 antibody response was observed in 7- to 13-week-old NOD mice, in which 30–70% of islets showed severe infiltration by mononuclear cells (68). The time course of humoral anti–hsp 60 immunity matches the time course of the development of hsp 60 autoreactive T cells (6,55,56). The T cells with hsp 60 specificity have been shown to transfer and induce insulitis and diabetes in naive mice (5,7).

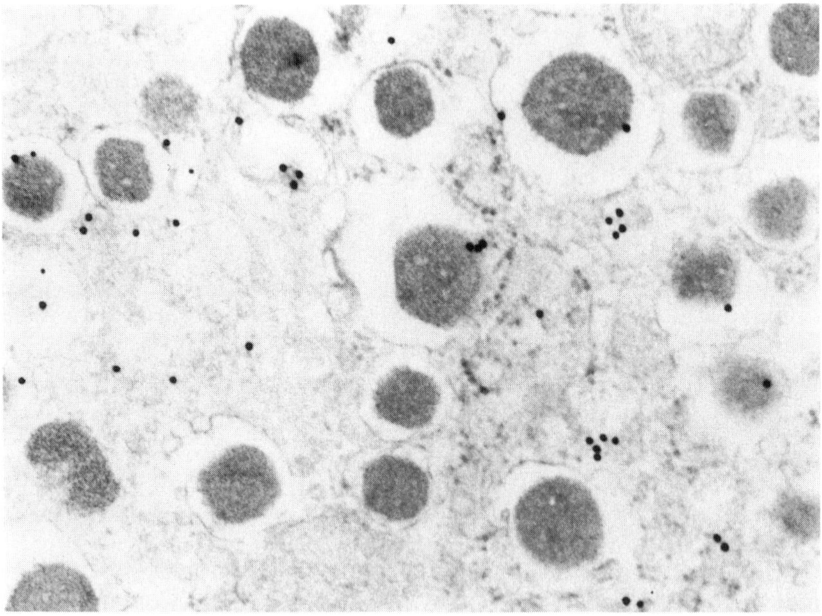

Figure 4 Immunoelectron microscopic localization of hsp 60–related proteins in NOD mouse islets infiltrated by mononuclear cells. Hsp 60 immunoreactivity is associated with β cell microvesicles.

These data are consistent with our view that cellular stress such as islet inflammation is the event that constitutes the first step leading to hsp 60 autoimmunity in NOD mice diabetes. It has to be emphasized, however, that immunity to endogenous hsp 60 is not the only cause of diabetes.

IV. POTENTIAL MECHANISMS FOR AUTOIMMUNITY TO HSP 60

A. Molecular Mimicry

Despite a growing amount of evidence on the involvement of hsp 60 in the immunopathology associated with inflammatory and autoimmune diseases, the mechanism by which hsp 60 elicits immune responses remains unknown. Antigen mimicry between pathogenic hsp 60 and endogenous hsp 60 has been proposed as a potential mechanism underlying hsp 60 autoimmunity in rheumatoid arthritis and diabetes (5,7,9,69,70). Investigations have therefore been focused on defining antigenic determinants shared by pathogenic hsp 60 and self-molecules, which influence T cells and antibody responses. Indeed, the data of Elias et al. (7) indicate that in NOD mice diabetes activation of a pathogenic T-cell clones occurred with peptide p277 of human hsp 60 which shared sufficient amino acid sequence homology to mycobacterial hsp 60. In contrast, arthritogenic T cells responded strongly to the epitope 180–188 of mycobacterial hsp 65, which did not show any homology to human hsp 60 (41). Nevertheless, in adoptive transfer experiments both arthritogenic T-cell clones specific to the peptide 180–188 as well as diabetogenic T-cell clones specific to the peptide p277 were able to induce autoimmune arthritis or diabetes, respectively (71,7). The development of the diseases were prevented by the administration of respective peptides or these T-cell clones. Thus, it seemed that in rat adjuvant arthritis and diabetes hsp 60 autoimmune responses to self-protein occurred, apparently irrespectively to the extent of sequence homology with pathogenic hsp 60. Therefore, if molecular mimicry is an explanation for hsp 60 immunity, one should predict the existence of hsp 60–related proteins, not yet identified, containing an amino acid sequence mimicking hsp 60 T-cell epitope (5). Our finding of multiple hsp 60 cross-reactive proteins in pancreatic β cells represents a good rationale for a search for such epitope.

It is assumed further that the stimulation of autoreactive T cells by antigen mimicking peptide may in turn lead to an autoantibody response against the native protein antigen.

B. Functional Mimicry

The essential biological function of hsp 60 molecular chaperons in protein folding and assembly is achieved through the interaction of its peptide-binding domain with target polypeptides. This peptide-binding ability of molecular chaperons resembles a peptide-binding function of MHC classes I and II molecules involved in antigen presentation. After antigen processing in the cell by ATP-dependent proteosomes, the MCH-peptide interaction occurs with those polypeptides, selected from a broad range of peptides, which topographically and chemically fit into the binding groove of MHC molecules (72). From amino acid sequence homology and secondary structures predictions, it appears that MHC molecules and hsp 70 chaperon bind peptides in a similar fashion (73,74). Although features of MHC class I–antigen interaction favor binding short polypeptides of 8–10 amino acid residue, class II molecules recognize longer polypeptide stretches in the range of 12–24 residues and also full-length proteins. Efficient formation of stable complexes with antigenic polypeptides by MHC class II molecules mimics the function of hsp 70. Since hsps are evolutionary older then MHC molecules, it raised the possibility that the peptide-binding domain of MHC molecules may have been donated by hsps or a hsp ancestor protein (73).

Hsps may also participate in the transport and presentation of antigenic peptides. Newly discovered molecular chaperon of endoplasmic reticulum—calnexin (calreticulin)—enables proper folding and assembly of class I heavy chain/β_2 microglobulin heterodimer that permits peptide biding (75). Studies on another heat shock protein, gp 96, revealed that it serves as a peptide acceptor and carrier of antigenic peptides to the endoplasmic reticulum and may also act as an accessory protein to peptide loading of MHC class I molecules (76,77).

The general capability of hsps to bind a wide array of polypeptides implicates them in a variety of cellular processes. It is not surprising that hsps also participate in the interactions with elements of the immune system. However, a functional mimicry between MHC molecules involved in antigen presentation and hsps is provocative. Are there any circumstances under which hsps would substitute or compensate for MHC function?

V. CONCLUSIONS

Understanding the molecular basis for hsp 60 autoimmunity in many organ-specific diseases emerges as a major focus in today's research aiming at the development of future prophylactive therapies. With more biochemical and biophysical information on the structure, function, and subcellular distribution of hsp 60, our perception of hsp 60 as an antigen is changing. In this new picture, hsp 60 appears as a complex molecule of characteristic double-donut structure with a high ability to bind a variety of polypeptides. It is a plastid-specific chaperonin. However, structurally and functionally similar proteins have been found outside these organelles. These hsp 60–related proteins could contribute to anti–hsp 60 autoimmune responses as potential additional antigenic targets. Some details of the cell biological events associated with islet inflammation, which lead to hsp 60 immunity in NOD mice diabetes, have been presented above. However, more studies combining molecular biology, cell biology, and immunology are needed to reveal the contribution of hsp 60 as a molecular chaperon and immunogen in the pathogenesis of autoimmune diseases. Ultimately, these could be expected to lead to the design and introduction of new strategies aimed at curing and/or preventing diabetes.

ACKNOWLEDGMENTS

The author is very grateful to Drs. R. S. Gupta, S. Jindal, and T. Mason for the generous supply of hsp 60 antibodies and Dr. K. Borkowski for critical reading of the manuscript. This work was supported by the Imperial Oil Funds for Diabetes Research and the Bickell Fundation.

REFERENCES

1. D. Young, R. Lathigra, R. Hendrix, D. Sweetser, and R.A. Young, Stress proteins are immune targets in leprosy and tuberculosis, *Proc. Natl. Acad. Sci. USA 85*:4267 (1988).
2. J.R. Lamb, V. Bal, P. Mendez-Samperio, A. Mehlert, A. So, J. Rothbard, S. Jindal, R.A. Young, and D.B. Young, Stress proteins mau provide a link between the immune response to infection and autoimmunity, *Int. Immunol. 1*:191 (1989).
3. R.A. Young, Stress proteins and immunology, *Ann. Rev. Immunol. 8*:401 (1990)
4. S.H.E. Kaufmann, B. Schoel, A. Wand-Wurttenbeger, U. Steinhoff, M.E. Munk, and T. Koga, T-cells, stress proteins and pathogenesis of mycobacterial infections, *Curr. Topics Microbiol. Immunol. 155*:125 (1990).
5. I.R. Cohen, Autoimmunity to chaperonins in the pathogenesis of arthritis and diabetes, *Annu. Rev. Immunol. 9*:567 (1991).
6. D. Elias, D. Markovits, T. Reshef, R. van der Zee, and I.R. Cohen, Induction and therapy of autoimmune diabetes in the non-obese diabetic (NOD/Lt) mouse by a 65-kDa heat shock protein, *Proc. Natl. Acad. Sci. USA 87*:1576 (1990).
7. D. Elias, T. Reshef, O.S. Birk, R. van der Zee, M.D. Walker, and I.R. Cohen, Vaccination against autoimmune mouse diabetes with a T cell epitope of the human 65 kDa heat shock protein, *Proc. Natl. Acad. Sci. USA 88*:3088 (1991).
8. K. Brudzynski, V. Martinez, and R.S. Gupta, Secretory granule autoantigen in insulin-dependent diabetes mellituis is related to 62 kDa heat-shock protein (HSP60), *J. Autoimmun. 5*:453 (1992).
9. W. Van Eden, J.E.R. Thole, R. van der Zee, R.A. Noordij, J.D.A. van Ebden, H.J. Hensen, and I.R. Cohen, Cloning of the mycobacterial epitope recognized by T lymphocytes in adjuvant arthritis, *Nature 331*:171 (1988).
10. P.C.M. Res, C.G. Schaar, F.C. Breedveld, W. van Eden, J.D.A. van Embden, and I.R. Cohen, Synovial fluid T cell reactivity against 65 KD heat shock protein of mycobacteria in early chronic arthritis, *Lancet 2*:305 (1988).
11. J.S.H. Gaston, P.F. Life, L.C. Bailey, and P.A. Bacon, In vitro responses to a 65-kilodalton mycobacterial protein by synovial T cells from inflammatory arthritis patients. *J. Immunol. 143*:2494 (1989).
12. M.G. Danieli, D. Markovits, A. Gabrielli, A. Corvetta, P.L. Giorgi, R. van der Zee, J.D.A. Embden, G. Danieli, and I.R. Cohen, Juvenile rheumatoid arthritis patients manifest immune reactivity to the mycobacterial 65 kD heat shock protein, to its 180–188 peptide and to partially homologous peptide of the proteoglycan link protein, *Clin. Immunol. Immunopathol. 64*:121 (1992).
13. E.R. DeGraeff-Meeder, R. van der Zee, G.T. Rijkers, H.-J. Schuurman, W. Kuis, W.J. Bijlsma, B.J.M. Zegers, and W. van Eden, Recognition of the human 60 kD heat shock protein by mononuclear cells from patients with juvenile chronic arthritis, *Lancet 337*:1368 (1991).
14. S.M. Hemmingsen, C. Woolford, S.M. van der Vies, K. Tilly, D.T. Dennis, C.P. Georgopoulos, R.W. Hendrix, and R.J. Ellis, Homologous plant and bacterial chaperone oligomeric protein assembly, *Nature 333*:330 (1988).
15. R.L. Hallberg, A mitochondrial chaperonin: genetic, biochemical and molecular characteristics, *Semin. Cell Biol. 1*:37 (1990).

16. C.P. Georgopoulos, W. Hendrix, S.R. Casjens, and A.D. Kaiser, Host participation in bacteriophage lambda head assembly, *J. Mol. Biol. 76*:645 (1973).
17. N. Sternberg, Properties of a mutant of Escherichia coli defective in bacteriophage lamda head formation (groEL). II. The propagation of phage lambda, *J. Mol. Biol. 76*:25 (1973).
18. R.J. Ellis, Molecular chaperones: The plant connection, *Science 250*:954 (1990).
19. T.W. McMullin and R.L. Hallberg, A highly evolutionarily conserved mitochondrial protein is structurally related to the protein encoded by the Escherichia coli groEL gene, *Mol. Cell. Biol. 8*:371 (1988).
20. D.S. Reading, R.L. Hallberg, and A.M. Myers, Characterization of the yeast hsp60 gene coding for a mitochondrial assembly factor, *Nature 337*:655 (1989).
21. J. Ostermann, A.L. Horwich, W. Neupert, and F.U. Hartl, Protein folding in mitochondria requires complex formation with hsp60 and ATP hydrolysis, *Nature 341*:125 (1989).
22. G.N. Chandrasekhar, K. Tilly, C. Woolford, R. Hendrix, and C. Georgopoulos, Purification and properties of the groES morphogenic protein of Escherichia coli, *J. Biol. Chem. 261*:12414 (1986).
23. T.H. Lubben, A.A. Gatenby, G.K. Donaldson, G.H. Lorimer, and P.V. Viitanen, Identification of a groES-like chaperonin in mitochondria that facilitates protein folding, *Proc. Natl. Acad. Sci. USA 87*:7683 (1990).
24. J. Martin, S. Geromanos, P. Tempst, and F.-U. Hartl, Identification of nucleotide-binding regions in the chaperonin proteins GroEL and GroES, *Nature 336*:279 (1993).
25. S. Landry, R. Jordan, R. McMacken, and L. Gierasch, Different conformations for the same polypeptide bound to chaperones DnaK and GroEL, *Nature 355*:455 (1992).
26. J. Martin, T. Langer, R. Boteva, A. Schramel, A.L. Horwich, and F.-U. Hartl, Chaperonin-mediated proteins folding at the surface of groEL through a "molten-globule"-like intermediate, *Nature 352*:36 (1991).
27. E.S. Bochkareva, N.M. Lissin, and A.S. Girshovich, Transient association of newly synthesized unfolded protein with the heat shock GroEL protein, *Nature 336*:254 (1988).
28. M.Y. Cheng, F.U. Hartl, J. Martin, R.A. Pollock, F. Kalousek, W. Neupert, E.M. Hallerberg, R.L. Hallerberg, and A.L. Horwich, Mitochondrial heat-shock protein hsp60 is essential for assembly of proteins imported in yeast mitochondria, *Nature 337*:620 (1989).
29. K. Braig, M. Simon, F. Furaya, J.F. Hainfield, and A.L. Horwich, A polypeptide bound by the chaperonin groEL is localized within a central cavity, *Proc. Natl. Acad. Sci. USA 90*:3978 (1993).
30. T. Langer, G. Pfeifer, J. Martin, W. Baumeister, and F.-U. Hartl, Chaperonin-mediated protein folding: groES binds to one end of the groEL cylinder which accomodates the protein substrate within its central cavity, *EMBO J. 11*:4757 (1992).
31. L.M. Silver, K. Artzt, and D. Bennett, A major testicular cell protein specified by a mouse T/t complex gene, *Cell 17*:275 (1979).

32. R.S. Gupta, Sequence and structural homology between a mouse t-complex protein TCP-1 and the chaperonin family of bacterial (groEL, 60-65 kDa heat shock antigen) and eukaryotic proteins, *Biochem. Int. 20*:833 (1990).
33. E. Mummert, R. Grimm, V. Speth, C. Eckerskorn, E. Schiltz, A.A. Gatenby, and E. Schaefer, A TCP-1-related molecular chaperone from plants refolds phytochrome to its photoreversible form, *Nature 363*:644 (1993).
34. D. Ursic and M.R. Culbertson, The yeast homolog to mouse TCP-1 affects microtubule-mediated processes, *Mol. Cell. Biol. 11*:2629 (1991).
35. V.A. Lewis, G.M. Hynes, D. Zheng, H. Saibil, and K. Willison, T-complex polypeptide-1 is a subunit of a heteromeric particle in the eukaryotic cytosol. *Nature 358*:249 (1992).
36. J. Frydman, E. Nimmesgern, H. Erdjument-Bromage, J.S. Wall, P. Tempst, and F.-U. Hartl, Function in protein folding of TriC, a cytosolic ring complex containing TCP-1 and structurally related subunits, *EMBO J. 11*:4767 (1992).
37. Y. Gao, I.E. Vainberg, R.L. Chow, and N.J. Cowan, Two cofactors and cytoplasmic chaperonin are required for the folding of alfa- and beta-tubulin, *Mol. Cell. Biol. 13*:2478 (1993).
38. M.B. Yaffe, G.W. Farr, D. Miklos, A.L. Horwich, M.L. Sternlicht, and H. Sternlicht, TCP1 complex is a molecular chaperone in tubulin biogenesis, *Nature 358*:245 (1992).
39. R. Melki, I.E. Vainberg, R.L. Chow, and N.J. Cowan, Chaperonin-mediated folding of vertebrate actin-related protein and gamma-tubulin, *J. Cell. Biol. 122*:1301 (1993).
40. A.K. Dudani and R.S. Gupta, Immunological characterization of a human homolog of the 65-kilodalton mycobacterial antigen, *Infect. Immun. 57*:2786 (1989).
41. S. Jindal, A.K. Dudani, B. Singh, C.B. Harley, and R.S. Gupta, Primary structure of a human mitochondrial protein homologous to the bacterial and plant chaperonins and to the 65 kilodalton mycobacterial antigen, *Mol. Cell. Biol. 9*:2279 (1989).
42. S. Makino, K. Kunimoto, Y. Muraoka, Y. Mizushima, K. Katagiri, and Y. Tochino, Breeding of a non-obese diabetic strain of mice, *Exp. Anim. 29*:1 (1980).
43. H. Kikutani and S. Makino, The murine autoimmune diabetes model: NOD and related strains, *Adv. Immunol. 51*:285 (1992).
44. J.A. Todd, T.J. Aitman, R.J. Cornall, S. Ghosh, J.R.S. Hall, C.M. Hearne, A.M. Knight, J.H. Love, M.A. McAleer, J.-B. Prins, N. Rordrigues, M. Lathrop, A. Pressey, N.H. DeLarato, L.B. Peterson, and L.S. Wicker, Genetic analysis of autoimmune type 1 diabetes mellitus in mice, *Nature 351*:542 (1991).
45. M. Prochazka, E.H. Leiter, D.V. Serreze, and D.L. Coleman, Three recessive loci required for insulin-dependent diabetes in nonobese diabetic mice, *Science 237*:286 (1987).
46. R.J. Cornall, J.-B. Prins, J.A. Todd, A. Pressey, N. DeLarto, L. Wicker, and L.B. Peterson, Type 1 diabetes in mice is linked to the interleukin-1 receptor and Lsh/Ity/Bcg genes on chromosome 1, *Nature 353*:262 (1991).
47. T. Fujita, R. Yui, Y. Kusumoto, Y. Serizawa, S. Makino, and Y. Tochino, Lymphocytic insulitis in a non-obese diabetic (NOD) strain of mice, an immunohistochemical and electron microscope investigation, *Biom. Res. 3*:429 (1982).

48. K. Haskins, M. Portas, B. Bergman, K. Lafferty, and B. Bradley, Pancreatic islet-specific T cell clones from nonobese diabetic mice, *Proc. Natl. Acad. Sci. USA 86*:8000 (1989).
49. M. Nagata, K. Yokono, M. Hayakawa, Y. Kawase, N. Hatamori, W. Ogawa, K. Yonezewa, K. Shii, and S. Baba, Destruction of pancreatic islet cells by cytotoxic T lymphocytes in nonobese diabetic mice, *J. Immunol. 143*:1156 (1989).
50. B.J. Miller, M.C. Appel, J.J. O'Neil, and L.S. Wicker, Both the Lyt-2+ and L3T4+ T-cell clones are required for transfer of diabetes in nonobese diabetic mice, *J. Immunol. 142*:52 (1988).
51. M. Nagate and J.-W. Yoon, Studies on autoimmunity for T-cell mediated beta cell destruction, *Diabetes 41*:998 (1992).
52. B.O. Roep, S.D. Arden, R.R.P. de Vries, and J.C. Hutton, T-cell clones from a type-1 diabetes patient respond to insulin secretory granule proteins, *Nature 345*:632 (1990).
53. B.O. Roep, A.A. Kallan, W.L.W. Hazenbos, G.J. Bruining, E.M. Bailyes, S.D. Arden, J.C. Hutton and R.R.P. de Vries, T-cell reactivity to 38 kD insulin-secretory-granule protein in patients with recent-onset type 1 diabetes, *Lancet 337*:1439 (1991).
54. J.-L. Diaz, J. Ways, and P. Hammonds, T-lymphocyte lines specific for glutamic acid decarboxylase (GAD) the 64 K beta-cell antigen of IDDM, *Diabetes 41*:118 (1992).
55. D.L. Kaufman, M. Clare-Salzier, J. Tian, T. Forsthuber, G.S.P. Ting, P. Robinson, M. Atkinson, E.E. Sercarz, A.J. Tobin, and P.V. Lehmann, Spontaneous loss of T-cell tolerance to glutamic acid decarboxylase in murine insulin-dependent diabetes, *Nature 366*:69 (1993)
56. R. Tisch, X.-D. Yang, S.M. Singer, R.S. Liblau, L. Fugger, and H.O. McDevitt, Immune responses to glutamic acid decarboxylase correlates with insulitis in nonobese diabetic mice, *Nature 366*:72 (1993).
57. M.A. Atkinson, L.A. Holmes, D.W. Scharp, P.E. Lacy, and N. Maclaren, No evidence for serological autoimmunnity to islet cell heat shock protein in insulin dependent diabetes, *J. Clin. Invest. 87*:721 (1991).
58. O. Kampe, L. Velloso, A. Andersson, and A. Karlsson, No role for 65 kD heat shock protein in diabetes, *Lancet 336*:1250 (1990).
59. D. Lachtman, No diabetes link to hsp65?, *Nature 356*:114 (1992) (Letter).
60. R.S. Gupta and A.K. Dudani, Mitochondrial binding of a protein affected in mutants resistant to the microtubule inhibitor podophylotoxin, *Eur. J. Cell Biol. 44*:278 (1987).
61. B. Singh and R.S. Gupta, Expression of human 60 kDa heat shock protein in E. coli and development and characterization of monoclonal antibodies to it, *DNA Cell Biol. 11*:489 (1992).
62. B.R. Johnson, K. Fearon, T. Mason, and S. Jindal, Cloning and characterization of the yeast chaperonin HSP60 gene, *Gene 84*:295 (1989).
63. J.M. Cuezva, A.I. Flores, A. Liras, J.F. Santaren, and A. Alconada, Molecular chaperones and the biogenesis of mitochondria and peroxisomes, *Biol. Cell 77*:47 (1993).

64. A. Luis, A. Alconada, and J.M. Cuezva, The alfa regulatory subunit of the mitochondrial F1-ATPase complex is a heat-shock protein. Identification of two highly conserved amino acid sequences among the alfa-subunits and molecular chaperones, *J. Biol. Chem. 265*:7713 (1990).
65. K. Brudzynski, V. Martinez, and R.S. Gupta, Immunocytochemical localization of heat-shock protein 60-related protein in beta-cell secretory granules and its altered distribution in non-obese diabetic mice, *Diabetologia 35*:316 (1992).
66. R.S. Gupta, T.K.W. Ho, M.R.K. Moffat, and R. Gupta, Podophyllotoxin resistant mutants of Chinese hamster ovary cells: alteration in a microtubule associated protein, *J. Biol. Chem. 257*:1071 (1982).
67. D.B. Jones, N.R. Hunter, and G.W. Duff, Heat-shock protein 65 as a beta cell antigen of insulin-dependent diabetes, *Lancet 336*:583 (1990).
68. K. Brudzynski, Insulitis-caused redistribution of heat-shock protein hsp60 inside beta-cells correlates with induction of hsp60 autoantibodies, *Diabetes 42*:908 (1993).
69. X.-D. Yang, J. Gasser, B. Riniker, and U. Feige, Treatment of adjuvant arthritis in rats: vaccination potential of a synthetic nonapeptide from the 65 kDa heat shock protein of mycobacteria, *J. Autoimmun. 3*:11 (1990).
70. X.-D. Yang, J. Gasser, and U. Feige, Prevention of adjuvant arthritis in rats by a nonapeptide from the 65 kD mycobacterial heat shock protein. II. Specificity and mechanism, *Clin. Exp. Immun. 87*:99 (1991).
71. J.M. Holoshitz, A. Matitiau, and I.R. Cohen, Arthritis induced in rats by cloned T lymphocytes responsive to mycobacteria but not to collagen type II, *J. Clin. Invest. 73*:211 (1984).
72. R.N. Germain, MHC-dependent antigen processing and peptide presentation: providing ligands for T lymphocyte activation, *Cell 76*:287 (1994).
73. F. Rippmann, W.R. Taylor, J.B. Rothbard, and N.H. Green, A hypothetical model for the peptide binding domain of hsp70 based on the peptide binding domain of HLA, *EMBO J. 10*:1053 (1991).
74. M.F. Flajnik, C. Canel, J. Kramer, and M. Kasahara, Which came first, MHC class I or class II?, *Immunogenetics 33*:296 (1991).
75. E. Degen, M.F. Cohendoyle, and D.B. Williams, Efficient dissociations of the p88 chaperone from major histocompatibility complex class I molecules requires both beta 2-microglobulin and peptide, *J. Exp. Med. 175*:1653 (1992).
76. P.K. Srivastava and R.G. Maki, Stress-induced proteins in immune response to cancer, *Curr. Top. Microbiol. Immunol. 167*:109 (1991).
77. P.K. Srivastava and M. Heike, Tumor-specific immunogenecity of stress-induced proteins: convergence of two evolutionary pathway of antigen presentation. *Semin. Immunol. 3*:57 (1991).

27

Expression of Stress Proteins in Diabetes Mellitus

Detection of a Heat Shock Protein 60–like Protein in Pancreatic RINm5F β-Cell Plasma Membranes

Burkhard Göke, Brigitte Lankat-Buttgereit, Hanna Steffen, and Rüdiger Göke
Philipps University of Marburg, Marburg, Germany

Friedrich Lottspeich
Max-Planck-Institute for Biochemistry, Martinsried, Germany

I. PUTATIVE ROLE FOR STRESS/HEAT SHOCK PROTEINS IN THE PATHOPHYSIOLOGY OF DIABETES

Cells utilize different strategies to respond to adverse changes in their environment (1–3). It is of high recent interest that many, although varying, deleterious treatments of cells result in rather uniform changes of gene expression (3). This response is today referred to as the stress response rather than the heat shock response, although the latter term was initially coined to describe the changes in cellular proteins observed in classic experiments which studied the effects of sudden increases in the normal growth temperature on the cells (4). It is believed that this stress response and, in particular, the stress proteins are essential for the survival of the cell confronted with a particular environmental insult. Although expressed at higher levels in the disturbed cell, most of the so-called stress proteins are in fact also synthesized under normal growth conditions and appear integral in a number of important cellular processes as the maturation of proteins (3,5,6).

Having their potential role as cellular safeguards in mind, it is somewhat puzzling that stress proteins presently receive so much attention because of their possible pathophysiological connections with immune responses and autoimmune disease (7,8). Recent studies address the role of stress proteins as targets for the immune response following infections (9). Obviously, a vigorous production of microbial stress proteins occurs after invasion into the warm-blooded host (9–11). Some of these proteins appear to represent immunodominant targets for both T- and B-cell responses. Furthermore, there are data supporting the concept that some autoimmune diseases may arise as a result of T- and/or B-cell activation, appearing specific for foreign stress proteins but also recognizing closely related proteins of the host (3,9). This is supported by the observations of antibodies to self-stress proteins in a variety of autoimmune disorders (3).

In a classic series of experiments with pancreatectomized dogs, von Mering and Minkowski in 1889 (12) showed that removal of the pancreas caused diabetes mellitus. Today, we know that the characteristic feature of type 1 diabetes is a specific loss or disappearance of the β cells from the islets of Langerhans (13). Simply described, type 1 diabetes represents nature's own β cytectomy (13). The molecular driving forces which govern the extermination of β cells still need to be defined. However, there is solid evidence that a precise and concerted activity of the immune system within an autoimmune reaction attacks certain molecules which among the various islet cells occur only on the β cells (13).

The sequence of events involved when the immune system mounts a response toward an antigen (foreign or autoantigen) is rather complex and not yet fully explained. Briefly, the antigen is presented to a T helper lymphocyte (T inducer cell) with a specific receptor for the antigen. The T helper cell recognizes the antigen only if it also recognizes an major histocompatibility complex (MHC) class II antigen simultaneously. This means that the antigen-presenting cell must have the same class II antigenic specificity as the T helper lymphocyte. Then the activated T helper lymphocyte promotes the proliferation and differentiation of cytotoxic T lymphocytes and/or B lymphocytes specifically. The latter carry receptors (membrane-bound antibody molecules) on their cell surfaces. The former cells identify and kill the target cell if this expresses the antigen on its cell surface. In order to verify the concept that an abnormal immune response toward the pancreatic β cell is responsible for the onset of type 1 diabetes mellitus, it is mandatory to identify and characterize relevant autoantigens (13).

We have good evidence today that diabetogenesis in both humans and in the nonobese diabetic (NOD) mouse model results from complex polygenic interactions between multiple alleles within the MHC and multiple susceptibility genes outside of the MHC (14). Furthermore, a multiplicity of protein antigens is likely to trigger the autoimmune reaction, which eventually causes β-cell destruction. Stress proteins are presently considered as highly interesting candidates in this connection (15–19). The immunogenicity of several stress protein constituents of bacteria and parasites or the host T-cell system makes them of central interest. Elias and Cohen's group has contributed significant evidence for a key antigenic epitope within a distinct heat shock protein (hsp 60/65) critical to the pathogenesis of diabetes, at least in the NOD mouse (20,21). The NOD mouse spontaneously develops diabetes resembling human insulin-dependent type I diabetes mellitus, which is characterized by a progressive lymphocytic infiltration of islets (insulitis) before the manifestation of hyperglycemia. Recent observations suggest that T cells recognize the stress protein hsp 65, carboxypeptidase H, peripherin, and further novel autoantigens which are involved in the early events of diabetes in addition to the conventional autoantigens, including GAD65, GAD67, ICA69 (22). Gelber et al. demonstrated that young NOD mice have peripheral autoreactive T cells that recognize these novel islet cell antigens even in the absence of antigen priming and, furthermore, that these autoreactive NOD T cells recognize human islet cell extracts, suggesting conservation of antigenic determinants between human and murine islet β cells (22). It has been suggested that there are multiple and distinct species of β-cell–specific antigens recognized in the early inflammatory events of NOD disease and that knowledge of the antigen(s) recognized by T cells before overt diabetes would be important for certain forms of effective immunotherapy (22,23). It was hypothesized that early recognition of novel β-cell–specific autoantigens leads to inflammatory insulitis with resultant β-cell destruction and the release of β-cell products whose recognition then results in the production of the more conventional autoantibodies and T-cell proliferative responses to antigens such as hsp 60 (22). However, very recently spectacular progress was achieved when Elias and Cohen applied a peptide therapy for diabetes in NOD mice which targeted diabetogenic T-cell clones (24). A peptide of the 60-kDa heat shock protein (hsp 60) prevented diabetes in NOD mice and even arrested the autoimmune process after it was far advanced (24).

II. FOCUS ON HSP 60

One prominent group of stress proteins, commonly referred to as hsp 60 or 60- to 65-kDa antigens, has achieved high attention (9,20,21,24), partly because recombinant mycobacterial hsp 60 was made easily available for investigators. Nevertheless, it obviously consists of determinants which are major antigens of a wide variety of pathogenic organisms (9). Hsp 60 was first described in *Tetrahymena* (9) and has meanwhile been identified, cloned, and sequenced from a number of eukaryotic organisms (25). Such analysis has revealed its homology to the previously characterized GroEL hsp of bacteria (26,27). The GroEL-related proteins are encoded by nuclear genes and are synthesized within the cytoplasma, and they are then translocated into the mitochondria (27). The homologues of hsp 60 have been shown to be present in all species ranging from bacteria and plants to humans (27). Their primary structure is highly conserved across all species. Recent data support that these hsps facilitate proper folding of newly synthesized polypeptide chains and their assembly into oligomeric protein complexes (28,29). Interestingly, although microbial and mammalian hsp 60/65 molecules are 50% identical in amino acid sequence and immunologically cross reactive, hsp 60/65 is a dominant antigen in infection (9). Immune responses against a 65-kDa mycobacterial hsp 60 have been implicated in the etiology of rat adjuvant arthritis and rheumatoid arthritis (8). Such an autoimmune response may be caused by molecular mimicry between mycobacterial and human hsps. Indeed, such a mechanism was also discussed in the etiology of type I diabetes (8).

There is evidence that a key antigenic epitope may be contained within the hsp 60 molecule (21,24). T cells specific for hsp 60 and for a contigous 24 amino acid peptide (p277) transferred acute diabetes to young prediabetic NOD mice (20). The insulitis lesions that preceded the symptoms of diabetes in the animals contained T lymphocytes reactive to hsp 60 and the p277 peptide (21). Furthermore, active immunization to entire hsp 60 induced acute diabetes (20). Vaccination with anti-p277 T lymphocytes or with p277 peptide led to decreased immunity to hsp 60 as well as to prevention of insulitis and insulin-dependent diabetes (21). In a very recent report, it was shown that subcutaneous injections of the p277 peptide into NOD mice with or without overt diabetes prevented or arrested the autoimmune process even after it was far advanced (24). The p277 substantially decreased the proportion of islets affected by insulitis, apparently by raising a population of anti-idiotypic T cells that inhibited the effector T cells, which recognize the offending portion of hsp 60. This finding is of importance, especially since it suggests that the immune system is responsive to manipulation by a specific signal even in the face of a virulent, full-blown autoimmune process.

Reports of the possible importance of hsp 60 in the etiology of diabetes, especially that β-cell autoantigen is cross reactive with mycobacterial hsp 60 and that mouse diabetogenic T-cell clones are stimulated by human hsp 60 (20, 21,30), have prompted studies which investigated the cellular localization of hsp 60 in pancreatic islet cells (31,32). It was speculated that although hsp 60 was reported to be a mitochondrial protein (27), quantitative differences in the expression of epitopes of this protein, or the putative expression of "new" hsp 60 molecules at intracellular locations other then the originally described mitochondrial one, may explain the tissue-restricted inflammatory disease in mammals (9,32). Recent studies revealed that, in addition to mitochondria, a hsp 60–like protein is present in secretory granules of insulin-secreting β cells (31) and that an increase of cytoplasm staining occurred (32,33) that is associated with some morphological evidence for an accumulation of the stress protein on the cell surface during progressive insulitis. Furthermore, NOD mice affected by insulitis contain an autoantibody directed against this secretory granule hsp 60 in their serum (33). In another investigation, hyperthermic incubation and cytokines such as interleukin-1β (IL-1β), interferon-gamma, and tumor necrosis factor induced synthesis of a 64-kDa protein in insulinoma RINm5F cells (30). This protein reacted with antibodies against mycobacterial hsp 60 and, interestingly, binding of such antibodies to cell lysates of interleukin-treated RIN cells was inhibited by prior addition of serum from patients with insulin-dependent diabetes (30).

We found observations especially intriguing which reported a cell surface location of hsp 60 in β cells of NOD mice with insulitis (33). In any case, it was argued that cell-mediated effector mechanisms did not need the expression of the full-size hsp 60 molecule on the cell surface, as hsp 60–derived peptides already associated with HLA antigens may be recognized by specific T cells (89). We have, nevertheless, readdressed the question whether a cell surface expression of a hsp 60–like molecule is detectable, since this expression would be an easy target for the immunogenic interaction with insulin-producing cells.

III. DETECTION OF A HSP 60–RELATED PROTEIN EXPOSED ON β-CELL MEMBRANES

Our studies were prompted by a previous study in rat RINm5F cells that suggested that hsp 60/65 may be a β-cell antigen and that autoreactivity to an epitope of hsp 60/65 may confer susceptibility to insulin-dependent diabetes mellitus (30). These studies utilized RINm5F cells, considering those rat islet cell preparations as valid model of human β-cell antigenicity (30). However, a characterization of the subcellular distribution of hsp 60/65 in RIN cells was not performed. By utilizing an antibody, which was previously characterized in the NOD mouse model against hsp 60 (32), we localized binding sites of this antibody to mitochondria, in cytosol, and weakly to the cell surface membrane of the RIN cell (Fig. 1). As mentioned earlier, some evidence raised by morphological means was recently reported for a redistribution of hsp 60 into the cell membrane of NOD mouse pancreatic β cell with insulitis (32,33). We asked two questions: (1) whether hsp 60 expression in cell membranes is a regular feature in RINm5F β cells, and (2) whether RINm5F cells under stress show an increased exposure of hsp 60 in the cell membranes.

Figure 2 demonstrates our experimental strategy. RINm5F cells were cultured as previously described (34). First, cells were labeled with [^{35}S]methionine for 1 hr at 37°C according to Welch et al. (35) and then heated in a waterbath at 42°C for 1 hr. Cells were then harvested and labeled proteins were analyzed by high-resolution two-dimensional (2D) gel electrophoresis (36) (Fig. 3). Ear-

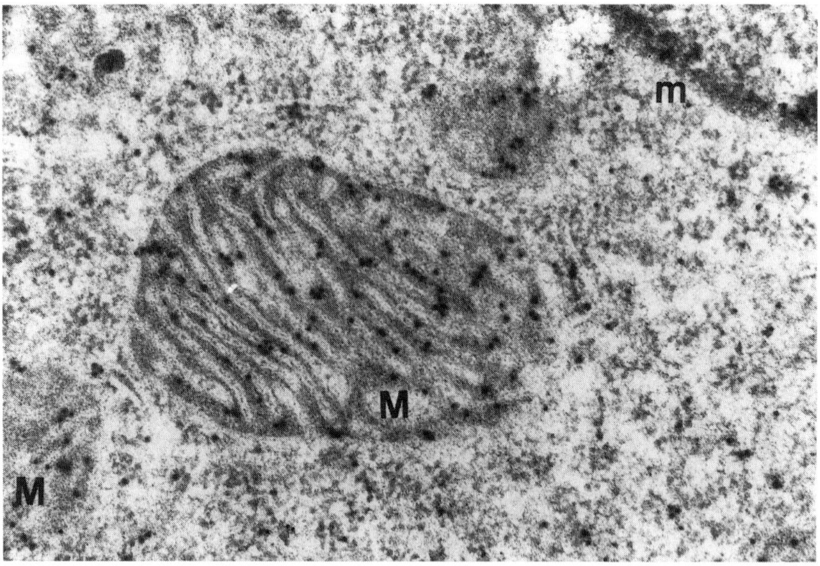

Figure 1 Localization of binding sites of antibodies to hsp 60 in RINm5F cells detected by immunogold technique (electron microscopy). Gold particles are concentrated over the mitochondria (M), diffusely found in cytosol, and few at the cell membrane (m). Control experiment with nonimmune rabbit serum did not reveal any immunogold staining at all (not shown).

lier, Welch et al. showed that the changes in the patterns of proteins being synthesized after stress appear very similar in different cell lines as HeLa cells, baby hamster kidney cells, fibroma cells, Chinese hamster ovary cells, and rat fibroblasts (37), which was corroborated by the patterns in RINm5F cells presently observed. Considering the molecular weights, among other protein species, a spot at ~60 kDa was visualized which became more prominent under heat stress (Fig. 3). This spot was detected independently whether the proteins were divided into soluble and particulate protein factions before subsequent 2D separation (not shown). The pH region where this spot at ~60 kDa appeared ranged between pIs of 7 and 8. We therefore decided to purify RINm5F cell membranes in order to solubilize the membrane proteins and to separate them by preparative isoelectric focusing.

To achieve this, cells were homogenized and the plasma membranes were obtained by sucrose gradient centrifugation (100,000*g* for 1 hr). The band at the interface between buffer and sucrose representing the plasma membranes was collected, washed, and marker enzymes were analyzed (38). Furthermore, electron microscopy was performed to ensure homogenity of the plasma membranes

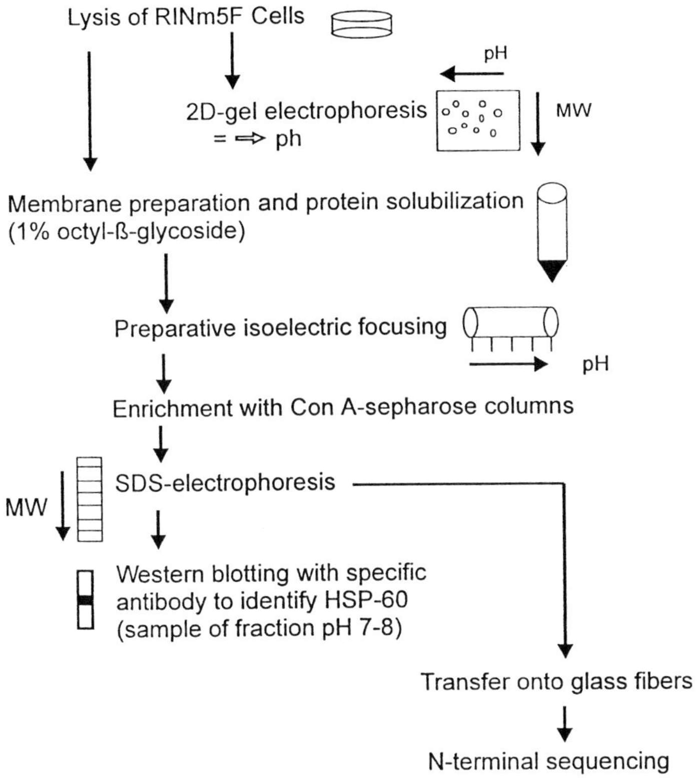

Figure 2 Purification and N-terminal amino acid sequence analysis of putative hsp 60 from RINm5F cell membranes: Experimental strategy.

and, especially, to exclude mitochondrial contamination. The proteins of the purified membranes were solubilized with 1% (w/v) octyl-β-glycoside (39) and thereafter subjected to preparative isoelectric focusing (IEF). Preparative IEF was facilitated utilizing a Rotofor cell from Biorad (Germany). The Rotofor cell was cooled down to 4°C, filed with 46 ml 0.5% (w/v) octyl-β-glycoside containing 2% (w/v) ampholytes (Bio-Lyte 3/10; Biorad), and prefocused for 1 hr at 12 W. Thereafter, 4 ml 0.5% octyl-β-glycoside containing 40 mg membrane protein was injected and focused at 12 W. After 4 hr of fractionation, samples were harvested. The pH and protein content was determined in every sample. Fractions with pH 7–8 were pooled and proteins were enriched with Con A Sepharose (Sigma, Germany) taking advantage of the glycoprotein structure of hsps. The purified membrane proteins were then subjected to SDS-polyacrylamide gel electrophoresis. Figure 4 shows silver staining of membrane proteins

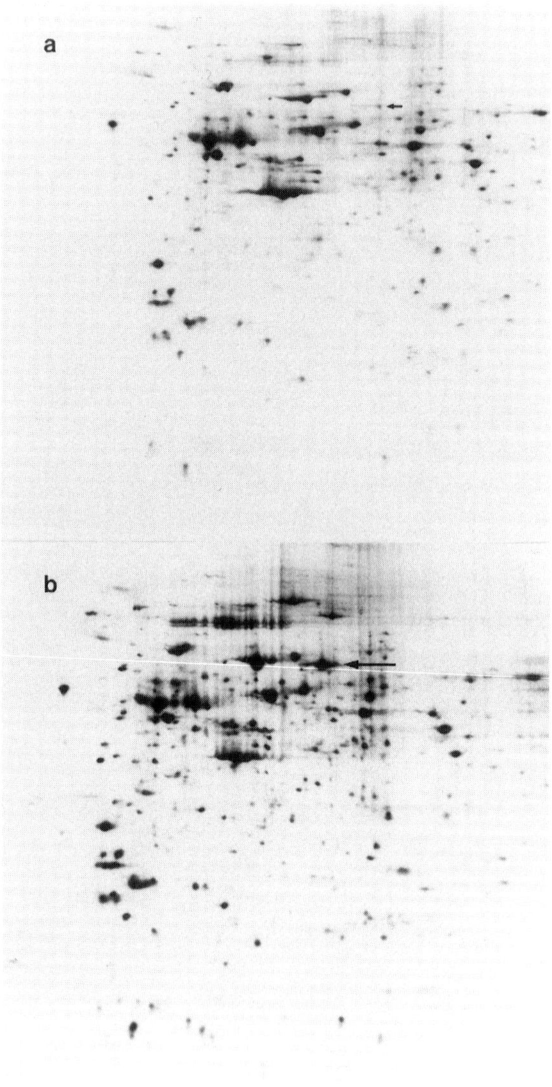

Figure 3 Analysis of RINm5F cell stress proteins by 2D gel electrophoresis. Cells were labeled with [^{35}S]methionine for 1 hr at 37°C, washed, then incubated at 37°C (control) or 42°C (heat stress), separated by 2D gels, and submitted to autoradiography. The first dimension (isoelectric focusing) extended over the pH range from 3.4 to 9.6 (left to right). The second dimension (sodium dodecyl sulfate gel electrophoresis) separated proteins according to molecular mass. Shown is the whole pattern of labeled proteins, respectively. (a) Control at 37°C; (b) heated RINm5F cells. Putative hsp 60 protein is indicated by an arrow.

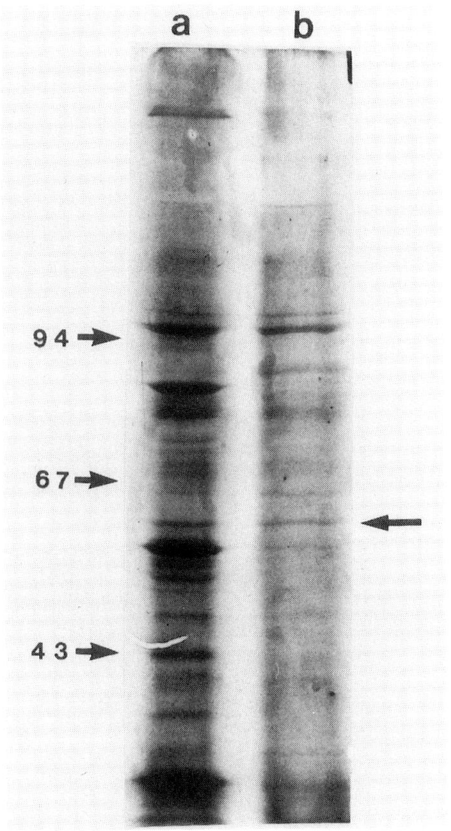

Figure 4 Silver staining of solubilized RINm5F cell membrane proteins after purification by preparative IEF (lane a) and subsequent enrichment with Con A column (lane b). Proteins were electrophoresed on a 10% SDS gel and stained with silver. Arrows with numbers indicate migration positions of corresponding molecular weight marker proteins. The arrow on the right side of lane b indicates the putative hsp 60 band which was further processed until sequencing.

after IEF (lane a) and subsequent Con A binding (lane b). The enrichment of a band at ~60 kDa is indicated by an arrow. When the proteins after SDS-PAGE were transfered to Immobilon membrane, an immunoreactive hsp 60–like protein was marked by employing a hsp 60-directed antibody, which was kindly provided by Dr. R. S. Gupta (Hamilton, Ontario) (Fig. 5). This antibody was earlier used for immunocytochemical localization of hsp 60-related protein in NOD mice (32). To further confirm the identity of the purified, putative hsp,

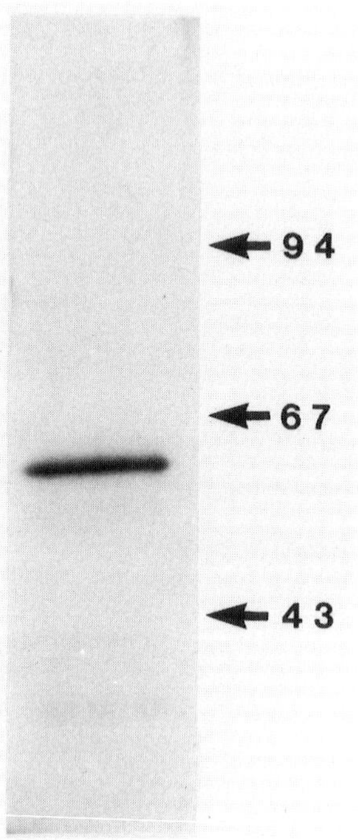

Figure 5 Western blot of Con A–enriched putative hsp 60. Membrane proteins were solubilized, purified by preparative IEF, and enriched by subsequent Con A–column treatment. The samples were then electrophoresed on a SDS-PAGE gel and transferred to Immobilon membranes. Specific hsp 60 protein was detected using a hsp 60 directed antibody and goat antirabbit antibody conjugated with horseradish peroxidase. Arrows indicate migration positions of corresponding molecular weight marker proteins.

the protein mixture was separated on a 10% SDS-polyacrylamide gel and electroblotted onto siliconized glass fiber (Glassybond, Biometra) according to Eckerskorn et al. (40). After staining the blot with Coomassie blue, the 60-kDa protein–containing band was excised and subjected to gas phase sequencing (ABI 477A, Applied Biosystems, Germany) according to the instructions of the manufacturer. Thereby we obtained 14 N-terminal amino acids of the purified protein yielding the following sequence: A K D V K F G A D A R A L M. This partial sequence is 100% homologue to the deduced N-terminus of rat hsp 60 which was recently cloned and sequenced from a cDNA rat kidney library (25). Therefore, we concluded that the RINm5F cell plasma membrane carries hsp 60 protein. Consecutively, we performed further functional studies. When

RINm5F cells were stressed with heat, an increased expression of the hsp 60–like protein became evident (Fig. 6).

Taken together, our data provide further proof that hsp 60 or a closely related protein is expressed in the β-cell membrane and that this protein is increasingly detected under cellular stress.

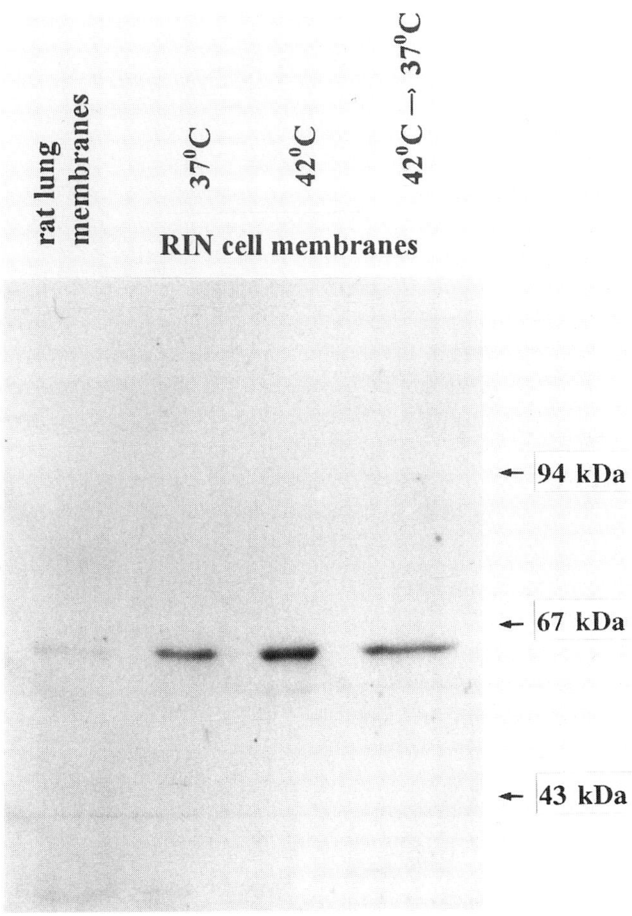

Figure 6 Western blot analysis of equal amounts of solubilized proteins from rat lung membranes and from RINm5F cell membranes, treated at 37°C, 42°C (1 hr), and 42°C, and thereafter 37°C (1 hr). Antibodies directed against hsp 60 showed only a weak band in lung membranes and a clearly stronger band in RINm5F cell membranes. After treatment at 42°C, signal intensity clearly increased and decreased again, when the cells were kept at 37°C for another hour after heat shock.

IV. ANOTHER PERSPECTIVE: HSPS AND REPAIR OF PANCREATIC ISLET β CELLS

Most studies focusing on hsps and the pathogenesis of diabetes mellitus have emphasized the immune assault against the insulin-producing β cells. It has been convincingly shown that classic cell-mediated and humoral immune responses to β-cell antigenic motifs are involved. Actually, the development of diabetes is predominantly seen by the light of immune effector cells. Keeping generally assigned biological function of stress proteins in mind, one should also consider that β cells are probably not passive to the immune attack but rather respond with an activation of cell-protective mechanisms. This is not unlikely, since we have learned that common to all organisms is such a highly conserved and exquisitely regulated cellular response to suboptimal physiological conditions. Actually, considerable evidence suggests that in response to injury, protective and/or repair mechanisms are activated within the β cells (41). This is supported by the clinical observation that several months after the initial diagnosis of insulin-dependent diabetes and subsequent treatment, most patients experience some recovery of β-cell function (42). In vitro data suggest that in prediabetic NOD mice, a population of still viable β cells exists which can use repair mechanisms to regain normal function (43). This can be seen when the islets are removed from their in vivo environment and placed in cell culture (43,44) or by eliminating the invading T cells with specific monoclonal antibodies (45) to arrest the immune assault.

When we analyzed RINm5F insulinoma cell stress proteins by high-resolution 2D gel electrophoresis, a dramatic regulation of the expression of the [^{35}S]methionine labeled protein pattern is seen (see Fig. 3). The enhanced expression of putative hsps can be visualized in such gels by their position on the gels, which is defined by molecular weights and isoelectric points. In fact, previous investigations have identified in lower resolution one-dimensional gels three proteins at ~32, 70–72, and 90 kDa molecular weight which were increasingly expressed in pancreatic islets exposed either to high temperatures (heat shock, 42°C) (15,16), the cytokine IL-1 (17–19), or the diabetogenic compound streptozotocin (18). The 32-kDa protein was identified as heme oxygenase (16), whereas the 70-kDa protein(s) resembled member(s) of the hsp 70 family (16–19).

Hsp 70 seems to be of considerable interest for the pathogenesis of diabetes. This is supported by several lines of evidence. Purified hsp 70 proteins introduced into pancreatic islets by the liposome technique protect β cells against the toxic effects of IL-1β (46). Hsp 70 genes have been localized within the HLA class III region of the MHC in humans (47), in the rat MHC (48), and in the mouse H-2 complex (49). It has been shown that the duplicated locus encoding the hsp 70 is located between the genes for complement and tumor necrosis factor (TNF) in humans (47). This part of the genome shows polymorphism, and differences in the frequencies of TNF-β and hsp 70 genotypes have been indicated between diabetic patients and controls (50,51). TNF enhances the deleterious effects of IL-1 on pancreatic β cells (52,53) . Since TNF may contribute to β-cell damage in early diabetes and hsp 70 may play a role in β-cell defense, it is conceivable that susceptibility to insulin-dependent diabetes is enhanced by certain unfavorable combinations of these genes (54).

V. CONCLUSIONS

We have discussed above the evidence that the exposure of hsp 60 in β-cell membranes may support the immune assault against the β cell. However, the observed redistribution of hsp 60 in prediabetic NOD mice (33) can be quite differently interpreted. An alternative concept would be to envisage that an immune-mediated β-cell dysfunction may be a signal for redistribution of the hsp 60, perhaps as part of activation of a defense mechanism that has a role in preventing protein denaturation both in cytosol and mitochondria (41).

Presently, the prevailing assumption is that hsps are pathogenetically significant targets in the immune response in β cells experiencing stress. As recent data point to the significance of hsps in cell repair mechanisms in other cell systems, it will be of interest to evaluate their presence and importance in β cells and to correlate their expression after cell damage with β-cell survival and functional outcome. It would be of interest to induce expression of hsps (e.g., in transfection studies or transgenic animals) to study the resulting resistance against immune-mediated β-cell destruction.

ACKNOWLEDGEMENT

This work was generously supported by the Deutsche Forschungsgemeinschaft (DFG).

V. REFERENCES

1. R.I. Morimoto, K.D. Sarge, and K. Abravaya, Transcriptional regulation of heat shock genes. A paradigm for inducible genomic responses, *J. Biol. Chem. 267*: 21987 (1992).
2. J.R. Subjeck and T.-T. Shyy, Stress protein systems of mammalian cells, *Am. J. Physiol. 250*:C1 (1986).
3. W.J. Welch, Mammalian stress response: cell physiology, structure/function of stress proteins, and implications for medicine and disease, *Physiol. Rev. 72*:1063 (1992).
4. F.M. Ritossa, A new puffing pattern induced by a temperature shock and DNP, *Drosophila. Experientia Basel 18*:571 (1962).

5. E.A. Craig, Chaperones: helpers along the pathways to protein folding, *Science 260*:1902 (1993).
6. R.J. Ellis and S.M. van der Vies, Molecular Chaperones, *Annu. Rev. Biochem. 60*:321 (1991).
7. C.J.P. Boog, E.R. de Graeff-Meeder, M.A. Lucassen, R. van der Zee, M.M. Voorhost-Ogink, P.J.S. van Kooten, H.J. Geuze, and W. van Eden, Two monoclonal antibodies generated against human hsp60 show reactivity with synovial membranes of patients with juvenile chronic arthritis, *J. Exp. Med. 175*:1805 (1992).
8. I.R. Cohen, Autoimmunity to chaperonins in the pathogenesis of arthritis and diabetes, *Ann. Res. Immunol. 367*:567 (1991).
9. R. Kiessling, A. Grönberg, J. Ivanyi, K. Söderström, M. Ferm, S. Kleinau, E. Nilsson, and L. Klareskog, Role of hsp60 during autoimmune and bacterial inflammation, *Immun. Rev. 121*:91 (1991).
10. B.S. Polla, A role for heat shock proteins in inflammation?, *Immunol. Today 9*:134 (1988).
11. S.H.E. Kaufmann, Heat shock proteins and immune response, *Current Topics in Microbiology and Immunology* (S.H.E. Kaufmann, ed.), Springer-Verlag, New York, 1991.
12. J. von Mering and O. Minkowski, Diabetes mellitus nach Pankreasextirpation, *Arch. Exp. Pathol. Pharmakol. 26*:371 (1889).
13. Ä. Lernmark, Molecular biology of type 1 (insulin-dependent) diabetes mellitus, *Diabetologia 28*:195 (1985).
14. D.V. Serreze, Autoimmune diabetes results from genetic defects manifest by antigen presenting cells, *FASEB J. 7*:1092 (1993).
15. M. Welsh, D.L. Eizirik, and E. Standell, Heat shock treatment of mouse pancreatic islets results in a partial loss of islet cells but no remaining among the surviving β-cells, *J. Mol. Endocrinol. 1*:27 (1988).
16. S. Helqvist, B.S. Polla, J. Johannesen, and J. Nerup, Heat shock protein induction in rat pancreatic islets by human recombinant interleukin-1β, *Diabetologia 34*:150 (1991).
17. S. Helqvist, B.S. Hansen, J. Johannesen, H.U. Andersen, J.H. Nielsen and J. Nerup, Interleukin-1 induces new protein formation in isolated rat islets of Langerhans, *Acta Endocrinol. 121*:136 (1989).
18. D.L. Eizirik, M. Welsh, E. Strandell, N. Welsh, and S. Sandler, Interleukin-1β depletes insulin messenger ribonucleic acid and increases the heat shock protein HSP70 in mouse pancreatic islets without impairing the glucose metabolism, *Endocrinology 127*:2290 (1990).
19. N. Welsh, M. Welsh, S. Lindquist, D.L. Eizirik, K. Bendtzen, and S. Sandler, Interleukin-1β increases the biosynthesis of the heat shock protein HSP70 and specifically decreases the biosynthesis of five proteins in rat pancreatic islets, *Autoimmunity 9*:33 (1991).
20. D. Elias, D. Markovits, T. Reshef, R. van der Zee, and I.R. Cohen, Induction and therapy of autoimmune diabetes in the non-obese diabetic (NOD/Lt) mouse by a 65-kDa heat shock protein, *Proc. Natl. Acad. Sci. USA 87*:1576 (1990).

21. D. Elias, T. Reshef, O.S. Birk, R. van der Zee, M.D. Walker, and I.R. Cohen, Vaccination against autoimmune mouse diabetes with a T-cell epitope of the human 65-kDa heat shock protein, *Proc. Natl. Acad. Sci. USA 88*:3088 (1991).
22. C. Gelber, L. Paborsky, S. Singer, D. McAteer, R. Tisch, C. Jolicoeur, R. Buelow, H. McDevitt, and C.G. Fathman, Isolation of nonobese diabetic mouse T-cells that recognize novel autoantigens involved in the early events of diabetes, *Diabetes 43*:33 (1994).
23. N. Shehadeh, F. Calcinaro, B.J. Bradley, I. Bruchlim, P. Vardi, and K.J. Lafferty, Effect of adjuvant therapy on development of diabetes in mouse and man, *Lancet 343*:706 (1994).
24. E. Dana and I.R. Cohen, Peptide therapy for diabetes in NOD mice, *Lancet 343*: 704 (1994).
25. T.J. Venner and R.S. Gupta, Nucleotide sequence of rat hsp60 (chaperonin, GroEL homolog) cDNA, *Nucleic Acids Res. 18*:5309 (1990).
26. R.S. Gupta, Sequence and structural homology between a mouse T-complex protein TCP-1 and the chaperonin family of bacterial (GroEL, 60-65 kDa heat shock antigen) and eukaryotic proteins, *Biochem. Int. 20*:833 (1990).
27. J. Zeilstra-Ryalls, O. Fayet, and C. Georgopoulos, The universally conserved GroE (Hsp60) chaperonins, *Annu. Rev. Microbiol. 45*:301 (1991).
28. H. Koll, B. Guiard, J. Rassow, J. Ostermann, A.L. Horwich, W. Neupert, and F.U. Hart, Antifolding activity of hsp60 couples protein import into the mitochondrial matrix with export to the intermembrane space, *Cell 68*:1163 (1992).
29. W.B. Pratt, The role of heat shock proteins in regulating the function, folding, and trafficking of the glucocorticoid receptor, *J. Biol. Chem. 268*:21455 (1993).
30. D.B. Jones, N.R. Hunter, and G.W. Duff, Heat-shock protein as a β cell antigen of insulin-dependent diabetes, *Lancet 336*:583 (1990).
31. K. Brudzynski, V. Martinez, and R.S. Gupta, Secretory granule autoantigen in insulin-dependent diabetes mellitus is related to 62 kDa heat-shock protein (hsp60), *J. Autoimmun. 5*:453 (1992).
32. K. Brudzynski, V. Martinez, and R.S. Gupta, Immunocytochemical localization of heat-shock protein 60–related protein in beta-cell secretory granules and its altered distribution in non-obese mice, *Diabetologia 35*:316 (1992).
33. K. Brudzynski, Insulitis-cause redistribution of heat-shock protein HSP60 inside β-cells correlates with induction of HSP60 autoantibodies, *Diabetes 42*:908 (1993).
34. R. Göke, M.E. Trautmann, E. Haus, G. Richter, H.C. Fehmann, R. Arnold, and B. Göke, Signal transmission after GLP-1 (7-36)amide binding in RINm5F cells, *Am. J. Physiol. 257*:G397 (1989).
35. W.J. Welch, J.I. Garrels, G.P. Thomas, J.J. Lin, and J.R. Feramisco, Biochemical characterization of the mammalian stress proteins and identification of two stress proteins as glucose- and Ca^{2+} ionophore-regulated proteins, *J. Biol. Chem. 258*: 7102 (1983).
36. B. Göke, J.A. Williams, M.J. Wishart, and R.C. DeLisle, Low molecular mass GTP-binding proteins in subcellular fractions of the pancreas: regulated phosphoryl G proteins, *Am. J. Physiol. 262*:C493 (1992).

37. W.J. Welch, L.A. Mizzen, and A.T. Arrigo, Structure and function of mamalian stress proteins, *Stress-Induced Proteins* (M.L. Purdue, J.R. Feramisco, and S. Lindquist, eds.), Liss, New York, 1988, p. 187.
38. R. Göke, T. Cole, and J.M. Conlon, Characterization of the receptor for glucagon-like peptide-1 (7-36)amide on plasma membranes from rat insulinoma-derived cells by covalent cross-linking, *J. Mol. Endocrinol. 2*:93 (1989).
39. R. Göke, B. Oltmer, S. Sheik, and B. Göke, Solubilization of active GLP-1 (7-36)amide receptors from RINm5F plasma membranes, *FEBS Lett. 300*:232 (1992).
40. C. Eckerskorn, W. Mewes, H. Goretzki, and F. Lottspeich, A new siliconized-glass fiber as support for protein-chemical analysis of electroblotted proteins, *Eur. J. Biochem. 176*:509 (1988).
41. D.L. Eizirik, S. Sandler, and J.P. Palmer, Repair of pancreatic β-cells: a relevant phenomenon in early IDDM, *Diabetes 42*:1383 (1993).
42. T. Agner, P. Damm, and C. Binder, Remission in IDDM: prospective study of basal C-peptide and insulin dose in 268 consecutive patients, *Diabetes Care 10*:164 (1987).
43. E. Strandell, D.L. Eizirik, and S. Sandler, Reversal of β-cell suppression in vitro in pancreatic islets isolated from nonobese diabetic mice during the phase preceding insulin-dependent diabetes mellitus, *J. Clin. Invest. 85*:1944 (1990).
44. D.L. Eizirik, E. Strandell, and S. Sandler, Prolonged exposure of islets isolated from prediabetic nonobese diabetic mice to a high glucose concentration does not impair β-cell function, *Diabetologia 34*:6 (1991).
45. E. Strandell, S. Sandler, C. Boitard, and D.L. Eizirik, Role of infiltrating T-cells for impaired glucose metabolism in pancreatic islets isolated from nonobese diabetic mice, *Diabetologia 35*:924 (1992).
46. B.A. Margulis, S. Sandler, D.L. Eizirik, N. Welsh, and M. Welsh, Liposomal delivery of purified heat shock protein HSP70 into rat pancreatic islets as protection against interleukin-1β-induced impaired β-cell function, *Diabetes 40:*1418 (1991)
47. C.A. Sargent, I. Dunham, J. Trowsdale, and R.D. Campbell, Human major histocompatibility complex contains genes for the heat shock protein HSP70, *Proc. Natl. Acad. Sci. USA 86*:1968 (1989).
48. W. Wurst, C. Benesch, B. Drabent, R. Rothermel, B.-J. Benecke, and E. Günther, Localization of heat shock protein 70 genes inside the rat major histocompatibility complex close to class III genes, *Immunogenetics 30*:46 (1989).
49. H.R. Gaskins, M. Prochazka, J.H. Nadeau, V.W. Henson, and E.H. Leiter, Localization of a mouse heat shock HSP70 gene within the H-2 complex, *Immunogenetics 32*:286 (1990).
50. F. Pociot, J. Mølvig, L. Wogensen, H. Worsaae, H. Dalbøge, L. Baek, and J. Nerup, A tumor necrosis factor β gene polymorphism in relation to monokine secretion and insulin-dependent diabetes mellitus, *Scand J. Immunol. 33*:37 (1991).
51. A. Pugliese, Z.L. Awdeh, A. Galluzzo, E.J. Yunis, C.A. Alper, and G.S. Eisenbarth, No independent association between HSP70 gene polymorphism and IDDM, *Diabetes 41*:788 (1992).

52. T. Mandrup-Poulsen, K. Bendtzen, C.A. Dinarello, and J. Nerup, Human tumor necrosis factor potentiates interleukin-1–mediated rat pancreatic β-cell cytotoxicity, *J. Immunol. 139*:4077 (1987).
53. D.L. Eizirik, Interleukin-1 induced impairment in pancreatic islet oxidative metabolism of glucose is potentiated by tumor necrosis factor, *Acta Endocrinol. 119*:321 (1988).
54. J. Mølvig, A model of pathogenesis of insulin-dependent diabetes, *Danish Med. Bull. 6*:509, (1992).

28

Stress Proteins in Atherogenesis

Qingbo Xu
Institute for Biomedical Aging Research, Austrian Academy of Sciences, Innsbruck, Austria

Georg Wick
Institute for General and Experimental Pathology, University of Innsbruck Medical School, Innsbruck, Austria

I. INTRODUCTION

Atherosclerosis, the principal cause of heart attack, stroke, and gangrene of the extremities, remains a major contributor to morbidity and mortality in the Western world. The etiology and pathogenesis of atherosclerosis have been difficult to elucidate, because this disease progresses slowly from childhood and does not become manifest until middle age or later. Based on the results of experimental and clinical investigations, many factors, for example, hypercholesterolemia, modified lipoproteins, smoking, hypertension, and diabetes mellitus, were identified to be involved in the development of atherosclerosis. However, owing to technological advances in cellular and molecular approaches to the study of cell interactions in the arterial wall as well as alterations of the lipid metabolism, new insights into atherosclerotic pathogenesis have been provided during the last decade, which have been broadly summarized in two main theories: (1) the response to injury and (2) the modified low-density lipoprotein hypotheses. The former hypothesis relies on the concept that the primary cause of atherosclerosis is an arterial endothelium injury induced by certain factors, such as mechanical stress, oxidized low-density lipoproteins, homocysteine, immunological events, toxins, and viruses (1). The latter theory hypothesizes that low-density lipoproteins oxidized by various factors, including endothelial cells, macrophages, and smooth muscle cells of the arterial wall, play a key role in the development of atherosclerosis (2). More recently, humoral and cellular immunological reactions have received increased attention as possible factors involved in atherogenesis (reviewed in refs. 3–5), but it is not yet known which event(s) or factor(s) initiate(s) disease development or which factor(s) are responsible for lesion perpetuation.

Heat shock proteins (hsps), or stress proteins, comprise a group of families of approximately two dozen proteins and cognates thereof that show highly homologous sequences between species ranging from bacteria to humans. Detailed accounts of the physiological and possible pathological roles of hsps are given in other chapters of this book. Recently, hsps have been identified to be involved in the pathogenesis of atherosclerosis not only in experimental animals but also in humans (6–11). This chapter does not focus on all of the various controversial aspects concerning the etiology and pathogenesis atherosclerosis, but rather on recent work from our laboratory pointing to the role of hsps in the process of atherosclerosis.

II. LESIONS OF ATHEROSCLEROSIS

The atherosclerotic lesion is defined by arterial intimal cell proliferation, lipid accumulation, and connective tissue deposition. Depending on size and composition, the lesions are usually divided into fatty streaks, which predominantly consist of lipid-rich macrophages and T lymphocytes within the innermost layer of the artery wall (14–17), and plaques, which are advanced stages of the lesions. The three major cellular components of the human atherosclerotic plaques are smooth muscle cells, which dominate the fibrous cap, the macrophages, which are the most abundant cell type in the lipid-rich core region, and lymphocytes, which have been mainly ascribed to the fibrous cap (18–21).

The relationship of fatty streaks to atherosclerotic plaques is controversial (22). Fatty streaks occur early in life and by the age of 10 years have covered approximately 10% of the aortic intimal surface, including areas not particularly susceptible to the plaque development later in life (23). However, fatty streaks also occur in coronary arteries at the same anatomical sites prone to later development of plaques (24). Evidence from animal experiments supports the concept of an evolution of fatty streaks into plaques. For instance, in a hypercholesterolemic nonhuman primate model, Faggiotto and Ross (25) showed that many of the advanced lesions occurred at sites where fatty streaks were originally present. Thus, it is conceivable that only some "specific" fatty streaks in young subjects progress into advanced lesions.

III. IMMUNE COMPONENTS IN ATHEROSCLEROTIC LESIONS

As mentioned above, it has been established that atherosclerotic lesions of humans and rabbits contain a large number of T lymphocytes (6,14–21,26). About half these T cells express major histocompatibility complex (MHC) class II antigens, and some also express interleukin-2 (IL-2) receptors (27), indicating a state of activation. Recent studies have shown that T lymphocytes in human atherosclerotic plaques are primarily memory cells expressing the low molecular weight form (CD45RO) of the leukocyte common antigen and the integrin very late activation antigen 1 (VLA-1). These T cells are polyclonal in origin based on the phenotype of their T-cell receptors (TCR), that is, α/β and γ/δ (28,29), respectively. The presence of T cells in atherosclerotic lesions could be important, since they can act as effector cells and secrete factors chemotactic for monocytes/macrophages and smooth muscle cells, as well as determine the differentiation and function of B cells and monocytes/macrophages.

In young subjects, the first lymphocytes infiltrating the atherosclerotic lesion are $CD4^+$ (16,30). In fatty streaks, they represent the predominant population of $CD3^+$ cells, but in atherosclerotic plaques, the $CD4^+$ to $CD8^+$ cell distribution reverts to a preponderance of the latter (7,16). This result led us to postulate that $CD4^+$ cells may play a role in initiating a process that entails a T-cell–mediated cytotoxic response in the arterial intima. Furthermore, most T lymphocytes involved in atherosclerosis bear the α/β TCR. However, in the earliest stage of atherogenesis, we found an average of 9.7% γ/δ T cells, the function of which is still elusive, although TCR γ/δ^+ cells have been proposed to constitute a first line of defense (31), and our results also point to a possible participation of these cells in the early stages of atherosclerosis (7).

The MHC class I molecules, HLA-A, HLA-B, and HLA-C, are essential for activation of $CD8^+$ T cells, and are expressed by all types of nucleated cells, whereas class II determinants are normally expressed mainly by cells of the immune system. The latter include HLA-DR, HLA-DQ, and HLA-DP molecules and are restriction elements in the process of $CD4^+$ T cell activation (32). However, MHC class II antigens can also be expressed by parenchymal and stromal cells in many inflammatory and autoimmune conditions. In atherosclerotic lesions, endothelial cells were HLA-DR^+ only above those regions of the intimal lesions where HLA-DR^+ and interferon gamma–producing T cells had accumulated. Furthermore, one third of smooth muscle cells within plaques are not only HLA-DR^+ but also express HLA-DR γ-chain, suggesting that HLA-DR protein is actively synthesized by these cells (16). In vitro, cultured endothelial and smooth muscle cells express HLA-DR when treated with interferon gamma (INF-γ) (33,34). MHC class II expression of endothelial and smooth muscle cells may be induced by INF-γ released by activated T cells in their vicinity within atherosclerotic lesions. It is conceivable that such HLA-DR^+ endothelial cells might then serve as antigen-presenting cells during the perpetuation of the immune response in the atherosclerotic pathogenesis.

Atherosclerotic lesions also contain immunoglobulin deposits not found in nonatherosclerotic arteries (35). With a similar tissue distribution, complement factors are found in the lesions, including the lytic C5b-9 complex (36–38) formed as the endproduct of the complement cascade and serving as a device for perforating cell membranes. The assembly of the C5b-9 complex from the C5b, C6, C7, C8, and C9 components creates conformational neoantigens not present on native proteins. These findings strongly suggest that complement activation occurs within the lesions (39).

IV. HOW WAS HSP CONSIDERED TO BE INVOLVED IN ATHEROSCLEROSIS?

As mentioned above, atherosclerotic lesions bear many similarities to chronic inflammatory conditions, the most striking of which are accumulation of macrophages and T lymphocytes (reviewed in refs. 3–5). It was not known, however, which pathogens and/or antigens might elicit the response of these immune cells in atherogenesis. Our theory that atherosclerosis might be immunologically mediated was aroused by the increasing reports in the literature, partially cited above, centering around the humoral and cellular immunological phenomena in this disease (14–21,26–39), including the involvement of antibodies to modified low-density lipoprotein (LDL) (40), and bacteria (41). Our approach to this issue was based on our previous experience with various animal models of experimentally induced and spontaneous autoimmune diseases (42), as well as the role of changes in lymphocyte lipid metabolism in the altered immune responsiveness during aging (43,44).

We initially performed a series of experiments immunizing rabbits with proteins extracted from human or rabbit atherosclerotic lesions and emulsified with Freund's complete adjuvant (FCA) to identify one or more autoantigens within atherosclerotic lesions that might be responsible for inducing an autoimmune response as the first step in the atherosclerotic pathogenesis. However, in contrast to our expectations, we observed that arteriosclerosis consistently developed when FCA, which is an adjuvant containing heat-killed mycobacteria, was included in the immunization mixture irrespective of the antigen itself, for example, total human or rabbit lesion proteins or ovalbumin. Since hsp 65 is the major antigenic component of *Mycobacterium tuberculosis*, we then immunized rabbits with this component alone and found it to be capable of inducing atherosclerosis in the same manner as FCA (6). This serendipitous finding led us to research further the possible role of hsps in atherogenesis.

V. HSP 65 ANTIBODIES IN ATHEROSCLEROSIS

Arteriosclerotic lesions can be induced by immunization with hsp 65 in animals. Obviously, hsp 65 may serve as antigen or pathogen to initiate a specific immune reaction in the development of the disease. We therefore wondered whether hsp 65 was also involved as an antigen to stimulate B-cell production of specific antibodies in humans with atherosclerosis. To investigate this issue, we participated in an epidemiological study carried out in Bruneck, a small town in South Tyrol (8). Of 13,534 clinically normal inhabitants, 867 volunteers, aged 40–79 years, were randomly selected for determination of serum antibodies against hsp 65, simultaneous sonographic assessment of carotid atherosclerotic lesions, and evaluation of established risk factors, that is, blood cholesterol, hypertension, smoking, diabetes mellitus, and obesity. Autoantibodies to nuclear antigens, thyroid antigens, and rheumatoid factors were determined for control purposes. Our data showed that serum hsp 65 antibodies were significantly ($p < 0.01$) increased in subjects with carotid atherosclerosis compared with those without lesions, and this increased antibody level was independent of other established risk factors. On the other hand, the incidence and titers of various autoantibodies (antinuclear antibodies, antibodies to thyroglobulin and thyroid microsomal antigens, and rheumatoid factors) did not correlate with carotid atherosclerotic lesions. These data provided the first strong correlation of hsp 65 antibodies with carotid atherosclerosis, suggesting that hsp 65 may be involved in the pathogenesis of human atherosclerosis (8).

Antibodies to hsp 65/60 can be induced by several different mechanisms: Infection with agents containing hsp proteins homologous to mammalian hsp 60 could induce an antiself response through molecular mimicry in susceptible individuals (45); viral infection might result in hsp 60 incorporation into the budding virus, rendering it immunogenic (46); the protein could become immunogenic owing to structural alteration or posttranslational modification resulting from metabolic changes or viral infection (47); other foreign or self-antigens could interact with hsp 60 to form immunogenic complexes in which B cells recognize hsp 60, and T cells direct their response at the associated antigen (48); and finally, a bona fide autoimmune reaction to hsp 60 might occur. We hypothesized that the increased levels of hsp 65 antibodies may be involved in the development of atherosclerosis by a cross-reactive response to lesion proteins or cell components.

On the other hand, increased reactivity to mycobacterial hsp 65 and homologous mammalian hsp 60 antigens was observed previously in several other diseases, such as adjuvant arthritis in rats (49), insulin-dependent diabetes mellitus in mice (51), and in humans with rheumatoid arthritis (50), schizophrenia (52), and systemic sclerosis (53). The hsp 65 molecule containing 573 amino acids could feature multiple lymphocyte and/or antibody epitopes, and it is conceivable that at least one is cross reactive with an antigen in atherosclerotic lesions, whereas another cross reacts with an antigen present in joints. Indeed, the hsp 65 epitope critical for adjuvant arthritis (amino acids 180–188) appears to be homologous to a sequence in the link protein of the cartilage proteoglycan (50). Hence, responses to different hsp 65 epitopes might lead either to arthritis or to atherosclerosis depending on genetically determined recognization of different epitopes. Therapeutically, vaccination with the nonapeptide amino acid sequence 180–188 of hsp 65 was demonstrated to reduce the incidence and severity of adjuvant arthritis in rats (54), and administration of hsp 65 can vaccinate against diabetes in mice (51). If different hsp 65 epitopes are confirmed to play a pathogenic role in different diseases, the use of various peptides may also be considered for therapeutic or preventive purposes in humans in the future. Furthermore, although increased concentrations of serum antibodies to hsp 65 are not specific for atherosclerosis, they may prove to be a new diagnostic marker for this disease (55).

The question arose as to whether human serum antibodies to mycobacterial hsp 65 could react specifically against autologous hsp 60 or another cellullar component of atherosclerotic lesions. We selected human sera with high or low titers to recombinant mycobacterial hsp 65 and investigated their reactivity with human arterial lesion components using immunoblotting and immunofluorescence. All high-titer sera against hsp 65 reacted with a 60-kDa band of atherosclerotic lesion proteins and human recombinant hsp 60 on Western blots (Fig. 1). Pooled sera with a low antibody titer to hsp 65, which was diluted in a manner similar to high-titer sera, showed no reactivity with atherosclerotic lesion and media proteins. Using immunohistochemistry and immunofluorescence with human immunoglobulin (IgG) isolated from different sera, labeled with biotin, and visualized with a streptavidin conjugate, positive staining was observed in sections of fatty streaks and atherosclerotic plaques of carotid arteries and weak staining in normal intima. Double immunofluorescence identified the majority of positively stained cells as macrophages, endothelial cells, and a few smooth muscle cells (9). Therefore, serum antibodies against hsp 65 cross react with the human 60-kDa homologue present in high levels in atherosclerotic lesions mainly associated with macrophages and endothelial cells, supporting our concept of a humorally mediated immune reaction against hsp 60 in atherogenesis.

VI. HSP EXPRESSION IN THE ARTERIAL INTIMA

Berberian et al. (10) have shown high levels of hsp 70 in human atherosclerotic lesions, particularly in the central portions of atheroma and at necrotic core sites, correlating in intensity with the severity of the lesion. Hsp 70 was also demonstrated to increase arterial cell survival, suggesting that these proteins are associated with a protective mechanism in normal and diseased arteries (11). It is, however, not clear if hsp 70, like hsp 60, is involved in atherogenesis.

We studied hsp 60 expression in human atherosclerotic lesions with specimens of aortae, carotid arteries, and internal mammary arteries and veins as well

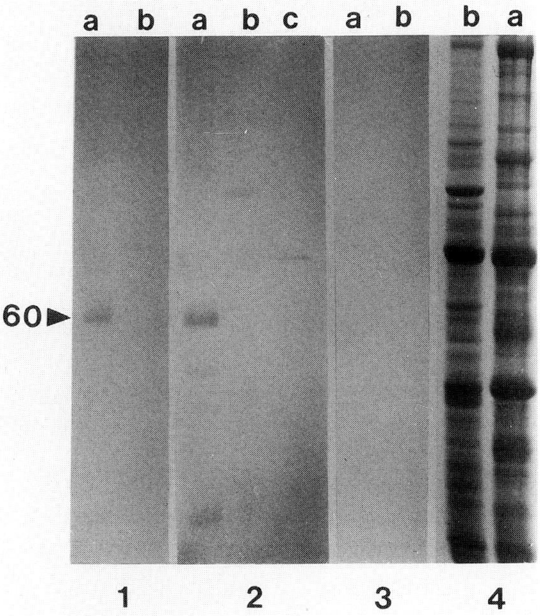

Figure 1 Western blots probed with sera from subjects with high and low titers against mycobacterial hsp 65. Homogenates of atherosclerotic lesions (a), arterial media (b), and purified recombinant mycobacterial hsp 65 (c), were separated on 10%T SDS-PAGE gels followed by transfer to a nitrocellulose membrane or Commassie staining (4). Blots were incubated with anti-hsp 65 high-titer sera (1,2) or pooled normal sera (3) for 1 hr at room temperature and the reaction visualized by diaminobenzidine/H_2O_2 after incubation with a rabbit anti-human Ig peroxidase conjugate. Note the 60-kDa band observed in atherosclerotic lesion proteins probed with high-titer anti-hsp 65 sera but not pooled normal serum.

as saphenous veins and venae cavae from 27 subjects, aged 23–80 years, using immunohistochemical and immunofluorescence techniques on serial frozen tissue sections (7). Hsp 60 was detected on the endothelium, smooth muscle cells, and/or mononuclear cells of all carotid and aortic specimens, whereas vessels of smaller diameters, serving as reference specimens for normal intima without atherosclerotic lesions and mononuclear infiltration, showed no detectable expression of this stress protein. In 30 specimens stained with a murine monoclonal antibody (ML30) recognizing an epitope common to mycobacterial hsp 65 and mammalian hsp 60 (56), no obvious association between superficial, that is, endothelial, staining and hsp 60 expression in macrophages, smooth muscle cells, and lymphocytes in deeper layers of the intima, was observed. From these results, we deduced that hsp 60 expression by endothelial cells depends on hemodynamic factors in large arteries, but its expression by mononuclear and smooth muscle cells seems to result from the inflammatory process itself. In detail, the factors leading to hsp 60 expression are numerous, but they can be collectively designated as influences threatening cellular integrity with possible lethal effects.

Interestingly, similar observations were obtained in an animal model of atherosclerosis (12), that is, in New Zealand White rabbits treated either by recombinant mycobacterial hsp 65 immunization or administration of a 0.2% cholesterol diet. Atherosclerotic lesions were assessed 16 weeks after immunization. Hsp 60 staining of aortas showed a heterogenous distribution and significantly increased staining intensity in atherosclerotic lesions compared with aortic media or adventitia. This abundantly expressed hsp 60 was observed in both atherosclerotic lesions induced by hsp 65 immunization and those induced by cholesterol-rich diet alone. Mechanisms leading to hsp 60 expression in aortic lesions may involve hemodynamic stress as well as hypercholesterolemia as primary stressors.

Although it is assumed that hsps are located in the cytoplasmic compartment to exert their function (57), evidence is emerging that surface expression of hsp 60 in mononuclear cells may also occur (58,59). To determine whether surface expression of hsps occurs on aortic endothelial cells, endothelial cells from rat aortae were cultured and stained with various monoclonal antibodies against hsp 60. Positive staining of endothelial cells on the surface was observed after stressing the cells with cytokine-containing medium, for example, tumor necrosis factor (TNF) α, and labeling with a specific monoclonal antibody (II-13) (60) against hsp 60, but not with another monoclonal antibody, ML30, recognizing a different epitope (56,57). FACS analyses showed over 80% of the stressed living endothelial cells to be positively surface stained with the anti–hsp 60 antibody II-13 (61).

The vascular surface of 1000 sqm represents an interface between humans and their inner environment. This surface is constantly exposed to aggressive attacks inherent in life, with an endless process of destruction by various stressors. Hsp 60 overexpression by endothelial cells may be one important response to these stresses in the course of physiological and pathological processes. For instance, stress factors such as mechanical stress (hypertension), shear stress (blood flow), hypercholesterolemia, free radicals, cytokines, infections, and surgical stress, may directly or indirectly stimulate hsp 60 expression by the arterial endothelium. The mechanisms of the different types of stress and the endothelial cell responses might be similar to that observed in other cells, as described in other chapters in this volume.

VII. HSP 60–SPECIFIC T CELLS IN ATHEROSCLEROSIS

In healthy humans, circulating T cells reacting with mycobacterial hsp 65 have been observed to cross react with shared epitopes of human hsps (57). All animals used in our studies showed circulating T cells reacting with hsp 65 at significantly higher levels than with ovalbumin, which was used as a nonrelevant control antigen (6,12). As described, hsps showed very high phylogenetic sequence homology, not only between bacterial genera but also throughout all higher forms of life. Life-long exposure to environmental microorganisms may partially explain the existence of certain basic levels of hsp-specific T cells in healthy subjects.

T cells in atherosclerotic lesions of rabbits constitute up to 20% of total lesion cells, most of which express MHC class II (Ia) antigens, that is, are activated (6,12,26). To determine whether these T cells specifically respond to hsp 65/60, we isolated and cultured T cells from atherosclerotic lesions and compared them with peripheral blood lymphocytes from the same animal (12). A population of the T lymphocytes isolated from all forms of atherosclerotic lesions specifically responded to hsp 65 in vitro. IL-2–expanded T-cell lines derived from atherosclerotic lesions showed significantly higher hsp 65 reactivity than those from peripheral blood of the same donor, supporting the hypothesis that T cells reacting to hsp 65/60 may be involved in the development of atherosclerosis.

There are different mechanisms by which activated T cells could emerge in the lesions, such as preferential recruitment of T cells already activated in regional lymph nodes and circulating in the blood, which is a possibility supported by the consisting finding of low numbers of activated T cells specifically reacting against hsp 65 in peripheral blood of rabbits (6,12) and humans (62) and reflecting previous contact with this antigen. Such cells adhere to the endothelial cells to a much higher degree than resting T cells (63). In addition, there is an increased expression of endothelial lymphocyte-adhesive molecules in atherosclerotic lesions (64), possibly resulting in preferential recruitment of activated T cells that could either maintain their activated state via local hsp 60 stimulation expressed in the lesions or disappear and become replaced by other activated T cells recruited from the blood.

VIII. INFECTIONS, HSP, AND ATHEROSCLEROSIS

Several lines of evidence point to infection as a possible contributor to atherosclerosis (66). Infection with Marek's disease virus (MDV) has been shown to induce atherosclerosis in chickens (67). Seroepidemiological studies have pointed to a higher prevalence of cytomegalovirus antibodies among several groups of patients with coronary heart disease as compared with controls (68), and immunohistological examination revealed the presence of herpesvirus in human atherosclerotic lesions (69).

With regard to bacterial infections, it has been demonstrated that serum antibody titers against *Chlamydia pneumoniae* posivitely correlate with coronary atherosclerosis. The antibody titers increase rapidly from the age of 5 to 20 years, and then increase more slowly into old age. Men have a higher prevalence of these antibodies than women, reflecting the higher rate of atherosclerosis in males (41,70). Interestingly, *C. pneumoniae* was detected within atherosclerotic lesions of coronary arteries (71), and prior infection with this microorganism positively correlated with the incidence of atherosclerosis (72). These studies suggest that *C. pneumoniae* infection may be a risk factor for the development of atherosclerosis.

It is also known that atherosclerotic lesions emerge in children as early as at 10–12 years of age (24), and the blood cholesterol concentration alone is not a plausible explanation, since most of these children have normal or even relatively low blood cholesterol levels. Alternatively, various microbial infections, such as those mentioned above, are frequent in children and may play a part in initiating atherosclerosis, although no causal relationship has yet been established between specific infections and atherosclerosis.

It is possible that a third variable may be associated with both infections and atherosclerosis, and that hsps might be the missing link between these conditions. For instance, infections can stimulate endothelial cells to overexpress hsps, directly or indirectly, by released cytokines and free radicals pathways. Infections trigger the immune system to induce T cells and antibodies specific for bacterial hsp 65, which in turn may bind to surface-expressed hsp 60 on the endothelium via cross reaction.

IX. LEUKOCYTE INTERACTION WITH ENDOTHELIUM

One of the earliest detectable events in atherosclerosis is the adherence of mononuclear cells to the endothelium and subsequent migration and accumulation in the intima where they form early lesions (1). Increased expression of vascular cell adhesion molecule 1 (VCAM-1) (64) and intercellular adhesion molecule 1 (ICAM-1) (73,74) have been found in arterial endothelial cells at sites prone to the emergence of atherosclerotic lesions in the intima. Recently, the induction of VCAM-1 on aortic endothelium was observed in rabbits as early as 1 week after feeding a cholesterol-rich diet (75). Alterations of adhesive properties of the vascular endothelial surface may contribute to leukocyte adherence, and depending on antigen specificity, to subsequent detachment or attachment and recruitment into subendothelial regions as a first stage in the atherosclerotic development (76).

To test the relationship between hsp 60 expression, ICAM-1, and leukocyte adhesion, we studied rat aortae after intravenous lipopolysaccharide (LPS) administration. Increased ICAM-1 expression on aortic endothelium was observed as early as 3 hr after LPS injection and persisted up to 72 hr, whereas elevated levels of hsp 60 were found between 6 and 48 hr. In vitro, cytokine-containing medium, H_2O_2, and high temperatures were found to stimulate endothelial expression of ICAM-1 and hsp 60. In this study, the predominant cells adhering to aortic endothelium after LPS administration were monocytes (80%) and T lymphocytes (8–20%). Interestingly, the areas with increased leukocyte adhesion to the aorta were identical to sites prone to the development of atherosclerotic lesions, that is, primarily on the aortic arch and arterial bifucation ostia. Most adherent cells were Ia^+ (Fig. 2). In organ cultures of rat aortae, LPS, cytokine-containing medium, and H_2O_2 evoked increased leukocyte adhesion to the endothelium, which was found to be a selective process, since adherent leukocytes were predominantly Ia^+ monocyte and T cells, that is, activated.

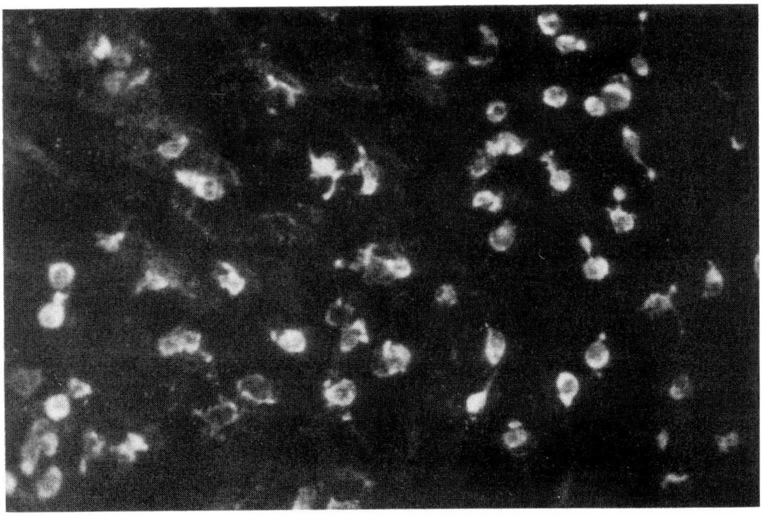

Figure 2 Ia^+ cells adhesion to arterial endothelium of rat in response to LPS. Aortic segments obtained from a rat 24-hr after LPS injection were mounted on glass slides, air-dried, fixed with ethanol for 20 min at room temperature, and labeled with monoclonal antibodies identifying Ia^+cells; for example, activated T cells and monocytes. Positive cells, visualized by rabbit anti-mouse Ig conjugated with TRITC, can be seen attaching to the endothelial surface, ×250.

Thus, increased expression of hsp 60 and ICAM-1 correlates with monocytes and T-cell adhesion to the aortic endothelium, observations that may elucidate the mechanisms of events initiating atherosclerosis (C. Seitz et al., submitted).

ICAM-1 expression in vivo could be induced by two distinct pathways: direct stimulation of endothelial cells by LPS or indirect stimulation via free radicals and cytokines released by macrophages, and even by endothelial cells themselves. Our in vitro data showed that H_2O_2, cytokine-containing medium and high temperature induced endothelial cell expression of ICAM-1, which further supports multiple pathways of endothelial cell stimulation by LPS. Similarly, hsp 60 expression of aortic endothelial cells in response to LPS stimulation may also involve the same factors. Thus, aortic endothelial cells injured by a single LPS administration can fully recover and replicate (77). Frequent repetition of this process during infection of lifelong exposure to environmental microorganisms may cause chronic inflammation in the arterial intima owing to humoral and/or cellular immune reactions to endothelial hsp 60.

X. CAUSAL INVOLVEMENT OF HSP 60 IN ATHEROGENESIS

Evidence for a causal involvement of hsp65 in atherogenesis is provided by our animal models induced by immunization or feeding a cholesterol-rich diet (6). As mentioned earlier in this chapter, we have immunized rabbits one or more times with various antigens, with or without adjuvants. The antigens and adjuvants, respectively, included human and rabbit atherosclerotic lesion proteins, ovalbumin, recombinant mycobacterial hsp 65, Freund's complete and incomplete adjuvants, and two hsp-free adjuvants, Ribi and lipopeptide. Arteriosclerotic lesions in the intima of the aortic arch developed only in animals immunized with antigenic preparations containing hsps, either as whole mycobacteria or purified recombinant hsp 65, irrespective of the addition of any further antigens and despite normal serum cholesterol levels (6,78). These results suggest that an (auto)immune response to hsp 65 may initiate the development of atherosclerosis.

Arteriosclerotic lesions induced by immunization with hsp 65–containing material were characterized by mononuclear cell infiltration, connective tissue deposition and smooth muscle cell proliferation, and a lack of foam cells. The cell compositions of these lesions showed distinct similarities to a chronic inflammation process. These lesions can regress after time in the absence of additional atherosclerotic risk factors (Q. Xu, et al., submitted). In addition, a combination of immunization with hsp 65–containing material and a cholesterol-rich diet led to the development of complicated atherosclerotic lesions similar to the classic human lesions (16).

XI. CONCLUSIONS

As detailed in this chapter, hsp expression in arterial endothelium can be induced or augmented by various types of stress, such as hypercholesterolemia, heat treatment, exposure to oxygen radicals or cytokines, ischemia, hemodynamic overload, surgical stress, and infections. Surface exposure of hsp 60 epitope(s) also emerged on endothelial cells. Thus, specific antibodies and T lymphocytes against hsp 60/65 preexisting in circulation could react with surface-exposed hsp 60 components or cross react with other proteins owing to sequence homologies (79) on aortic endothelium, including antibody-complement–mediated cell lysis, antibody-dependent cellular cytotoxicity (ADCC), and/or specific T-cell–mediated responses, to cause endothelial injury and subsequently initiate development of early lesions. Furthermore, antibodies could penetrate the injured endothelium of early atherosclerotic lesions and specifically bind hsp 60 expressed in high levels on macrophages/foam cells, resulting in the cell lysis that may contribute to the formation of the necrotic core promoting the lesions into advanced stages. In this process, a great number of cytokines may be released by endothelial cells, macrophages, and smooth muscle cells and participate in the formation of atherosclerosis via regulation of cell chemotaxis and proliferation. This hypothesis is summarized in Figure 3.

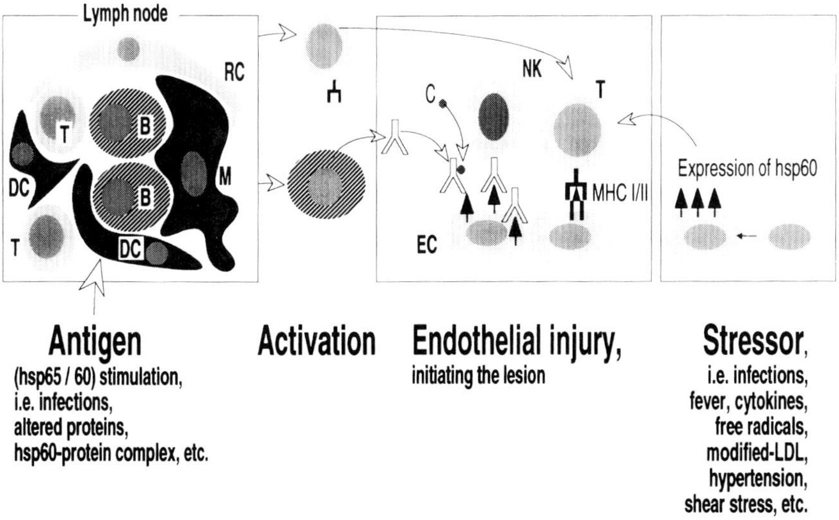

Figure 3 Schematic figure representing the stress-induced immune response hypothesis for the initiation of atherosclerosis. T, T cells; B, B cells; M, macrophages; RC, reticular cells; DC, dendritic cells; NK, NK cells mediating antibody-dependent cellular cytotoxicity (ADCC), C, complement.

ACKNOWLEDGMENTS

This work has been supported by grants from Austrian Research Council (projects no. 8925 and 10677 G.W.) and the Austrian Society for Geriatrics and Gerontology (Walter Doberauer Stipend for Aging Research, Q.X.).

We thank A. Mair, R. Kleindienst, C.S. Seitz, and H. Dietrich, who, together with many other collaborators, played a critical role in the work summarized in this review; and I. Atzinger for the preparation of photographs.

REFERENCES

1. R. Ross, The pathogenesis of atherosclerosis: a perspective for the 1990s, *Nature 362*:801 (1993).
2. D. Steinberg and J.L. Witztum, Lipoproteins and atherogenesis: current concepts, *J.A.M.A. 264*:3047 (1990).
3. G.K. Hansson, L. Jonasson, P.S. Seifert, and S. Stemme, Immune mechanisms in atherosclerosis, *Arteriosclerosis. 9*:567 (1989).
4. P. Libby and G.K. Hansson, Involvement of the immune system in human atherogenesis: current knowledge and unanswered questions. *Lab. Invest. 64*:5 (1991).
5. G. Wick, R. Kleindienst, H. Dietrich, and Q. Xu, Is atherosclerosis an autoimmune disease? *Trends Food Sci. Technol. 3*:114 (1992).
6. Q. Xu, H. Dietrich, H.J. Steiner, A.M. Gown, B. Schoel, G. Mikuz, S.H.E. Kaufmann, and G. Wick, Induction of arteriosclerosis in normocholesterolemic rabbits by immunization with heat shock protein 65, *Arterioscler. Thromb. 12*:789 (1992).
7. R. Kleindienst, Q. Xu, J. Willeit, F. Waldenberger, S. Weimann, and G. Wick, Immunology of atherosclerosis: Demonstration of heat shock protein 60 expression and T-lymphocytes bearing α/β and γ/δ receptor in human atherosclerotic lesions, *Am. J. Pathol. 142*:1927 (1993).
8. Q. Xu, J. Willeit, M. Marosi, R. Kleindienst, F. Oberhollenzer, S. Kiechl, T. Stulnig, G. Luef and G. Wick, Association of serum antibodies to heat shock protein 65 with carotid atherosclerosis, *Lancet 341*:255 (1993).
9. Q. Xu, G. Luef, S. Weimann, R.S. Gupta, H. Wolf, and G. Wick, Staining of endothelial cells and macrophages in atherosclerotic lesions with human heat-shock protein reactive atisera, *Arterioscler. Thromb. 13*:1763 (1993).
10. P.A. Berberian, W. Myers, M. Tytell, V. Challa, and M.G. Bond Immunohistochemical localization of heat shock protein-70 in normal appearing and atherosclerotic specimens of human arteries, *Am. J. Pathol. 136*:71 (1990).
11. A.D. Johnson, P.A. Berberian, and M.G. Bond, Effect of heat shock proteins on survival of isolated aortic cells from normal and atherosclerotic cynomolgus macaques. *Atherosclerosis 84*:111 (1990).
12. Q. Xu, R. Kleindienst, W. Waitz, H. Dietrich, and G. Wick, Increased expression of heat shock protein 65 coincides with a population of infiltrating T lymphocytes in atherosclerotic lesions of rabbits specfically responding to heat shock protein 65. *J. Clin. Invest. 91*:2693 (1993).

13. A.D. Johnson, P.A. Berberian, M. Tytell, and M.G. Bond, Atherosclerosis alters the localization of hsp70 in human and macaque aortas, *Exp. Mol. Pathol. 58*:155 (1993).
14. J.M. Munro, J.D. van der Walt, C.S. Munro, J.A.C. Chambers, and E.L. Cox, An immunohistochemical analysis of human aortic fatty streaks. *Hum. Pathol. 18*:375 (1987).
15. E.E. Emeson and A.L. Robertson, Jr., T lymphoctyes in aortic and coronary intimas: their potential role in atherogenesis, *Am. J. Pathol. 130*:369 (1988).
16. Q. Xu, G. Oberhuber, M. Gruschwitz, and G. Wick, Immunology of atherosclerosis: cellular composition and major histocompatibility complex class II antigen expression in aortic intima, fatty streaks, and atherosclerotic plaques in young and aged human specimens, *Clin. Immunol. Immunopathol. 56*:344 (1990).
17. S. Katsuda, H.C. Boyd, C. Fligner, R. Ross, and A.M. Gown, Human atherosclerosis: III. Immunocytochemical analysis of the cell composition of lesions of young adults. *Am. J. Pathol. 140*:907 (1992).
18. L. Jonasson, J. Holm, O. Skalli, G. Bondjers, and G.K. Hansson, Regional accumulations of T cells, macrophages and smooth muscle cells in the human atherosclerotic plaque, *Arteriosclerosis 6*:131 (1986).
19. A.M. Gown, T. Tsukada, and R. Ross, Human atherosclerosis: II. Immunocytochemical analysis of the cellular composition of human lesions, *Am. J. Pathol. 125*:191 (1986).
20. G.K. Hansson, L. Jonasson, B. Lojsthed, S. Stemme, O. Kocher, and G. Gabbiani, Localization of T lymphocytes and macrophages in fibrous and complicated human atherosclerotic plaque, *Atherosclerosis 72*:135 (1988).
21. A.C. van der Wal, P.K. Das, D.B. van de Berg, C.M. van der Loos, and A.E. Becher, Atherosclerotic lesions in humans: *in situ* immunophenotypic analysis suggesting an immune mediated response, *Lab. Invest. 61*:166 (1989).
22. H.C. McGill HC, Jr., Persistent problems in the pathogenesis of atherosclerosis, *Arteriosclerosis 4*:443 (1984).
23. H.C. McGill Jr., Fatty streaks in the coronary arteries and aorta, *Lab. Invest. 18*:560 (1968).
24. H.C. Stary, Evolution and progression of atherosclerotic lesions in coronary arteries of children and young adults. *Arteriosclerosis 9 (Suppl.)*:19 (1989).
25. A. Faggiotto and R. Ross, Studies of hypercholesterolemia in the nonhuman primate. II. Fatty streak conversion to fibrous plaque, *Arteriosclerosis 4*:341 (1984).
26. G.K. Hansson, P.S. Seifert, G. Olsson, and G. Bondjers, Immunohistochemical detection of macrophages and T lymphocytes in atherosclerotic lesions of cholesterol-fed rabbits, *Arterioscler. Thromb. 11*:745 (1991).
27. G.K. Hansson, J. Holm, and L. Jonasson, Detection of activated T-lymphocytes in the human atherosclerotic plaque, *Am. J. Pathol. 135*:169 (1989).
28. S. Stemme, J. Holm, and G.K. Hasson, T lymphocytes in human atherosclerotic plaques are memory cells expressing CD45RO and the integin VLA-1. *Arterioscler. Thromb. 12*:206 (1992).
29. S. Stemme, L. Rymo, and G.K. Hansson, Polyclonal origin of T lymphocytes in human atherosclerotic plaques, *Lab. Invest. 65*:654 (1991).

30. Q. Xu, G. Oberhuber, and G. Wick, Immunology of atherosclerosis: cell composition and MHC class II antigen expression of aortic intima, (abstr.), *Arteriosclerosis 9*:703a (1989).
31. R.L. O'Brien, M.P. Happ, A. Dallas, E. Palmer, R. Kubo, and W.K. Born, Stimulation of a major subset of lymphocytes expressing T cell receptor gamma/delta by an antigen derived from mycobacterium tuberculosis, *Cell 57*:668 (1989).
32. M.M. Davis and P.J. Bjorkman, T-cell antigen genes and T-cell recognition, *Nature 334*:395 (1988).
33. J.S. Pober, and M.A. Gimbrone Jr., Expression of Ia-like antigens by human vascular endothelial cells in inducible *in vitro*, *Proc. Natl. Acad. Sci. USA 79*:6641 (1982).
34. G.K. Hansson, J. Holm, L. Jonasson, M.M. Clowes, A.W. Clowes, γ-Interferon regulates vascular smooth muscle proliferation and Ia expression *in vivo* and *in vitro*, *Cir. Res. 63*:712 (1988).
35. W. Hollander, M.A. Colombo, B. Kirkpatrick, and J. Paddock, Soluble proteins in the human atherosclerotic plaque, *Atherosclerosis 38*:391 (1979).
36. R. Vlaicu, F. Niculescu, H.G. Rus, and A. Cristea, Immunohistochemical localization of the terminal C5b-9 complement complex in human aortic fibrous plaque, *Atherosclerosis 57*:163 (1985).
37. P.S. Seifert and G.K. Hansson, Complement receptors and regulatory proteins in human atherosclerotic lesions, *Arteriosclerosis 9*:802 (1989).
38. P.S. Seifert, F. Hugo, G.K. Hansson, and S. Bhakdi, Prelesional complement activation in experimental atherosclerosis: termianl C5b-9 complement deposition coincides with accumulation in aortic intima of hypercholesterolemic rabbits, *Lab. Invest. 6*:747 (1989).
39. P.S. Seifer, F. Hugo, J. Tranum-Jensen, U. Zahringer, M. Muhly, and S. Bhakdi, Isolation and characterization of a complement-activating lipid extracted from human atherosclerotic lesion, *J. Exp. Med. 172*:547 (1991).
40. J.T. Salonen, S. Ylä-Herttuala, R. Yamamoto, S. Butler, H. Korpela, R. Salonen, K. Nyyssönen, W. Palinski, and J.L. Witztum, Autoantibody against oxidised LDL and progression of carotid atherosclerosis, *Lancet 339*:883 (1992).
41. P. Saikku, M. Leinonen, K. Mattila, M.R. Ekman, M.S. Nieminen, P.H. Mäkelä, J.K. Huttunen, and V. Valtonen, Serological evidence of an association of a novel chlamydia, TWAR, with chronic coronary heart disease and acute myocardial infarction, *Lancet 2*:983 (1988).
42. G. Wick, H.P. Brezinschek, K. Hala, H. Deitrich, H. Wolf, and G. Kroemer, The Obese strain of chickens: an animal model with spontaneous autoimmune thyroiditis, *Adv. Immunol. 47*:433 (1989).
43. K.N. Traill, L.A. Huber, G. Wick, and G. Jürgens, Lipoprotein interactions with T cells: an update, *Immunol. Today 11*:411 (1990).
44. G. Wick, L.A. Huber, Q. Xu, E. Jarosch, D. Schönitzer, and G. Jürgens, The decline of the immune response during aging: The role of an altered lipid metabolism, *Ann. N.Y. Acad. Sci. 621*:277 (1991).
45. R.I. Cohen, A heat shock protein, molecular mimicry and autoimmunity, *Israel J. Med. Sci. 26*:673 (1990).

46. R.M. Zinkernagel, S. Cooper, J. Chambers, R.A. Lazzarini, H. Hengartner, and H. Arnheiter, Virus-induced autoantibody response to a transgenic viral antigen, *Nature 345*:68 (1990).
47. A. Schattner and B. Rager-Zisman, Virus-induced autoimmunity, *Rev. Infect. Dis. 12*:204 (1990).
48. P.G. Coulie and J. van Snick, Rheumatoid factor production during anamnestic responses in the mouse, *J. Exp. Med. 161*:88 (1985).
49. W. van Eden, J.E.R. Thole, R. van der Zee, A. Noordzij, J.D.A. van Embden, E.J. Hensen, and I.R. Cohen, Cloning of the mycobacterial epitope recognized by T lymphocytes in adjuvant arthritis, *Nature 331*:171 (1988).
50. P.C.M. Res, C.G. Schaar, F.C. Breedveld, W. Van Eden, J.D.A. Van Embden, I.R. Cohen, and R.R.P. de Vries, Synovial fluid T cell reactivity against the 65 kD heat-shock protein of mycobacteria in early onset of chronic arthritis, *Lancet 2*:478 (1988).
51. D. Elias, D. Markovits, T. Reshef, R. van der Zee, and I.R. Cohen, Induction and therapy of autoimmune diabetes in the non-obese diabetic (NOD/Lt) mouse by a 65-kDa heat shock protein, *Proc. Natl. Acad. Sci. USA 87*:1576 (1990).
52. K. Kilidireas, N. Latov, D.H. Strauss, A.D. Gorig, G.A. Hashim, J.M. Gorman, and S.A. Sadiq, Antibodies to the human 60 kDa heat-shock protein in patients with schizophrenia, *Lancet 340*:569 (1992).
53. M.G. Danieli, M. Candela, A.M. Ricciatti, R. Reginelli, G. Danieli, I.R. Cohen, and A. Gabrielli, Antibodies to mycobacterial 65 kDa heat shock protein in systemic sclerosis (scleroderma), *J. Autoimmun. 5*:443 (1992).
54. X.D. Yang, J. Gasser, B. Riniker, and U. Feige, Treatment of adjuvant arthritis in rats: vaccination potential of a synthetic nonapeptide from the 65 kDa heat shock protein of mycobacteria, *J. Autoimmun. 3*:11 (1990).
55. G.K. Hansson, Immunological markers of atherosclerosis, *Lancet 341*:278 (1993).
56. D.J. Evans, P. Norton, and J. Ivanyi, Distribution in tissue sections of the human GroEL stress protein homologue, *A.P.M.I.S. 98*:437 (1990).
57. R. Kiessling, A. Grönberg, J. Ivanyi, K. Söderstrom, M. Ferm, S. Kleinau, E. Nilsson, and L. Klareskog, Role of hsp60 during autoimmune and bacterial inflammation, *Immunol. Rev. 12*:91 (1991).
58. A.W. Wurttenberg, B. Shoel, J. Ivanyi, and S.H.E. Kaufmann, Surface expression by mononuclear phagocytes of an epitope shared with mycobacterial heat shock protein 60, *Eur. J. Immunol. 21*:1089 (1991).
59. S.D. Cesare, F. Poccia, A. Mastino, and V. Colizzi, Surface expressed heat-shock proteins by stressed or human immunodeficiency virus (HIV)–infected lymphoid cells represent the target for antibody-dependent cellular cytotoxicity, *Immunology 76*:341 (1992).
60. B. Singh, and R.S. Gupta, Expression of human 60-kD heat shock protein (HSP 60 or P1) in *Escherichia coli* and the development and characterization of corresponding monoclonal antibodies, *DNA Cell. Biol. 11*:489 (1992).
61. Q. Xu, and G. Wick, Co-expression of heat shock protein 60 and ICAM-1 in aotic endothelial cells after stress (abstr.), *Eur. J. Cell. Biol. 60 (Suppl. 37)*:73 (1993).

62. M.E.B. Munk, B. Schoel, and S.H.E. Kaufmann, 1988. T cell responses of normal individuals towards recombinant protein antigens of *Mycobacterium tuberculosis*. *Eur. J. Immuno. 18*:1835 (1988).
63. N.E. Damle and L.V. Doyle, Ability of human T lymphocytes to adhere to vascular endothelial cells and to augment endothelial permeability to macromolecules is linked to their state of post-thymic maturation, *J. Immunol. 144*:1233 (1990).
64. M.I. Cybulsky and M.A. Gimbrone Jr., Endothelial expression of a mononuclear leukocyte adhesion molecule during atherogenesis, *Science 251*:788 (1991).
65. Q. Xu and G. Wick, Surface expression of heat shock protein 60 on endothelial cells (abstr.), *Immunobiol. 189*:131 (1993).
66. V.V. Valtonen, Infection as a risk factor for infarction and atherosclerosis, *Ann. Med. 23*:539 (1991).
67. D.P. Hajjar, C.G. Fabricant, C.R. Minick, and J. Fabricant, Virus-induced atherosclerosis, *Am. J. Pathol. 122*:62 (1986).
68. E. Adam, J.L. Melnick, J.L. Probtsfield, B.L. Petrie, J. Burek, K.R. Bailey, C. H. McCollum, and M.E. DeBakey, High levels of cytomegalovirus antibody in patients requiring vascular surgery for atherosclerosis, *Lancet 2*:291 (1987).
69. H.M. Yamashiroya, L. Ghosh, R. Yang, and A.L. Jr. Robertson, Herpesviridae in coronary vessels and aorta of young trauma victims, *Am. J. Pathol. 130*:71 (1988).
70. D.H. Thom, S.P. Wang, T. Grayston, D.S. Siscovick, D.K. Stewart, R.A. Kronmal, and N.S. Weiss, *Chlamydia pneumoniae* strain TWAR antibody and angiographically demonstrated coronary artery disease, *Arterioscler. Thromb. 11*:547 (1991).
71. C.-c. Kuo, A. Shor, L.A. Cambell, H. Fukushi, D.L. Patton, and J.T. Grayston, Demonstration of *Chlamydia pneumoniae* in atherosclerotic lesions of coronary arteries, *J. Infect. Dis. 167*:841 (1993).
72. D.H. Thom, J.T. Grayston, D.S. Siscovick, S.P. Wang, N.S. Weiss, and J.R. Daling, Association of prior infection with Chlamydia pneumoniae and angiographically demonstrated coronary artery disease, *J.A.M.A. 268*:68 (1992).
73. R.N. Poston, D.O. Haskard, J.R. Coucher, N.P. Gall, and R.R. Johnson-Tidey, Expression of intercellular adhesion molecule-1 in atherosclerotic plaques, *Am. J. Pathol. 140*:665 (1992).
74. K.M. Wood, M.D. Cadogan, A.L. Ramshaw, and D.V. Darums, The distribution of adhesion molecules in human atherosclerosis, *Histopathology 22*:437 (1993).
75. H. Li, M.I. Cybulsky, M.A. Gimbrone, Jr., and P. Libby, An atherogenic diet rapidly induces VCAM-1, a cytokine-regulatable mononuclear leukocyte adhesion molecule, in rabbit aortic endothelium, *Arterioscler. Thromb. 13*:197 (1993).
76. G. Wick, R. Kleindienst, C. Seitz, and Q. Xu, Atherosclerosis is initiated by an immune reaction against heat shock protein 65/60 (abstr.), *J. Immunol. 150*:76a (1993).
77. M.A. Reidy, A reassessment of endothelial injury and arterial lesion formation, *Lab. Invest. 53*:513 (1985).

78. G. Wick G, and Q. Xu, Heat shock protein 65 as an antigen in atherogenesis (abstr.), *Arterioscler. Throm. 11*:1526a (1991).
79. D.B. Jones, A.F.W. Coulson, and G.W. Duff, Sequence homologies between hsp60 and autoantigens, *Immunol. Today 14*:115 (1993).

29

Stress Proteins and Myocardial Protection

Michael R. Gralinski and Benedict R. Lucchesi
University of Michigan Medical School, Ann Arbor, Michigan

Shawn C. Black
Merck Frosst, Pointe-Claire, Quebec, Canada

I. INTRODUCTION

Research involving the putative function of stress proteins as they relate to the cardiovascular system has increased within the past decade. The cardiovascular literature contains a growing number of articles hypothesizing that stress, or heat shock, proteins play a conspicuous role in protecting the heart from detrimental events that may compromise the functioning myocardium. Since the primary cause of death in many countries continues to involve the cardiovascular system, investigators from many disciplines carry on the search for innovative methods to confer a "cardioprotective" state on this system.

The early cardiovascular literature dealing with stress proteins is embodied in an article by Currie that appeared in 1981 (1). Currie demonstrated that hyperthermic stress induced the in vivo synthesis of a novel protein, which he designated P71, in many rat tissues, including the myocardium. This stress, or heat shock, protein, he purported, may be induced in vivo by stimuli other than hyperthermia. Since this report, investigators have utilized means other than hyperthermia to induce the expression of stress proteins, including the 71-kDa protein. Li and colleagues demonstrated that sodium arsenite, ethanol, and hypoxia induced both increased protein synthesis (MW 70,000 and 87,000) and transient thermotolerance in plateau-phase Chinese hamster fibroblast (HA-1) cells (2). Administering high doses of isoproterenol, which produces myocardial necrosis, can induce the synthesis and accumulation of a 71,000-dalton protein (SP71) in rodent tissues, including the heart and aorta (3). These investigators found that "SP71" was identical to the aforementioned protein found to be synthesized in the heart in response to heat shock and tissue slicing (4). Detectable amounts of SP71 were no longer seen 5 days after the isoproterenol injection, suggesting that this protein accumulates as part of a cellular response to stress and subsequently degrades during tissue repair (3). Even the imposition of a volume or pressure overload has been implicated as an impetus for stress protein expression. Delcayre and colleagues (5) used isolated rat heart myocytes incubated with [^{35}S]methionine to study the expression of proteins during the first 2 weeks after either pressure or volume overload. In both models, an early (2–4 days) and transient expression of three major stress proteins (heat shock proteins [hsps] 70, 68, and 58) was observed together with an increased synthesis of putative ribosomal proteins. Only traces of ^{35}S-labeled hsps were detected in controls and sham-operated animals. These prefatory results suggesting a protective role for heat stress proteins in the cardiovascular system become more interesting on examination of the circumstances under which most deleterious cardiovascular events occur.

II. MYOCARDIAL ISCHEMIC AND REPERFUSION INJURY

In the clinical scenario, the antecedent of most myocardial damage continues to be a combination of two phenomena termed ischemia and reperfusion. A regional ischemic event occurs in the heart when the blood supply to an area of functioning myocardium is interrupted, usually due to the formation of a thrombus within the lumen of a coronary vessel. If the thrombus reaches a critical mass, an occlusion forms, depriving the viable tissue of nutrients. Since the ischemic tissue attempts to function without incoming blood flow, detrimental metabolic by-products accumulate within the tissue. For example, high-energy phosphate-producing glycolytic pathways lead to the formation of lactic acid under anaerobic conditions. On restoration of flow to the ischemic area, oxygen is delivered into the tissue. The reintroduction of molecular oxygen to a previously ischemic region of the heart can lead to profoundly damaging effects (6). This damage results, in part, from the generation of oxygen-derived free radicals released during the initial reperfusion period. Sources of oxygen-derived free radicals (notably superoxide) include xanthine oxidase (7), mitochondria (8), and infiltrating neutrophils (9). Contributing to the damage caused by the reintroduction of oxygen, and hence superoxide, is an ischemia-related reduction in cellular antioxidant defenses (10). After superoxide-mediated cellular damage, the intracellular environment is not able to maintain a normal ionic gradient with the extracellular milieu. Calcium enters the cell to a greater than normal extent and overwhelms the sarcoplasmic reticulum which is involved in the regulation of intracellular calcium and excitation-contraction coupling (11–13). The "paralyzed" sarcoplasmic reticulum results in a progressive loss of cardiac function.

III. CARDIOPROTECTIVE EFFECTS OF HSP INDUCTION—IN VITRO

As indicated, myocardial ischemia plays a major pathophysiological role in the genesis of cardiovascular disorders. Interestingly, ischemia has been shown to induce the expression of one or more "stress proteins" in mammalian cells and tissues (14–19). During a less severe form of oxygen depletion (hypoxia), hearts accumulate messenger RNA for at least two polypeptides at substantially elevated levels (20–22). Howard and colleagues (20) demonstrated that the molecular weights of proteins, 85 kDa and 95 kDa, are similar to those reported for other mammalian stress proteins or glucose-regulated proteins. Time course experiments suggest that mRNAs for these species increase continuously for up to 16 hr of treatment, whereas mRNA for 71- and 79-kDa polypeptides are elevated early in the treatment but later decrease to control values. The results demonstrate that rodent cardiac tissue is capable of mounting a cellular stress-like response when exposed to moderately stressful conditions, including reduced oxygen supply.

Since myocardial ischemia is a major concern, investigators began to expand the scope of their search for cardioprotection. If the expression of the "protective" stress protein could be increased by subjecting the animal to hyperthermia before an episode of myocardial ischemia, any putative cardioprotective properties of the stress protein should be revealed. An early report addressing this hypothesis was by Karmazyn and colleagues (23) who demonstrated that induction of the heat shock phenomenon alters the response of isolated hearts to ischemia and reperfusion. Anesthetized male rats were exposed to 15 min of hyperthermia (internal temperature of 42°C) and then allowed to recover for 0, 24, 48, 96, or 192 hr. The hearts were isolated from control and the hyperthermia-treated rats and perfused by the Langendorff method. After 30 min of global ischemia, improved functional recovery was observed in 48-hr post–heat shock hearts as compared with control hearts. During reperfusion there was an increased recovery of force, +dF/dt, and −dF/dt in the heat shocked hearts when compared with controls (Fig. 1). Creatine kinase efflux during reperfusion was reduced by 75% in the 24-hr post–heat shock hearts. The results suggested a correlation between the acquisition and decay of the enhanced ex vivo postischemic ventricular recovery and the hyperthermic induction of the heat shock response indicated by the accumulation of the heat shock protein hsp 71. Other

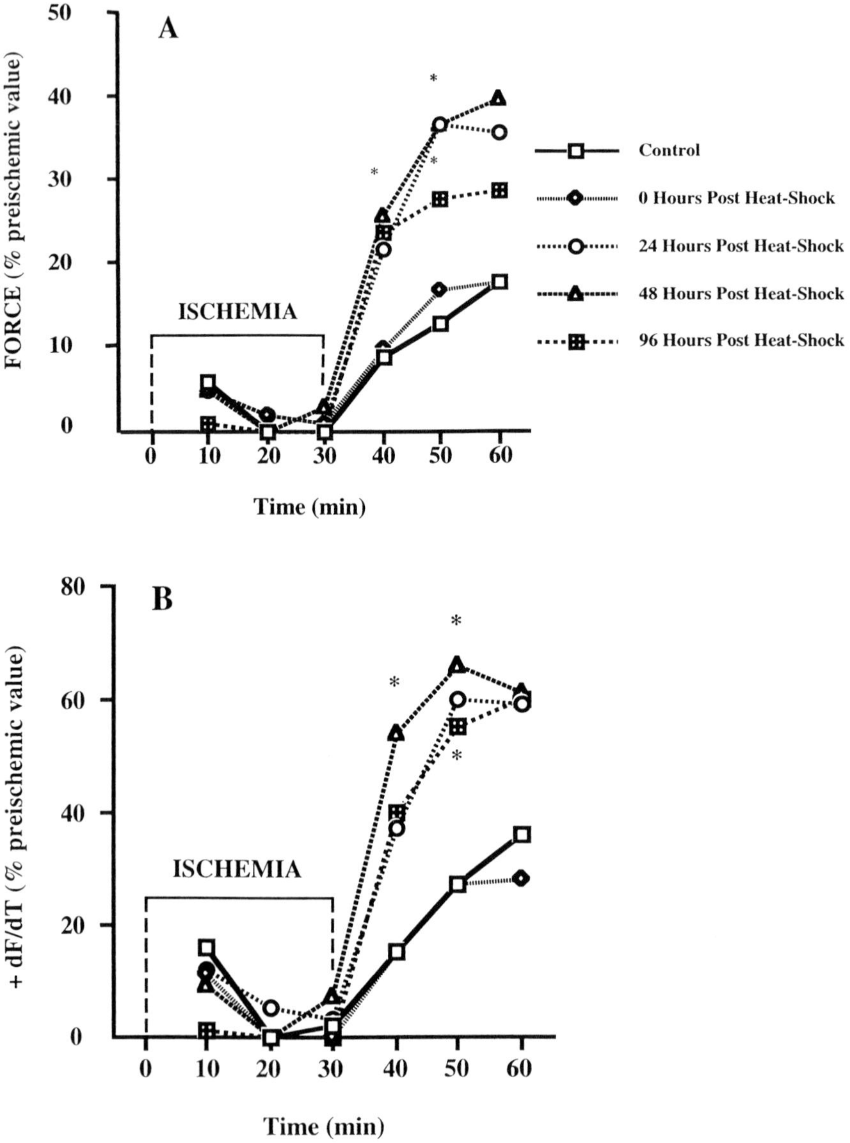

Figure 1 Contractile responses of isolated hearts from control and 0-, 24-, 48-, and 96-hr posthyperthermia treated rats. Data were normalized to preischemic contractile values. Each point represents the mean. Ischemia (I, vertical dotted lines) was initiated at 0 min and maintained for 30 min.* $p<0.05$ from control values using Student's *t*-test. (A) Force. (B) +dF/dT (rate of contraction). (C) -dF/dT (rate of relaxation). (Adapted from Ref. 22.)

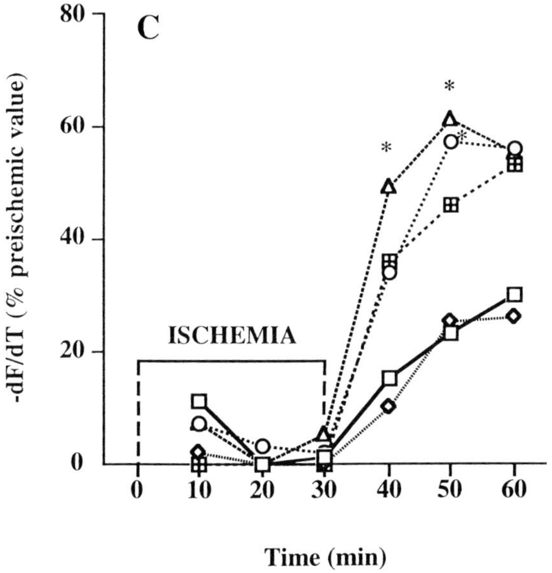

Figure 1 Continued.

investigators have since demonstrated the beneficial effects of whole-body hyperthermia prior to exposing the isolated heart to ischemia and reperfusion in the rat and rabbit (24–26).

IV. CARDIOPROTECTIVE EFFECTS OF HSP INDUCTION—*IN VIVO*

Since positive data were obtained in the ex vivo models of myocardial ischemia, many laboratories have chosen to explore if heat shock protection could be extended to an in vivo paradigm. Donnelly et al. (27) described the possible role for stress proteins to improving myocardial salvage after ischemia and reperfusion in the rat heart. To test the hypothesis that the heat shock response is associated with improved myocardial salvage after ischemia and reperfusion, rats were subjected to whole-body hyperthermia and allowed 24 hr to recover. The rats were then subjected to 35 min of left coronary artery (LCA) occlusion and 120 min of reperfusion. Non–heat stressed controls were also subjected to LCA occlusion and reperfusion. Cardiac muscle specimens from rats subjected to hyperthermia showed induction of hsp 72. Control rats showed no significant presence of myocardial hsp 72. Infarct size (amount of necrosis resulting from 35 min LCA occlusion and 2 hr reperfusion) was significantly reduced in heat-shocked rats compared with controls (8.4 ± 1.7% vs 15.5 ± 1.9%); infarct mass/left ventricular mass × 100) (Fig. 2). There were no significant differences in hemodynamic variables between heat-shocked and control animals during the ischemic period. From these findings, they concluded that heat shock was associated with significantly improved myocardial salvage after 35 min of LCA occlusion and reperfusion. The improved salvage was correlated with marked hsp 72 induction and was independent of the hemodynamic determinants of myocardial oxygen supply and demand during the ischemic period. Other laboratories have since reported that whole-body hyperthermia is cardioprotective using analogous in vivo protocols of ischemia and reperfusion (28,29). The enigma regarding the hypothetical cardioprotective properties of stress proteins continues even among contributors in the field. Some researchers who have published data stating that stress proteins are indeed cardioprotective have also been associated with other reports that deny the protective nature of these proteins albeit the use of two different periods of coronary artery occlusion (29,30). It is apparent, therefore, that cardioprotection conferred by whole-body heat shock in vivo is dependent on the intensity of the applied ischemic insult.

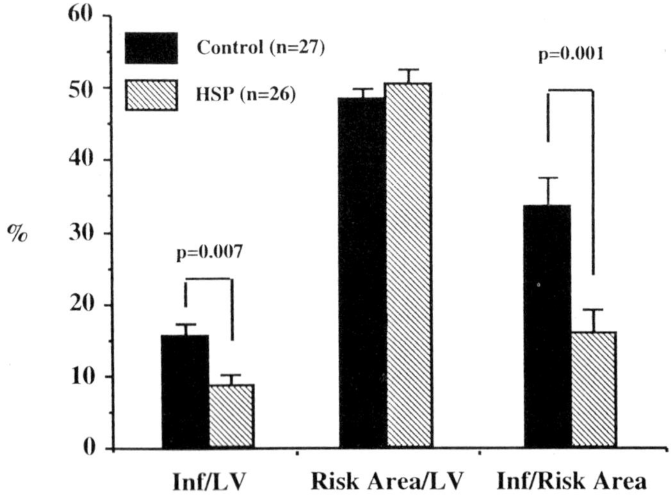

Figure 2 Bar graph shows infarct size in heat-shocked and control rats after 35 min. of ischemia and 2 hr of reperfusion. In comparison with control animals, heat-shocked rats demonstrated a significant reduction in infarct size measured as a percentage of total left ventricular mass and a percentage of the ischemic risk area. Inf, infarct mass; LV, left ventricular mass; risk area, mass of ischemic risk area; HSP, heat-shocked rats. (Adapted from Ref. 26.)

V. MYOCARDIAL PRECONDITIONING AND HSP INDUCTION

Thus far, we have focused primarily on the involvement of stress proteins as a means of conferring cardioprotection. It is readily apparent that the effort expended on discovering additional factors that can confer protection against myocardial ischemia has gained increased attention. The idea of "preconditioning" the myocardium against ischemia was first introduced in 1986 by Murry and colleagues (31). As previously discussed, an extended period of ischemia is detrimental to the functioning myocardium. However, a brief ischemic period (5 min) followed by reperfusion (5 min), although itself not causing any permanent damage, has been demonstrated to be beneficial in the protection of myocytes against such incidents. Murry et al. (31) demonstrated that by subjecting the canine myocardium to a preconditioning protocol immediately before a 90-min period of ischemia followed by reperfusion, the resulting infarction or amount of necrosis due to the extended ischemic event was significantly reduced when compared with animals that were not preconditioned.

Murry and colleagues were the first to raise the possibility that the synthesis of new, protective proteins could mediate the preconditioning phenomenon (31). This idea fell out of favor for a number of years after a publication describing the involvement of preconditioning in myocardial protection despite in vivo pretreatment with the protein synthesis inhibitors cyclohexamide and actinomycin D (32). As in the majority of previous studies, ischemia was produced by occluding a branch of the left coronary artery in open-chest anesthetized rabbits. All groups were subjected to 30 min of ischemia followed by 3 hr of reperfusion. Along with noninhibited controls, additional groups were pretreated with cyclohexamide or actinomycin D to examine the involvement of protein synthesis in myocardial preconditioning. The percentage of the ischemic zone infarcted was small and similar in all preconditioned groups (preconditioning with two 5-min ischemic periods, each followed by 10 min of reperfusion before the 30-min ischemic period). All nonpreconditioned groups had large infarcts with no differences among groups. Murry et al. concluded that whereas neither cycloheximide nor actiomycin D could prevent protection afforded by preconditioning, it seemed unlikely that synthesis of a protective protein is the mechanism by which preconditioning reduces myocardial injury due to ischemia/reperfusion.

Presently, a definitive mechanism by which preconditioning protects the myocardium is unknown. Without a viable explanation for this phenomenon, cardiovascular investigators reexamined a putative role for stress proteins. The preconditioning stress protein literature revival occurred in 1992 when investigators contemplated a possible relationship between the two. Das et al. (33) utilized a model of "stunning" to examine the possible biochemical mechanism for the myocardial preservation afforded by preconditioning. Swine hearts were subjected to four episodes of 5 min of preconditioning by occluding a left anterior coronary artery (LAD), followed by 10 min of reperfusion after each preconditioning period. The hearts were then made regionally ischemic for 60 min by LAD occlusion, followed by 6 hr of reperfusion. Control hearts were treated identically, but without preconditioning. Although the study confirmed that ischemic preconditioning did reduce infarct size, it also indicated that a number of new proteins were expressed after preconditioning the heart. Among these moieties were some oxidative stress-related proteins and the 72-kDa heat shock protein.

In 1993, Marber and colleagues (29) demonstrated that both heat stress and ischemic preconditioning induce the increased expression of stress proteins 24 hr after the preconditioning period. Rabbits were preconditioned by brief ischemic insults with four 5 min episodes of coronary ligation separated by 10 min of reperfusion. The corresponding control group underwent surgical preparation without coronary ligation. Twenty-four hours later, hearts were removed for determination of tissue hsps and assessment of infarct size. Myocardial hsp 72 content assessed by Western blot analysis was increased by both ischemic and thermal pretreatments compared with the corresponding control groups. Hsp 60 was increased preferentially by ischemic pretreatment. It was concluded that myocardial stress protein induced by either sublethal thermal treatment or ischemic injury is associated with myocardial salvage. They also suggested that increased stress protein expression, rather than the nonspecific effects of thermal or ischemic stress, may be responsible for the myocardial protection seen in this model.

VI. CORRELATION BETWEEN HSP INDUCTION AND REDUCTION OF MYOCARDIAL NECROSIS

Since many of the aforementioned studies have demonstrated that heat shock treatment results in the induction of hsp 72 and a reduction of infarct size after subsequent ischemia and reperfusion, Hutter and colleagues tested the hypothesis that the degree of protection from ischemic injury in heat-shocked animals correlates with the degree of previous hsp 72 induction (34). Rats were pretreated with 40, 41, or 42°C of whole-body hyperthermia followed by 24 hr recovery. Control rats were subjected to an identical surgical technique without the hyperthermic treatment. Hearts from these groups were assessed quantitatively for the presence of myocardial hsp 72 by optical densitometry of Western blots and a primary antibody that is specific for hsp 72 and a tertiary antibody labeled with ^{125}I. Although rats heat shocked to 40°C had no significant induction of myocardial hsp 72, rats heat stressed to 41 and 42°C demonstrated progressively increased amounts of myocardial hsp 72 compared with controls. Separate groups of rats heat stressed to 40, 41, and 42°C with 24 hr of recovery, and controls were subjected to 35 min of left coronary artery occlusion and 120 min of reperfusion. Compared with control and rats heated to 40°C, there was a progressive reduction in infarct size in rats that were heat stressed to 41 and 42°C. Hutter et al. demonstrated a correlation between the amount of hsp 72 induced and a reduction in infarct size (Fig. 3). The results suggest that the improved salvage after heat stress pretreatment may be related to the amount of hsp 72 induced before application of the more prolonged ischemic insult and reperfusion.

An issue that has not been addressed sufficiently is the effect of whole-body heat stress on extracardiac cells and the possible contribution of such effects to cardioprotection. The use of an in vivo ischemia/reperfusion protocol introduces the possibility that whole-body hyperthermia affects noncardiac cells, tissue, or organ function that may impinge on cardiac susceptibility to ischemia and reperfusion induced damage. Although it is established that whole-body heat stress protects the heart from ischemia/reperfusion damage in vitro (indicative of a direct myocardial effect), the complexity of the in vivo milieu and the pancellular nature of the heat stress response imply that additional mechanisms must be considered. Altered neutrophil function is one possibility. In response to heat stress, neutrophil hsp expression is increased and neutrophil NADPH oxidase activation is inhibited, indicating that neutrophil superoxide production may be attenuated concomitant with increased hsp expression (35). Since neutrophil

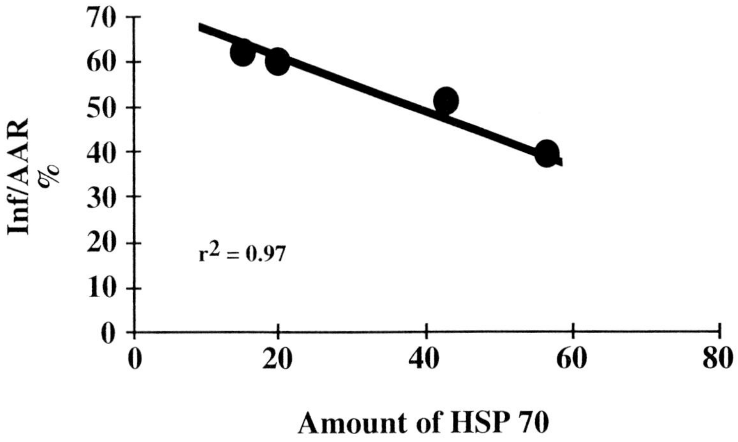

Figure 3 Graph showing correlation between the amount of left ventricular hsp 72 induced and infarct size expressed as a percentage of the ischemic risk area. There was a linear correlation between the amount of hsp 72 induced and infarct size reduction. Inf/RA indicates infarct size/ischemic risk area. (Adapted from Ref. 33.)

ablation reduces infarct size (36), whole-body heat stress may confer cardioprotection in vivo in part by influencing neutrophil function. However, it is possible that whole-body heat stress exerts negative effects because the duration of cardioprotection is much less in vivo than in vitro (23). Several studies have demonstrated that heat stress has failed to protect the heart in vivo (30).

Differentiation of the effect of heat stress on coronary artery endothelial cells versus cardiac myocytes has not been addressed in studies that have shown cardioprotection by heat stress. The vascular endothelium plays a significant role in cardiac pathophysiology, and heat stress has been shown to increase hsp expression in vascular endothelial cells (37). The relevance, if any, of a putative increase in endothelial hsp expression to cardioprotection is unknown. Furthermore, it remains to be determined if heat stress mediates a change in the expression of endothelial cell surface adhesion molecules. In view of hsp's role in the cellular protection and trafficking of proteins (38), and of the time course of both hsp expression and endothelial cell receptors (39) during reperfusion of the previously ischemic heart, a relationship may exist between hsp and cell surface adhesion receptor expression. As alluded to above, such a relationship may be beneficial.

VII. CONCLUSIONS

The approach of conferring cardioprotection against ischemia by heat stress is of particular interest because it speaks to the possibility of a novel mechanism by which the heart may be preconditioned to withstand ischemic episodes. It is apparent that cardiac hsp research will continue to define the mechanism by which a sublethal insult confers protection against a subsequent noxious ischemic insult. Issues that remain to be determined include definition of the role and mechanism of stress proteins in the cardioprotective process and determining whether other mechanisms are contributory in these models. Ultimately, the question remains whether a pharmacological intervention could be designed to target the responsible element and stimulate the mechanism(s) responsible for cardioprotection. Clearly, this possibility should stimulate further research in the area of stress proteins and their relationship to the cardiovascular system.

REFERENCES

1. R.W. Currie and F.P. White, Trauma-induced protein in rat tissues: a physiological role for a "heat shock" protein?, *Science 214*:72 (1981).
2. G.C. Li and Z. Werb, Correlation between synthesis of heat shock proteins and development of thermotolerance in Chinese hamster fibroblasts, *Proc. Natl. Acad. Sci. USA 79*:3218 (1982).
3. F.P. White and S.R. White, Isoproterenol induced myocardial necrosis is associ-

ated with stress protein synthesis in rat herat and thoracic aorta, *Cardiovacs. Res.* *20*:512 (1986).

4. F.P. White, Differences in protein synthesized in vivo and in vitro by cells associated with the cerebral microvasculature. A protein synthesized in response to trauma?, *Neuroscience 5*:1793 (1980).
5. C. Delcayre, J.L. Samuel, F. Marotte, M. Best-Belpomme, J.J. Mercadier, L. Rappaport, Synthesis of stress proteins in rat cardiac myocytes 2–4 days after imposition of hemodynamic overload, *J. Clin. Invest. 82*:460 (1988).
6. D.J. Hearse, S.M. Humphrey, W.G. Nayler, A. Slade, and D. Border, Ultrastructural damage associated with reoxygenation of the anoxic myocardium, *J. Mol. Cell. Cardiol. 7*:315 (1975).
7. J.V. Bannister, W.H. Bannister, H.A. Hill, and P.J. Thornalley, Enhanced production of hydroxyl radicals by the xanthine-xanthine oxidase reaction in the presence of lactoferrin, *Biochim. Biophys. Acta 715*:116 (1982).
8. A. Boveris and B. Chance, The mitochondrial generation of hydrogen peroxide. General properties and effect of hyperbaric oxygen, *Biochem. J. 134*:707 (1973).
9. J.C. Fantone and P.A. Ward, Role of oxygen-derived free radicals and metabolites in leukocyte-dependent inflammatory reactions, *Am. J. Pathol. 107*:395 (1982).
10. M. Dikshit, M.H. Van Oosten, S. de Graff, and R.C. Srimal, Free radical scavenger mechanisms in experimentally induced ischemia in the rabbit heart and protective effect of verapamil, *Arch. Int. Pharmacodyn. Ther. 318*:55 (1992).
11. S. Krause and M.L. Hess, Characterization of cardiac sarcoplasmic reticulum dysfunction during short-term, normothermic, global ischemia, *Circ. Res. 55*:176 (1984).
12. N.S. Dhalla, V. Panagia, P.K. Singal, N. Makino, I.M. Dixon, and D.A. Eyolfson, Alterations in heart membrane calcium transport during the development of ischemia-reperfusion injury, *J. Mol. Cell. Cardiol. 2*:3 (1988).
13. P. Kaplan, M. Hendrikx, M. Mattheussen, K. Mubagwa, and W. Flameng, Effect of ischemia and reperfusion on sarcoplasmic reticulum calcium uptake, *Circ. Res. 71*:1123 (1992).
14. H.B. Mehta, B.K. Popovich, and W.H. Dillmann, Ischemia induces changes in the level of mRNAs coding for stress protein 71 and creatine kinase M, *Circ. Res. 63*:512 (1988).
15. R.W. Currie, Protein synthesis in perfused rat hearts after in vivo hyperthermia and in vitro cold ischemia, *Biochem. Cell. Biol. 66*:13 (1988).
16. F.Z. Meerson and I. Malyshev, [Adaptation to stress limits the reperfusion injury of the heart after total ischemia and increases its resistance to the heat shock], *Dokl. Akad. Nauk. SSSR. 313*:750 (1990).
17. T. Nowak, Jr., Protein synthesis and the heart shock/stress response after ischemia. *Cerebrovasc. Brain Metab. Rev. 2*:345 (1990).
18. M.L. Entman, L. Michael, R.D. Rossen, et al., Inflammation in the course of early myocardial ischemia, *F.A.S.E.B. J. 5*:2529 (1991).
19. A.A. Knowlton, P. Brecher, and C.S. Apstein, Rapid expression of heat shock protein in the rabbit after brief cardiac ischemia, *J. Clin. Invest. 87*:139 (1991).
20. G. Howard and T.E. Geoghegan, Altered cardiac tissue gene expression during acute hypoxic exposure, *Mol. Cell. Biochem, 69*:155 (1986).

21. F.Z. Meerson, I. Malyshev, and A.V. Zamotrinskii, [Differences in the development of the phenomenon of the adaptive stabilization of the DNA in myocardial nuclei during adaptation to stress and altitude hypoxia: the role of heat shock proteins], *Dokl. Akad. Nauk. SSSR, 317*:1503 (1991).
22. K. Iwaki, S.H. Chi, W.H. Dillmann, and R. Mestril R, Induction of HSP70 in cultured rat neonatal cardiomyocytes by hypoxia and metabolic stress, *Circulation 87*:2023 (1993).
23. M. Karmazyn, K. Mailer, and R.W. Currie, Acquisition and decay of heat-shock-enhanced postischemic ventricular recovery, *Am. J. Physiol, 259(Pt. 2)*:H424 (1990).
24. D.M. Yellon, E. Pasini, A. Cargnoni, M.S. Marber, D.S. Latchman, and R. Ferrari, The protective role of heat stress in the ischaemic and reperfused rabbit myocardium, *J. Mol. Cell. Cardiol, 24*:895 (1992).
25. D.M. Walker, E. Pasini, S. Kucukoglu S, et al., Heat stress limits infarct size in the isolated perfused rabbit heart, *Cardiovasc. Res, 27*:962 (1993).
26. M. Amrani, N.J. Allen, J. O'Shea, et al., Role of catalase and heat shock protein on recovery of cardiac endothelial and mechanical function after ischemia, *Cardioscience 4*:193 (1993).
27. T.J. Donnelly, R.E. Sievers, F.L. Vissern, W.J. Welch, and C.L. Wolfe, Heat shock protein induction in rat hearts. A role for improved myocardial salvage after ischemia and reperfusion?, *Circulation 85*:769 (1992).
28. R.W. Currie, R.M. Tanguay, and J. Kingma, Jr., Heat-shock response and limitation of tissue necrosis during occlusion/reperfusion in rabbit hearts, *Circulation 87*:963 (1993).
29. M.S. Marber, D.S. Latchman, J.M. Walker, and D.M. Yellon, Cardiac stress protein elevation 24 hours after brief ischemia or heat stress is associated with resistance to myocardial infarction, *Circulation 88*:1264 (1993).
30. D.M. Yellon, E. Iliodromitis, D.S. Latchman, et al., Whole body heat stress fails to limit infarct size in the reperfused rabbit heart, *Cardiovasc. Res, 26*:342 (1992).
31. C.E. Murry, R.B. Jennings, and K.A. Reimer, Preconditioning with ischemia: a delay of lethal cell injury in ischemic myocardium, *Circulation 74*:1124 (1986).
32. J. Thornton, S. Striplin, G.S. Liu, et al., Inhibition of protein synthesis does not block myocardial protection afforded by preconditioning, *Am. J. Physiol, 259*: H1822 (1990).
33. D.K. Das, M.R. Prasad, D. Lu, and R.M. Jones, Preconditioning of heart by repeated stunning. Adaptive modification of antioxidative defense system, *Cell. Mol. Biol, 38*:739 (1992).
34. M.M. Hutter, R.E. Sievers, V. Barbosa, and C.L. Wolfe, Heat-shock protein induction in rat hearts. A direct correlation between the amount of heat-shock protein induced and the degree of myocardial protection, *Circulation 89*:355 (1994).
35. I. Maridonneau-Parini, J. Clerc, and B.S. Polla, Heat shock inhibits NADPH oxidase in human neutrophils, *Biochem. Biophys. Res. Commun, 154*:179 (1988).
36. J.L. Romson, B.G. Hook, S.L. Kunkel, G.D. Abrams, M.A. Schork, and B.R. Lucchesi, Reduction of the extent of ischemic myocardial injury by neutrophil depletion in the dog, *Circulation 67*:1016 (1983).

37. L. Jornot, M.E. Mirault, and A.F. Junod, Differential expression of hsp70 stress proteins in human endothelial cells exposed to heat shock and hydrogen peroxide, *Am. J. Respir. Cell. Mol. Biol, 5*:265 (1991).
38. W.J. Welch, H.S. Kang, R.P. Beckmann, and L.A. Mizzen, Response of mammalian cells to metabolic stress; changes in cell physiology and structure/function of stress proteins, *Curr. Top. Microbiol. Immunol, 167*:31 (1991).
39. C.W. Smith, D.C. Anderson, A.A. Taylor, R.D. Rossen, and M.L. Entman, Leukocyte adhesion molecules and myocardial ischemia, *Trends Cardiovasc. Med, 1*:167 (1991).

30

Heat Shock Proteins in Eosinophilic Inflammation

Pandora Christie and Muriel R. Jacquier-Sarlin
University Hospital, Geneva, Switzerland

Anne Janin
Calmette Hospital, Lille, France

Jean Bousquet
Arnaud de Villeneuve Hospital, Montpellier, France

Barbara S. Polla
UFR Cochin Port Royal, Paris, France

I. INTRODUCTION

Much of the work supporting a role for heat shock proteins (hsps) in inflammation, in particular eosinophilic inflammation, results from observations of the stress response in human monocytes-macrophages with relationship to generation of reactive oxygen species (ROS) by these cells during the phagocytic process.

We suggested for the first time a role for hsp in inflammation (1). We showed that exposure of human macrophages to exogenous H_2O_2 induces a heat shock response and in particular the synthesis of hsp 70 (2) and established that phagocytosis of red blood cells by human macrophages, including alveolar macrophages (AMs), induces in these latter cells a complex stress response which was related to the endogenous generation of ROS as occurs during phagocytosis (3). Since AMs from patients with inflammatory lung diseases produce excess amounts of ROS (4), we extended these studies to these cells. We found that AMs recovered from patients suffering from certain types of pulmonary inflammatory diseases associated with alveolar eosinophilia (a sign of gravity in these diseases) spontaneously synthesized stress proteins (5) (Fig. 1) (see color plate). Heme oxygenase (HO), an oxidation specific stress protein (6), was induced along with the classic hsp.

Much of this chapter deals with the possible mechanism(s) by which eosinophils have the ability to induce hsp in neighbor cells. Whatever the mechanism(s), we suggest that the increased expression of hsp in inflammation serves protective functions.

On the other hand, hsps are immunodominant antigens. Another link we will consider between hsps and eosinophils is the possibility that hsps would be TH2-stimulating allergens and parasites' antigens as well as TH1-stimulating bacterial antigens, leading to hsp-specific immunoglobulin E (IgE) synthesis and IgE-driven accumulation of eosinophils at inflammatory sites such as bronchi in asthma.

II. EOSINOPHILS AND HSP INDUCTION IN Mϕ

A. Eosinophils

Eosinophils are so-called because of their strong affinity for the acidic dye eosin, which is related to their content in highly toxic basic proteins. Major basic protein (MBP) is located in the core of the granules, whereas eosinophil cationic proteins (ECPs), eosinophil-derived neurotoxin (EDN), and eosinophil peroxidase (EPO) are located in their matrix (Fig. 2). MBP is highly toxic to bronchial epithelium and may contribute to the respiratory epithelial damage of patients with severe asthma (7). The granule components are also powerful toxins to multicellular parasites. As a likely reflection of those properties, the eosinophils are involved in allergic and parasitic diseases. An interesting link between these proteins and hsp has been found by Rosenberg et al., who reported the specific and high-affinity binding of GroEL (the prokaryote homologue of hsp 60) to recombinant ECP and EDN, suggesting that hsp may chaperone these proteins in vivo (8).

Eosinophils are phagocytic cells and produce high amounts of lipid mediators of inflammation and of ROS, with the latter being generated both via activation of the respiratory burst enzyme NADPH oxidase and via EPO (see Fig. 2). Eosinophils also are immunocompetent cells: they express major histocompatibility class II antigens when activated, produce a number of cytokines (see Fig. 2), and express receptors for these as well as other cytokines (9). Many of the mediators produced by eosinophils may be involved in the induction of hsp, particularly in eosinophilic inflammation and when eosinophils are phagocytosed by neighbor cells (Fig. 3).

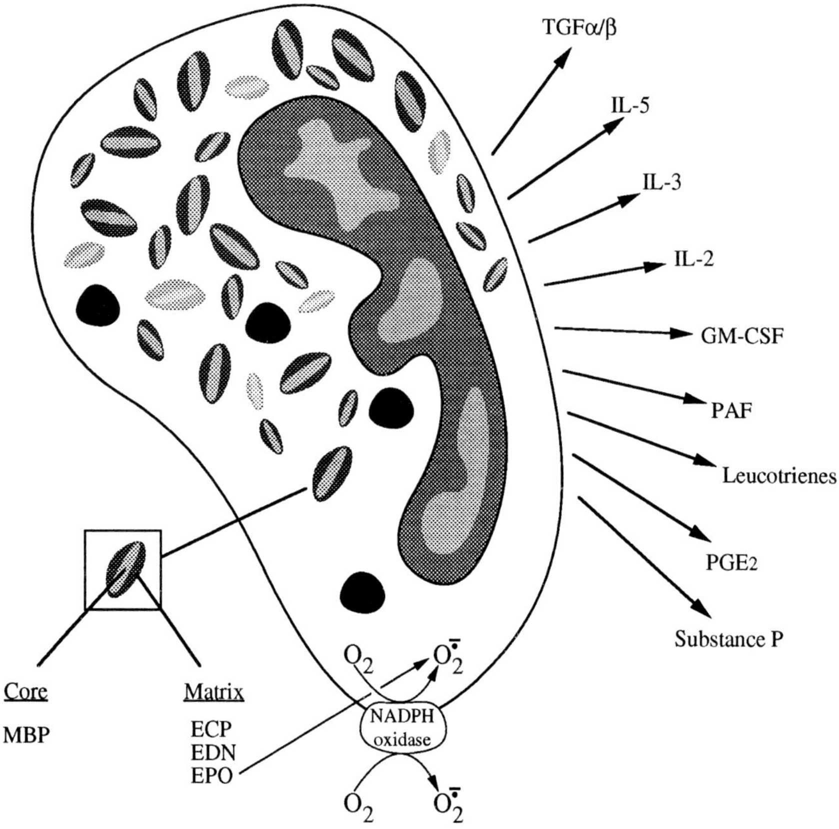

Figure 2 Mediators of the eosinophil. Inflammatory mediators include leukotrienes, platelet-activating factor (PAF), prostaglandins (PG) E_2, and substance P. Cytokines include transforming growth factor (TGF) TGF-α/β, interleukins (IL) IL-3, IL-5, IL-2, and granulocyte-macrophage colony-stimulating factor (GM-CSF). The eosinophil granules consist of a central core containing major basic protein (MBP) and a matrix containing eosinophil cationic protein (ECP), eosinophil-derived neurotoxin (EDN), and eosinophil peroxidase (EPO). Superoxide (O_2^-) is generated via activation of the respiratory burst enzyme NADPH oxidase and ROS are amplified by EPO.

B. ROS and Phagocytosis

Although the induction of HO has been closely related to oxidative stress, and more specifically hydroxyl radicals (6), unpublished studies from our laboratory indicate that it is unlikely that the induction of stress proteins during eosinophilic inflammation solely relates to the generation of high levels of ROS (Jacquier-Sarlin et al., in preparation).

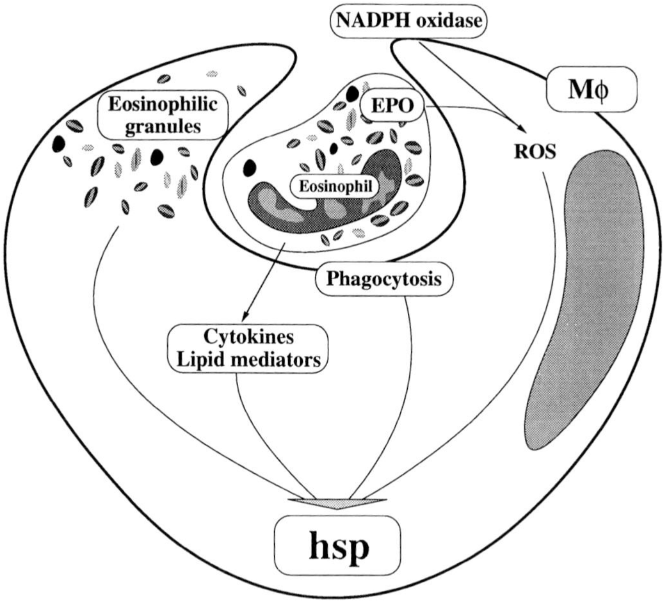

Figure 3 Phagocytosis of the eosinophil by monocyte-macrophage (mϕ) and mechanisms for induction of hsp. Toxic proteins of the eosinophil granules, eosinophil mediators (lipids, cytokines), and ROS may all contribute to the induction of hsp during phagocytosis of eosinophils by macrophages (Mϕ).

In order to sort out which one(s) of the above-mentioned possible mechanisms for hsp induction was most important, we took a number of approaches. We first addressed the issue as to whether other lung pathologies associated with eosinophilic infiltrates were characterized by an increased hsp expression. It was found that both AMs and epithelial cells form patients with asthma expressed increased levels of hsp 70 and that there was a close correlation between the levels of hsp 70 and the severity of asthma as established by AAS score (10), whereas eosinophilia also correlates with the severity of asthma (11). These data suggested that hsp could be a marker of eosinophilic inflammation in asthma. We next investigated whether this was also the case in another pathology associated with eosinophils, that is, chronic eosinophilic pneumonia (Carrington's disease), a disease in which phagocytosis of eosinophils by the AMs is a characteristic feature (12). As shown in Figure 4, using immunoelectron microscopy and human hsp 70–specific monoclonal antibodies, no hsp 70 expression was observed in pulmonary tissue from patients with Carrington's disease, whereas eosinophilic proteins and Charcot-Leyden crystals were identified within AMs phagocytosing eosinophils, using the same technique, but distinct antibodies.

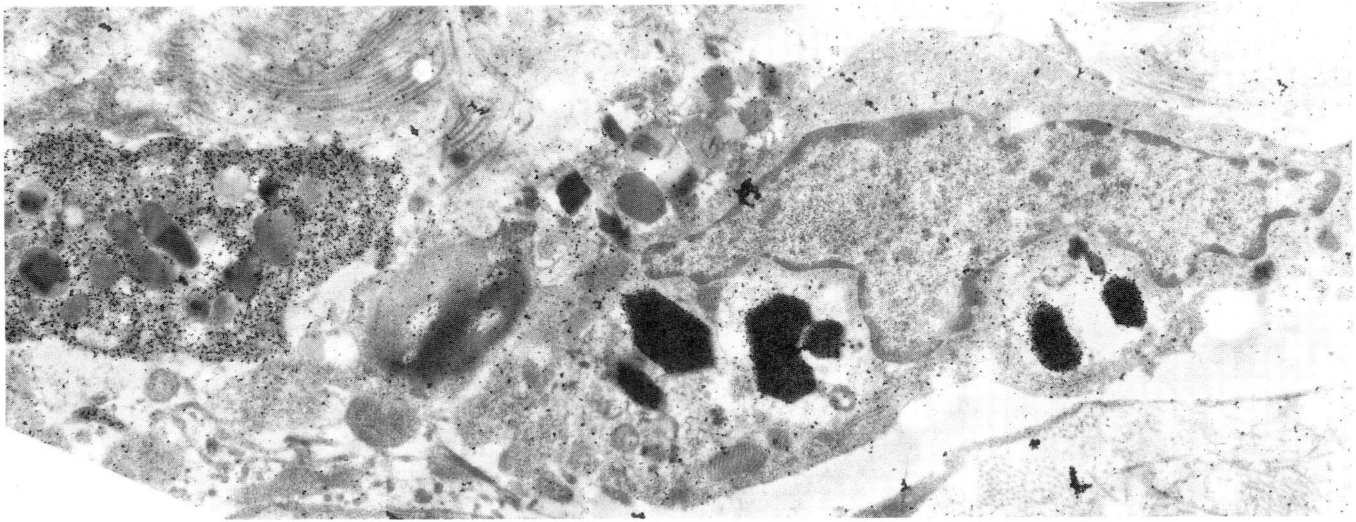

(a)

Figure 4 Eosinophil phagocytosis in Carrington's disease is not associated with hsp synthesis. Cells obtained by bronchoalveolar lavage from patients with chronic eosinophilic pneumonia (Carrington's disease) were prepared for electron microscopy as described in detail in ref. 12. Ultrathin sections on nickel grids were incubated with rabbit polyclonal antibody to human MBP, mouse monoclonal antibody to EDP, or mouse monoclonal antibody to hsp 70. The grids were rinsed and incubated with gold-conjugated goat antirabbit or mouse IgG, washed, and sections subjected to silver enhancement. Lowicryl sections were stained with uranyl acetate and lead citrate before examination with a Philips EM 420 microscope. In the four cases of chronic eosinophilic pneumonia examined, BAL showed similar features for macrophages and eosinophil ultrastructure, redistribution of MBP, and ECP into macrophage cytoplasmic compartments and absence of hsp labeling. The cytoplasmic vacuoles containing granular material were close to Charcot-Leyden crystals (a). Analysis of the macrophages staining with the three different antibodies showed that MBP was only found on pseudomyelinic structure when ECP selectively concentrated in vacuoles with granular content. No gold deposit was observed within macrophages with the antibody directed against the inducible form of hsp 70 (b and c, see pp. 484 and 485).

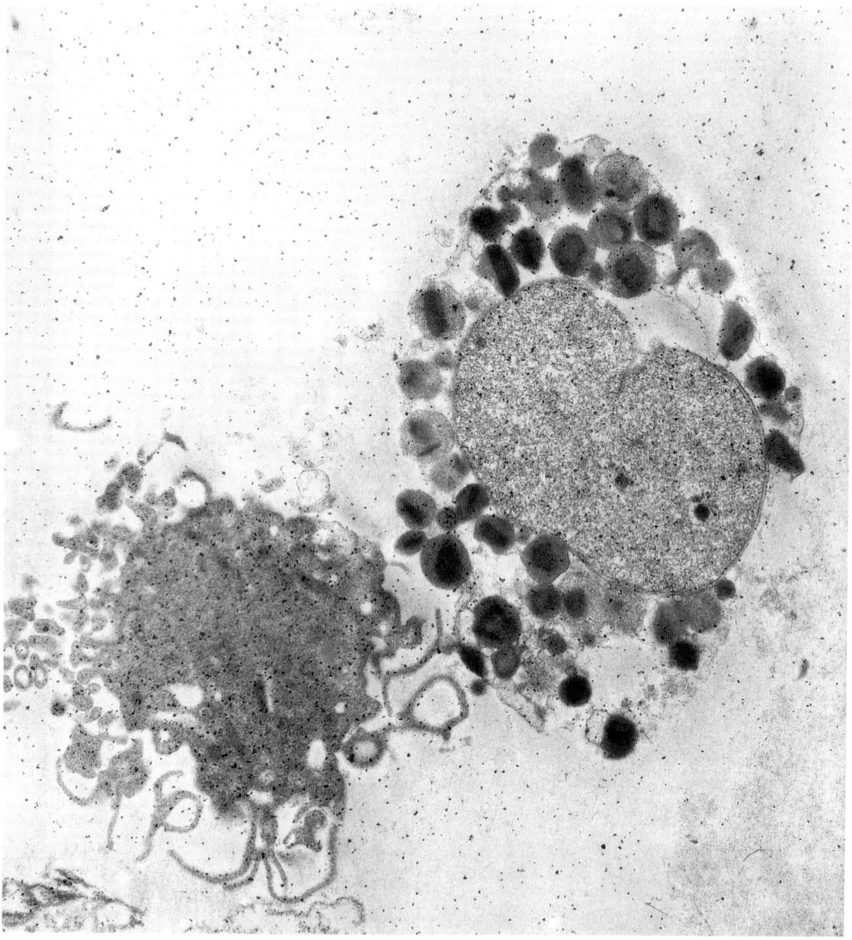

(b)

Figure 4 Continued.

We then assessed whether AMs exposed in vitro to the purified eosinophilic proteins increased their stress protein expression. Exposure of AMs to the purified eosinophil-derived proteins ECP, EDN, MBP, and/or EPO, alone or in combination, did not induce hsp synthesis even under conditions in which there was toxicity to the AMs as assessed by morphological criteria (AM vacuolization) and by a decrease in total normal protein synthesis (5). To assess the uptake of EPO by AMs, peroxidase staining was performed which confirmed internalization of EPO. We suggest that the induction of stress proteins within the AMs requires other events than the sole presence of eosinophils or eosinophilic proteins.

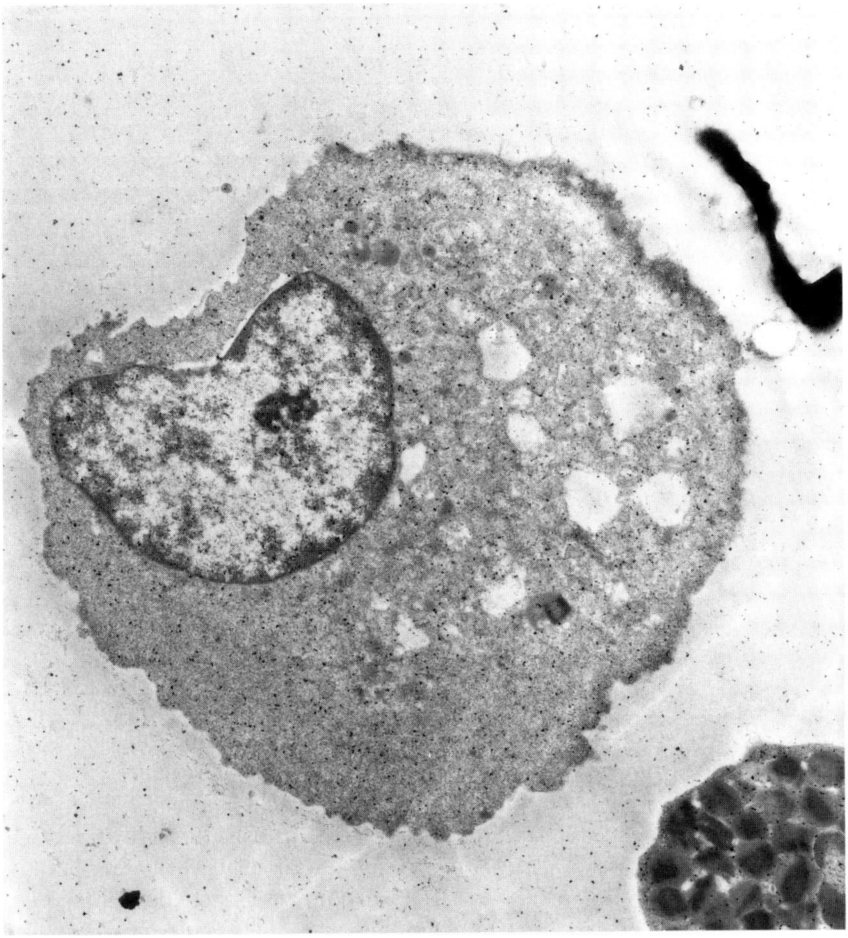

(c)

Figure 4 Continued.

The increased hsp expression we observed in some cases of interstitial diseases thus appears to strictly relate neither to phagocytosis nor to the effects of the toxic proteins of eosinophils. Induction of hsp may be the consequence of exposure to other specific products of activated eosinophils such as lipid mediators and/or cytokines (see Figure 3).

C. Lipid Mediators of Inflammation

The metabolism of arachidonic acid generates the lipid mediators of inflammation, including platelet-activating factor (PAF acether), and, via the cyclooxygenase and the lipo-oxygenase pathways, respectively, prostaglandins (PGs), thromboxanes, and leukotrienes. PAF acether activates the respiratory burst in human macrophages but fails to induce hsp synthesis, although it likely induces the glucose-regulated protein (grp) grp 78 (13). The cyclopentone prostaglandin PGA_1 induces hsp 70 in nonstress situations (14). Since the induction of thermotolerance by PGA_1 is associated in human cells with hsp 70 synthesis, this activity of PGA_1 could be part of a protective control mechanism during fever (15). Furthermore, another lipid mediator, 12-hydroxyeicosatetraenoic acid (12-HETE) generated via 12-lipooxygenase, also induces the expression of hsps (hsp 65 and hsp 83) in human leukocytes (16).

D. Cytokines

Among the cytokines produced by the eosinophil, transforming growth factor β (TGF-β) is a candidate of particular interest. Indeed, TGF-β increases the expression of members of the hsp 70 and hsp 90 families in cultured chicken embryo cells (17). It has been suggested that the induction of hsps by TGF-β is mediated by the increase in protein synthesis which enhances the requirement for chaperoning. TGF-β released by eosinophils in asthma may thus be the mediator for hsp induction. In Carrington's disease, eosinophils express TGF-α (18). However, it has not yet been determined whether TGF is differentially secreted in the two eosinophil-mediated pulmonary diseases mentioned above (asthma and Carrington's disease) or whether TGF-α has the same effects on hsp synthesis than TGF-β.

Other cytokines may also be involved in this upregulation. Interleukin-1 (IL-1) selectively induces hsp synthesis in pancreatic β cells, which may relate to the ability of IL-1 to induce oxidative stress in these cells (19). Platelet derived growth factor (PDGF), an important factor in cell proliferation and activation, may also modulate hsp synthesis. Hightower et al. observed the induction of hsp by PGDF in chicken and rat embryo cells (20). We could not, however, reproduce these data using recombinant PDGF and human macrophages (13). Platelets are involved in the pathogenesis of both asthma and pulmonary fibrosis but not of Carrington's disease, whereas similar levels of PDGF have been detected in patients with asthma, chronic bronchitis, and in normal subjects (unpublished data).

TH2 cytokines such as IL-4, a crucial cytokine for induction of IgE synthesis, as well as IL-6, have no effect on hsp synthesis in human macrophages. This contrasts with the ability of IL-6 to induce acute phase proteins and metallothioneins, which belong to the stress protein families. Although the role of IL-5 for eosinophil survival and proliferation is well established, there are no data so far regarding a direct effect of IL-5 on hsp expression.

III. HSPS AS TH2-ACTIVATING ANTIGENS AND ALLERGENS

Another link between eosinophilic inflammation and hsps concerns the complex processes involved in antigen and/or allergen uptake and presentation, activation of specific immune responses (T cells and antibodies) and inflammation. We are currently considering the possibility that hsps, which are immunodominant antigens of a still growing number of bacteria (21), also are IgE-stimulating parasites' antigens and/or plant allergens. Such a possibility would provide a link between parasitic diseases and allergy, both of which are characterized by eosinophilic inflammation and elevated IgE. The possibility of plant stress proteins to act as allergens has been established for profilins (22). Hsp 70 is an immunodominant antigen in many species and grass pollens and other allergens are likely to contain hsp 70 (23,24). Hsps may have the potential to activate both TH1 and TH2 cells, leading to the production by B cells of hsp-specific IgE isotype antibodies as well as of IgG and IgM isotype antibodies. This hypothesis is schematically depicted in Figure 5.

The role of hsps in parasitic disease has been reviewed elsewhere (25). The initial change in temperature experienced by the parasite may induce hsp synthesis (26). Hsps may interfere with parasite recognition and infectivity (27), differentiation (28), and resistance (29). On the one hand, hsps are major parasite antigens, and on the other hand, they are involved in differentiation and protection of these organisms. Thus, one has to consider the possibility that in parasitic diseases (as in the pulmonary inflammatory diseases mentioned above), hsps would be upregulated, not only in host cells but in parasites as well. There may be differential upregulation of hsps in host cells and in pathogens, leading to preferential protection of either one. For example, infection of human and murine macrophages with *Leishmania major* is associated with parasite hsp synthesis, but there is no detectable host stress response (29a).

The hypothesis presented here deserve further studies, among which the analysis of the possible presence of hsp-specific IgE antibodies in the sera of patients with parasitic disease or allergy. Such investigations are currently underway.

IV. HSP, EOSINOPHILS, AND ATHEROSCLEROSIS

ROS may be involved in the development of atherosclerotic plaques via oxidation of low-density lipoproteins deposited under endothelial cells of arterial walls. Lipid peroxidation is toxic to endothelial cells and may be a key event in the formation of atherosclerotic plaques (reviewed in ref. 30). Although there is evidence that the macrophage is the major source of oxidative stress in athero-

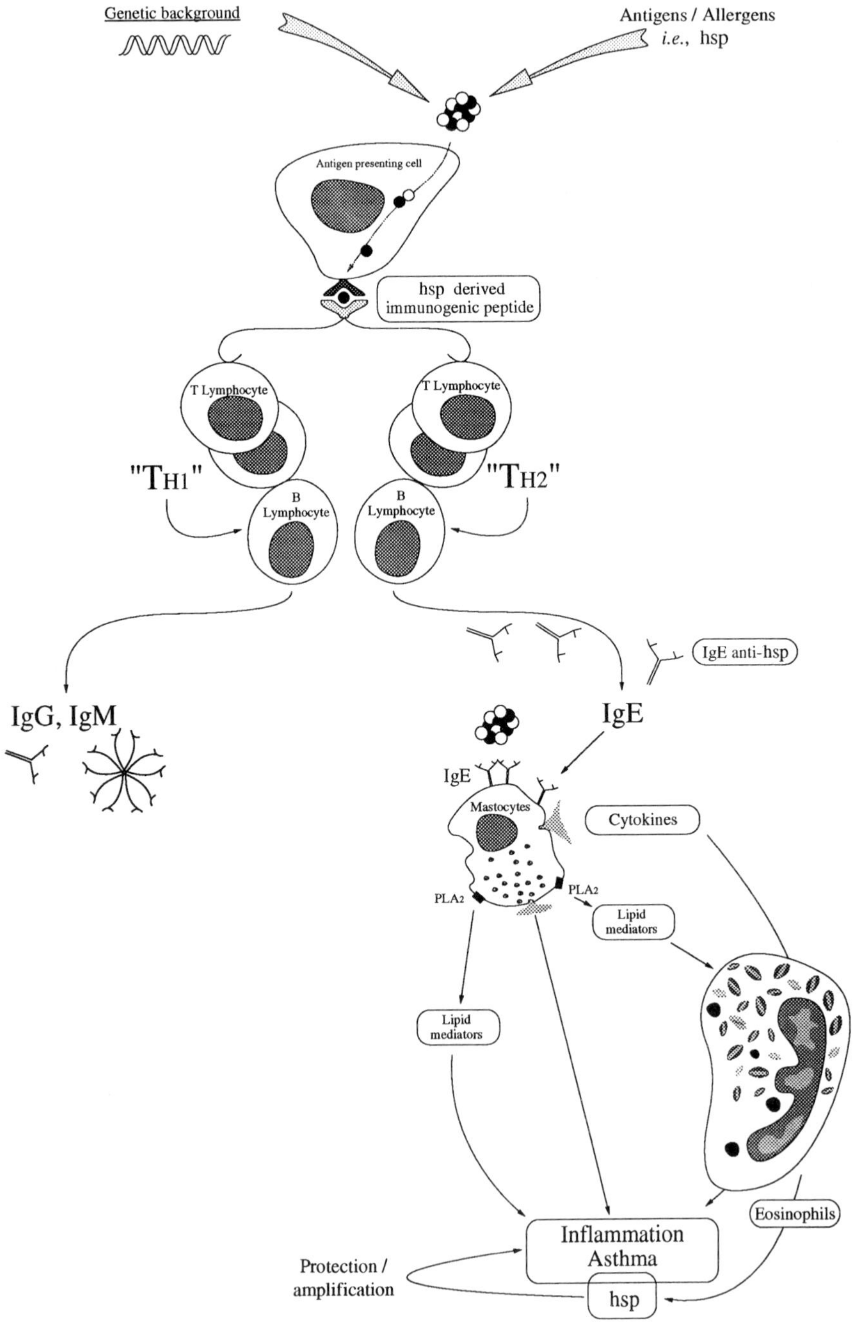
Genetic background
Antigens / Allergens
i.e., hsp
Antigen presenting cell
hsp derived immunogenic peptide
T Lymphocyte
T Lymphocyte
"TH1"
"TH2"
B Lymphocyte
B Lymphocyte
IgE anti-hsp
IgG, IgM
IgE
IgE
Mastocytes
Cytokines
PLA2
PLA2
Lipid mediators
Lipid mediators
Eosinophils
Inflammation Asthma
Protection / amplification
hsp

sclerosis, an intriguing observation is the presence of eosinophilia in atherosclerosis associated with high levels of cholesterol (31,32). The toxic products of eosinophils may thus contribute to oxidative stress in atherosclerosis. The underlying mechanism for increased hsp synthesis observed in atherosclerotic lesions (33) and the relationship between this observation, eosinophilia, and oxidative stress requires further work. As for asthma, the induction of hsp in atherosclerosis may both indicate injury and provide protection (34).

V. PROTECTION BY HSPS FROM EOSINOPHIL-MEDIATED TISSUE DAMAGE

As mentioned above, hsp synthesis within the lung may represent an autoprotective mechanisms for this organ against the toxic effects of ROS to which the lung is often exposed. Stress proteins have a general function in cellular protection from injury, and we have previously reported that hsps have the potential to protect macrophages from oxidative injury (reviewed in ref. 35). The induction of hsps as observed in eosinophilic inflammation within the lung likely deserves such protective functions. HO may contribute to the these protective effects: indeed, HO was found to be induced along with the classic hsp in the AMs of patients with eosinophilic inflammation (see Fig. 1) and has an important potential for protection from oxidative stress (5,6,36), although the hypothesis of a specific protective effect of HO against eosinophil-mediated cell and tissue damage has not been addressed yet.

An interesting specific protective mechanism by which hsps may prevent eosinophil-mediated damage derives from the observation by Rosenberg et al. (8) that hsp 60/65 has the property to chaperone ECP. Hsps induced during exposure of other cells to eosinophils may bind ECP and prevent further damage, inasmuch the chaperoning of ECP alters the function of the cationic protein.

One example where protection by hsps from eosinophil-mediated oxidative tissue damage may occur in vivo is bleomycin-induced pulmonary inflamma-

---

Figure 5 Are hsp TH2-activating as well as TH1-activating antigens/allergens? Hsp are major immunogens and may activate both TH1 an TH2 cells leading to the production by B cells of IgG and IgM or IgE, respectively. Hsp-specific IgE bind high-affinity IgE receptors on mast cells. Cross linking by subsequent antigen (hsp) exposure leads to the release by the mast cells of inflammatory mediators and cytokines which recruit eosinophils (Il-5). Eosinophilic inflammation of the bronchi increases host hsp expression, which may either provide protection from the inflammatory lesions or amplify the immune response. A polymorphism in hsp genes may be part of the genetic basis which determines the response to allergens.

tion and fibrosis. Bleomycin is a chemotherapeutic agent with a high affinity for DNA, which generates ·OH by entering oxidoreduction cycles in the presence of iron. Although we did not observe detectable hsp 70 synthesis in cells exposed to bleomycin (unpublished results), the drug is able to activate the HS promotor region in COS-1 transfected cells (37). While inducing pulmonary eosinophilia, intraperitoneal administration of bleomycin primes the respiratory burst enzyme NADPH oxidase for ROS production in macrophages (38). The presense of eosinophils in bleomycin-induced lung injury may be an additional source of ROS and other factors participating in the development of pulmonary fibrosis (39). On the other hand, heat shock protects cells from in vitro bleomycin-mediated oxidative injury (40; B.S. Polla et al., in preparation). In vivo protection by hsps of eosinophil-mediated tissue damage should be further examined using similar protocols as described by Villar et al. (41) but inducing agents specifically leading to eosinophilic inflammation.

VI. CONCLUSIONS

The induction of hsps during some, but not other, forms of eosinophilic inflammation and the links between eosinophilia, the severity of inflammation, and the presence of hsps, suggest that hsps could represent new clinical markers (both diagnostic and prognostic) for eosinophilic inflammation. The possibility that hsps, acting as parasites' antigens or plant allergens, induce hsp-specific IgE antibodies and subsequent mastocyte-derived IL-5 release, leading to eosinophil accumulation in parasitic diseases and in allergy, would provide a missing link between these two conditions. The general protective properties of hsps during cellular stress, together with the ability of hsp 60/65 to chaperone ECP, indicate that there may be specific mechanisms by which induction of hsps may provide protection from eosinophil-mediated tissue damage.

AKNOWLEDGMENTS

Work mentioned here was supported in part by the Swiss National Science Foundation grant Nb.32-028645.90 to BSP. PC is supported by grants from UCB and the CISC. MRJS is supported by grants from ARC and from INSERM. AJ is supported by grants from the University and CHRU of Lille. The authors kindly aknowledge Yves R.A. Donati for help in graphical work.

REFERENCES

1. B.S. Polla, A role for heat shock proteins in inflammation? *Immunol. Today 9*:134 (1988).

2. B.S. Polla, A.M. Healy, W.C. Wojno, and S.M. Krane, Hormone 1α, 25-dihydroxyvitamin D_3 modulates heat shock response in monocytes, *Am. J. Physiol.* *252*:C640 (1987).
3. M. Clerget and B.S. Polla, Erythrophagocytosis induces heat shock protein synthesis by human monocytes-macrophages, *Proc. Natl. Acad. Sci. USA.* *87*:1081 (1990).
4. R.G. Crystal, P.B. Bitterman, S.I. Rennard, A.J. Hance, and B.A. Keogh, Interstitial lung diseases of unknwon cause: disorders characterized by chronic inflammation of the lower respiratory tract, *N. Engl. J. Med.* *310*:154 (1984).
5. B.S. Polla, S. Kantengwa, G.J. Gleich, M. Kondo, C.M. Reimert, and A.F. Junod, Spontaneous heat shock protein synthesis by alveolar macrophages in interstitial lung disease associated with phagocytosis of eosinophils, *Eur. Respir. J.* *6*:483 (1993).
6. S.M. Keyse and R.M. Tyrrell, Heme oxygenase is the major 32-kD stress protein induced in human skin fibroblasts by UVA radiation, hydrogen peroxide, and sodium arsenite, *Proc. Natl. Acad. Sci. USA* *86*:99 (1989).
7. S. Motojima, E. Frigas, D.A. Loegering, and G.J. Gleich, Toxicity of eosinophil cationic proteins for guinea pig tracheal epithelium *in vitro*, *Am. Rev. Respir. Dis.* *139*:801 (1989).
8. H.F. Rosenberg, S.J. Ackerman, and D.G. Tenen, Characterization of a distinct binding site for the prokaryotic chaperone, GroEL, on a human granulocyte ribonuclease, *J. Biol. Chem.* *268*:4499 (1993).
9. G.J. Gleich, C.R. Adolphson, M.S, and K.M. Leiferman, The biology of the eosinophilic leukocyte, *Annu. Rev. Med.* *44*:85 (1993).
10. A.M. Vignola, P. Chanez, B.S. Polla, P. Vic, F.B. Michel, P. Godard, and J. Bousquet, Heat shock proteins (HSP 65 and 70) expression on airway cells in asthma, *Am. Rev. Respir. Dis.* *147*:A518 (1993).
11. J. Bousquet, P. Chanez, J.Y. Lacoste, G. Barnéon, N. Ghavanian, I. Enander, P. Venge, S. Ahlstdt, J. Simony-Lafontaine, P. Godard, and F-B. Michel, Eosinophilic inflammation in asthma, *N. Engl. J. Med.* *323*:1033 (1990).
12. A. Janin, G. Torpier, P. Courtin, M. Capron, L. Prin, A.-B. Tonnel, P.-Y. Hatron, and B. Gosselin, Segregation of eosinophil proteins in alveolar macrophage compartments in chronic eosinophilic pneumonia, *Thorax* *48*:57 (1993).
13. B.S. Polla and S. Kantengwa, Heat shock proteins and inflammation, *Current Topics in Microbiology and Immunology* (S.H.E. Kaufmann, ed.), Springer-Verlag Berlin, 1991, p. 93.
14. M.G. Santoro, E. Garaci, and C. Amici, Prostaglandins with antiproliferative activity induce the synthesis of a heat shock protein in human cells, *Proc. Natl. Acad. Sci. USA.* *86*:8407 (1989).
15. C. Amici, T. Palamara, and M.G. Santoro, Induction of thermotolerance by prostaglandin A in human cells, *Exp. Cell. Res.* *207*:230 (1993).
16. M. Köller, and W. König, Arachidonic acid metabolism in heat shock treated human leukocytes, *Immunology* *70*:458 (1990).
17. I.M. Takenaka and L.E. Hightower, Transforming growth factor-β1 rapidly induces HSP70 and HSP90 molecular chaperones in cultured chicken embryo cells, *J. Cell. Physiol.* *152*:568 (1992).

18. D.T.W. Wong, P.F. Weller, S.J. Galli, A. Elovic, T.H. Rand, G.T, Gallagher, T. Chiang, M.Y. Chou, K. Matossian, J. McBride, and R. Todd, Human eosinophils express transforming growth factor α, *J. Exp. Med. 172*:673 (1990).
19. S. Helqvist, B.S. Polla, J.Johannesen, and J. Nerup, Heat shock protein induction in rat pancreatic islets by recombinant human interleukin 1β, *Diabetologia 34*:150 (1991).
20. L.E. Hightower and F.P. White, Cellular responses to stress: comparison of a family of 71-73-kilodalton proteins rapidly synthesized in rat tissue slices and canavanin-treated cells in culture, *J. Cell. Physiol. 108*:261 (1981).
21. D.B. Young, Heat-shock proteins: immunity and autoimmunity, *Curr. Opin. Immunol. 4*:396 (1992).
22. R. Valenta, M. Duchêne, K. Pettenburger, C. Sillaber, P. Valent, P. Bettelheim, M. Breitenbach, H. Rumpold, D. Kraft, and O. Scheiner, Identification of Profilin as a novel pollen allergen; IgE autoreactivity in sensitized individuals, *Science 253*:558 (1991).
23. S.H.E. Kaufmann, Heat shock proteins and the immune response, *Immunol. Today 11*:129 (1990).
24. S. Vrtala, M. Grote, M. Duchene, R. Van Ree, D. Kraft, O. Schiener, and R. Valenta, Properties of tree and grass pollen allergens: reinvestigation of the linkage between solubility and allergenicity, *Int. Arch. Allergy Immunol. 102 (2)*:160 (1993).
25. B.S. Polla, Heat shock proteins in host-parasite interactions, *Immunol. Today 12*: A38 (1991).
26. F. Lawrence, and M. Robert-Gero, Induction of heat shock and stress proteins in promastigotes of three *Leishmania species*, *Proc. Natl. Acad. Sci. USA. 82*:4414 (1985).
27. R.M. Smejkal, R. Wolff, and J.G. Olenick, *Leishmania braziliensis panamensis*: increased infectivity resulting from heat shock, *Exp. Parasitol. 65*:1 (1988).
28 L.H.T. Van Der Ploeg, S.H. Giannini, and C.R. Cantor, Heat shock genes: regulatory role for differentiation in parasitic protozoa, *Science 228*:1443 (1985).
29. J.H. Zarley, B.E. Britigan, and M.E. Wilson, Hydrogen peroxide–mediated toxicity of *Leishmania donovani chagasi promastigotes*: role of hydroxyl radical and protection by heat shock, *J. Clin. Invest. 88*:1511 (1991).
29a. S. Kantengwa, I. Müller, J. Louis, and B.S. Polla, Infection of human and murine macrophages with *Leishmania major* is associated with early parasite heat shock protein synthesis but fails to induce a host cell stress response. *Immunol. Cell Biol. 73*:73 (1995).
30. M. Perin-Minisini, M.-J. Richard, and B.S. Polla, Radicaux libres de l'oxygène: rôle pathogénique et cibles thérapeutiques dans l'athérosclérose, *STV*, (1994).
31. B.S. Kasinath and E. J. Lewis, Eosinophilia as a clue to the diagnosis of atheroembolic renal disease, *Arch. Intern. Med. 147*:1384 (1987).
32. D.K. Young, M.F. Burton, and J.H. Herman, Multiple cholesterol emboli syndrome simulating systemic necrotizing vasculitis, *J. Rheumatol. 13*:423 (1986).
33. P.A. Berberian, W. Myers, M. Tytell, V. Challa, and M. Gene Bond, Immunohistochemical localization of heat shock protein-70 in normal-appearing and atherosclerotic specimens of human arteries, *Am. J. Pathol. 136*:71 (1990).

34. B.S. Polla, L. Bornman, M. Perin, Heat shock proteins, oxidative stress and atherosclerosis, *Biologie prospective* (M.M. Glateau, G. Siest, and J. Henny eds.), John Libbey, Paris, 1993, p. 223.
35. B.S. Polla, N. Mili, and S. Kantengwa, Heat shock and oxidative injury in human cells, *Heat shock* (B. Maresca and S. Lindquist, eds.), Springer-Verlag, Berlin, 1991, p. 279.
36. M. Perin-Minisini, S. Kantengwa, and B.S. Polla, DNA damage and stress protein synthesis induced by oxidative stress proceed independently in the human premonocytic line U937, *Mutat. Res. 315*:169 (1994).
37. P.L. Moseley, S.J. York, and J. York, Bleomycin induces the hsp 70 heat shock promotor in cultured cells, *Am. J. Respir. Cell. Mol. Biol. 1*:89 (1989).
38. D.O. Slosman, P.M. Costabella, M. Roth, G. Werlen, and B.S. Polla, Bleomycin primes monocytes-macrophages for superoxide production, *Eur. Respir. J. 3*:772 (1990).
39. P.F. Piguet, M.A. Collart, G.E. Grau, Y. Kapanci, and P. Vassali, Tumor necrosis factor/cachectin plays a key role in bleomycin-induced pneumopathy and fibrosis, *J. Exp. Med. 170*:655 (1989).
40. B.S. Polla, N. Pittet, and D.O. Slosman, Heat shock (HS) protects the human monocytic line U937 from bleomycin (BLM)-mediated oxidative injury, *Eur. Respir. J. 3*:335s (1990).
41. J. Villar, J.D. Edelson, M. Post, B.M. Mullen, and A.S. Slutzky, Induction of heat stress proteins is associated with decreased mortality in an animal model of acute lung injury, *Am. Rev. Respir. Dis. 147*:177 (1993).

31

Heat Shock Proteins in Duchenne Muscular Dystrophy and Other Muscular Diseases

Liza Bornman
University of Pretoria, Pretoria, South Africa, and UFR Cochin Port Royal, Paris, France

Barbara S. Polla
UFR Cochin Port Royal, Paris, France

I. INTRODUCTION

Both stressful and nonstressful stimuli, including elevated temperatures, oxidative stress, tissue trauma, embryonic development, and cell proliferation, are associated with the overexpression of heat shock/stress proteins (hsps). Hsps are essential for cell survival during and after stress as well as during normal cell metabolism. Indeed, these "molecular chaperones" contribute to the correct folding or refolding of unfolded or misfolded (denatured) proteins (1). Although the requirements for chaperoning are high during active protein synthesis, they increase during stress (2). Hsps exert important protective functions during and after heat shock, and there is increasing evidence that they do so as well during cellular injury in general and oxidative stress in particular both in vitro (reviewed in ref. 3) and in vivo (4). Hsps also contribute to the induction of the immune response both by acting as potent immunogens and by contributing to antigen processing/presentation (5). Furthermore, both αβ and γδ hsp-specific T cells may contribute to the amplification of an immune response regardless of whether primary recognition is of autologous or heterologous hsp.

Our long-lasting interests in the interactions between oxidative injury and hsps have led us to consider the possibility that they might play a role in oxidative-stress associated muscular diseases. Hsps (hsp 70 in particular) can be induced by oxidative stress in human cells. Other stress proteins, including heme oxygenase (HO) (6) and scavenging enzymes (superoxide dismutase [SOD], glutathione peroxidase [GSHP], and catalase), some of which are classified among the hsps, are oxidation-specific stress proteins in human cells. In contrast, glucose-regulated proteins (grps) are modulated by alterations in protein glycosylation as well as modifications in the intracellular calcium homeostasis.

A number of myopathies, and Duchenne's muscular dystrophy (DMD) in particular, display a variety of pathological processes suggestive of oxidative stress which could potentially lead to stress-related gene activation (7,8). The induction of stress proteins in vivo during reperfusion injury has been documented in a number of animal models (for review, see ref. 9). The generation of reactive oxygen species (ROS) secondary to ATP depletion is the likely inducer of hsp under these conditions. These findings closely relate to muscle disease, inasmuch ATP depletion is a common feature of altered muscle cell metabolism. In cultured myogenic cells, ATP depletion has been shown to induce activation of heat shock transcription factor 1 (HSF-1) (10).

We investigated the expression of hsp and oxidation-specific stress proteins in DMD based on the hypothesis that oxidative stress would be a major feature in muscle wasting and necrosis in this disease. We then extended these studies to other muscular disorders and to other stress proteins such as grps. In this chapter, we review available data on stress protein expression in DMD and establish comparisons with other myopathies, including polymyositis-dermatomyositis (PM-DM) (as examples of inflammatory myopathies) and mitochondrial myopathies (MM).

II. HSP AND ANTIOXIDANT ENZYME EXPRESSION IN DMD

DMD is one of the most frequent and severe inherited human muscular dystrophies (11). It is caused by mutations of the dystrophin gene at Xp21. The protein product of the defective gene, dystrophin, has been discovered by reverse genetics and localized to the cytoplasmic face of the plasma membrane in all muscle tissue (12,13). Dystrophin is associated with an integral membrane glycoprotein complex in the sarcolemma which anchors it to the plasmalemma and provides a link to the extracellular matrix. Inspite of a fixed mutation which is present from fetal life onward, DMD is a progressive disorder, and the identification of the gene defect has not allowed us yet to characterize fully the biochemical basis of DMD. Defective membrane stability and calcium regulation have been particularly considered (14,15). The finding that a similar dystrophin deficiency in the *mdx* mouse model of DMD does not lead to progressive muscle wasting should help us identify the metabolic defects eventually leading to muscle necrosis, since they are present in humans but not in the mouse (16).

There are a number of factors that provide evidence in favor of an imbalance between oxidants and antioxidants in DMD. A number of studies have examined the antioxidant status in DMD and found either decreased or increased (as well as normal) levels of antioxidants in the blood or the muscle, respectively, of patients suffering DMD (reviewed in ref. 17) (Table 1). When the products of lipid peroxidation were examined either in the blood or in the muscle, all studies (6/6) indicated increased lipid peroxidation (17). However, among 15 studies evaluating the results of antioxidant therapy (e.g., selenium, vitamin E, orgotein, allopurinol), only 3 showed a significant benefit in DMD (17).

Using Western blotting and monospecific antibodies, we observed an increased expression of GSHP and catalase in muscle homogenates of certain patients with DMD (L. Bornman et al., submitted). Our results thus provide further evidence in favor of an oxidative load in DMD muscle fibers. The controversial results reported in the literature may relate the progressive nature of DMD, patients in one experimental group usually being at different stages of the disease associated with specific pathological reactions; the oxidant-antioxidant balance may vary between patients in different stages of the disease.

Table 1 Antioxidant Status in Duchenne Muscular Dystrophy

	Results		
	Studies nb	Normal	Decreased
Blood			
Vitamin E	2	1	1
Selenium	2	1	1
SOD	3	2	1
GSH	1	1	
	Results		
	Studies nb	Normal	Increased
Muscle			
Catalase	2	1	1
SOD	2		2
GSH	3	1	2

Source: Adapted from Ref. 17.

We then examined the expression of hsps and grps using computer-assisted image processing of immunohistochemically stained cryosections. We found a significant induction of hsp 70 and hsc 70 (the constitutive, cognate form of hsp 70), hsp 60, and ubiquitin in the hypercontracted fibers, a distinctive feature of DMD often associated with necrotic fiber segments (Fig. 1) (see color plate), and of hsp 90 and ubiquitin in regenerating fibers and in macrophage-invaded necrotic fibers of DMD muscle (18) (Fig. 2) (see color plate) (as well as in inflammatory myopathies, see below). Subsequent Western blot analysis in an expanded group of patients with DMD revealed the confinement of hsp 72 upregulation to a subgroup of patients with a more advanced stage of muscle degeneration which also overexpressed the oxidation-specific stress proteins GSPH and catalase (our unpublished results).

In contrast, the expression of grp 75 and hsp 60 were elevated in all biopsies analyzed, probably relating to the elevated intracellular calcium levels which are prominent throughout the whole disease process and are known to induce the grp in several cell types, including macrophages (14; L. Bornman et al., unpublished data). The differential expression of hsp and grp indicates that although a link between calcium-mediated oxidative stress and hsps has been suggested (19), calcium is unlikely the sole inducer of hsp 70. Thus, we conclude at this time that both the toxicity of ROS, potentiated by calcium, and the presence of damaged proteins within the hypercontracted fibers, are the most likely inducers of the above-mentionned stress proteins in DMD (2,20).

The induction of hsp 90 and ubiquitin in regenerating fibers appears to result from a selective regulation linked to features distinct from the disease process. Myogenesis as a differentiation process is regulated by various factors, including hormones and regulatory kinases. Hsp 90 interacts with several viral oncogene products that display tyrosine kinase activity, associates with tubulin and actin, stimulates the activity of eukaryotic initiation factor 2a subunit–specific protein kinase, and forms stable complexes with steroid receptors (reviewed in ref. 21). Furthermore, the transcription factor MyoDI, a myogenic determination DNA-binding protein which plays a major role in myogenesis, is conformationally activated to its DNA binding form by hsp 90 (22). Comparison of hsp 90 expression in normal and DMD muscle of both infants and adults leads us now to propose that in DMD, the upregulation of hsp 90, in contrast to hsp 70, essentially relates to muscle fiber regeneration (L. Bornman et al., *Muscle and Nerve*, in press).

III. PROPOSED MODEL FOR STRESS PROTEIN INDUCTION IN DMD

We propose the following model for the mechanisms involved in the induction of a stress response in DMD (Fig. 3) and postulate that ROS play a key role in

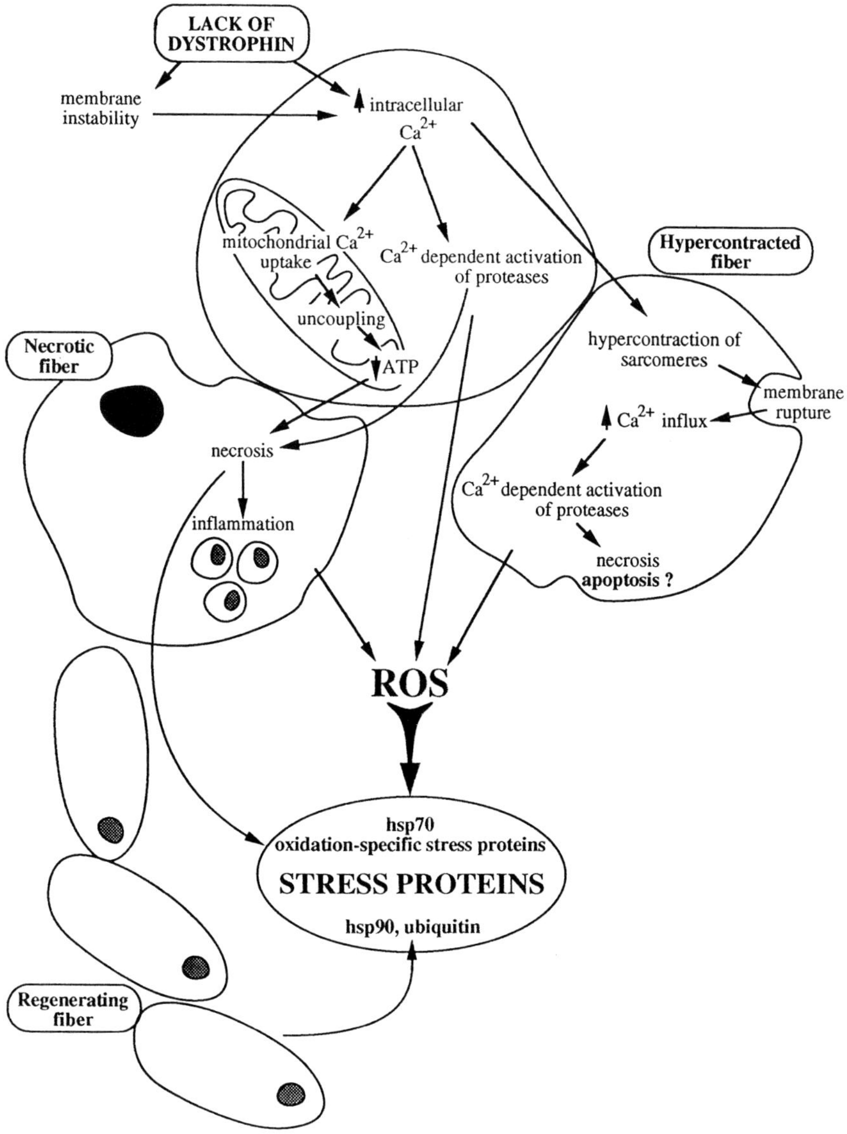

Figure 3 Proposed model for selective stress protein induction in DMD. See text for details.

the induction of hsp 70 and oxidation-specific stress proteins. Hsp 70 may be directly upregulated by ROS which lead, in human cells, to translocation and binding of HSF-1 to the heat shock promotor (heat shock element, HSE) (Jacquier-Sarlin and Polla, submitted). Alternatively, oxidatively altered cellular proteins may represent the signal for such induction, whereas fiber regeneration appears to be the signal for the induction of hsp 90 and ubiquitin.

In DMD, oxidative damage is potentiated by calcium overloading of mitochondria which is associated with a consequent drop in ATP levels. Muscle fibers deprived of energy become necrotic and infiltrated by inflammatory cells. Calcium-dependent proteases directly damage intracellular proteins and may convert xanthine-dehydrogenase to xanthine-oxidase, which results in a further sustained production of superoxide anions (23). Sarcomere hypercontraction induces membrane rupture, which further elevates intracellular calcium level with the consequences described above. We propose that the upregulation of grp 75 and hsp 60 closely relates to the uptake of calcium by the endoplasmic reticulum and the mitochondria, respectively, and to its above-mentioned consequences in terms of calcium-dependent proteases and sarcomere hypercontraction.

IV. THE POLYMYOSITIS AND DERMATOMYOSITIS SYNDROMES

The inflammatory myopathies are a heterogeneous group of disorders comprising a number of entities which display symptoms of connective tissue diseases as well as characteristic immunological alterations (reviewed in ref. 24). Immune-mediated mechanisms are central to the pathogeny of these diseases: For example, an antibody- or immune complex–mediated response against a vascular component plays a key role in DM. The C5b-9 membranolytic attack complex induces an intramuscular microangiopathy, leading to a loss of capillaries, muscle ischemia, muscle fiber necrosis, and perifascicular atrophy. Inflammatory changes are present in both skin and muscle in DM, whereas T-cell–mediated destruction of muscle fibers is uncommon.

In contrast, a T-cell–mediated response against muscle fibers is central in the etiology of polymyositis (PM) (25). Autoagressive $CD8^+$ cytotoxic T cells that recognize MHC-I restricted muscle antigens initiate fiber injury, resulting in phagocytosis and muscle necrosis. The vast majority of $CD8^+$ cells in PM use the common $\alpha\beta$ T-cell receptor to recognize antigens, whereas $\gamma\delta$ T cells are rare or absent in PM muscle (26,27). In a distinct subtype of PM, however, Holfeld et al. (26,27) described the initiation of necrosis by autoagressive $\gamma\delta$ T cells surrounding and invading nonnecrotic muscle fibers expressing HLA-class I antigen and the 65-kDa hsp.

Certain pathological features are shared by all types of PM and dermatomyositis (DM) syndromes and are required for diagnosis. These features include varying stages of muscle necrosis accompanied with macrophage and lymphocyte infiltration (24). Muscle fibers at various stages of regeneration are frequently observed, together with a variation in fiber diameter which becomes more prominent along with the chronicity of the disease. PM and DM are both associated with an increased amount of connective tissue (24).

We investigated the expression of hsps in muscle from patients with PM and DM by means of immunohistochemistry and image analysis (18) and observed the induction of hsp 60 and of ubiquitin in necrotic and regenerating muscle fibers. A similar induction was found for hsp 90, which was, however, less significant in necrotic fibers. Other hsps were not induced. Subsequent Western blot analysis in a large group of patients with PM as well as patients with DM revealed the increased expression of SOD (Mn) and GSHP but not of SOD (Cu/Zn), catalase, or HO. This latter finding is in contrast with rhabdomyolysis, where HO has been found to be induced and suggested to exert protective functions (28,29).

The induction of a stress response in inflammatory conditions is reported in many other chapters in this book (see in particular Natvig, Yu, Rook, Petermans, Bielefldt-Ohmann, Polla). Whether under these conditions the immune response to self-hsp (mostly described for hsp 60/65) is detrimental or beneficial remains a matter of controversy, although the most recent evidence from a number of laboratories rather favors the latter possibility (D. Young, personal communication).

V. MITOCHONDRIAL MYOPATHIES

Mitochondrial myopathies (MM), which give rise to a large group of clinical syndromes, are associated with mutations in the mitochondrial or nuclear genome affecting mitochondrial function (see Chapter 32). MM can affect multiple systems, but they cause chiefly muscle defects, since muscle tissue is the most susceptible to mitochondrial dysfunction. Cells from individuals with MM usually contain mitochondria with both normal and mutant genomes (heteroplasmy) and mitochondrial dysfunction occurs when mutant mitochondria predominate over a certain threshold. The amount and distribution among and within the patient's tissue of the mitochondria harboring a mutated genome together with mitochondrial DNA mutations distinct from the primary mutation account for the heterogeneity in the phenotypic expression of MM. A characteristic histological feature of these myopathies is the massive proliferation of mitochondria beneath the plasma membrane and between the myofibrils showing intense staining with the trichrome technique, commonly referred to as "ragged-red fibers." This proliferation is considered to be an attempt to compensate for insufficient energy production from oxidative phosphorylation.

The ragged-red fibers characteristic of MM have previously been associated with an enhanced expression of hsp 60, ubiquitin and αβ-crystallin, another member of the small hsp family (29, 30). Tubular aggregates in subsarcolemmal areas deficient in myofibrillar structures are occasionally found in various muscle disorders and have also been associated with the increased expression of hsp 70 (31). Hsp 70 and hsp 60 play important roles in mitochondrial protein import (reviewed in ref. 32). Although the cytosolic hsp 70 maintains precursor proteins in a transport-competent conformation, the mitochondrial hsp 70 and hsp 65 assist translocation across mitochondrial membranes and folding of imported proteins in the mitochondrial matrix. The mitochondrial hsp 70 and hsp 60 have been implicated in the maintenance of ATP levels and protection against uncoupling of oxidative phosphorylation during thermotolerance (33,34; B.S. Polla et al., submitted). The importance of hsp 60/65 in mitochondrial protein import and enzyme assembly to maintain functional mitochondria is underlined by the recent report by Agsteribbe et al. (35) of a fatal, systemic mitochondrial disease associated with a deficiency in hsp 60 and with decreased mitochondrial enzymes activities and abnormal ultrastructure of the mitochondria. Mitochondria, which were mainly concentrated around the nuclei, appeared swollen with locally disintegrated innermembranes and large vacuoles. Although cytoplasmic hsp 70 was normal, hsp 60 levels were one fifth of that found in normal controls.

We investigated the expression of hsp 90, hsp 70, hsc 70, and hsp 60, and ubiquitin in three patients with MM. Although immunohistochemistry did not reveal the induction of any of these hsps in cryosections from the patients with MM investigated, not even in the ragged-red fibers, Western blot analysis indicated an increased expression of hsp 70 and hsp 60, whereas all other stress proteins and scavengers examined remained unchanged. Our results are in contrast with those of Sparaco et al. (30), who observed the immunolocalization of hsp 60 and ubiquitin in ragged-red fibers of patients with mitochondrial encephalomyopathies. The basis of this dissimilarity is uncertain, but the variable induction of a heat shock response in MM may relate to the heterogeneity of these syndromes. Iwaki et al. (31) observed increased immunoreactivity for αβ-crystallin in ragged-red fibers of MM.

The fact that selected stress proteins are induced in DMD and not in MM suggest that other factors than oxidative stress (which most likely occurs in both), factors which may include type, subcellular location and amount of ROS produced, are involved in the regulation of hsp expression in these diseases.

VI. HSP FUNCTION IN MUSCULAR DISEASES

We propose that the induction of stress proteins deserves protective functions in muscle diseases. This hypothesis is based on observations from many different groups and in various models, both in vitro and in vivo, for a protective function of stress proteins, as well as from direct evidence showing that *mdx* mice fed an iron-free diet display simultaneously less muscle necrosis and less hsp expression (L. Bornman et al., submitted).

Potential mechanisms for protection by stress proteins in muscle pathology include:

1. Chaperoning of unfolded proteins
2. Increased proteolysis of denatured proteins
3. Prevention of inhibition of oxidative phosphorylation
4. Regulation of the production of ROS
5. Other direct or indirect protection mechanisms against the toxic effects of ROS such, for example, as direct scavenging or DNA repair

Members of both the hsp 60 and hsp 70 families act as molecular chaperones and may facilitate the repair of damaged proteins by preventing aggregation and assisting the refolding of unfolded proteins to their native conformation (36). Chaperones also facilitate the turnover of denatured proteins by solubilizing aggregates of denatured proteins, whereas ubiquitin targets denatured proteins for nonlysosomal degradation and is often observed in conditions where denatured proteins need to be removed. There is increasing evidence that hsp induction protects mitochondrial ATPase activity. Patriarca and Maresca (33) illustrated the positive effect of a heat shock response on the ability of cells to maintain respiration coupled to ATP production in *Saccharomyces cerevisiae*. We observed a similar effect in mitochondria from thermotolerant rat hearts exposed to hydrogen peroxide during reperfusion (L. Bornman et al., submitted). On the other hand, heat shock prevents the production of superoxide by NADPH oxidase (37).

HO and ferritin are stress proteins with an interesting potential for radical scavenging. Degradation of heme by HO results in the production of the antioxidant bile pigments bilirubin and biliverdin (29). Subsenquently, ferritin is induced by the iron released during heme catabolism, leading to an improved iron storage and thus decreasing iron-catalyzed free radical reactions. Nath et al. (30) provided evidence that the induction of HO, coupled to ferritin synthesis during myoglobin degradation, is a rapid effective antioxidant response which reduces mortality in rhabdomyolysis. HO and/or ferritin may play essential roles in protecting DNA from oxidatively induced strand breaks or fragmentation (39). Although our results on HO expression in the myopathies examined are as for now unconclusive, we have yet to determine HO activity and ferritin expression. The important role iron may play in DMD by potentiating oxidative injury (reviewed in ref. 40) is supported by our above-mentionned data on the effects of an iron-free diet in *mdx* mice.

Finally, the overexpression of hsp 90 may modulate the cellular responses to steroids. The anti-inflammatory effects of steroids are well established both in DMD and in the inflammatory myopathies, with possible targets belonging to either the immune system or the muscle itself. Indeed, glucocorticoids enhance myoblast proliferation and promote muscle regeneration. Hsp 90, by forming a complex with the steroid receptors, may mediate these effects. Furthermore, hsp 90 may also activate MyoDI, a major protein involved in myogenesis (41).

VII. CONCLUSIONS

Having reviewed available data on the expression of stress proteins in DMD, DM, PM, and MM, we propose that the stress response observed in myopathies with diverse etiologies relates to the biochemical and pathological features such as, respectively, an imbalance in the oxidant/antioxidant status or alterations in intracellular calcium, or hypercontraction, necrosis, regeneration, and/or ragged-red fibers, rather than to the disease itself. The selective upregulation of various stress proteins in distinct myopathies may thus have diagnostic as well as prognostic value.

Furthermore, since stress proteins likely represent an autoprotective mechanism, manipulation of the stress response prior to, or during, muscle injury may reduce subsequent, chronic damage. Further insight into the role of stress proteins in muscular disorders may contribute to the understanding of the biochemical basis of these diseases and to the development of new avenues to therapy.

AKNOWLEDGMENTS

Experimental work presented here was supported by the Medical Research Council of South Africa and the Swiss National Research Foundation (Grant no. 32.28645.90 to B.S.P.) We thank G. S. Geriche for continuing support and J. C. Kaplan for stimulating discussions.

REFERENCES

1. R.J. Ellis and S.M. Van der Vies, Molecular chaperones, *Annu. Rev. Biochem. 60*:321 (1991).
2. L.E. Hightower, Heat shock, stress proteins, chaperones and proteotoxicity, *Cell 66*:191 (1991).
3. B.S. Polla, N. Mili, and S. Kantengwa, Heat shock and oxidative injury in human cells, *Heat Shock* (B. Maresca and S. Lindquist eds.), Springer-Verlag, Berlin, 1991, p. 279.
4. M.R. Jacquier-Sarlin, K. Fuller, A.T. Dinh-Xuan, M.-J. Richard, and B.S. Polla, Protective effects of hsp70 in inflammation, *Experientia* (1994).

5. E. Mariéthoz, F. Tacchini-Cottier, M.R. Jacquier-Sarlin, F. Sinclair, and B.S. Polla, Exposure to heat shock does not increase class II expression but modulates antigen-dependent T cell response, *Int. Immunol. 6*:925 (1994).
6. S.M. Keyse and R.M. Tyrrell, Heme oxygenase is the major 32-kDa stress protein induced in human skin fibroblasts by UVA radiation, hydrogen peroxide, and sodium arsenite, *Proc. Natl. Acad. Sci. USA 86*:99 (1989).
7. L. Austin, M. De Niesc, A. McGregor, H. Arthur, A. Gurusinghe, and M.K. Gould, Potential oxyradical damage and energy status in individual muscle fibers from degenerating muscle diseases, *Neuromusc. Disord. 2*:27 (1992).
8. M.E. Murphy, and J.P. Kehrer, Free radicals: a potential pathogenic mechanism in inherited muscular dystrophy, *Life Sci. 39*:2271 (1986).
9. D.M. Yellon, and M.S. Marber, Hsp70 in myocardial ischemia, *Experientia* (1994).
10. I.J. Benjamin, S. Horie, M.L. Greenberg, R.J. Alpern, and R.S.Williams, Induction of stress proteins in cultured myogenic cells: molecular signals for the activation of heat shock transcription factor during ischemia, *J. Clin. Invest. 89*:1685 (1992).
11. M. Koenig, E.P. Hoffman, C.J. Bertelson, A.P. Monaco, C. Freener, and L.M. Kunkel, Complete cloning of the Duchenne muscular dystrophy (DMD) cDNA and preliminary genomic organization of the DMD gene in normal and affected individuals, *Cell 50*:509 (1987).
12. E.P. Hoffman, R.H. Brown, and L.M. Kunkel, Dystrophin: the protein product of the Duchenne muscular dystrophy locus, *Cell 51*:919 (1987).
13. E.P. Hoffman, and L. Schwarts, Dystrophin and disease, *Mol. Aspects Med. 12*:175 (1991).
14. J.S.H. Tay, P.S. Lai, P.S. Low, W.L. Lee, and G.C. Gan, Pathogenesis of Duchenne muscular dystrophy: the calcium hypothesis revisited, *J. Paediatr. Child Health 28*:291 (1992).
15. B.P. Lotz and A.G. Engel, Are hypercontracted muscle fibers artifacts and do they cause rupture of the plasma membrane?, *Neurology 37*:1466 (1987).
16. G. Bulfield, W.G. Siller, P.A. Wight, and K.J. Moore, X chromosome-linked muscular dystrophy (mdx) in the mouse, *Proc. Natl. Acad. Sci. USA 79*:1189 (1984).
17. M.J. Jackson, and R.H.T. Edwards, Free radicals and trial of antioxidant therapy in muscle diseases, *Antioxidant in Therapy and Preventive Medicine* (I. Emerit et al., eds.), Plenum Press, New York, 1990, p. 485.
18. L. Bornman, B.S. Polla, B.P. Lotz, and G.S. Gericke, Expression of heat-shock/stress proteins in Duchenne muscular dystrophy, *Muscle Nerve 18*:23 (1995).
19. S. Kantengwa, A.M. Capponi, J.V. Bonventre, and B.S. Polla, Calcium and the heat shock response in the human monocytic line U937, *Am. J. Physiol. 259 (Cell Physiol. 28)*:C77 (1990).
20. C.D. Malis and J.V. Bonventre, Mechanism of calcium potentiation of oxygen free radical injury to renal mitochondria. A model for post-ischemic and toxic mitochondrial damage, *J. Biol. Chem 261*:14201 (1986).
21. B.S. Polla, M. Perin, and L. Pizurki, Regulation and functions of stress proteins in allergy and inflammation, *Clin. Exp. Allergy 23*:548 (1993).

22. R. Shaknovich, G. Shue, and D.S. Kohtz, Conformational activation of a basic helix-loop-helix protein (MyoD1) by the C-terminal region of murine HSP 90 (HSP 84), *Mol. Cell. Biol. 12*:5059 (1992).
23. J.M. Mc Cord, Oxygen-derived free radicals in postischemic tissue injury, *N. Engl. J. Med. 312*:159 (1985).
24. R.R. Heffner Jr., Inflammatory myopathies. A review, *J. Neuropathol. Exp. Neurol. 52*:339 (1993).
25. I.N. Targoff, Polymyositis, *Systemic autoimmunity* (M. Reichlin, P. Bigazzi, eds.), Marcel Dekker, New York, 1991, p. 201.
26. R.H. Hohlfeld, A.G. Engel, K. Li, and M.C. Harper, Polymyositis mediated by T lymphocytes that express the γ/δ receptor, *N. Engl. J. Med. 324*:877 (1991).
27. R. Hohlfeld, A.G. Engel, The rold of gamma-delta T lymphocytes in inflammatory muscle disease, *Heat Shock Proteins and Gamma-Delta T Cells*, (C.F. Brosnan, ed.), Karger, Basel, 1992, p. 75.
28. K.A. Nath, G. Balla, G.M. Vercellotti, J. Balla, H.S. Jacob, M.D. Levitt, and M.E. Rosenberg, Induction of heme oxygenase is a rapid protective response in rhabdomyolysis in the rat, *J. Clin. Invest. 90*:267 (1992).
29. R. Stocker, Induction of haem oxygenase as a defence against oxidative stress, *Free Rad. res. Commun. 9*:101 (1990).
30 . M. Sparaco, G. Rosoklija, K. Tanji, M. Sciacco, S. DiMauro,and E. Bonilla, Immunolocalization of heat shock proteins in ragged-red fibers of patients with mitochondrial encephalomyopathies, *Neuromuscular Dis. 3*:71 (1993).
31. T. Iwaki, A. Iwaki, and J.E. Goldman, $\alpha\beta$-Crystallin in oxidative muscle fibers and its acculation in ragged-red fibers: a comparative immunohistochemical and histochemical study in human skeletal muscle, *Acta Neuropathol. 85*:475 (1993).
31. J.E. Martin, K. Mather, M. Swash, and A.B. Gray, Expression of heat shock protein epitopes in tubular aggregates, *Muscle Nerve 14*:219 (1991).
32. R.A. Stuart, D.M. Cyr, and W. Neupert, Hsp 70 in mitochondrial biogenesis: from chaperoning nascent polypeptide chains to facilitation of protein degradation, *Experientia* (1994).
33. E.J. Patriarca and B. Maresca, Acquired thermotolerance following heat shock protein synthesis prevents impairment of mitochondrial ATPase activity at elevated temperatures in Saccharomyces cerevisiae, *Exp. Cell. Res. 190*:57 (1990).
34. B.S. Polla and J.V. Bonventre, Heat shock protects cells dependent on oxidative metabolism from inhibition of oxidative phosphorylation, *Clin. Res. 35*:555A (1987).
35. E. Agsteribbe, A. Huckriede, M. Veenhuis, M.H.J. Ruiters, K. E. Niezen-Koning, O.H. Skjedal, K. Skullerud, R.H. Gupta, R. Hallberg, O.P. van Diggelen, and H.R. Scholte, A fatal, systemic mitochondrial disease with decreased mitochondrial enzyme activities, abnormal ultrastructure of the mitochondria and deficiency of heat shock protein 60, *Biochem. Biophys. Res. Comm. 193*:146 (1993).
36. F.U. Hartl, and J. Martin, Protein folding in the cell: the role of molecular chaperones Hsp70 and Hsp60, *Ann. Rev. Biophys. Biomol. Struct. 21*:293 (1992).
37. I. Maridonneau-Parini, S.E. Malawista, H. Stubbe, F. Russo-Marie, and B.S. Polla, Heat-shock proteins in human neutrophils: protection of NADPH oxidase activity, *J. Cell Phsyiol. 156*:204 (1993).

38. I.A. Clark, Proposed treatment of Duchenne muscular dystrophy with desferrioxamine, *Med. Hyp. 13*:153 (1984).
39. M.-J. Perin-Minisini, Kantengwa and B.S. Polla, Heat shock protein synthesis and DNA damage proceed independently in the human premonocytic line U937. *Mut. Res. (DNA Repair) 15*:169 (1994).
40. B.G. Atkinson, Synthesis of heat-shock proteins by cells undergoing myogenesis, *J. Cell. Biol. 89*:666 (1981).
41. V. Guerriero, and J.R. Florini, Dexamethasone effects on myoblast proliferation and differentiation, *Endocrinology 106*:1198 (1980).

32

Mitochondrial Neuromuscular Disease Associated with Partial Deficiency of Heat Shock Protein 60

Etienne Agsteribbe and Anke Huckriede
University of Groningen, Groningen, The Netherlands

I. INTRODUCTION

Human mitochondrial diseases are caused by substantially decreased activities of one or more of the many enzymes of mitochondrial metabolism. The enzyme deficiencies are due to mutations in genes encoding enzyme subunits or they may reflect a defect in the biosynthesis of mitochondria. In the latter case, several enzymes will be affected simultaneously. In this paper, mitochondrial disorders will be presented which are associated with a partial deficiency of heat shock protein (hsp) 60. This stress protein is located in the mitochondrial matrix where it is needed for the folding of newly imported proteins and their subsequent assembly into multimeric enzyme complexes. As proper folding and assembly are essential for the realization of functional enzymes, a defect in one of these processes can be defined as a disorder of mitochondrial biosynthesis.

II. MITOCHONDRIAL BIOSYNTHESIS

Mitochondria contain their own genetic information and protein synthesizing system, including ribosomes, tRNAs, and the enzymes needed for transcription and translation. The human mitochondrial genome comprises genes for rRNAs and tRNAs and for mRNAs encoding 13 polypeptides, all of which are subunits of enzymes of oxidative phosphorylation (1).

The vast majority of proteins required for oxidative phosphorylation as well as for all other mitochondrial processes are, however, encoded by nuclear genes. These proteins are synthesized in the cytosol and imported into the mitochondria. Mitochondrial protein import is a multistep process which is mediated by a number of nuclearly encoded proteins. A major role in the regulation of mitochondrial protein import is reserved for chaperone proteins of the classes of heat shock proteins 60 and 70, respectively (2–7). These heat shock proteins catalyze the folding of the mitochondrial protein into conformations required at the different import steps. Cytosolic heat shock protein 70 (chsp 70) binds to newly synthesized precursor polypeptides to maintain them in an unfolded conformation necessary for membrane translocation. After binding to the protein receptor, translocation of the polypeptides through the outer and inner membrane contact sites proceeds as a result of interaction with mitochondrial heat shock protein 70 (mhsp 70). Inside, in the mitochondrial matrix, the N-terminal signal sequence, which is needed for receptor recognition, is removed by proteolytic processing. Hsp 60 mediates the folding of processed polypeptides in order to facilitate their assembly into multimeric enzymes.

The elucidation of the process of mitochondrial protein import is mainly the result of studies on the lower eukaryotes *Saccharomyces cerevisiae* (baker's yeast) and the mold Neurospora crassa. However, there is good experimental evidence that for higher eukaryotes, including humans, mitochondrial protein import occurs in a similar manner.

III. MITOCHONDRIAL NEUROMUSCULAR DISORDERS

Mitochondrial neuromuscular disorders are in clinical practice commonly referred to as mitochondrial encephalomyopathies (8–12). Biochemically, these diseases are characterized by impaired synthesis of ATP due to defects of pyruvate metabolism, the citric acid cycle, and/or oxidative phosphorylation. Clinical manifestation of these defects can differ widely among patients but generally encompasses hypotonia and exercise intolerance and occasionally cardiomyopathy and brain damage. Lactic acidosis is typical of all mitochondrial encephalomyopathies. In muscular tissue, the accumulation of morphologically aberrant mitochondria is frequently observed. Genetic analysis revealed that defects of oxidative phosphorylation are often associated with point mutations or deletions of mitochondrial DNA (13,14). Mutations in nuclear DNA giving rise to mitochondrial encephalomyopathies have rarely been identified. The nuclear mutations described so far can be divided into mutations affecting nuclearly encoded subunits of mitochondrial enzymes (15,16), mutations affecting the fidelity of mitochondrial DNA replication (17,18), and mutations impairing the proper import of proteins into the mitochondria (19,20), respectively. In the latter case, mutations were detected in the signal sequences of specific proteins, which resulted in decreased import of these proteins. Recently, we have shown that in the mitochondria of two patients with a mitochondrial disease concentrations of hsp 60 were significantly lower than in mitochondria from healthy individuals (21). Owing to the protein's role in mitochondrial protein import and assembly, deficiency of hsp 60 will affect the maturation of most mitochondrial proteins that have to be imported into the mitochondria. Below, the further characterization of the partial hsp 60 deficiency is documented and the implications of such a deficiency for the understanding of mitochondrial disease are discussed.

IV. PATIENTS WITH A SEVERE AND SYSTEMIC MITOCHONDRIAL DISORDER AND PARTIAL DEFICIENCY OF hsp 60

Deficiency of hsp 60 was discovered in cultured skin fibroblasts from two patients with mitochondrial encephalomyopathy. The two patients had a number of symptoms in common, but they also presented with differences in the clinical manifestation of the mitochondrial defect. Patient 1 died shortly after birth of heart failure (21). Patient 2 reached the age of 4.5 years, was mentally and physically retarded, and died from acute, severe acidosis (22). Both children had facial dysmorphic features, a symptom rarely encountered in patients with a mitochondrial encephalomyopathy, and they showed directly at birth a severe lactic acidosis and hypotonia. Pathological examination revealed cardiac and lung abnormalities, hepatomegaly, and brain tissue damage. Metabolically, the children differed with respect to the organic acids excreted in the urine. Patient 1 accumulated only lactate in the urine; patient 2 also other organic acids like pyruvate, ketone bodies, and 2-hydroxy-glutarate.

Levels of hsp 60 in the fibroblasts from the two patients were decreased to about 20% of those in control cells (Fig. 1). The low amount of hsp 60 was not a secondary effect common to disorders of mitochondrial metabolism, since fibroblast cultures from eight other patients with systemic presentation of a mitochondrial defect showed hsp 60 levels similar to those in control fibroblasts. Furthermore, levels of chsp 70 and mhsp 70 also proved to be normal in the hsp 60–deficient patient cells.

Functional characterization of the two hsp 60–deficient fibroblast cultures revealed strongly decreased activities of mitochondrial enzymes of pyruvate metabolism, the citric acid cycle, oxidative phosphorylation, fatty acid transport and oxidation, and amino acid degradation (Table 1). The affected enzymes are all located either in the mitochondrial inner membrane or the mitochondrial matrix. In contrast, activity of the outer membrane enzyme carnitine palmitoyl transferase I appeared to be normal. Cytosolic enzymes and enzymes of other organelles like lysosomes and peroxisomes also exhibited normal activities.

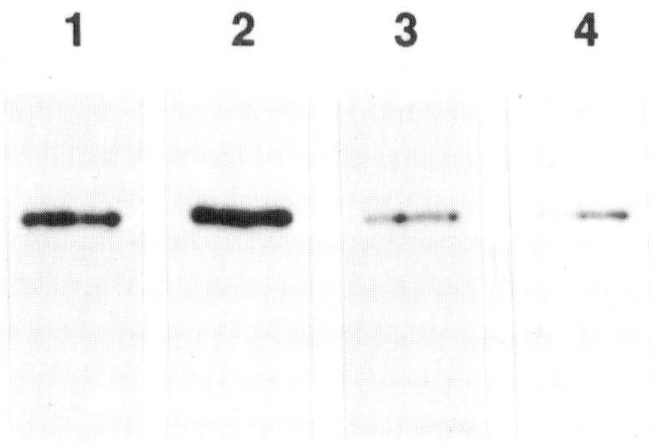

Figure 1 Immunodetection of hsp 60 in lysates of fibroblasts from healthy individuals (lanes 1 and 2) and from patients 1 and 2 (lanes 3 and 4, respectively).

Fluorescence visualization of the mitochondria in patient fibroblasts showed clustering of swollen mitochondria around the nucleus (Fig. 2) and an unusual patch-like distribution of mitochondrial enzymes; for example, medium chain acyl-CoA dehydrogenase (Fig. 3). Mitochondrial inner membranes were locally disintegrated (Fig. 4).

Table 1 Mitochondrial Enzyme Activities in Fibroblasts from Patient 1

		Activity (nmol/min.mg)	
Metabolic pathway	Enzyme	patient	control
Pyruvate metabolism	Pyruvate dehydrogenase	0.32	0.92;122
	Pyruvate carboxylase	0.061	0.17–0.85
Citric acid cycle	2-Ketoglutarate dehydrogenase	1.90	6.30;7.07
Oxidative phosphorylation	Succinate cytochrome *c* reductase	0.85	6.81–13.7
	Cytocrome *c* oxidase	0.16	0.94–2.90
	Mg^{2+}-ATPase	8.70	13.33
Fatty acid transport and oxidation	Carnitine palmitoyl transferase I	0.57	0.43–1.46
	Carnitine palmitoyl transferase II	0.30	0.54–0.75
	Medium-chain acyl-CoA dehydrogenase	0.59	0.98–2.56
Amino acid catabolism	Propionyl-CoA carboxylase	0.21	0.25–1.20
	Leucine decarboxylation	2.4	14.2–16.4

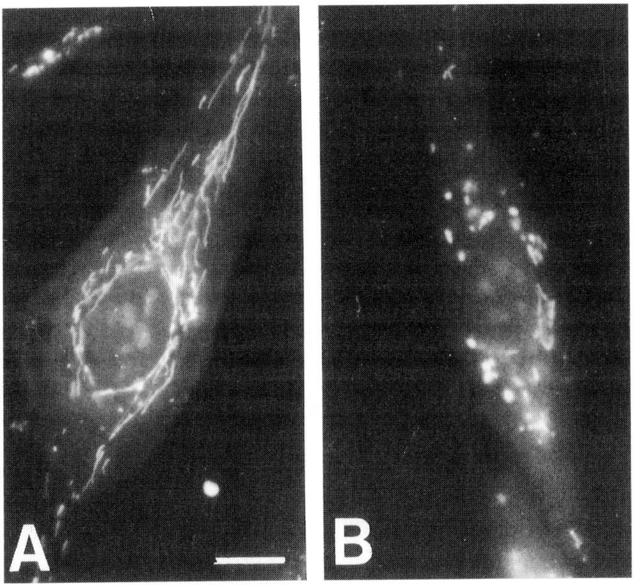

Figure 2 Visualization of mitochondria in fibroblasts from a healthy control (A) and from patient 1 (B) by in vivo staining with the fluorescence dye 2-[4-(dimethylamino)-styryl]-1-methylpyridinium iodide (26). Scale bar = 10 μm.

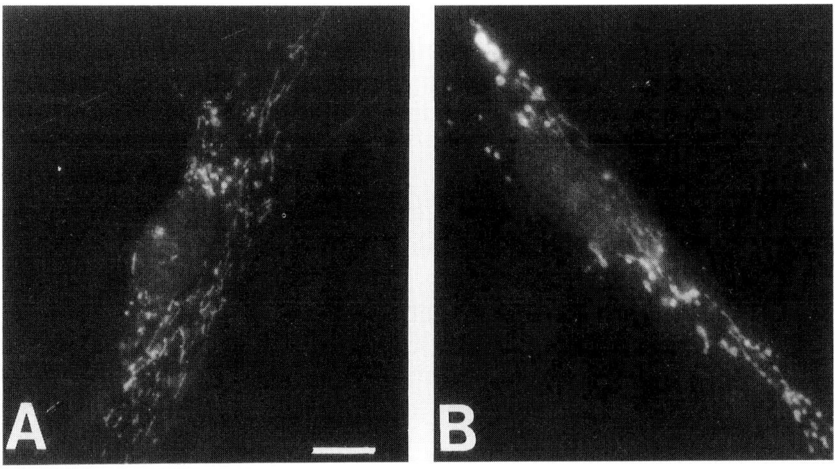

Figure 3 Immunofluorescence visualization of medium chain acyl-CoA dehydrogenase in fibroblasts from a healthy control (A) and from patient 1 (B). Scale bar = 10 μm.

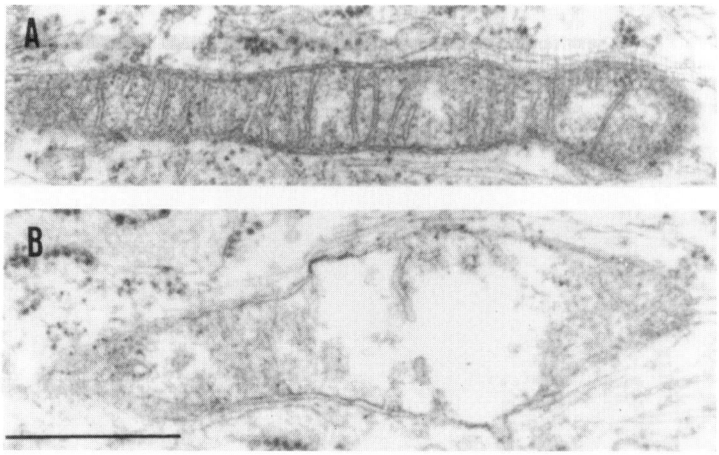

Figure 4 Electron micrographs of fibroblast mitochondria from a healthy control (A) and patient 2 (B). Scale bar = 0.5 μm.

Synthesis as well as maturation of hsp 60 appeared to be depressed in the patients' cells (23). Hsp 60 is synthesized in the cytosol as a precursor with an N-terminal extension, which is cleaved off after transport into the mitochondria. The result of the processes of synthesis and maturation can be monitored by in vivo labeling of the cells with [^{35}S]methionine, immunoprecipitation of hsp 60 from the cell lysate, and visualization of labeled hsp 60 on electrophoresis gels by autoradiography. As shown in Figure 5, less hsp 60 was synthesized in patients' cells than in control cells after a labeling period of 5 minutes. Moreover, as can be deduced from the precursor-product ratios, maturation of hsp 60 proceeds at a lower rate in patients' cells. Experiments designed to test the functional properties of the residual hsp 60 revealed that it behaved as hsp 60 derived from control fibroblasts (23). The monomers assembled into the native 14-mer complex, this complex bound to denatured protein with high affinity, and it was located inside the mitochondria.

V. DISCUSSION AND PROSPECTS

We have described the intriguing observation that some cases of mitochondrial encephalomyopathy are associated with reduced amounts of the mitochondrial chaperonin hsp 60. Characteristic findings in cultured skin fibroblasts from the two patients detected thus far comprised decreased activities of a large number of functionally unrelated mitochondrial enzymes and striking changes in morphology and ultrastructure of the mitochondria. It is this kind of phenotype that

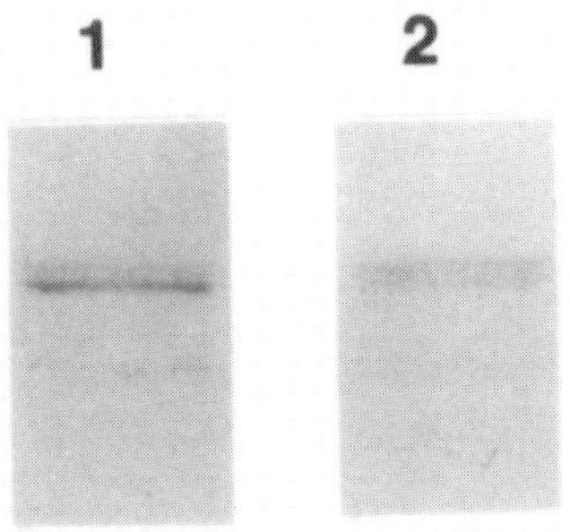

Figure 5 Autoradiographs of immunoprecipitated, ^{35}S-labeled precursor (upper bands) and mature hsp 60 (lower bands) from fibroblasts of a healthy control (lane 1) and patient 1 (lane 2).

has been described for hsp 60–deficient mutants of lower eukaryotes (24,25). This congruity invites to the hypothesis that the quantitative deficiency of hsp 60 is responsible for the development of the diseased phenotype of the patient: The residual hsp 60 is not sufficient to ensure the proper folding of newly imported enzyme subunits which gives rise to severely decreased activities of several metabolically independent enzymes. The resulting accumulation of harmful metabolic intermediates and the ATP deficiency then cause the typical symptoms like lactic acidosis and severe muscle weakness. The normal activity of carnitine palmitoyl transferase I is also in line with this hypothesis: As an outer membrane enzyme, its subunits do not have to be imported into the mitochondrial matrix and thus do not have to interact with hsp 60 to ensure proper enzyme maturation. The low amounts of hsp 60 may be caused by mutations affecting either the rate of synthesis of the hsp 60 mRNA or of the protein itself or its maturation. However, genetic analysis is needed to establish unequivocally whether or not a mutant allele of the hsp 60 gene is responsible for these disorders. This will comprise base sequence determination of the hsp 60 gene and the rescue of the normal phenotype by transformation of the patient cells with a wild-type hsp 60 gene.

Disorders of biosynthesis of mitochondria due to deficiency of proteins mediating mitochondrial protein import and/or assembly constitute a new group of mitochondrial diseases. Common characteristics of the patients will by systemic manifestation of the defect, multiple mitochondrial enzyme deficiencies, and , since the proteins in question are nuclearly encoded, Mendelian transmittance of the defect. For the correct and fast diagnosis of mitochondrial disorders possibly associated with a defect of mitochondrial protein import, it is essential that a survey of prevalent clinical and biochemical symptoms becomes available soon. To that end, more patients have to be screened and the underlying defect has to be characterized.

ACKNOWLEDGMENTS

This work was made possible by a grant from the Prinses Beatrix Fonds. Fibroblasts from patients 1 and 2 were kindly provided by Dr. O.H. Skjeldal (Oslo, Norway) and Dr. P. Briones (Barcelona, Spain), respectively.

REFERENCES

1. G. Attardi and G. Schatz, Biogenesis of mitochondria, *Annu. Rev. Cell Biol. 4*:257 (1988).
2. B. Glick and G. Schatz, Import of proteins into mitochondria, *Annu. Rev. Genet. 25*:331 (1990).
3. R.J. Ellis, Molecular chaperones, *Annu. Rev. Biochem. 60*:321 (1991).
4. M.J. Gething and J. Sambrook, Protein folding in the cell, *Nature 355*:33 (1992).
5. N. Pfanner, J. Rassow, I.J. van der Klei, and W. Neupert, A dynamic model of the mitochondrial protein import machinery, *Cell 68*:990 (1992).
6. F.-U. Hartl, J. Martin, and W. Neupert, Protein folding in the cell: the role of molecular chaperones Hsp70 and Hsp60, *Annu. Rev. Biophys. Biomol. Struct. 21*:293 (1992).
7. J.P. Hendrick and F.-U. Hartl, Molecular chaperone functions of heat-shock proteins, *Annu. Rev. Biochem. 62*:349 (1993).
8. A. Lombes, E. Bonilla, and S. DiMauro, Mitochondrial encephalomyopathies, *Rev. Neurol. (Paris) 145*:671 (1989).
9. L.A. Bindoff and D.M. Turnbull, Defects of the respiratory chain, *Baillière's Clin. Endocrinol. Metab. 4*:583 (1990).
10. M.H. Tulinius, E. Holme, B. Kristiansson, N.-G. Larsson, and A. Oldfors, Mitochondrial encephalomyopathies in childhood. I. Biochemical and morphological investigations, *J. Pediatr. 119*:242 (1992).
11. M.H. Tulinius, E. Holme, B. Kristiansson, N.-G. Larsson, and A. Oldfors, Mitochondrial encephalomyopathies in childhood. II. Clinical manifestations and syndromes, *J. Pediatr. 119*:251 (1992).
12. S. DiMaura and D.C. Devivo, Diseases of carbohydrate, fatty acid and mitochondrial metabolism, *Basic Neurochemistry: Molecular, Cellular, and Medical Aspects* (G.J. Siegel, ed.), Raven Press, New York, 1994, p. 723.
13. A. Harding, The other genome, *Br. Med. J. 303*:377 (1991).
14. D.C. Wallace, Diseases of the mitochondrial DNA, *Annu. Rev. Biochem. 61*:1175 (1992).
15. X. Zheng, J.M. Shoffner, M.T. Lott, A.S. Voljavec, N.S. Krawieck, K. Winn, and D.C. Douglas, Evidence in a lethal infantile mitochondrial disease for a nuclear mutation affecting respiratory complexes I and IV, *Neurology 33*:1203 (1989).
16. H.-J. Tritschler, E. Bonilla, A. Lombes, F. Andreetta, S. Servidei, B. Schneider,

A.F. Miranda, E.A. Schon, B. Kadenbach, and S. DiMauro, Differential diagnosis of fatal and benign cytochrome *c* oxidase–deficient myopathies of infancy: an immunochemical approach, *Neurology 41*:300 (1991).
17. M. Zeviani, N. Bresolin, C. Gellera, A. Bordoni, M. Pannacci, P. Amati, M. Moggio, S. Servidei, G. Scarlato, and S. DiDonato, Nucleus-driven multiple large scale deletions of the human mitochondrial genome: a new autosomal dominant disorder, *Am. J. Hum. Genet. 47*:904 (1990).
18. C.T. Moraes, S. Shanske, H.-J. Tritschler, J.R. Aprille, F. Andreetta, E. Bonilla, E.A. Schon, and S. DiMauro, mtDNA depletion with variable tissue expression: A novel genetic abnormality in mitochondrial diseases, *Am. J. Hum. Genet. 48*:492 (1991).
19. F.D. Ledley, R. Jansen, S.-U. Nham, W.A. Fenton and L.E. Rosenberg, Mutation eliminating mitochondrial leader sequence of methylmalonyl-CoA mutase causes *mut*^{o} methylmalonic acidemia, *Proc. Natl. Acad. Sci. USA 87*:3147 (1990).
20. A.H.V. Schapira, J.M. Cooper, J.A. Morgan-Hughes, D.N. Landon, and J.B. Clark, Mitochondrial myopathy with a defect of mitochondrial protein import, *N. Engl. J. Med. 323*:37 (1990).
21. E. Agsteribbe, A. Huckriede, M. Veenhuis, M.H.J. Ruiters, K.E. Niezen-Konings, O.H. Skjeldal, K. Skullerud, R.S. Gupta, R. Hallberg, O.P van Diggelen, and H.R. Scholte, A fatal, systemic mitochondrial disease with decreased mitochondrial enzyme activities, abnormal ultrastructure of the mitochondria and deficiency of heat shock protein 60, *Biochem. Biophys. Res. Commun. 193*:146 (1993).
22. P. Briones, M.A. Vilasera, A. Ribes, A. Vernet, M. Lluch, A. Huckriede and E. Agsteribbe, A new case of multiple mitochondrial enzyme deficiencies with decreased amount of heat shock protein 60, *Eur. J. Pediatr.* (submitted).
23. A. Huckriede and E. Agsteribbe, Characterization of heat shock 60-deficiency in a patient with a mitochondrial encephalomyopathy, *Biochim. Biophys. Acta 1227*: 200 (1994).
24. M.Y. Cheng, F.-U. Hartl, J. Martin, R.A. Pollock. F. Kalousek, W. Neupert, E.M. Hallberg, R.L. Hallberg, and A.L. Horwich, Mitochondrial heat-shock protein hsp60 is essential for assembly of proteins imported into yeast mitochondria, *Nature 337*:620 (1989).
25. E.M. Hallberg, Y. Shu, and R.L. Hallberg, Loss of mitochondrial hsp60 function: nonequivalent effects on matrix-targeted and intermembrane-targeted proteins, *Mol. Cell. Biol. 13*:3050 (1993).
26. M.H.J. Ruiters, E.A. van Spronsen, O.H. Skjeldal, P. Stromme, H.R. Scholte, H. Pzyrembel, G.P.A. Smit, W. Ruitenbeek, and E. Agsteribbe, Confocal scanning laser microscopy: A possible tool in the diagnosis of mitochondrial disorders, *J. Inher. Metab. Dis. 14*:45 (1991).

33

Heat Shock Proteins as Chaperones of Unique and Shared Antigenic Epitopes of Human Cancers

A Novel Approach to Vaccination

Pramod K. Srivastava
Fordham University, Bronx, New York

I. INTRODUCTION

Heat shock proteins (hsps) gp 96/grp 94, hsp 90, and hsp 70 have been reported to be associated with antigenic peptides (1–3). We have proposed that the peptides are chaperoned by a relay line of hsps within and outside the endoplasmic reticulum during antigen processing and presentation by major histocompatibility complex (MHC) class I molecules (4). This aspect of the role of hsps has been discussed at length elsewhere (3,4) and will not be addressed here. Instead, in keeping with the title of this book, I focus solely on the implications of our observations on immunotherapy of human cancers.

II. BACKGROUND: HSP–PEPTIDE COMPLEXES AS VACCINES AGAINST CANCERS AND INFECTIOUS DISEASES

Immunization of mice and rats with hsp peptide complexes generated in vivo has been shown to elicit specific tumor immunity and cytotoxic T lymphocyte (CTL) responses against tumors from which the hsp-peptide complexes were isolated but not to antigenically distinct tumors (2,5–10). Further pursuit of these observations has indicated that the hsps are not antigenic per se but act as carriers of antigenic peptides (3). The idea that hsps chaperone not only tumor-specific peptides but indeed all peptides generated within a cell, including viral peptides generated during infection, was developed and successfully tested (10; N.E. Blachere and P.K. Srivastava, unpublished results). These results have led to the possibility of development of hsp-peptide complexes as the basis of a new generation of specific vaccines.

It is important to distinguish our studies from *apparently* similar studies where a foreign hsp transfected into a tumor renders the tumor highly immunogenic (11). In such systems, the hsp is an antigen by itself by virtue of being bacterial and therefore foreign. Transfection of a murine tumor with *any* bacterial or other foreign gene (not only an hsp gene) will render murine tumors highly immunogenic in mice. The observation that mice immunized with tumor cells transfected with the foreign hsp gene are also resistant to the untransfected tumors is a classic observation made in a number of other systems and is unrelated to the transfected gene being an hsp (e.g., see ref. 12). Such xenogenized systems have little relevance for tumor immunity and are excluded from the present discussion.

Hsp-peptide complexes offer unique and unprecedented advantages over other types of vaccines against cancers, infectious and transforming viruses, intracellular bacteria, and protozoa:

1. Knowledge of the antigenic epitopes which elicit immunity is a prerequisite for all forms of vaccination. Hsp-peptide based vaccination circumvents this necessity, as hsps are naturally complexed with the repertoire of peptides generated in a cell. For this reason, it is an ideal means for vaccination against infections for which the protective epitopes are yet undefined, or where a single epitope may not be sufficient for eliciting immunity, or where the infectious agent is so highly variable (in a population-, season-, or individual-specific manner) that the prospect of identifying the immunogenic epitopes for each variant is simply impractical.
2. Even in the case of infectious diseases where the relevant antigenic epitopes *have been* defined, hsp-based vaccination offers a unique advantage: Hsp-peptide complexes can be readily stripped of their natural peptides and these "empty" hsps can be reconstituted with known, synthetic peptide epitope(s) (H. Udono and P.K. Srivastava, unpublished results). The in vitro reconstituted noncovalent hsp-peptide complexes elicit potent *T-cell response* to the complexed peptide. Further, if the peptide is conjugated covalently to the hsp, the hsp-peptide complexes elicit potent *antibody responses* to the complexed peptide (13,14).
3. One of the major conceptual difficulties in cancer immunotherapy has been the possibility that human cancers, like cancers of experimental animals, are antigenically distinct (15,16). The prospect of identification

of immunogenic antigens of individual tumors from patients with cancer is daunting to the extent of being impractical. The ability of hsps to chaperone the entire repertoire of antigenic peptides of the cells from which they are derived circumvents this extraordinary hurdle. Thus, patients can be vaccinated with hsp-peptide complexes derived from their own tumors, or cell lines derived from them, without any need for identification of the antigenic epitopes of the patient's tumor. In light of the emerging evidence for existence of "shared" tumor antigens (17–19), this particular point of advantage for hsp-based vaccines may at first sight appear less profound than originally imagined; however, closer scrutiny suggests that even if human cancer antigens are cross reactive rather than individually distinct, hsp-peptide complexes offer a uniquely effective method of vaccination. This premise is elucidated in detail in Sections III–IV.

4. As hsps are nonpolymorphic (i.e., show no allelic diversity, although there are several hsp families), they bind the entire spectrum of peptides regardless of the MHC haplotype of a cell. Thus, an hsp-peptide complex isolated from cells of a given haplotype may be used to vaccinate individuals of other haplotypes (4).
5. Vaccination with hsp-peptide complexes elicits $CD8^+$ T cells without the use of live (attenuated or otherwise) agents and in spite of exogenous administration (20).
6. Hsp-peptide based vaccines are inherently mutlivalent because hsps chaperone, not one or a few but the entire repertoire of epitopes generated in a cell.
7. As hsp-peptide complexes can be purified easily to apparent homogeneity, vaccination with such preparations circumvents the risks associated with vaccination with attenuated organisms or undefined biological extracts which contain transforming DNA and immunosuppressive factors such as transforming growth factor β (TGF-β).

III. HUMAN TUMOR ANTIGENS—INDIVIDUALLY DISTINCT OR CROSS REACTIVE?

Tumor antigens of experimental cancers have been historically considered to be antigenically distinct. Early experiments by Prehn and Main (21), Old et al. (22), Globerson and Feldman (15), and Basombrio (16) led to definition of transplantation antigens of methylcholanthrene-induced sarcomas of inbred mice as being antigenically individually distinct, and this became the defining paradigm for cancer immunity. However, recent studies on identification of tumor epitopes recognized by cytotoxic T lymphocytes against human melanomas have begun to challenge this paradigm. So far, at least five cytotoxic T-lymphocytes (CTL) epitopes have been identified, and in four of these, the CTL epitope is derived from a melanocytic differentiation antigen (23–30). Further, in each case, the amino acid sequence of the melanoma CTL epitope is identical to the corresponding sequence in normal melanocytes. Thus, in contrast to the tumor antigens of mutagenized tumors (31–33), no mutations are detected in CTL epitopes of human melanomas. These results are generating a paradigm shift in our understanding of cancer immunity. It has long been assumed that cancers would contain a repertoire of mutations, some of which will be transforming and some antigenic. The possibility that some of the transforming mutations, for example, those in *ras* and p53, may themselves be antigenic has also been considered (34–37). The multimutation model of cancers also fits neatly with our understanding of the multistep nature of familial and environmentally induced cancers (38).

These assumptions regarding an inherent mutational repertoire of cancer cells are now being called into question. First, the identified antigenic epitopes are not mutated; hence the idea that cancers are recognized by the immune system because of genetic alterations stands unsubstantiated. Second, immunogenicity of unmutated epitopes raises questions regarding the state of tolerance to self-antigens, particularly differentiation antigens as all known epitopes (with one exception, see refs. 39 and 40) are derived from melanocyte-differentiation antigens. This observation is a reincarnation of one of the earliest speculations on the nature of tumor antigens (41,42) as differentiation antigens. The question of immune response to unaltered antigens also arises with respect to immune response to the HER2/neu oncogene product, which is not altered in tumor cells but is overexpressed in a proportion of breast cancers (43). Finally, and for the purposes of this chapter, most significantly, the presence of unaltered CTL epitopes derived from differentiation antigens suggests that human melanomas possess shared or cross reactive immunogenic antigens. A particular tissue has only a finite set of differentiation antigens, and if genes for these antigens are not mutated in tumor cells, all tumors of a given histological origin can possess only a finite set of immunogenic antigens. Thus, the recent structural studies on CTL epitopes indicate that human melanomas possess shared antigens. Earlier, nonstructural studies have also indicated this cross reactivity (17–19).

IV. USE OF HSP–PEPTIDE COMPLEXES AS CANCER VACCINES IN VIEW OF EXISTENCE OF SHARED MELANOMA ANTIGENS

One of the advantages of hsp-peptide complexes for vaccination against human cancers lies in the possibility that human cancers, like their murine counterparts, are antigenically diverse and individually distinct (see item number 3 in Section II). In that scenario, it would be practically impossible to identify the antigenic epitopes of individual cancer patients. If, however, human tumors are antigenically cross reactive (as outlined in the preceding section), antimelanoma vaccines could be designed simply on the basis of peptide epitopes of known cross reactive melanoma antigens instead of hsp-peptide complexes isolated from individual melanomas. It is the premise of this chapter that hsp-peptide complexes provide a uniquely effective method of vaccination regardless of antigenic individuality or cross reactivity of human tumor antigens. This premise is elaborated as follows.

First, the antigenic cross reactivity among human melanomas suggests that once a number of shared melanoma antigens are identified, patients can be immunized with synthetic peptides corresponding to the relevant epitopes and that the vaccinated patients will elicit a $CD8^+$ CTL response. Vaccination with peptides under suitable conditions has indeed been shown to elicit $CD8^+$ CTLs in a number of systems (37,44). Such conditions usually include the use of incomplete Freund's adjuvant along with large quantities (~ 100 μg peptide for a 20-gr mouse) of peptide. This is clearly incompatible with human use. Alternative approaches, such as addition of a lipophilic tail to the peptides, have been employed successfully (45) and could be potentially suited for human applications. In this context, vaccination with in vitro reconstituted complexes of human hsps with the relevant antigenic peptides offers an economical and technologically simple method of vaccination. The ability of hsp-peptide complexes, *administered in saline and without any adjuvants*, to prime naive $CD8^+$ CTLs in vivo has been reported recently (10,20). The use of human hsps and synthetic peptides is also attractive from the point of view of circumventing hazards associated with the vaccination of patients with chimeric molecular constructs of unknown toxicity.

Second, vaccination with a given peptide will be effective only for patients with a given HLA allele. If different epitopes from a single molecule are recognized by different HLA alleles (as appears to be the case in case of tyrosinase; see refs. 23 and 24), a cocktail of peptides will have to be used for vaccination of a general population. Even for a given patient, a cocktail may have to be used, as humans are outbred and possess several restriction elements. A far more effective and simpler alternative will be to isolate hsp-peptide complexes from human cell lines transfected with the relevant gene under the control of a high-expression promoter. The hsp-peptide complexes purified from such transfectants will consist of the entire repertoire of antigenic peptides derived from that particular protein. As hsp-peptide binding is proximal to HLA-peptide binding during antigen processing, there is no HLA restriction in the hsp-bound peptides. Peptides capable of binding to all possible HLA alleles will be represented among the hsp-peptide complexes.

Third, the methodology of identification of CTL epitopes of human cancers suggests that these epitopes may represent a significantly biased sample of the antigenic repertoire of human cancers. Generation of cell lines is an essential prerequisite for isolation of CTLs, and only a very small proportion of human cancers (less than 2% for breast cancers to about 30% for melanomas) lend themselves to it. Of the tumors from which cell lines are developed, only a small proportion permit generation of CTLs. Thus, the CTL epitopes being identified may represent an atypical and sparse sampling of the cancer antigenic repertoire. Immunogenicity of cancers represents, in all likelihood, the sum total of immunogenicity of a large number of immunogenic epitopes, and effective anticancer vaccines should include this antigenic multiplicity. Hsp-peptide complexes are such a multicomponent, multivalent vaccine. If cancer antigens are shared and not individually indistinct, the use of hsp-peptide complexes isolated from tumors becomes even simpler, such that these complexes need not be isolated from cancers of individual patients. Instead, they could be purified from a mixture of human melanomas and be used to vaccinate allogeneic melanomas. The lack of HLA restriction of hsp-bound peptides (discussed in the preceding section) is a key advantage in this regard. Vaccination with multicomponent, multivalent vaccines rather than single or oligocomponent vaccines is also necessary for protection against antigenic escape or preexisting antigenic heterogeneity of human cancers.

These three major considerations indicate that regardless of the cross-reactive or individually distinct nature of human cancer antigens, hsp-peptide complexes offer unique and unprecedented advantages over other existing methods in vaccination against human cancer. In the final section of this chapter, the emerging thesis that human cancer antigens are cross reactive is examined critically.

V. HIERARCHY OF IMMUNOGENIC HUMAN TUMOR ANTIGENS: THE UNIQUE, THE SHARED, AND THE ONCOGENIC

Although the emerging data on CTL epitopes of human melanomas indicate that human tumor antigens may be shared, rumors of the demise of unique antigens on human cancers may be premature (apologies to Mark Twain). The studies carried out so far leave open a significant possibility that they are uncovering an unrepresentative sample of the repertoire (as discussed in Section III). The existence of shared antigens is not in question, but the absence of unique antigens is. After all, shared or cross-protective antigens can be detected readily even in methylcholanthrene-induced mouse sarcomas—the quintessential models of nonshared or individually distinct cancer antigens (46,47). However, it is the unique antigens which are the stronger, the most reproducibly observed, and the ones which elicit tumor rejection (48,49).

Tumors appear to contain a hierarchy of antigens which constitute a continuum with respect to strength of immunogenicity (Fig. 1). The unique antigens may form the strongest end of this spectrum. It is expected that they are generated by mutations accompanying malignant transformation but not causative to it. The randomness of mutations would generate an individually distinct mutational repertoire in individual cancers. We refer to such mutations as constituting the "incidental mutational repertoire" of tumors (3). It is puzzling, however, that such mutated, unique antigens have not been detected so far among the CTL-defined epitopes of human melanomas. It is conceivable that such unique epitopes, by virtue of being subtle modifications of self proteins, may elicit lower affinity CTLs, which are effective in vivo but do not survive the stringent conditions of in vitro selection of CTLs. (That such CTL epitopes are detected in tumor mutants may result from the fact that tumor mutants contain an unusually large repertoire of mutations such that selection of higher affinity CTLs is more likely.) Further, tumors arising in lungs, colon, or other tissues acutely exposed to carcinogenic insult may have a higher frequency of the

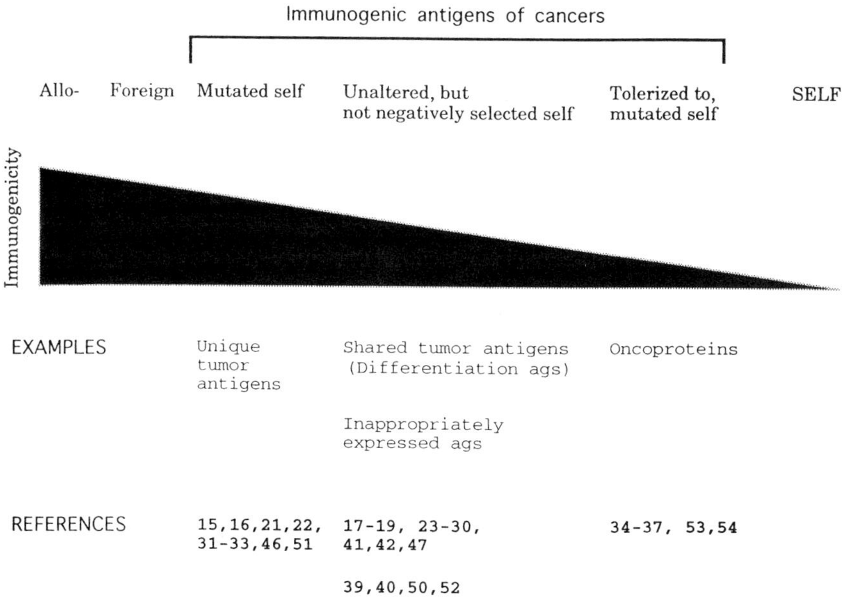

Figure 1 Schematic representation of one possible view of the universe of antigens, with special emphasis on immunogenic antigens of cancers.

mutated unique antigens. Characterization of CTL epitopes of human cancers other than melanomas will address this question.

Although the CTL epitopes of human melanomas and one mouse mastocytoma P815 are unmutated (23–30, 50), the recently reported CTL-defined epitope of the murine Lewis lung carcinoma 3LL is mutated with respect to the normal sequence (51). It may be of interest to contemplate in this regard that although the epitopes of human melanomas and of the P815 are not unique but are shared with other tumors, the epitope of the 3LL carcinoma is indeed unique. This is consistent with our premise (see above) that shared antigens will turn out to be unmutated, whereas the unique antigens will be altered with respect to their normal counterparts.

The shared antigens of tumors may form the center of the spectrum of immunogenicity. The shared tumor antigens identified so far are unmutated, differentiation antigens, and in hindsight, it is reasonable that shared antigens should be differentiation antigens rather than housekeeping proteins. Tolerance to housekeeping proteins is established efficiently during thymic development, as housekeeping proteins are by definition common to the thymic epithelium and the rest of the tissues. Tolerance to differentiation antigens on the other hand may result from peripheral mechanisms. Tumors may modulate this peripheral tolerance by factors secreted by them, and this may lead to development of immune responses to differentiation antigens. In this context, it is also possible to conceive of shared antigens which are not classic differentiation antigens but which are also not ubiquitous housekeeping proteins. The melanoma antigen MAG (39,40) and the mucins associated with breast, pancreatic, and other cancers (52) may belong to this category.

The shared antigens are followed in the spectrum of strength of immunogenicity by molecules which cause malignant transformation. These consist of *ras*, p53, HER2/neu, and other mutated or inappropriately expressed oncogenic proteins. We have suggested earlier that the immune system is tolerant to these antigens simply because they are the best tumor antigens and tolerance to them is one mechanisms of escape from immunosurveillance (53). Tolerance to this group of antigens, like tolerance to the differentiation antigens, must develop peripherally during the early phase of tumorigenesis. Although there are occasional examples of immune response to p53 and HER2/neu in tumor-bearing mice and humans (54,55), the immune system does not generally recognize this category of antigens. The lack of significant cross reactivity among chemically induced cancers, which carry the same mutations in *ras* and p53, attests to the lack of immunogenicity of these molecules. However, it is equally clear that immune response to oncoproteins can be elicited under certain conditions; that is, by vaccination with large quantities of relevant peptides or the use of appropriate adjuvants (34–37). Among the three groups of cancer antigens discussed, the therapeutic possibilities of this last group of antigens are perhaps the least clear.

Admittedly, this article is restricted to antigens recognized by T cells. Although there is reasonable evidence that the cellular arm of the immune response is primarily involved in cancer immunity, we cannot not entirely ignore the possibility that we are relearning through the T-cell arm what we have learned earlier through the serological analysis of cell surface antigens of cancer cells (56).

VI. CONCLUSIONS

Immunogenicity of tumors results from the sum total of a number of immunogens in each of the categories discussed above—the unique, the shared, and the oncogenic. Characterization of this large and complex antigenic repertoire is a worthy and difficult goal. No doubt, the remarkable ongoing advances in cellular immunity and peptide chemistry will move us rapidly towards this objective. However, regardless of these future advances, it is safe to predict that the antigenic complexity, plasticity, and heterogeneity of tumors and their ability to interact dynamically with the immune system will make it impractical to develop the kind of structurally defined anticancer vaccines which have been so effective in the war against infection. Hsp-peptide complexes representing the total antigenic repertoire of human cancers, cut through the antigenic complexity of cancers, and offer a unique opportunity for vaccination against human cancers independent of the structural analyses of the immunogenic antigens.

REFERENCES

1. Z. Li and P.K. Srivastava, Tumor rejection antigen Gp96/Grp94 is an ATPase: implications for protein folding and antigen presentation, *EMBO J. 12*:3143 (1993).
2. H. Udono and P.K. Srivastava, Heat shock protein 70–associated peptides elicit specific cancer immunity. *J. Exp. Med. 178*:1391 (1993).
3. P.K. Srivastava, Peptide-binding heat shock proteins in the endoplasmic reticulum: role in immune response to cancer and in antigen presentation. *Adv. Cancer Res. 62*:153 (1993).
4. P.K. Srivastava, H. Udono, N.E. Blachere, and Z. Li, Heat shock proteins transfer peptides during antigen processing and CTL priming, *Immunogenetics 39*:93 (1994).
5. P.K. Srivastava and M.R. Das, Serologically unique surface antigen of a rat hepatoma is also its tumor-associated transplantation antigen, *Int. J. Cancer 33*:417 (1984).
6. P.K. Srivastava, A.B. DeLeo, and L.J. Old, Tumor rejection antigens of chemically induced sarcomas of inbred mice, *Proc. Natl. Acad. Sci. USA 83*:3407 (1986).
7. M.A. Palladino, P.K. Srivastava, H.F. Oettgen, and A.B. DeLeo, Expression of

a shared tumor-specific antigen by two chemically induced BALB/c sarcomas. I. Detection by a cloned cytotoxic T cell line. *Cancer Res. 47*:5047 (1987).
8. A.M. Feldweg and P.K. Srivastava, Molecular heterogeneity of tumor rejection antigen/heat shock protein gp96. *Int. J. Cancer* 1995 (in press).
9. S. Janetzki, N.E. Blachere, and P.K. Srivastava, Cognate heat shock protein gp96 preparations from two antigenically distinct UV-induced carcinomas elicit tumor-specific rejection and cytotoxic T lymphocyte response, *J. Exp. Med.* (1994).
10. N.E. Blachere NE, H. Udono, S. Janetzki, Z. Li, M. Heike, and P.K. Srivastava, Heat shock protein vaccines against cancer. *J. Immunother. 14*:352 (1993).
11. K.V. Lukacs, D.B. Lowrie, R.W. Stokes, and M.J. Colston, Tumor cells transfected with a bacterial heat shock protein gene lose tumorigenicity and induce protection against tumors. *J. Exp. Med. 178*:343 (1993).
12. E.R. Fearon, T. Itaya, B. Hunt, B. Vogelstein, and P. Frost, Induction in a murine tumor of immunogenic variants by transfection with a foreign gene. *Cancer Res. 48*:2975 (1988).
13. A.R. Lussow, C. Barrios, J. van Embden, R. Van der Zee, A.S. Verdini, A. Pessi, J.A. Louis, P.H. Lambert, and G. Del Giudice, Mycobacterial heat shock proteins as carrier molecules. *Eur. J. Immunol. 21*:2297 (1991).
14. C. Barrios, A.R. Lussow, J. van Embden, R. Van der Zee, R. Rappuoli, P. Constantino, J.A. Louis, P.H. Lambert, and G. Del Guidice, Mycobacterial heat shock proteins as carrier molecules. II:The use of the 70 kda mycobacterial heat shock protein as carrier for conjugated vaccines can circumvent the need for adjuvants and BCG priming. *Eur. J. Immunol. 22*:1365 (1992).
15. A. Globerson and M. Feldman, Antigenic specificity of benzo(a)pyrene induced sarcomas. *J. Natl. Cancer Inst. 32*:1229 (1964).
16. M.A. Basombrío, Search for common antigenicities among 25 sarcomas induced by methylcholanthrene. *Cancer Res. 30*:2458 (1970).
17. T.L. Darrow, C.L. Slingluff, and H.F. Siegler, The role of HLA I antigens in recognition of melanoma cells by tumor-specific CTLs: evidence for shared tumor antigens. *J. Immunol. 142*:3329 (1989).
18. Y. Kawakami, R. Zakut, S.L. Topalian, H. Stotter, and S.A. Rosenberg, Shared human melanoma antigens: recognition by tumor-infiltrating lymphocytes in HLA A2.1-transfected melanomas. *J. Immunol. 148*:638 (1992).
19. S.S. Hom, S.L. Topalian, T. Simonis, M. Mancini, and S.A. Rosenberg, Common expression of melanoma tumor-associated antigens recognized by human tumor-infiltrating lymphocytes: analysis by HLA-restriction. *J. Immunother. 10*:153 (1991).
20. H. Udono H, D.L. Levey DL, and P.K. Srivastava, Definition of T cell sub-sets mediating tumor-specific immunogenicity of cognate heat shock protein gp96, *Proc. Natl. Acad. Sci. USA 91*:3077 (1994).
21. R.T. Prehn RT, and J.M. Main, Immunity to methylcholanthrene-induced sarcomas, *J. Natl. Cancer. Inst. 18*:769 (1957).
22. L.J. Old, E.A. Boyse, D.A. Clarke, and E.A. Carswell, Antigenic properties of chemically induced tumors, *Ann N.Y. Acad. Sci. 101*:80 (1962).
23. V. Brichard, A. van Pel, T. Wolfel, E. DePlaen, B. Lethe, P. Coulie, and T. Boon,

The tyrosinase gene codes for an antigen recognized by autologous cytolytic T lymphocytes on HLA-A2 melanomas, *J. Exp. Med. 178*:489 (1993).
24. T. Wolfel, A. van Pel, V. Brichard, J. Schneider, B. Seliger, K.H. Meyer zum Buschenfeld and T. Boon, Two tyrosinase nonapeptides recognized on HLA-A2 melanomas by autologous CTLs. *Eur. J. Immunol. 24*:759 (1994).
25. P.G. Coulie, V, Brichard, A, van Pel, T. Wolfel, J. Schneider, C, Traversari, E, DePlaen, C. Lurquin, J.P. Szikora, J.C. Renauld, and T. Boon, A new gene coding for a differentiation antigen recognized by autologous cytolytic T lymphocytes on HLA-A2 melanomas, *J. Exp. Med. 180*:35 (1994).
26. Y. Kawakami, S. Eliyahu, K. Sakaguchi, P.F. Robbins, L. Rivoltini, J.R. Yannelli, E. Appella, and S.A. Rosenberg, Identification of the immunodominant peptides of the MART-1 human melanoma antigen recognized by the majority of HLA-A2 restricted tumor infiltrating lymphocytes, *J. Exp. Med. 180*:347 (1994).
27. Y. Kawakami, S. Eliyahu, C.H. Delgado, P.F. Robbins, L. Rivoltini, S.L. Topalian, T. Miki, and S.A. Rosenberg, Cloning of the gene coding for a shared human melanoma antigen recognized by autologous T cells infiltrating into tumor, *Proc. Natl. Acad. Sci. USA 91*:3515 (1994).
28. Y. Kawakami, S. Eliyahu, C.H. Delgado, P.F. Robbins, K. Sakaguchi, E. Appella, J.R. Yannelli, G.J. Adema, T. Miki, and S.A. Rosenberg, Identification of a human melanoma antigen recognized by tumor-infiltrating lymphocytes associated with in vivo tumor rejection. *Proc. Natl. Acad. Sci. USA 91*:6458 (1994).
29. A.B.H. Bakker, M.W.J. Schreurs, A.J. de Boer, Y. kawakami, S.A. Rosenberg, G.J. Adema, and C.G. Figdor, Melanocyte linage-specific antigen gp100 is recognized by melanoma-derived-tumor-infiltrating lymphocytes, *J. Exp. Med. 179*:1005 (1994).
30. A.L. Cox, J. Skipper, Y. Chen, R.A. Henderson, T.L. Darrow, J. Shabonowitz, V.H. Engelhardt, D.F. Hunt, and C.L. Slingluff, Jr., Identification of a peptide recognized by five melanoma-specific human cytotoxic T cell lines, *Science 264*:716 (1994).
31. J.P. Szikora, A. VanPel, V. Brichard, M. Andre, N. VanBaren, P. Henry, and T. Boon, Structure of the gene of tumor transplantation antigen P35B:presence of a point mutation in the antigenic allele, *EMBO J. 9*:1041 (1990).
32. C.A. Lurquin, A. VanPel, B. Mariame, E. DePlaen, J.P. Szikora, C. Janssens, J. Reddehase, J. Lejeune, and T. Boon, Structure of the gene coding for tumor transplantation antigen P91A. A peptide encoded by the mutated exon is recognized with L^d cytolytic T cells. *Cell 58*:293 (1989).
33. C. Sibille, P. Chomez, C. Wildmann, A. Van Pel, E. DePlaen, J. Maryanski, V. de Bergeyck, and T. Boon, Structure of the gene of tumor transplantation antigen P198:a point mutation generates a new antigenic peptide, *J. Exp. Med. 172*:35 (1990).
34. D.J. Peace, W. Chen, H. Nelson, and M.A. Cheever, T cell recognition of transforming proteins encoded by mutated ras protooncogenes. *J. Immunol. 146*:2059 (1991).
35. M.L. Disis, E. Calenoff, G. McLaughlin, A.E. Murphy, W. Chen, B. Groner, M. Jeschke, N. Lydon, E. McGlynn, R.B. Livingston, R. Moe, and M.A. Cheever,

Existent T cell and antibody immunity to HER2/neu protein in patients with breast cancer, *Cancer Res. 54*:16 (1994).
36. M. Yanuck, D.P. Carbone, D. Pendleton, T. Tsukui, S.F. Winter, J.D. Minna, and J.A. Berzofsky, A mutant p53 tumor suppressor protein is a target for peptide-induced CD8+ cytotoxic T cells, *Cancer Res. 53*:3257 (1993).
37. Y. Noguchi, Y.T. Chen, and L.J. Old, A mouse mutant p53 product recognized by CD4+ and CD8+ T cells, *Proc. Natl. Acad. Sci. USA 91*:3171(1994).
38. P.C. Nowell, Molecular events in tumor development, *N. Engl. J. Med. 319*:575 (1988).
39. P. van der Bruggen, C. Traversari, P. Chomez, C. Lurquin, E. DePlaen, B. van den Eynde, A. Knuth, and T. Boon, A gene encoding an antigen recognized by cytolytic T cells on a human melanoma, *Science 254*:1643 (1991)
40. B.B. Gaugler, B. van den Eynde, P. van der Bruggen, P. Romero, J.J. Gaforio, E. DePlaen, B. Lethe, F. Brasseur, and T. Boon, Human MAGE-3 codes for an antigen recognized on a human melanoma by autologous cytotoxic T lymphocytes, *J. Exp. Medicine 179*:921 (1994).
41. L.J. Old and E.A. Boyse, Some aspects of normal and abnormal cell surface genetics, *Ann Rev. Genet. 3*:269 (1969).
42. E.A. Boyse and L.J.Old, Immunogenetics of differentiation in the mouse, *Harvey Lect. 71*:23 (1978).
43. D.J. Slamon, G.M. Clark, S.G. Wong, W.J. Levin, A. Ullrich, and W.L. McGuire, Human breast cancer: correlation of relapse and survival with amplification of the HER-2/neu oncogene, *Science 235*:177 (1987).
44. M. Schulz, R. Zinkernagel, and H. Hengartner, Peptide-induced antiviral protection by cytotoxic T cells, *Proc. Natl. Acad. Sci. USA 88*:991 (1991).
45. K. Deres, H. Schild, K-H Wiesmuller, G. Jung, and H.-G. Rammensee, In vivo priming of virus-specific cytotoxic T lymphocytes with synthetic lipopeptide vaccines, *Nature 342*:561 (1989).
46. P.K. Srivastava and L.J. Old, Individually distinct transplantation antigens of chemically induced mouse tumors, *Immunol. Today 9*:78 (1988).
47. J.W. Rohrer, S.D. Rohrer, A. barsoum, and J.H. Coggin, Jr., Differential recognition of murine tumor-associated oncofetal transplantation antigen and individually specific transplantation antigens by syngeneic cloned BALB/c and RFM mouse T cells, *J. Immunol. 152*:754 (1994).
48. J.H. Coggin, Jr., Shared cross-protective TATA on rodent sarcomas, *Immunol. Today 10*:76 (1989).
49. P.K. Srivastava and L.J. Old, Reply to Coggin's letter. *Immunol. Today 10*:78 (1989).
50. B. Van den Eynde, B. Lethe, A. VanPel, E. DePlaen, and T. Boon, The gene coding for a major tumor rejection antigen of tumor P815 is identical to the normal gene of syngeneic DBA/2 mice, *J. Exp. Med. 173*:1373 (1991).
51. O. Mandelbolm, G. Berke, M. Fridkin, M. Feldman, M. Eisenstein. and L. Eisenbach, CTL induction by a tumor-associated antigen octapeptide derived from a murine lung carcinoma, *Nature 369*:67 (1994).
52. D.L. Barnd, M.S. Lan, R.S. Metzgar, and O.J. Finn, Specific major histocom-

patibility complex–unrestricted recognition of tumor-associated mucins by human cytotoxic T cells, *Proc. Natl. Acad. Sci. USA 86*:7159 (1989).
53. M. Heike M, N.E. Blachere N, and P.K. Srivastava, Cellular immune response to spontaneous breast cancers of the v-Ha-ras transgenic mice, *Immunobiology 190*:411 (1994).
54. A.B. DeLeo, G. Jay, E. Appella, G.C. DuBois, L.W. Law, and L.J. Old, Detection of a transformation-related antigen in chemically induced sarcomas and other transformed cells of the mouse, *Proc. Natl. Acad. Sci. USA 76*:2420 (1979).
55. I. Yoshino, G.E. Peoples, P.S. Goedegebuure, R. Maziarz, and T.J. Eberlein, Association of HER2/neu expression with sensitivity to tumor-specific CTL in human ovarian cancer, *J. Immunol. 152*:2393 (1994).
56. A.N. Houghton, Cancer antigens: immune recognition of self and altered self, *J. Exp. Med. 180*:1 (1994).

34

In Vivo Carrier Effect of Heat Shock Proteins in Conjugated Vaccine Constructs

Giuseppe Del Giudice
World Health Organization—Immunology Research and Training Centre and Institute of Biochemistry, University of Lausanne, Switzerland

I. INTRODUCTION

Vaccines represent the most cost-effective approach to control infectious diseases. Major efforts are currently being devoted to the improvement of existing vaccines and/or the development of new vaccines. To this aim, several strategies are being followed; some of these foresee the production of vaccine constructs consisting of defined microbial antigens or epitopes obtained by classic purification procedures or by genetic engineering and peptide synthesis (1,2). Such an approach is particularly appropriate when the preparation of vaccines via inactivation or attenuation of the microorganisms is not achievable, or when vaccines produced in these classic manners do not prove to be efficacious. However, several orders of qualitative and quantitative problems can be encountered when recombinant and/or synthetic antigens are considered as immunogens in the development of potential vaccines.

To be recognized by T cells, antigens must be taken up by specialized cells (antigen-presenting cells, APCs) able to enzymatically process them and subsequently present the processed parts (T-cell epitopes) at their surface in the context of major histocompatibility complex (MHC) molecules. Thus, the T-cell recognition will be dictated by the nature of the epitope formed after cleavage of the antigen and by the presence of appropriate MHC restriction elements.

However, some epitopes, potentially important in the development of vaccine constructs against some infectious agents, do not have the ability to bind to MHC molecules (B-cell epitopes), because they lack the structural elements allowing this binding: This is the case of conformational epitopes or polysaccharide antigens. The MHC control of the T-cell recognition has been extensively investigated in experimental (mouse) models in vivo, and the problem has often been circumvented by the chemical linking of the epitope with carrier molecules; that is, containing a source of T-cell epitopes able to be expressed on the surface of APCs and induce the carrier-specific T-cell help necessary for an optimal B-cell priming. In animal models, carrier molecules such as, for example, keyhole limpet hemocyanin, ovalbumin, bovine serum albumin, and tetanus toxoid, have been widely utilized, linked to synthetic or recombinant peptides via glutaraldehyde or other cross linkers. Tetanus and diphtheria toxoids are also largely being employed in the development of conjugated vaccines for human use (3–5).

One problem which may be associated with the use of carrier molecules for the induction of antibody responses specific for conjugated B-cell epitopes is the so-called epitope-specific suppression. This is represented by the possibility that anticarrier antibodies induced by previous immunization(s) with the carrier alone (e.g., with tetanus toxoid as a vaccine) may interfere with the induction of an effective antibody response to a given epitope, when this is given as an immunogen conjugated to the same carrier molecule (6).

Although carrier molecules offer a solution to the qualitative problems of antigen presentation and cell activation, they only partially solve the quantitative problems involved in the degree of immune responses induced by a particular vaccine. In fact, most of the synthetic and recombinant constructs recently developed and envisaged as potential vaccines show very limited immunogenicity when given in the absence of strong adjuvants. So far, aluminum salts (alum) remain the only adjuvants admitted for human use. Despite its wide use and acceptance, alum faces a series of problems; Not all proteins and peptides are equally well adsorbed onto it (7); alum-adjuvanted vaccines can not be lyophilized; and aluminum-associated pathology and induction of vaccine-specific immunoglobulin (IgE) antibodies have been reported (8).

II. PPD AS A CARRIER MOLECULE IN ANIMALS SENSITIZED WITH BCG

Bacille Calmette-Guérin (BCG, attenuated *Mycobacterium tuberculosis* var. *bovis*, used as a vaccine against tuberculosis) has been used in the past to nonspecifically enhance the immune response to antigens coinjected with it (9). In addition to this, Lachmann and coworkers originally reported (10) that mice first primed with BCG and then immunized in the absence of adjuvants with B-cell epitopes conjugated to purified protein derivative (PPD, tuberculin, currently used for diagnostic skin testing) mounted antibody responses against the B-cell epitopes; such a response was absent in mice not primed with BCG (10,11).

The same strategy was applied in our laboratory to investigate whether it was possible to enhance the immunogenicity of defined peptides from antigens of malaria parasites. A few years ago, we had shown that the immunogenicity of a synthetic polypeptide $(NANP)_{40}$, reproducing the entire repetitive sequence of the *Plasmodium falciparum* circumsporozoite (CS) protein and envisaged as potential malaria vaccine (5), was under strict MHC control, with only mouse haplotype ($H\text{-}2^b$) being able to recognize it as both a B- and T-cell epitope (12,13). The MHC-controlled immune response to the $(NANP)_{40}$ peptide was overcome when it was coupled with carrier molecules (12). However, both the peptide alone and the peptide conjugated to the carrier required strong (Freund's) adjuvants in order to induce detectable responses.

When strains of mice of different H-2 haplotypes were first primed with BCG and then immunized with PPD-$(NANP)_{40}$ conjugate in the absence of adjuvants, it was found that anti-(NANP) IgG antibodies were detectable at titers comparable with those obtained when the Freund's complete adjuvant was used. Such antibodies persisted for a long time (at least 20 weeks) and were boosted by reinjection of the conjugate (14). Furthermore, the antipeptide antibodies recognized the natural sporozoite antigen, as shown by immunofluorescence and Western blot experiments, and functionally inhibited the penetration of sporozoite into target liver cells in vitro and their further development (14). More recent data from our laboratory have shown that this model of immunization can also be applied to nonhuman primates: Indeed the same adjuvant-free carrier effect was observed in squirrel monkeys (*Saimiri sciureus guyanensis*) sensitized with BCG and then immunized subcutaneously with the PPD-$(NANP)_{40}$ conjugate in the absence of adjuvants (15).

This model of immunization presents several advantages. First, through PPD it supplies the T-cell epitopes necessary, for a wide proportion of individuals respond to peptide vaccine candidates, and at the same time it avoids the need for adjuvants. Second, after priming with BCG, no antimycobacterial antibodies are detectable; thus one would not expect to observe the phenomenon of epitope-specific suppression, which has been supposed to have the same role in the poor reactivity to the (NANP) epitope observed in some vaccinees when tetanus toxoid was used as a carrier (16). Third, this system may not lead to a secondary immune response on the natural exposure to the microorganisms (17), but it presents the advantage of inducing substantial levels of specific antibodies lasting for long time. Finally, the main constituents for this systems of immunization, BCG and PPD, are being widely used for vaccination against tuberculosis and for skin testing, respectively, to make one think about their use in increasing the immunogenicity of peptidic constructs in humans.

III. HEAT SHOCK PROTEINS AS CARRIER MOLECULES FOR CONJUGATED B-CELL EPITOPES AVOID THE NEED FOR ADJUVANTS

PPD is composed of a plethora of mycobacterial epitopes, and repeated analysis of the preparation have failed to characterise it chemically. (10). In the definition of which mycobacterial antigen(s) present in the PPD were mediating the carrier effect observed, three observations were made. The PPD portion of the conjugate could not be replaced by T-cell epitopes without homologies with known mycobacterial antigens. Second, only mice previously primed with live BCG, but not those that had received heat-killed or sonicated BCG, produced anti-(NANP) antibodies after immunization with the PPD-peptide conjugate in the absence of adjuvants. Finally, antipeptide IgG antibodies could also be induced following immunization with the PPD-peptide conjugate without adjuvants in mice sensitized with other live intracellular parasites of the macrophage, such as *Salmonella typhimurium* and *Leishmania tropica* (18). Taken together, these data suggested that the priming was being mediated by mycobacterial antigens, which were more abundant during active infection, and that there were common antigen(s) shared between microorganisms of phylogenetically diverse origins. A group of molecules which could fulfill these criteria were the heat shock proteins (hsps).

In fact, fragments of the 65-kDa (GroEL type) hsp were known to be present in the PPD (19). Furthermore, expression of bacterial hsps has been shown to increase after microorganisms have been taken up by macrophages (20). Finally, hsps are among the best conserved molecules across evolution (21).

To test the hypothesis that mycobacterial hsps were indeed mediating the in vivo helper effect observed with PPD in the absence of adjuvants, purified recombinant *Mycobacterium bovis* hsp 65 (hsp R65) (22) and *M. tuberculosis* 70 kDa (hsp R70, DnaK type) (23) were conjugated with the $(NANP)_{40}$ synthetic peptide, and the conjugates were administered to mice in the absence of adjuvants. Mice previously primed with BCG and then immunized with the hsp R65– or hsp R70–$(NANP)_{40}$ conjugates mounted a strong antipeptide IgG antibody response similar to that previously observed when PPD was used as a carrier (18). Furthermore, no epitope-specific suppression was detectable in mice having high titers of anti-hsp antibodies following immunization with hsp alone, unlike groups of mice receiving tetanus toxoid (TT)–peptide conjugates and previously immunized with the TT alone (24).

In experiments designed to investigate the requirement for a previous BCG priming in the induction of the adjuvant-free helper effect of mycobacterial hsps, it was observed that priming with BCG was required when the 65-kDa hsp was used as a carrier molecule. However, priming with BCG was not required when using the 70-kDa hsp. In fact, mice not previously sensitized with live BCG and immunized with the peptide conjugated to the 70-kDa hsp mounted an antipeptide IgG antibody response similar to that observed in BCG-primed mice (24). It is interesting to note that when hsp R65–based conjugates were used as immunogens, the priming with live BCG could be replaced by injection with the hsp R65 alone, suggesting that a priming of hsp R65–specific T cells after both pretreatments was necessary for an effective in vivo helper effect on the induction of antipeptide antibody responses after immunization with the hsp R65–based conjugates (24).

The in vivo helper effect mediated by mycobacterial hsps in the absence of adjuvants was also applicable to other conjugated antigens and to other animal species. In fact, mice immunized with conjugates consisting of hsp and oligosaccharide from group C *N. meningitidis* produced antioligosaccharide IgG antibodies at titers similar to or even higher than those obtained after immunization with a conjugate containing diphtheria toxoid and given together with alum as an adjuvant (24). Furthermore, an adjuvant-free helper effect in vivo of mycobacterial hsp R65 and hsp R70 was also observed in *Saimiri sciureus* (squirrel) monkeys after subcutaneous immunization with the conjugates. Interestingly, as observed in mice and also in monkeys, a previous priming with BCG was required when hsp R65–based conjugates were used as immunogens; however, this was not required when monkeys were immunized with hsp R70–based conjugates (15).

Previous results from our laboratory have shown that not all mycobacterial hsps appear to mediate the adjuvant-free helper effect observed in vivo with the 65- and 70-kDa hsp conjugated with peptides or oligosaccharides. In fact, such an effect was not observed in mice immunized with nonadjuvanted conjugates consisting of the *M. leprae* 18-kDa hsp (25) and the $(NANP)_{40}$ synthetic peptide (18). On the other hand, similarly negative results were obtained in mice immunized with non-hsp from *M. tuberculosis* conjugated to peptide, irrespective of a previous priming with live BCG, unless incomplete Freund's adjuvant was used as an adjuvant (25a). These results suggest that the in vivo helper effect in the absence of adjuvants is a characteristic of hsp, and of 65- and 70-kDa hsp in particular. This hypothesis is supported by the recent finding from our laboratory that purified GroEL and DnaK hsp from *Escherichia coli* exhibit a behavior similar to that observed with their mycobacterial homologues, in terms of both strong helper effect for the production of antibodies to conjugated peptides in the absence of adjuvants, and the requirement of previous priming with live BCG for GroEL-peptide, but not for DnaK-peptide conjugates (25a). It is thus reasonable to hypothesize that similar effects could be observed with GroEL-like and DnaK-like hsps from other microorganisms.

The exact mechanisms through which bacterial hsps mediate their strong helper effect in vivo in the absence of adjuvants remains to be fully understood.

Based on the different requirements for previous sensitization of animals with live BCG, one may predict that such mechanisms would be different for GroEL- and DnaK-like hsps. It could be that infection of mice with live BCG (or immunization with 65-kDa hsp alone [24]) primes T cells specific for the 65-kDa hsp at high frequency; these cells would then undergo clonal expansion on immunization with conjugates containing this hsp; thus providing the help to B cells necessary for production of antibodies specific for peptides or oligosaccharides conjugated to it. This sequence of events would be in agreement with findings by other investigators showing a remarkably high frequency of T-cell precursors specific for the mycobacterial 65-kDa hsp (26,27), and also with findings showing that hsp R65–specific T-cell lines and clones can be originated from *M. leprae*– and BCG-vaccinated individuals which recognize this molecule within the context of most of the HLA-DR molecules tested (28). Preliminary results in our laboratory have shown that infection of mice with BCG induces a high frequency of mycobacterial hsp R65–specific T-cell precursors, which increases after immunization with hsp R65–based conjugates in the absence of adjuvants.

The most simplistic explanation for the adjuvant-free helper effect of mycobacterial hsp R70, which does not require any previous priming of the animals with live BCG, would be a nonspecific polyclonal stimulation of B lymphocytes through bacterial lipopolysaccharides (LPS) (29), possibly contaminating the preparation of the purified mycobacterial hsp R70. Several lines of evidence speak, however, against this hypothesis. First, immunization of animals with adjuvant-free hsp R70–based conjugates induced high titers of peptide- or oligosaccharide-specific IgG antibodies (with only negligible amounts of IgM being detectable), which were boosted on reimmunization (24). Second, the in vivo helper effect of hsp R70 in the absence of adjuvants for the induction of antibody responses to conjugated peptides was not observed in athymic *nu/nu* mice; on the contrary, it was observed in LPS-resistant C3H/HeJ mice (24). Finally, the mycobacterial hsp R70 exerted its adjuvant-free helper effect also in *Saimiri sciureus* (squirrel) monkeys (15), the B lymphocytes of which do not respond to LPS stimulation (30). Taken together, these findings support the concept that the in vivo helper effect of the mycobacterial hsp R70 in the absence of adjuvants depends on the presence of an intact T-cell compartment, which would be specifically stimulated (or restimulated) following immunization with hsp R70 conjugates. This hypothesis is further supported by more recent data showing that in order to mediate such an in vivo helper effect, mycobacterial hsps require to be covalently linked to antigens (25a), a requirement commonly met in systems where, after immunization with hapten-carrier conjugates, the induction of hapten-specific antibody responses relies on the activation of carrier-specific T cells (31).

One explanation for the carrier effect of the 70-kDa hsp in the absence of a previous priming with BCG or hsp may be that a priming of hsp 70–specific T cells had already naturally occurred in the animals at the moment of the immunization with hsp-based conjugates; for example, through hsp derived from the saprophytic intestinal flora. This hypothesis could be easily tested in germ-free animals, and would be in agreement with recent findings by others showing that intestinal intraepithelial T lymphocytes strongly respond to the mycobacterial hsp of 70 kDa but not to that of 65 kDa (32). It is not known whether the mycobacterial 70-kDa hsp could have an activity similar to that of hsp-like molecules (belonging to the same hsp family) shown to bind immunogenic peptides in APCs, possibly facilitating the association between processed antigens and MHC class II (33) and class I molecules (34). One could also hypothesize that the peculiar carrier effect of this hsp might be exerted through some of the specific functions of this class of stress proteins, that is, their molecular chaperone function (35), after endocytosis of hsp-peptide and hsp-oligosaccharide conjugates. Recent data from our laboratory, however, do not appear to support this hypothesis (25a).

IV. PROKARYOTIC HSP IN VACCINE CONSTRUCTS: RISK OF AUTOIMMUNITY?

Just opposite of what one would have expected, owing to the high degree of homologies existing between mycobacterial and human hsp (21), it appears that hsps are among the most powerful immunogens of mycobacteria and that T cells at a high frequency are able to respond to them (26,27).

The findings that murine and human TCR-$\alpha\beta$ and $\gamma\delta^+$ T cells that recognize the mycobacterial 65-kDa hsp can also recognize the host's homologue, that some T-cell clones from autoimmune animals and humans proliferate in vitro in the presence of mycobacterial hsps, and that anti-hsp antibodies are detectable in the sera and in the sites of tissue damage of some autoimmune patients have suggested a role for microbial hsps (and particularly for the mycobacterial 65-kDa hsp) in the induction of autoimmune phenomena (36,37).

In this respect, it would logical to consider the potential risk that active immunization with microbial hsps in conjugated constructs given as vaccines may trigger immune responses possibly cross reactive with self-homologous hsp or with other proteins sharing some homologies with prokaryotic hsps (38). In fact, in our model of immunization using hsp-based constructs in the absence of adjuvants, anti-hsp antibodies are usually detected, as well as T lymphocytes that can be specifically restimulated in vitro by mycobacterial hsp (18,24; unpublished results). However, preliminary results obtained using serum samples from mice immunized several times with hsp R65–peptide conjugates without adjuvants (38a) have shown that anti–hsp R65 IgG antibodies cross reacted with *E.*

coli GroEL hsp but not with the homologous recombinant human 60-kDa hsp (39), nor with the 60-kDa hsp expressed by murine cells, which is almost identical to its human counterpart (40). These data suggest that, at least for the mycobacterial hsps of 65 kDa and at least for the antibody response induced after immunization with conjugates containing this hsp, the frequency of hsp-specific B-cell clones induced after immunization with hsp-containing constructs, and potentially cross-reactive with self-homologues, should be low or, in any case, the antibodies produced should have low affinities, since antibodies cross-reacting with human and mouse hsp were not detectable. It remains to be determined whether or not hsp-specific T cells cross reacting with homologous human hsp are induced following our model of immunization. However, even in the case cross-reactive T cells would be detected, still their participation in the induction of autoimmune phenomena would need to be formally proven.

Hsps are expressed by all living organisms, and a "natural" priming to hsp is known to occur following a variety of stimuli; for example, after bacterial, fungal, or parasitic infections (41,42) and after vaccinations with whole-cell vaccines. T cells reacting with the 65- and 70-kDa hsps of *M. tuberculosis* and *M. leprae* have been found in several individuals vaccinated against these pathogens (28). One of the most utilized whole-cell vaccines throughout the world is the inactivated pertussis vaccine, which is usually given in association with tetanus and diphtheria toxoid (DTP) at a very young age. Furthermore, *Bordetella pertussis* is known to possess a hsp of 65 kDa recognized by sera of mice immunized with DTP (43). By following up for almost 1 year the appearance and the persistence of anti-hsp antibodies in serum samples taken from infants before and after vaccination with DTP, we found that IgG antibodies to the mycobacterial 65- and 70-kDa hsp appeared in almost 90% of infants after two immunizations with DTP. The titers of these antibodies were boosted by a third injection of DTP. Antibodies and lymphocytes reactive to the mycobacterial 65-kDa hsp were also found in mice immunized with DTP (44). The priming of the immune response to hsp was mediated by the whole-cell pertussis component of the DTP vaccine, since it was not observed in infants receiving DT plus an acellular recombinant pertussis vaccine (45). Interestingly, whole-cell pertussis vaccine–induced anti-hsp antibodies cross reacted with the *Escherichia coli* GroEL hsp but not with the human 65-kDa hsp homologue (44).

Taken together, all these data suggest that priming of the immune system to microbial hsp is a common event which can even occur very early in life. Furthermore, if most healthy individuals can exhibit a priming to microbial hsp early in life following vaccination against pertussis, or subsequently following other stimuli, it becomes questionable whether such a priming could contribute to the induction of autoimmune disorders, which are relatively rare, any time in life unless other factors concomitantly operate in their induction.

On the other hand, if the immune response to microbial hsps in some experimental models and in certain human diseases has been postulated to represent a possible link with autoimmunity, immunization of animals with the mycobacterial 65-kDa hsp has been shown to prevent the appearance of autoimmune disorders, such as arthritis or diabetes mellitus (37). Interestingly, immunization of Lewis rats with the human 60-kDa expressed in recombinant vaccinia virus has been reported to have a beneficial effect against adjuvant arthritis despite antibody and T-cell responses induced against this hsp, which is virtually identical to the rat 60-kDa hsp homologue (46).

V. CONCLUSIONS

The use of hsps as carrier molecules for the induction of immune responses to conjugated antigens offers several advantages. The procedure of immunization with hsp-based conjugates does not require any adjuvant, the antibody titers obtained are the same as those obtained with conventional strong (e.g., Freund's) adjuvants, and are long-lasting. The conjugation procedure, via glutaraldehyde, is commonly used for the preparation of protein-polysaccharide conjugated vaccines for human use. This model of immunization is easily applicable to higher animals (monkeys) by subcutaneous injection of the conjugates (15). BCG is commonly and widely used very early in life (mainly in developing countries) as a vaccine against tuberculosis. Furthermore, the use of BCG for priming can turn out to be unnecessary if the 70-kDa hsps had to be used in conjugated constructs. No epitope-specific suppression has been observed in mice with very high titers of anti-hsp antibodies at the moment of immunization with the hsp-peptide conjugates (24). Several T-cell epitopes have been identified on the mycobacterial 65-kDa (47) and the 70-kDa hsps (48), and the entire proteins have been demonstrated to be immunogenic in vivo in a wide variety of mouse strains (47) and in vitro with T-cell lines from individuals with different HLA-DR haplotypes (28). This would speak against a possible restriction of the immune response in a large population owing to the MHC class II polymorphism. Furthermore, beyond their possible use in individuals already sensitized to mycobacteria (via vaccination or natural contact), the hsps may also be able to interact with hsp-specific T-cell populations that are found in apparently healthy individuals (49,50), and that may reflect a widespread and common sensitization to microbial hsps.

ACKNOWLEDGMENTS

I wish to thank all the colleagues who generously provided reagents and the PhD students and technicians at the WHO-IRTC, Department of Pathology, Geneva, who participated in the work discussed here.

Work from the author's laboratory received support from WHO Programme for Vaccine Development, and from IRIS, Siena, Italy.

REFERENCES

1. G.L. Ada, Vaccines, *Fundamental Immunology* (W.E. Paul, ed.), Raven Press, New York, 1989, p. 985.
2. D.R. Milich, Synthetic T and B recognition sites: implications for vaccine development, *Adv. Immunol. 45*:195 (1989).
3. J.B. Robbins and R. Schneerson, Polysaccharide-protein conjugates: a new generation of vaccines, *J. Infect., Dis. 161*:821 (1990).
4. J. Ward, Prevention of invasive *Haemophilus influenzae* type b disease: lessons from vaccine efficacy trials, *Vaccine 9* (Suppl.):S17 (1991).
5. P. Romero, Malaria vaccines, *Curr. Opin. Immunol. 4*:432 (1992).
6. M.P. Schutze, C. Leclerc, M. Jolivet, F. Audibert, and L. Chedid, Carrier-induced epitopic suppression, a major issue for synthetic vaccines, *J. Immunol. 135*:2319 (1985).
7. S.J. Seeber, J.L. White, and S.L. Hem, Predicting the adsorption of proteins by aluminum-containing adjuvants, *Vaccine 9*:201 (1991).
8. R.K. Gupta and E.H. Relyveld, Adverse reactions after injections of adsorbed diphtheria-pertussis-tetanus (DPT) vaccines are not only due to pertussis organisms or pertussis components in the vaccine. *Vaccine 9*:699 (1991).
9. D. Frommel, and P. Lagrange, BCG: a modifier of immune responses to parasites, *Parasitol. Today 5*:188 (1989).
10. P.J. Lachmann, L. Strangeways, A. Vyakarmam, and G. Evans, Raising antibodies by coupling peptides to PPD and immunizing BCG-sensitized animals, *Ciba Found. Symp. 119*:25 (1986).
11. P.J. Lachmann, Purified protein derivative (PPD), *Springer Semin. Immunopathol. 10*:279 (1988).
12. G. Del Giudice, J.A. Cooper, J. Merino, A.S. Verdini, A. Pessi, A.R. Togna, H.D. Engers, G. Corradin, and P.-H. Lambert, The antibody response in mice to carrier-free synthetic polymers of *Plasmodium falciparum* circumsporozoite repetitive epitope is I-A^b restricted: possible implications for malaria vaccines, *J. Immunol. 137*:2952 (1986).
13. A.R. Togna, G. Del Giudice, A.S. Verdini, F. Bonelli, A. Pessi, H.D. Engers, and G. Corradin, Synthetic *Plasmodium falciparum* circumsporozoite peptides elicit heterogenous $L3T4^+$ T cell proliferative responses in H-2^b mice, *J. Immunol. 137*:2956 (1986).
14. A.R. Lussow, G. Del Giudice, L. Renia, D. Mazier, J.P. Verhave, A.S. Verdini, A. Pessi, J.A. Louis, and P.-H. Lambert, Use of tuberculin purified protein derivative-Asn-Ala-Asn-Pro conjugate in bacillus Calmette-Guérin primed mice overcomes H-2 restriction of the antibody response and avoids the need for adjuvants, *Proc. Natl. Acad. Sci. USA 87*:2960 (1990).
15. R. Perraut, A.R. Lussow, S. Gavoille, O. Garraud, H. Matile, C. Tougne, J. van Embden, R. van der Zee, P.-H. Lambert, J. Gysin, and G. Del Giudice, Successful

primate immunization with peptides conjugated to purified protein derivative or mycobacterial heat shock proteins in the absence of adjuvants, *Clin. Exp. Immunol. 93*:382 (1993).

16. H.M. Etlinger, A.M. Felix, D. Gillessen, E.P. Heimer, M. Just, J.R.L. Pink, F. Sinigaglia, D. Stürchler, B. Takacs, A. Trzeciak, and H. Matile, Assessment in humans of a synthetic peptide-based vaccine against the sporozoite stage of the human malaria parasite, *Plasmodium falciparum*, *J. Immunol. 140*:626 (1988).
17. G. Del Giudice, Q. Cheng, D. Mazier, J. Meuwissen, G. Grau, G. Corradin, A. Pessi, A.S. Verdini, and P.-H. Lambert, Primary and secondary responses to (NANP) by *Plasmodium falciparum* sporozoites in various strains of mice, *Scand J. Immunol. 29*:555 (1989).
18. A.R. Lussow, C. Barrios, J. van Embden, R. van der Zee, A.S. Verdini, A. Pessi, J.A. Louis, P.-H. Lambert, and G. Del Giudice, Mycobacterial heat shock proteins as carrier molecules, *Eur. J. Immunol. 21*:2297 (1991).
19. J.E. Thole, H.G. Dauwerse, P.K. Das, D.G. Groothius, L.M. Schouls, and J.D. van Embden, Cloning of *Mycobacterium bovis* BCG DNA and expression of antigens in *Escherichia coli, Infect. Immun. 50*:800 (1985).
20. N.A. Buchmeier, and F. Heffron, Induction of *Salmonella* stress proteins upon infection of macrophages, *Science 248*:730 (1990).
21. S. Lindquist, and E. Craig, The heat shock proteins, *Annu. Rev. Genet. 22*:631 (1988).
22. J.E. Thole, W.C. van Schooten, W.J. Keulen, P.W. Hermans, A.A. Janson, R.R. Vries, A.H. Kolk, and J.D. van Embden, Use of recombinant antigens expressed in *Escherichia coli* K-12 to map B-cell and T-cell epitopes on the immunodominant 65-kilodalton protein of *Mycobacterium bovis* BCG. *Infect. Immun. 56*:1633 (1987).
23. A. Mehlert, and D.B. Young, Biochemical and antigenic characterization of the *Mycobacterium tuberculosis* 71 kD antigen, a member of the 70 kD heat shock protein family, *Mol. Microbiol. 3*:125 (1989).
24. C. Barrios, A.R. Lussow, J. van Embden, R. ven der Zee, R. Rappuoli, P. Constantino, J.A. Louis, P.-H. Lambert, and G. Del Giudice, Mycobacterial heat shock proteins as carrier molecules. II: The use of the 70-kDa mycobacterial hsat shock protein as a carrier for conjugated vaccines can circumvent the need for adjuvants and Bacillus Calmette-Guérin priming, *Eur. J. Immunol. 22*:1365 (1992).
25. F.I. Lamb, N.B. Singh, and M.J. Colston, The specific 18-kilodalton antigen of *Mycobacterium leprae* is present in *Mycobacterium habana* and functions as a heat-shock protein, *J. Immunol. 144*:1922 (1990).

25a. C. Barrios, C. Georgopoulos, P.H. Lambert, and G. Del Giudice, Heat shock proteins as carrier molecules: in vivo helper effect mediated by *Escherichia coli* GroEL and DnaK proteins required cross-linking with antigen. *Clin. Exp. Immunol. 98*:229 (1994).

26. S.H.E. Kaufmann, U. Väth, J.E.R. Thole, J.D.A. van Embden, and F. Emmrich, Enumeration of T cells reactive with *Mycobacterium tuberculosis* organisms and specific for the recombinant mycobacterial 64 kilodalton protein, *Eur. J. Immunol. 17*:351 (1987).

27. S. Kaufmann, B. Schoel, A. Wand-Württenberger, U. Steinhoff, M. Munk, and T. Koga, T cells, stress proteins, and pathogenesis of mycobacterial infections, *Curr. Top. Microbiol. Immunol. 155*:125 (1990).
28. A.S. Mustafa, K.E. Lundin, and F. Oftung, Human T cells recognize mycobacterial heat shock proteins in the context of multiple HLA-DR molecules: studies with healthy subjects vaccinated with *Mycobacterium bovis* BCG and *Mycobacterium leprae, Infect. Immun. 61*:5294 (1993).
29. J.M. Chiller, B.J. Skidmore, D.C. Morrison, and W.O. Weigle, Relationship of the structure of bacterial lipopolysaccharides to its function in mitogenesis and adjuvanticity, *Proc. Natl. Acad. Sci. USA 70*:2129 (1973).
30. O. Garraud, R. Perraut, D. Blanchard, P. Chouteau, E. Bourreau, C. Le Scanf, B. Bonnemains, and J.C. Michel, Squirrel monkey (*Saimiri sciureus*) B lymphocytes: secretion of IgG directed to *Plasmodium falciparum* antigens, by primed blood B lymphocytes restimulated in vitro with parasitized red blood cells, *Res. Immunol. 144*:407 (1993).
31. A. Lanzavecchia, Receptor-mediated antigen uptake and its effect on antigen presentation to class II-restricted T lymphocytes, *Annu. Rev. Immunol. 8*:773 (1990).
32. K.W. Beagley, K. Fujihashi, C.A. Black, A.S. Lagoo, M. Yamamoto, J.R. McGhee, and H. Kiyono, The *Mycobacterium tuberculosis* 71-kDa heat-shock protein induces proliferation and cytokine secretion by murine gut intraepithelial lymphocytes, *Eur. J. Immunol. 23*:2049 (1993).
33. A. vanBuskirk, B.L. Crump, E. Margoliash, and S.K. Pierce, A peptide binding protein having a role in antigen presentation is a member of the hsp70 heat shock family, *J. Exp. Med. 170*:1799 (1989).
34. P.K. Srivastava, H. Udono, N.E. Blachere, and Z. Li, Heat shock proteins transfer peptides during antigen processing and CTL priming, *Immunogenetics 39*:93 (1994).
35. F.-U. Hartl, R. Hlodan, and T. Langer, Molecular chaperones in protein folding: the art of avoiding sticky situations, *TIBS 19*:20 (1994).
36. P. Res, J. Thole, and R. de Vries, Heat-shock proteins and autoimmunity in humans, *Springer Semin. Immunopathol. 13*:81 (1991).
37. U. Feige, and I.R. Cohen, The 65-kDa heat-shock protein in the pathogenesis, prevention and therapy of autoimmune arthritis and diabetes mellitus in rats and mice, *Springer Semin. Immunopathol. 13*:99 (1991).
38. D.B. Jones, A.F.W. Coulson, and G.W. Duff, Sequence homologies between hsp60 and autoantigens, *Immunol. Today 14*:115 (1993).

38a. C. Barrios, C. Tougne, B.S. Polla, P.H. Lambert, and G. Del Giudice, Specificity of antibodies induced after immunization of mice with the mycobacterial heat shock protein of 65 kD. *Clin. Exp. Immunol. 98*:224 (1994).

39. S. Jindal, A.K. Dudani, B. Singh, C.B. Harley, and R.S. Gupta, Primary structure of a human mitochondrial protein homologous to the bacterial and plant chaperonins and to the 65-kilodalton mycobacterial antigen, *Mol. Cell. Biol. 9*:2279 (1989).
40. T.J. Venner, and R.S. Gupta, Nucleotide sequence of mouse HSP60 (chaperonin, GroEL homolog) cDNA, *Biochem. Biophys. Acta 1087*:336 (1990).

41. D.B. Young, A. Mehlert, and D.F. Smith, Stress proteins and infectious diseases, *Stress Proteins in Biology and Medicine* (R.I. Morimoto, A. Tissières, and C. Georgopoulos, eds.), Cold Spring Harbor Laboratory, Cold Spring Harbor, NY, 1990, p. 131.
42. D. Mazier, and D. Mattei, Parasite heat-shock proteins and host responses: the balance between protection and immunopathology, *Springer Semin. Immunopathol. 13*:37 (1991).
43. D.L. Burns, J.L. Gould-Kostka, M. Kessel, and J.L. Arciniega, Purification and immunological characterization of a GroEL-like protein from *Bordetella pertussis, Infect. Immun. 59*:1417 (1991).
44. G. Del Giudice, A. Gervaix, P. Constantino, C.-A. Wyler, C. Tougne, E.R. de Graeff-Meeder, J. van Embden, R. van der Zee, L. Nencioni, R. Rappuoli, S. Suter, and P.-H. Lambert, Priming to heat shock proteins in infants vaccinated against pertussis, *J. Immunol. 150*:2025 (1993).
45. A. Podda, E. Carapella De Luca, L. Titone, A.M. Casadei, A. Cascio, S. Peppoloni, G. Volpini, I. Marsili, L. Nencioni, and R. Rappuoli, Acellular pertussis vaccine composed of genetically inactivated pertussis toxin: safety and immunogenicity in 12-24 and 2-4 month old children, *J. Pediatr. 120*:680 (1992).
46. J.A. Lopez-Guerrero, J.P. Lopez-Bote, M.A. Ortiz, R.S. Gupta, E. Paez, and C. Bernabeu, Modulation of adjuvant arthritis in Lewis rats by recombinant vaccinia virus expressing the human 60-kilodalton heat shock protein, *Infect. Immun. 61*: 4225 (1993).
47. S.J. Brett, J. Lamb, J. Cox, J. Rothbard, A. Mehlert, and J. Ivanyi, Differential pattern of T cell recognition of the 65-kDa mycobacterial antigen following immunization with the whole protein or peptides, *Eur. J. Immunol. 19*:1303 (1989).
48. E. Adams, W.J. Britton, A. Morgan, A.L. Goodsall, and A. Basten, Identification of human T cell epitopes in the *Mycobacterium leprae* heat shock protein 70-kD antigen, *Clin. Exp. Immunol. 94*:500 (1993).
49. M.E. Munk, B. Schoel, and S.H.E. Kaufmann, T cell responses in normal individuals towards recombinant antigens of *Mycobacterium tuberculosis, Eur. J. Immunol. 18*:1835 (1988).
50. M.E. Munk, B. Schoel, S. Modrow, R.W. Karr, R.A. Young, and S.H.E. Kaufmann, Cytolytic T lymphocytes from healthy individuals with specificity to self epitopes shared by mycobacterial and human 65-kDa heat shock protein, *J. Immunol. 143*:2844 (1989).

35

Major Histocompatibility Complex Class Ib Molecules

A Role in the Presentation of Heat Shock Proteins to the Immune System?

Farhad Imani
The Johns Hopkins University School of Medicine, Baltimore, Maryland

Thomas M. Shinnick
Centers for Disease Control and Prevention, Atlanta, Georgia

Mark J. Soloski
The Johns Hopkins University School of Medicine, Baltimore, Maryland

I. INTRODUCTION

Class Ib molecules encoded within the major histocompatibility complex (MHC) have been shown to serve "specialized" roles in antigen presentation to T cells (1). It has also been speculated that class Ib molecules serve as antigen presentation structures for γ/δ T cells (2). Given that γ/δ T cells have been shown to recognize heat shock proteins (hsps), it seems reasonable to propose that class Ib molecules may play a role in the presentation of hsps to γ/δ T cells. In this chapter, we first review the novel cellular and molecular properties of class Ib proteins and then outline the evidence that argues for or against a role for class Ib molecule in the presentation of hsps to the immune system.

II. OVERVIEW OF CLASS IB MOLECULES

Class I molecules function to present peptide fragments derived from endogenous (self), viral or bacterial proteins to $CD8^+$ T cells (3). Our understanding of this function is largely based on studies addressing the role of the murine class I molecules H-2K, H-2D, and H-2L in T-cell antigen recognition. However, murine class I genes represent a large multigene family located on the 17th chromosome, and the majority of class I genes map telomeric to H-2K, H-2D, and H-2L in the Q, T, and M chromosomal regions (formerly called Qa, Tla, Hmt; ref. 4). The Q, T, and M genes encode polypeptides of 39–48 kDa that are noncovalently associated with β_2-microglobulin. Based on overall structural homologies to H-2K/D/L, the α_1-α_2 structural domains of these polypeptides are predicted to form a peptide binding pocket (4). Interestingly, the protein products of the Q, T, and M class I genes display several properties which distinguish them from H-2K/D/L. First, these proteins serve as weak transplantation antigens. Second, the class Ib molecules display little polymorphism. Finally, many of these gene products display tissue-specific expression. Based on these distinctive properties, the Q, T, and M class I genes/proteins have been termed class Ib genes/proteins.

The class Ib molecule Qa-2 molecule is perhaps the best characterized class Ib protein. Qa-2 is encoded by two genes, Q7 and Q9, which differ by a single base (5,6). In some mouse strains, a single Qa-2 encoding gene (Q7 or Q9) is present, and these mice display lower levels of Qa-2 expression (7). Also in the Q region, the genes Q4 and Q10 encode the class Ib proteins Qb-1 and Q10, respectively (8,9). For the other Q-region genes (Q1,Q2,Q3,Q5,Q6, and Q8), partial sequence information is available, much of which indicates that these genes should be expressed and functional (4,10).

Depending on the strain, the T region can contain up to 18 class I genes. Based on homologies, these genes have been designated T1–T24 (4,11). Within this cluster is found the genes that encode Qa-1 (T23) and TL (T3 and T13), as well as a ligand for a self-reactive γ/δ T cell (T22) (4,12–16).

Recently, a cluster of class Ib genes have been identified in the M (formerly Hmt) region (15). This region is telomeric to Tla and has seven (M1–M7) identified class Ib genes. Within this group is the gene (M3) which encodes the Mta alloantigen (16).

Several class Ib molecules have novel structural features. For example, although the majority of class Ib molecules are cell surface glycoproteins, the products of genes Q4 and Q10 encode secreted class Ib molecules (8,9). The Qa-2 class Ib molecule is a 40-kDa cell surface polypeptide which is attached to the cell membrane via a phosphatidylinositol-bearing glycolipid anchor and can be expressed as a secreted from (6,16–20). At present, it is unclear how these structural features relate to physiological function.

Class Ib molecules can be expressed in a tissue-specific fashion. The secreted gene product Q10 is expressed only in liver, whereas the expression of TL is confined to thymocytes and intestinal epithelium (9,21). Qa-2 transcripts are widespread, yet cell surface expression is found only on hematopoietic cells with the highest levels on T cells (7). Qa-2 expression is regulated during thymocyte development (22). In contrast, the protein products of the Qa-1 (T23) and Mta (M3) genes are expressed in a wide variety of cell types (23,24). Although the reasons these genes have evolved distinct expression patterns is not clear, it is quite likely that these patterns are reflective of the physiological settings in which they function.

In contrast to other MHC loci, class Ib loci are largely nonpolymorphic (4). Protein and nucleic acid studies have failed to detect significant Qa-2 structural polymorphism among Qa-2$^+$ inbred or wild mouse substrains that are geographically indigenous and have been isolated from other strains for 2–3 million years (25,26). Similarly, the Hmt encoding the M3 gene and Q10 appear to be largely invariant (1,4). Serological or cytotoxic T lymphocyte (CTL) analysis has detected modest allelic polymorphism for Qa-1 (five alleles) and TL (five alleles) (4,27). It is not completely appreciated why class Ib genes have not evolved polymorphism. It is likely that class Ib and class I molecules were subjected to different evolutionary pressures. Considering that class I MHC molecules bind and present peptide fragments to T cells, it has possible that the lack of polymorphism in class Ib molecules is mirroring the chemical nature of their peptide ligands.

The function of MHC class I and class II molecules is to bind to and present peptide antigens to T lymphocytes. This fundamental role serves to shape the T-cell repertoire and ultimately dictates the ability of the immune system to respond to a foreign antigen. Given that class Ib and "conventional" class I molecules are structurally homologous, it is likely that class Ib molecules also present peptide ligands to T cells. Several lines of evidence support this. First, T-cell recognition or surface expression of Qa-2, Qa-1, and Hmt have been shown to be effected by mutations in the peptide transporter gene Tap-2 (28,29). Second, the class Ib molecules Qa-2, Qa-1, Hmt, and recently TL have all been demonstrated to serve as weak transplantation antigens and can be recognized by alloreactive CD8$^+$ CTLs (4,27,30). Although recent studies have shown that some alloreactive T cells have complex recognition properties (31), it appears that in many different experimental settings, the bulk of alloreactive T-cell receptors (TCRs) recognize a complex of non-self MHC and peptide (31,32). Thus, it is likely that many class Ib–specific alloreactive CTLs are likewise recognizing a complex of class Ib plus peptide. Finally, peptides have been isolated from the class Ib molecule Qa-2 (33,34).

A novel role for class Ib molecule in antigen presentation has been revealed during investigations of the murine maternally transmitted antigen (Mta; ref. 1).

Mta has been defined as a alloantigen which is a complex of the class Ib molecule M3 and a peptide derived from the NH_2-terminus of mitochondrial NADH dehydrogenase (ND1) (35). The binding of the ND1 peptide to M3 is dependent on the presence of the N-formyl group (35,36). Since all prokaryotes initiate protein synthesis with N-formyl methionine, it was speculated that M3 may be a specialized peptide binding structure tailored to bind and present N-formyl peptides derived from microbial antigens (1,35,36). Indeed, M3 can present N-formyl bacterial antigens derived from the intracellular pathogen *Listeria* to *Listeria*-specific $CD8^+$ effector T cells (37,38).

Studies on Mta led to the hypothesis that M3 gene, and perhaps other class Ib molecules, is a specialized binding structure that evolved to exploit microbe-unique chemical structures (1,36–38). This novel feature could explain the failure of class Ib molecules to evolve polymorphism and their low levels of tissue expression.

III. THE MURINE CLASS IB MOLECULE QA-1(T23$^{b/d}$) AND HSPS

Previous studies from our laboratory have suggested that the murine class Ib molecule Qa-1 can bind peptides derived from stress proteins. The two observations that led to this hypothesis are as follows: First, mouse L fibroblasts transfected with the Qa-1 encoding class Ib gene T23 express low levels of Qa-1 on the cell surface, whereas the predominant Qa-1 species was present as a immature intracellular form which is rapidly degraded. The expression of mature cell surface Qa-1 was dramatically and selectively increased following heat shock (39). Second, "empty" Qa-1 molecules transported to the cell surface at 25°C could be stabilized when shifted to 37°C with a tryptic digest of mycobacterial hsp 65. Similar digests of hsp 70 or bovine serum albumin (BSA) failed to stabilize Qa-1 expression.

To identify which peptide(s) derived from hsp 65 binds to Qa-1, a panel of mycobacterial hsp 65 overlapping synthetic 15 amino acid peptides were used in a direct binding assay. Peptides were radiolabeled using the Bolton-Hunter reagent and added to L-g37 cells which had been incubated overnight at 25°C to increase the levels of "empty" class I molecules in the cell surface. Peptides were incubated with cells for 30 min at 37°C in the presence of β_2-microglobulin, and the cells were washed, lysed in NP-40, and the lysate immunoprecipitated with control, anti–H-2K^k, or polyclonal anti Qa-1 COOH peptide serum. Immunoprecipitates were analyzed by SDS-PAGE. Specific binding was scored by the presence of radiolabeled peptide. Typical results are displayed in Figure 1. Four peptides (Table 1) did display specific binding to Qa-1 in L cells. An identical analysis using the RMA-S cell line found that only peptide 181–

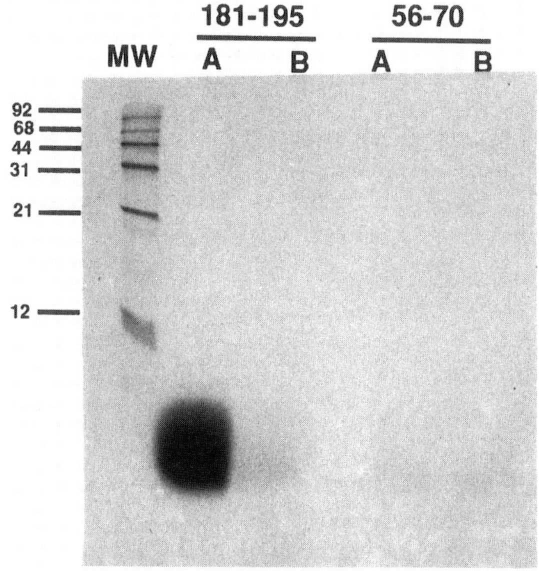

Figure 1 Mycobacterial hsp 65 peptides bind to Qa-1b. A series of 15-mer overlapping peptides spanning hsp 65 were radiolabeled via the Bolton-Hunter reagent. Peptides were added to Lg37 cells that had been incubated overnight at room temperature to upregulate "empty" Qa-1 molecules. Peptides were bound at 37°C for 30 min., and cells washed, lysed, and immunoprecipitated with anti–QA-1^b (lanes A) or anti–H-$2K^k$ (lanes B). Immunoprecipitates were analyzed by SDS-PAGE. Displayed are typical results showing peptide 181–195, which bound to Qa-1^b but not H-$2K^k$, and peptide 56–70, which bound to neither.

195 displayed specific binding to Qa-1. Since L-g37 cells express the $T23^d$ gene from BALB/c mice, whereas RMA-S is derived from C57BL/6 and expresses $T23^b$, this difference could be due to sequence differences between the two.

Peptide 181–195 has a number of interesting properties which may bear on the physiological significance of the Qa-1 molecule. First, this peptide has been

Table 1 Mycobacterial Hsp 65 Peptides that Bind to Qa-1^b

Peptide	Sequence	L-g37	RMA-S
11–25	ARRGLERGLNALADA	++	–
31–45	GPKGRNVVLEKKWGA	++	–
394–408	IEDAVRNAKAAVEEG	++	–
181–195	FGLQLELTEGMRFDK	++	++

shown to be recognized by several non–H-2–restricted hsp 65–reactive γ/δ T cells (40). Second, this peptide can be recognized by arthritogenic T cells in rats (41). It is also interesting to note that all the peptides that were found to bind Qa-1^b have at least one glutamic acid residue. A synthetic polymer of glutamic acid and tyrosine (poly GT) has been shown to be recognized by GT-specific γ/δ T cell hybrids in a Qa-1^b restricted fashion (42)

Collectively, the above observation suggest that Qa-1^b can form a complex with peptides derived from the stress protein hsp 60/65. However, it is not clear at this time that this interaction involves the peptide binding site nor is it resolved if hsp + Qa-1^b–specific T cells exist.

IV. TL AND HSPS

There are several studies in which γ/δ T cells reactive with syngeneic T cells have been proposed to recognize a complex between a class Ib gene product and a peptide derived from a stress protein.

A γ/δ T-cell hybridoma has been derived from C57BL/6 mice which recognizes syngeneic cells (14). The T-cell receptor expressed by the hybridoma is known to recognize the protein product of the $T22^b$ gene. It has been proposed that T22 is complexed with a self-peptide derived from a stress protein. The involvement of a self-peptide is supported by the observation that mutagenesis of amino acid residues in T22 that are unlikely to be TCR contacts but would be involved in peptide binding dramatically alters the ability of T22 to be recognized. However, the identity of the self-peptide bound to T22 has not been determined (43).

Recently, a novel Vγ1.1–expressing T-cell population has been recovered from intestinal epithelium that recognized the self class Ib TL protein encoded by the T3 gene (44). Again, it has been suggested that the peptide antigen is derived from a self–stress protein based on the observation that such cells are not normally activated in vivo and are only activated when cultured with explanted intestinal epithelial cells.

Thy1^+ dendritic epidermal T cells (dECs) expressing a narrow repertoire of γ/δ TCRs have been identified that react with syngeneic as well as allogeneic kerotinocytes (45). Although the molecular nature of the antigen recognition unit recognized by these TCRs has not been determined, the lack of MHC restriction has been interpreted as suggestive evidence for the involvement of nonpolymorphic antigen-presenting (class Ib?) molecules. In addition, the investigators have argued that such cells may perform an immune surveillance function, perhaps through the recognition of stress protein antigens.

V. SUMMARY AND CONCLUSIONS: GUILT BY ASSOCIATION OR REASONABLE DOUBT?

Clearly the notion that class Ib molecules are involved in the presentation of hsp-derived peptides is an attractive one. A compelling case can be made for this hypothesis given that class Ib molecules have been proposed to be "specialized" presentation structures involved in microbial immunity, the fact that hsps are highly conserved structures among prokaryotes and eukaryotes, the observation thatγ/δ cells that recognize stress proteins have been readily identified and that class Ib molecules have been implicated in γ/δ T-cell recognition. Thus, one can postulate a scenario where stress proteins synthesized in "sick" (infected) cells are degraded and peptides loaded onto newly synthesized class Ib molecules. When these molecular complexes arrive at the cell surface, they serve as a "flag" providing a early signal to the immune system. γ/δ T cells have been proposed to be involved in the early steps of a T-cell immune response. If γ/δ T cells are the predominate population that initially recognizes such "sick" cells, then their activation may potentiate (via cytokines) the subsequent antigen-specific $\alpha\beta$ T-cell response that ultimately clears pathogens.

However, one should keep in mind that the collective observations outlined above provide, at best, suggestive evidence that class Ib molecules are involved in the presentation of peptides derived from stress proteins. To date, studies have not evolved to a point where a given T-cell receptor (α/β or γ/δ) has been shown to recognize a specific complex of a stress protein–derived peptide bound to a class Ib molecule. A piece of the puzzle seems always to be unavailable. It is possible that class Ib + hsp-reactive T cells do not exist, and we have been seduced by an alluring hypothesis. Alternatively, the tissue compartmentalization, growth properties, and/or low frequency of such cells and the low levels of class Ib expression have made the identification and characterization difficult. Only the future will tell.

REFERENCES

1. K. Fischer Lindahl, E. Hermel, B. Loveland, and C.R. Wang, Maternally transmitted antigen of mice: a model transplantation antigen, *Ann. Rev. Immunol. 9*:351 (1991).
2. J. Strominger, The T cell receptor and class Ib MHC-related proteins: enigmatic molecules of immune recognition, *Cell 57*:895 (1989).
3. R. Germain, MHC-dependent antigen processing and peptide presentation: providing ligands for T cell activation, *Cell 76*:287 (1994).
4. I. Stroynowski, Molecules related to class I-major histocompatibility complex antigens, *Ann. Rev. Immunol. 8*:501 (1990).
5. M.J. Soloski, L. Hood, and I. Stroynowski, Q-region class I gene expression: iden-

tification of a second Qa-region gene encoding a Qa-2 polypeptide, *Proc. Natl. Acad. Sci. USA 85*:3100 (1989).
6. G. Waneck, D. Sherman, P. Kincade, M. Low, and R. Flavell, Molecular mapping of signals in the Qa-2 antigen required for attachment of the phosphtidylinositol membrane anchor, *Proc. Natl. Acad. Sci. USA 85*:577 (1989).
7. H. Tian, F. Imani, F., and M.J. Soloski, Biophysical and molecular genetic analysis of Qa-2 gene expression, *Mol. Immunol. 28*:845 (1990).
8. P. Robinson, Qb-1, a new class I polypeptide encoded by the Qa region of the mouse H-2 complex, *Immunogenetics 22*:285 (198).
9. W.L. Maloy, J. Coligan, Y. Barra, and G. Jay, Detection of a secreted form of the murine H-2 class I antigen with an antibody against its predicted carboxyl terminus, *Proc. Natl. Acad. Sci. USA 81*:1216 (1984).
10. S. Watts, A.C. Davis, B. Gaut, C. Wheeler, L. Hill, and R. Goodenow, Organization and structure of the Qa genes of the major histocompatibility complex of the C3H mouse: implications for Qa function and class I evolution, *EMBO J. 8*:1749 (1989).
11. J. Klien, C. Benoist, C.S. David, P. Demant, K. Fischer-Lindahl, L. Flaherty, R.A. Flavell, U. Hammerling, L. Hood, S.W. Hunt III, P.P. Jones, P. Kourilsky, H.O. McDevitt, D. Meruelo, D. Murphy, S.G. Nathenson, D.H. Sachs, M. Steinmetz, S. Tonegawa, E.K. Wakeland, and E.H. Weiss, Revised nomenclature of mouse H-2 genes, *Immunogenetics 32*:147 (1990).
12. P. Wolf, R. Cook, The TL region gene 37 encodes a Qa-1 antigen, *J. Exp. Med. 172*:1795 (1990).
13. Y.T. Chen, Y. Obata, E. Stockert, T. Takahashi, and L. Old, Tla-region genes and their products, *Immunol. Res. 6*:30 (1987).
14. K. Ito, L.V. Kaer, M. Bonneville, S. Hsu, D. Murphy, and S. Tonegawa, Recognition of the product of a novel MHC TL region gene (27^b) by a mouse $\gamma\delta$ T cell receptor, *Cell 62*:549 (1990).
15. S. Richards, M. Bucan, K. Brorson, M.C. Kiefer, S.W. Hunt, M. Lehrach, M., and K. Fisher Lindahl, Genetic and molecular mapping of the *Hmt* region of the mouse, *EMBO J. 8*:3749 (1989).
16. C. Wang, B. Loveland, and K. Fischer Lindahl, H2-M3 encodes the MHC class I molecule presenting the maternally transmitted antigen of the mouse, *Cell 66*:335 (1991).
17. I. Stroynowski, M.J. Soloski, M. Low, and L.H. Hood, A single gene encodes soluble and membrane-bound forms of the major histocompatibility Qa-2 antigen: anchoring of the product by a phospholipid tail, *Cell 50*:759 (1987).
18. N. Ulker, K. Lewis, L. Hood, and I. Stroynowski, Activated T cells transcribe an alternatively spliced mRNA encoding a soluble form of the Qa-2 antigen, *EMBO J. 9*:3839 (1990).
19. J. Stiernberg, M.G. Low, L. Flaherty, and P.W. Kincade, Removal of lymphocyte surface molecules with phosphatidylinositol-specific phospholipase C: effects on mitogen responses and evidence that ThB and certain Qa antigens are membrane-anchored via phosphatidylinositol, *J. Immunol. 38*:3877 (1987).
20. P.J. Robinson, Two different pathways for the secretion of Qa region associated class I antigens by mouse lymphocytes, *Proc. Natl. Acad. Sci. USA 84*:527 (1987).

21. M. Wu, L. Kaer, S. Itohara, and S. Tonegawa, Highly restricted expression of the thymus leukemia antigens on intestinal epithelial cells, *J. Exp. Med. 174*:213 (1991).
22. J. Vernachio, M. Li, A. Donnenberg, and M. Soloski, Qa2 antigen expression in the adult thymus: a marker for thymocytes with a mature phenotype, *J. Immunol. 142*:48 (1989).
23. C. Aldrich, J. Rodgers, and R.R. Rich, Regulation of Qa-1 expression and determinant modification by an H-2D linked gene Qdm, *Immunogenetics 28*:334 (1988).
24. K. Fischer Lindahl, E. Hermal, E. Grigorenko, E. Jones, C. Wang, Molecular definition of the maternally transmitted antigen, MTA, *J. Cell. Biochem. 16D*:O005 (1992).
25. J. Michaelson, L. Flaherty, B. Hutchinson, and H. Yudkowitz, Qa-2 does not display structural genetic polymorphism detectable on isoelectric-focusing gels, *Immunogenetics 16*:363 (1982).
26. J. Tine, A. Walsh, D. Rathbun, L. Leonard, E. Wakeland, R. Dilwith, and L. Flaherty, Genetic polymorphisms of Q-region genes from wild-derived mice: implicatins for Q region evolution, *Immunogenetics 31*:315 (1990).
27. L. Flaherty, E. Elliott, J. Tine, A. Walsh, and J. Water, Immunogenetics of the Q and TL regions of the mouse, *Crit. Rev. Immunol. 10*:131 (1990).
28. E. Hermel, E. Gigorenko, and K. Fischer Lindahl, Expression of medial class I histocompatibility antigens on RMA-S mutant cells, *Int. Immunol. 3*:407 (1991).
29. M. Attaya, S. Jameson, C. Martinez, E. Hermel, C. Aldrich, J. Forman, K. Fischer Lindahl, M. Bevan, and J. Monaco, Ham-2 corrects the class I antigen-processing defect in RMA-S cells, *Nature 355*:647 (1992).
30. A. Morita, T. Takahashi, E. Stockert, E. Nakayama, T. Tsuji, Y. Matsudaira, L. Old, and Y. Obata, TL antigen as a transplantation antigen recognized by TL-restricted cytotoxic T cells, *J. Exp. Med. 179*:777 (1994).
31. O. Rotzschke, K. Falk, S. Faath, and H.G. Rammensee, On the nature of peptides involved in T cell alloreactivity, *J. Exp. Med. 174*:1059 (1991).
32. W.R. Heathm, M.E. Hurd, F.R. Carbone, and L.A. Sherman, Peptide-dependent recognition of H-2K^b by alloreactive cytotoxic T lymphocytes, *Nature 341*:749 (1989).
33. O. Rotzschke, K. Falk, S. Stefanovic, B. Grahovac, M.J. Soloski, G. Jung, and H.G. Rammensee, Qa-2 molecules are peptide receptors of higher stringency than ordinary class I molecules, *Nature 361*:642 (1992)
34. S. Joyce, P. Tabaczewski, R.H. Angeletti, S.G. Nathenson, and I. Stroynowski, A non-polymorphic MHC class Ib molecule binds a large array of diverse self peptides, *J. Exp. Med. 179*:579 (1994).
35. B. Loveland, C.R. Wang, H. Yonekawa, E. Hermal, and K. Fischer Lindahl, Maternally transmitted histocompatibility antigen of mice: a hydrophobic peptide of a mitochondrially encoded protein, *Cell 60*:971 (1990).
36. S. Shawar, R. Cook, J. Rodgers, and R.R. Rich, Specialized functions of MHC class I molecules, *J. Exp. Med. 171*:897 (1990).
37. R. Kurlander, S. Shawar, M. Brown, and R.R. Rich, Specialized role for a murine class I-b MHC molecule in prokaryotic host defenses, *Science 257*:678 (1992).
38. E. Pamer, C.R. Wang, L. Flaherty, K. Fischer-Lindahl, and M. Bevan, H2-M3

presents a Listeria monocytogenes peptide to cytotoxic T lymphocytes, *Cell 70*:215 (1992).
39. F. Imani and M.J. Soloski, Heat shock proteins can regulate expression of the T1a region encoded class Ib molecule, Qa1, *Proc. Natl. Acad. Sci. USA 88*:10475 (1991).
40. W. Born, L. Hall, A. Dallas, J. Boymel, T. Shinnick, D. Young, P. Brennan, and R. O'Brien, Recognition of a peptide antigen by heat-shock reactive γδ T lymphocytes, *Science 249*:67 (1990).
41. W., Van Eden, J.E. Thole, R. van der Zee, A. Noordzij, J.D. van Embden. E. Hensen, and I.R. Cohen, Cloning of the mycobacterial epitope recognized by T lymphocytes in adjuvant arthritis, *Nature 331*:171 (1988)
42. D. Vidovic and Z Dembic, Qa-1 Restricted γδ T Cells can Help B Cells, *Curr. Top. Microbiol. Immunol. 173*:239 (1991).
43. S. Moriwaki, B. Korn, Y. Ichikawa, L. Kaer, and S. Tonegawa, Amino acid substitutions in the floor of the putative antigen-binding site of H-2T22 affect recognition by a γδ T-cell receptor, *Proc. Natl. Acad. Sci. USA 90*:11396 (1993).
44. P. Eghtesady, C. Panwala, and M. Kronenberg, Recognition of the thymus Leukemia antigen by Intestinal γδ T Lymphocytes, *Proc. Natl. Acad. Sci. USA* (1994)
45. W. Havaran, L. Chien, and J. Allison, Recognition of self antigens by skin-derived T cells with invariant γδ antigen receptors, *Science 52*:1430 (1991).

36

Polyclonal Responses of γδ T Cells to Heat Shock Proteins

Willi Born, Mary Ann DeGroote, Yang-Xin Fu, Christina Ellis Roark, Kent Heyborne, Harshan Kalataradi, Katherine A. Kelly, Christopher Reardon, and Rebecca O'Brien
National Jewish Center for Immunology and Respiratory Medicine and University of Colorado Health Sciences Center, Denver, Colorado

I. INTRODUCTION

T lymphocytes expressing γδ T-cell receptors (γδ T cells) were found 9 years ago, some time after the discovery of T-cell receptor γ genes (1,2). Populations of these cells have since been detected in many species. In the meantime, it has become quite clear that γδ T cells play an important role in the body's defenses against a variety of infectious pathogens (3). The details of their contribution to host resistance are still a matter of speculation. Responses of γδ T cells have been noted under various pathological conditions, including malignancies, autoimmunity, and tissue transplantation, as well as in bacterial, viral, and parasitic infections (4). In experimental infections of mice with *Listeria monocytogenes, Salmonella typhimurium,* or *Plasmodium yoelii*, host resistance was found to depend on endogenous γδ T cells (5–8; unpublished observations).

Little is known about how γδ T cells mediate protection. They are capable of producing a variety of cytokines, perhaps most notably interferon gamma but they are also capable of producing interleukin-2 (IL-2), IL-3, IL-4, IL-6, IL-10, granulocyte-macrophage colony-stimulating factor, tumor necrosis factor α (TNF-α), and transforming growth factor β (TGF-β), suggesting that they interact with other components of the immune system (9; our unpublished observations). Our recent studies suggest that such interactions also involve inflammatory cells, aid in the regulation of the inflammatory response, and result in reduced concomitant tissue destruction (our unpublished data). Others have reported evidence for interactions between γδ and αβ T cells (10–12), as well as between γδ T cells and B lymphocytes (13).

Most γδ T cells express polymorphic T-cell receptor (TCR) molecules, which in many ways resemble those of αβ T cells (14). A reasonable assumption then seems to be that γδ T-cell functions are triggered through specific receptor-ligand interactions. To identify and characterize such interactions, however, has been exceedingly difficult. In vivo immunizations with soluble antigens readily eliciting αβ T-cell responses have usually failed to produce γδ T-cell reactivity, with occasional exceptions (15,16). Alloimmunization has led to the isolation of major histocompatibility complex (MHC) class I/II–reactive γδ T-cell clones (17,18), but these cells appear to be less frequent than alloreactive αβ T cells (19) both in absolute number and relative to the population of γδ T cells, and they do not seem to recognize the same portion of the MHC molecule as do αβ T cells (20). Therefore, the deduction that γδ T cells, like αβ T cells, typically rely on MHC-encoded molecules as presenters of antigen is likely to be incorrect.

Lacking an efficient strategy for the identification of γδ T-cell ligands, researchers have searched for responses of γδ T cells to defined soluble antigens. Reports of antigens stimulating clonal responses have been scarce, and the biological significance of the responses questionable (4). On the other hand, polyclonal responses to a variety of stimuli have been noted in both murine and human γδ T cells (19,21–25). These responses differ from those of αβ T cells to conventional antigens in that they involve γδ T-cell subsets defined by expression of certain TCR-γ chains or γ/δ chain pairs. Stimulators of such polyclonal responses include cultured cell lines, killed bacteria, and crude bacterial antigen preparations, as well as purified molecules such as recombinant heat shock proteins, synthetic peptides, and nonproteinaceous substances of low molecular weight (4). Because the polyclonal responses have been observed among both murine and human γδ T cells, indicating their probable evolutionary conservation, they are unlikely to merely represent a laboratory artifact. However, evidence for a biological role of the polyclonal reactivity is still missing.

II. RESPONSES OF γδ T CELLS TO 60- AND 70-KDA HEAT SHOCK PROTEIN

Two γδ T-cell subsets in mice express nearly monomorphic TCRs. One of these subsets resides as intraepithelial T lymphocytes of the skin and the other of the uterus and tongue (26,27). To account for the lack of receptor diversity in these cells, instead of recognizing polymorphic foreign antigens like other T cells, they have been hypothesized to respond to invariant antigens such as stress signals of autologous cells, which can be induced by trauma, malignancy, or infection (28). Providing experimental proof of this attractive idea has been difficult, although a recent study has shown that epidermal $\gamma\delta$ T cells can respond to culture- or arsonate-stressed keratinocytes (29). In a different experimental system, however, it became evident much sooner that $\gamma\delta$ T cells can respond to stress associated proteins. Testing randomly generated collections of $\gamma\delta$ T-cell hybridomas derived from cells of newborn mouse thymus, adult mouse spleen, and various other tissues, we found that the majority of $V\gamma1^+$ cells could be stimulated to cytokine secretion in the presence of recombinant mycobacterial heat shock protein 60 (hsp 60) or synthetic peptides derived from this stress protein (19,30–33). The response is specific, and its dependence on the TCR has been confirmed by TCR gene transfection. $\gamma\delta$ T cells exhibiting polyclonal reactivity with hsp 60 are present in major lymphoid and many nonlymphoid tissues, including liver (34), placenta (35), small intestine (36), and even skin (37). We have found such hsp 60–reactive $\gamma\delta$ T cells, which bear polyclonal receptors but are invariably $V\gamma1^+$, in all mouse strains tested so far with the exception of AKR mice. The absence of reactivity in this mouse strain is based on a genetic polymorphism within Vγ1-J4C4 (our unpublished observations).

Several lines of evidence indicate that murine hsp 60–reactive $\gamma\delta$ T cells are triggered in vivo under various pathological conditions. We found that after repeated immunization with a peptide covering amino acids 180–196 of mycobacterial hsp 60 (p180–196), $V\gamma1^+$ T cells increased in relative frequency (among $\gamma\delta$ T-cell hybridomas) and showed changes in TCR composition (32). Specifically, whereas most $V\gamma1^+$ hsp 60–reactive cells in normal mice (C57BL/10) expressed Vδ6.3, the majority of such cells in immunized mice expressed Vδ6.1. Notably, immunization with the intact hsp 60 expanded no $\gamma\delta$ T cells recognizing other portions of the hsp 60 molecule, although $\alpha\beta$ T cells specific for various portions of hsp 60 were triggered. Thus, either murine $\gamma\delta$ T cells recognize only a single region of this molecule, or one region of the molecule easily triggers the polyclonal response of an already prevalent $\gamma\delta$ T-cell subset. That hsp 60–reactive $\gamma\delta$ T cells become activated during experimental listeriosis seems likely, because others have reported that $\gamma\delta$ T cells from infected mice showed enhanced reactivities with mycobacterial hsp 60 (5). Whether these $\gamma\delta$ T cells are initially stimulated by pathogen- or by host-derived hsp 60, or perhaps by another ligand, has still to be resolved. However, several observations suggest that autologous hsp 60 can stimulate murine $\gamma\delta$ T-cell reactivity, both in vitro and in vivo. Responses of $\gamma\delta$ T cells to stressed autologous lymph node cells have been noted, but no specific antigen has been implicated as stimulator (38). Hsp 60–reactive $\gamma\delta$ T-cell hybridomas show spontaneous autoreactivity (19), and this TCR-dependent response can be increased by a mild heat stress (D. Young, unpublished data). Finally, we have noted a weak response of the hsp 60–reactive hybridomas toward a peptide covering the portion of human/mouse hsp 60 equivalent to amino acids 180–196 of mycobacterial hsp 60 (31).

An hsp 60–reactive $\gamma\delta$ T-cell clone (291-H4) has also been isolated from a *P. yoelii*–infected mouse (6). This clone is of particular interest because it conveys partial protection against this parasite to adoptive transfer recipients. In contrast, another $\gamma\delta$ T-cell clone from the same source but not reactive with hsp 60 failed to protect against the parasite. The clones have not yet been further characterized, and it is still unclear whether the hsp 60–reactive clone bears a TCR of the type previously described for the hsp 60–reactive subset.

In humans, the first observations of $\gamma\delta$ T-cell responses to hsp 60 were made with $\gamma\delta$ T-cell lines. In one study, a $CD4^-8^-$ $\gamma\delta$ T-cell clone isolated from the synovial fluid of a patient with rheumatoid arthritis was found to respond to mycobacterial hsp 60 (39). In another study, a $CD4^-8^-$ $\gamma\delta$ T-cell line from a healthy but BCG-immune individual also reacted with mycobacterial hsp 60 in the presence of autologous accessory cells (40). A clone later derived from this line also responded to a recombinant human homologue of mycobacterial hsp 60; again requiring the presence of autologous accessory cells (41). In a third study by another group, human $\gamma\delta$ T cells, like mouse $\gamma\delta$ T cells, were shown to respond to hsp 60 polyclonally (42,43). In a large collection of human $\gamma\delta$ T-cell clones isolated from peripheral blood lymphocytes of healthy individuals by cell sorting and limiting dilution cloning on autologous feeder cells and Epstein-Barr virus (EBV)–transformed lymphoblastoid B-cell lines, $V\gamma9/V\delta2^+$ clones were found specifically to lyse the Burkitt's lymphoma cell line Daudi. The response to Daudi cells could be inhibited with a anti-TCR monoclonal antibody (MAb). The response could also be inhibited with both a polyclonal rabbit antiserum reactive with human hsp 60 and with the monoclonal mouse antibody II-13, which was raised against recombinant human mitochondrial hsp 60. II-13 also specifically stained Daudi cells. Both reagents precipitated from these cells molecules which had electrophoretic mobility patterns consistent with hsp 60. A similar observation had been earlier reported by others examining five $\gamma\delta$ T-cell clones isolated from peripheral blood of patients with active lupus nephritis (44). Responses of these cells to autologous EBV-transformed B-cell lines could be blocked by two anti–hsp 60 MAbs, IIH-9 and IIIE-9. Strong inhibition was found with all $\gamma\delta$ T-cell clones tested, whereas only one of two $CD4^-8^-$ $\alpha\beta$ T-cell clones and one of six $CD4^+$ $\alpha\beta$ T-cell clones were partially inhibited in their response to the same B lymphoma. However, none of the B lymphoma–reactive cells responded to soluble mycobacterial hsp 60.

Unlike the polyclonal response of murine $V\gamma1^+$ cells, the polyclonal response of human $V\gamma9/V\delta2^+$ cells appears to be triggered mainly by the autologous hsp 60 homologue (42,43), whereas human clones reactive with mycobacterial hsp 60 are comparatively infrequent (45). This was unexpected given that both types of $\gamma\delta$ T cells respond polyclonally to crude mycobacterial antigen preparations (19,42). The polyclonal responses to the mycobacterial antigen preparations now seem likely to be stimulated by some other molecule, at least in the case of human γδ T cells (46,47). This may pertain only to the Vγ9/Vδ2$^+$ subset, because a study with human synovial T-cell populations suggests that Vδ1$^+$ γδ T cells, frequent at this site and infrequent in peripheral blood, do respond to mycobacterial hsp 60 (48). Despite their differences, the response patterns of human and murine hsp 60–reactive γδ T cells appear to be similar. In each case, polyclonal responses to a single antigen by an entire subset of γδ T cells are the hallmark, but clonally distributed specificities may also exist. On the other hand, these subset responses of both murine and human γδ T cells are different from any of the antigen responses observed with αβ T cells (49) and thus should attract special attention (see below).

More recently, γδ T cells from murine and human sources were also discovered to respond to 70-kDa heat shock proteins (25,50). So far, comparatively little is known about these responses. In a study on murine intestinal intraepithelial lymphocytes (IELs), proliferative responses to mycobacterial hsp 70 of both αβ and γδ T cells were noted (50). Because these responses could be detected with cell populations that had been purified by sorting, they are likely to be independent of each other. We have recently observed a response of cloned γδ T cells to mycobacterial hsp 70 (C.R., unpublished data); thus furthering the idea that the γδ T-cell response to hsp 70 is independent of the presence of αβ T cells.

The response of intestinal IELs to mycobacterial hsp 70 was detected in three different mouse strains (C3H/HeN, BALB/c, and C57BL/6), although purified γδ T cells were only tested from one mouse strain, C3H/HeN. The responses could not be ascribed to residual endotoxin, because the level within the hsp 70 preparation was low, and far larger amounts of *Escherichia coli*–derived lipopolysaccharide (LPS) induced comparatively weak proliferative reactivity. The response of intestinal IELs to hsp 70 is remarkable also, because these cells fail to respond to other stimuli such as concanavalin A, phorbol ester, and calcium ionophore or TCR–cross-linking MAbs. Based on the strength of the proliferation in primary stimulation cultures, one can assume that responses of both αβ and γδ T cells to hsp 70 are polyclonal. The observed reactivity may thus be different from previously described conventional antigen responses of αβ T cells to mycobacterial hsp 70. In fact, for both polyclonal αβ and γδ T-cell responses to hsp 70, their TCR dependency remains to be shown.

The observation that human γδ T cells can respond to hsp 70 was made in a study on human leishmaniasis (25). Patients with cutaneous, mucosal, and visceral leishmaniasis were found to have elevated levels of γδ T cells in their peripheral blood. These responses parallel polyclonal responses of murine γδ T cells to experimental infections with *Leishmania major*. The responses of human γδ T cells appeared to be driven only by certain antigens of the parasites, because blood lymphocyte cultures from infected patients stimulated with promastigote lysates of *L. amazoniensis* and *L. braziliensis* contained up to 60%

αβ T cells. In contrast, two surface antigens of *L. amazonensis*, gp 63 and gp 42, did not expand γδ T cells, although both antigens elicited strong γδ T-cell responses. γδ T cells enriched by cell sorting from a *Leishmania*-specific T-cell line responded to *L. amazonensis* lysate, suggesting that their reactivity is independent of αβ98 T cells. Most interestingly, γδ T cells derived form peripheral blood mononuclear cells of a patient with mucosal leishmaniasis were found to respond to stimulation with a recombinant hsp 70 derived from *L. chagasi*. Again, this stimulation experiment was carried out with sorted cell populations, suggesting that the response of human γδ T cells to hsp 70, like that of murine γδ T cells, is independent of αβ T cells. Human γδ T cells reactive with hsp 70 or other leishmanial antigens have not yet been characterized in terms of TCR expression. However, in mice, experimental infections with *L. major* led to selective expansion of $V\delta4^+$ cells (51), suggesting that the γδ T-cell responses to leishmanial antigens are TCR dependent.

III. MECHANISM OF THE γδ T-CELL RESPONSES TO HSPS

The precise mechanisms of any of the antigen responses of γδ T cells await further clarification. In the case of the polyclonal antigen responses, it seems clear that the γδ T-cell reactivity has different requirements than those of either conventional antigen or superantigen responses of αβ T cells. These include requirements on the TCR, the stimulators (antigens), and the accessory cells. The polyclonal response of murine γδ T cells to hsp 60 is still the best-studied example. The response has now been clearly shown to be TCR dependent by TCR gene transfer (52). Moreover, a recent TCR chain shuffling experiment has confirmed that the reactivity is dependent on the TCR-γ chain (our unpublished data). The portion of TCR-γ required for hsp 60 reactivity has not yet been unequivocally determined but indirect evidence suggests that the constant portion of the TCR chain or junctional sequences are not critical. Instead, certain amino acid positions within Vγ1 are likely to be essential. The role played by TCR-δ is not yet established, but the lesser constraints on TCR-δ expression in hsp 60–reactive cells suggest a comparatively minor contribution of this TCR chain (53).

The requirements for the polyclonal response of human γδ T cells to hsp 60 have not yet been studied in such detail, but the available data indicate again similarities with the murine response. Thus, all reactive cells express Vγ9 suggestive of a strict dependency on TCR-γ (42). Human Vγ9 is typically coexpressed with Vγ2, reminiscent of the preferential expression of Vδ6 in association with Vγ1 in the case of murine hsp 60–reactive cells. However, as is the case with murine cells, the requirement for TCR-δ in human cells coes not seem to be strict, because responses to hsp 60 have also been noted with $V\delta1^+$ cells (48). A possible difference between Vδ2 and $V\delta1^+$ subsets may be that the former respond to only the human hsp 60 (expressed, e.g., on the surface of Daudi cells), whereas the latter are also stimulated by mycobacterial hsp 60. Although all of these data indicate that γδ T-cell responses to hsp 60 require the TCR, they do not yet address the question of TCR ligand interactions.

The molecular form of hsp 60 stimulating polyclonal γδ T-cell responses is still unknown. With murine γδ T cells, we found that a small synthetic peptide corresponding to amino acids 181–187 of mycobacterial hsp 60 (p181–187) could substitute for the whole molecule inasmuch as the peptide stimulated the same γδ T cells (52). In analogy to the antigens of αβ T cells, this could be taken to indicate that hsp 60 is recognized by γδ T cells in the form of processed and presented protein fragments. In this regard, it is noteworthy that the responses to the peptide could be both diminished and enhanced by appropriate amino acid substitutions, and individual hybridomas of the hsp 60–reactive type responded differently to these changes. Such clonal differences in reactivity patterns are consistent with direct cognate interactions between the TCR and the hsp 60–derived peptide and an involvement of TCR junctional sequences. However, whether the peptide is recognized either bound to an antigen-presenting molecule or by itself, perhaps representing a surface-exposed epitope of hsp 60 and thus mimicking the intact molecule, is unclear. Finally, since direct binding of the peptide or of hsp 60 to the γδ TCR has not yet been demonstrated, these stimulators may possibly merely facilitate cognitive interactions with other, unknown ligands.

So far, evidence that human γδ T cells polyclonally recognize hsp 60 is based mostly on antibody-blocking experiments (42–44). Several anti–hsp 60 MAbs could inhibit γδ T-cell responses to B lymphoma cells. This is most easily explained assuming direct cognate interactions between cell surface–expressed native hsp 60 and TCR-γδ. Similar to the response of human γδ T cells to B lymphomas, murine hsp 60–reactive γδ T-cell hybridomas respond to trophoblasts and trophoblast cell lines (our unpublished data) in a TCR-dependent fashion (verified by TCR gene transfection). Furthermore, the response seems to be directed against an evolutionarily conserved determinant, because a human trophoblast cell line is also recognized. Blocking of the trophoblast response using anti–hsp 60 antibodies have not yet been carried out. Despite this, similarities between the mouse and the human system are tangible. By extrapolation, we would therefore predict that peptides stimulating hsp 60–reactive human γδ T-cell clones will also be found. So far, no such data have been reported, but this could be due to other still ill-defined requirements. We have found that although murine γδ T-cell hybridomas readily respond to these stimulators, normal cells do not respond unless they are already activated in vivo (32). In contrast, normal murine hsp 60–reactive cells appear to respond to trophoblast cells (35). Similarly, direct stimulation of normal human T cells or clones with peptides or hsp 60 may be difficult if these stimulators can only be "seen" in the context of certain other markers of cell activation or stress. Such limitations may be less stringent for some γδ T cells, however, because human Vγ1$^+$ populations and occasional clones did respond to soluble hsp 60 (6,39,40,48).

A common property of all polyclonal responses to hsp 60 appears to be the lack of requirements for MHC class I/II–dependent antigen presentation. In the case of hsp 60–reactive murine γδ T-cell hybridomas, we have so far not found any convincing evidence of antigen presentation (19; unpublished data). The hybridomas responded to hsp 60 in the absence of any accessory cells. The responders themselves did not express MHC class II. Using a number of different anti–MHC class I antibodies, we have thus far failed to inhibit the reactivity of the hybridomas. Reagents used included antibodies against Qa-1 and TL. Although it is conceivable that other anti–MHC class I antibodies would have been more effective, we also found that the hsp 60–reactive hybridomas, and no other γδ T-cell hybridomas, responded to trophoblast cells (unpublished data). This TCR-dependent response occurred even if the trophoblasts, owing to a mutation in β_2-microglobulin, fail to express class I. This observation is very similar to another with human hsp 60–reactive γδ T cells (see below). Finally, we have been able to isolate hsp 60–reactive γδ T cells (in the form of hybridomas) from both neonatal thymus and adult spleen of β_2-microglobulin knockout mice, indicating that hsp 60–reactive murine γδ T cells can develop in the absence of all β_2-microglobulin–associated class I molecules (unpublished). These findings are consistent with the general absence of gross changes in γδ T-cell populations of MHC class I–deficient or class II–deficient mice, as reported by others (54,55). The majority of γδ T cells thus appear to develop without the need of MHC-dependent selection.

Similarly to the findings with murine cells, no MHC restriction of human hsp 60–reactive γδ T cells has been detected (42,56). Thus, Vγ9/Vδ2$^+$ T-cell clones of individuals with different MHC types all responded to Daudi cells by proliferation and cytotoxicity and without prior exposure to these cells. Daudi cells do not express HLA class I molecules because of a mutation of the β_2-microglobulin mRNA initiation codon. The response to the Daudi cells was not affected under serum-free conditions; a step undertaken to avoid the possibility that serum β_2-microglobulin restored the transort of HLA class I molecules to the cell surface. Lack of class I expression, a marker for susceptibility to natural killer cell lysis, was not the trigger of the γδ T-cell response, because Daudi cells expressing a transfected β_2-microglobulin gene were still stimulatory. The response of Vγ9/Vδ2–positive cells to the Daudi lymphoma appears to be also independent of MHC class II expression, because anti–class II MAbs were not inhibitory (43), whereas strong inhibition was noted with anti hsp 60 reagents.

In sum, the available data suggest a unique mechanism for polyclonal responses of γδ T-cell subsets to hsp 60 that differs from the responses of αβ T cells to both conventional antigens and superantigens. Further investigations should clarify whether hsp 60 is recognized during these responses in its native conformation or in a somehow processed and perhaps presented form. Requirements for γδ T-cell responses to hsp 70 have not yet been analyzed in similar detail. However, the polyclonality of both the γδ T-cell responses to hsp 60 and hsp 70 may suggest that they are subject to similar rules.

IV. FUTURE DIRECTIONS

Although we begin to develop a sense for the unique quality of the polyclonal antigen responses of γδ T-cell subsets and the responses to hsp 60 and hsp 70 in particular, what if any biological functions are associated with this type of reactivity remains unclear. Experimental manipulation of hsp 60–reactive γδ T-cell populations under controlled conditions in vivo, for example, in mice expressing the transgenic γδ TCR of a hsp 60–reactive clone, may be helpful in studying effector cell responses. Another issue in need of resolution is the terms and conditions of cells' surface expression of hsp 60 and hsp 70, especially with regard to the polyclonal γδ T-cell responses to these antigens, in which recognition of processed and presented peptides does not seem to pay a role. Here, short of understanding the physiological mechanisms of hsp surface expression, one could artificially direct expression to the cell surface and test for the immunological consequences, in particular with regard to γδ T cells. As the evidence for a contribution of γδ T cells to immune protection against infectious diseases in increasing, these questions become ever more attractive research problems.

REFERENCES

1. H. Saito, D.M. Kranz, Y. Takagaki, A. Hayday, H. Eisen, and S. Tonegawa, Complete primary structure of a heterodimeric T-cell receptor deduced from cDNA sequences, *Nature 309*:757 (1984).
2. M.B. Brenner, J. McLean, D.P. Dialynas, J.L. Strominger, J.A. Smith, F.L. Owen, J.G. Seidman, S. Ip, F. Rosen, and M.S. Krangel, Identification of a putative second T-cell receptor, *Nature 322*:145 (1986).
3. W.K. Born, K. Harshan, R.L. Modlin, and R.L. O'Brien, The role of γδ T lymphocytes in infection, *Curr. Opin. Immunol. 3*:455 (1991).
4. W. Born, R.L. O'Brien, and R. Modlin, Antigen specificity of γδ T lymphocytes, *F.A.S.E.B. J. 5*:2699 (1991).
5. K. Hiromatsu, Y. Yoshikai, G. Matsuzaki, S. Ohga, K. Muramori, K. Matsumoto, J.A. Bluestone, and K. Nomoto, A protective role of γ/δ T cells in primary infection with Listeria monocytogenes in mice, *J. Exp. Med. 175*:49 (1992).
6. M.Tsuji, P. Mombaerts, L. Lefrancois, R.S. Nussenzweig, F. Zavala, and S. Tone-

gawa, γδ T cells contribute to immunity against the liver stages of malaria in αβ T-cell-deficient mice, *Proc. Natl. Acad. Sci. USA 91*:345 (1994).
7. M.J. Skeen, and H.K. Ziegler, Induction of murine peritoneal γ/δ T cells and their role in resistance to bacterial infection, *J. Exp. Med. 178*:971 (1993).
8. P. Mombaerts, J. Arnoldi, F. Russ, S. Tonegawa, and S.H.E. Kaufmann, Different roles of αβ and γδ T cells in immunity against an intracellular bacterial pathogen, *Nature 365*:53 (1993).
9. C.T. Morita, S. Verma, P. Aparicio, C. Martinez-A, H. Spits, and M.B. Brenner, Functionally distinct subsets of human γ/δ T cells. *Eur. J. Immunol. 21*:2999 (1991).
10. D.A. Ferrick, S.R. Sambhara, W. Ballhausen, A. Iwamoto, H. Pircher, C.L. Walker, W.M. Yokoyama, R.G. Miller, and T.W. Mak, T cell function and expression are dramatically altered in T cell receptor Vγ1.1Jγ4Cγ4 transgenic mice, *Cell 57*:483 (1989).
11. S.H.E. Kaufmann, C. Blum, and S. Yamamoto, Crosstalk between α/β T cells and γδ T cells in vivo: activation of α/β T-cell responses after γ/δ T-cell modulation with the monoclonal antibody GL3, *Proc. Natl. Acad. Sci. USA 90*:9620 (1993).
12. M.J. Skeen, and H.K. Ziegler, Intercellular interactions and cytokine responsiveness of peritoneal α/β and γ/δ T cells form Listeria-infected mice: synergistic effects of interleukin 1 and 7 on γ/δ T cells, *J. Exp. Med. 178*:985 (1993).
13. P. Quere, M.D. Cooper, and G.J. Thorbecke, Characterization of suppressor T cells for antibody production by chicken spleen cells. I. Antigen-induced suppressor cells are $CT8^+$, $TcR1^+$ (γδ) T cells, *Immunology 71*:517 (1990).
14. W. Born, K.A. Kelly, and R.L. O'Brien, γδ T cells *Handbook of B and T Lymphocytes* (E.C. Snow, ed.), San Diego, 1994, pp. 179–214.
15. D. Vidovic, M. Roglic, K. McKune, S. Guerder, C. MacKay, and Z. Dembic, Qa-1 restricted recognition of foreign antigen by a γδ T-cell hybridoma, *Nature 340*:646 (1989).
16. R.M. Johnson, D.W. Lancki, A.I. Sperling, R.F. Dick, P.G. Spear, F.W. Fitch, and J.A. Bluestone, A murine $CD4^-$, $CD8^-$ T cell receptor–γδ T lymphocyte clone specific for Herpes simplex virus glycoprotein I, *J. Immunol. 148*:983 (1992).
17. L.A. Matis, R. Cron, and J.A. Bluestone, Major histocompatibility complex–linked specificity of γδ receptor-bearing T lymphocytes, *Nature 330*:262 (1987).
18. L.A. Matis, A.M. Fry, R.Q. Cron, M.M. Cotterman, R.F. Dick, and J.A. Bluestone, Structure and specificity of a class II alloreactive γδ T cell receptor heterodimer, *Science 245*:746 (1989).
19. R.L. O'Brien, M.P. Happ, A. Dallas, E. Palmer, R. Kubo, and W.K. Born, Stimulation of a major subset of lymphocytes expressing T cell receptor γδ by an antigen derived from Mycobacterium tuberculosis, *Cell 57*:667 (1989).
20. H. Schild, N. Mavaddat, C. Litzenberger, E.W. Ehrich, M.M. Davis, J.A. Bluestone, L. Matis, R.K. Draper, and Y.-H. Chien, The nature of major histocompatibility complex recognition by γδ T cells, *Cell 76*:29 (1994).
21. P. Fisch, M. Malkovsky, E. Braakman, E. Sturm, R.L.H. Bolhuis, A. Prieve, J.A. Sosman, V.A. Lam, and P.M. Sondel, γ/δ T-cell clones and natural killer–cell clones mediate distinct patterns of non-major histocompatibility complex–restricted cytolysis, *J. Exp. Med. 171*:1567 (1990).
22. G. Panchamoorthy, J. McLean, R.L. Modlin, C.T. Morita, S. Ishikawa, M.B.

Brenner, and H. Band, A predominance of the T cell receptor of Vγ2/Vδ2 subset in human mycobacteria-responsive T cells suggests germline gene encoded recognition, *J. Immunol. 147*:3360 (1991).

23. P. DePaoli, D. Gennari, P. Martelli, V. Cavarzerani, Comoretto, and G. Santini, γδ T cell receptor-bearing lymphocytes during Epstein/Barr virus infection, *J. Infect. Dis. 161*:1013 (1990).
24. P. DePaoli, D. Gennari, P. Martelli, G. Basaglia, M. Crovatto, S. Battistin, and G. Santini, A subset of γδ lymphocytes is increased during HIV-1 infection, *Clin. Exp. Immunol. 83*:187 (1991).
25. D.M. Russo, R.J. Armitage, M. Barral-Netto, A. Barral, K.H. Grabstein, and S.G. Reed, Antigen-reactive γδ T cells in human leishmaniasis, *J. Immunol. 151*:3712 (1993).
26. D.M. Asarnow, W.A. Kuziel, M. Bonyhadi, R.E. Tigelaar, P.W. Tucker, and J.P. Allison, Limited diversity of γδ antigen receptor genes of Thy-1$^+$ dendritic epidermal cells, *Cell 55*:837 (1988).
27. S. Itohara, A.G. Farr, J.J. Lafaille, M. Bonneville, Y. Takagaki, W. Haas, and A. Tonegawa, Homing of a γδ thymocyte subset with homogeneous T-cell receptors to mucosal epithelia, *Nature 343*:754 (1990).
28. C.A. Janeway, Jr., B. Jones, and A. Hayday, Specificity and function of T cells bearing γδ receptors, *Immunol. Today 9*:73 (1988).
29. W.L. Havran, Y.-H. Chien, and J.P. Allison, Recognition of self antigens by skin-derived T cells with invariant γδ antigen receptors, *Science 252*:1430 (1991).
30. M.P. Happ, R.T. Kubo, E. Palmer, W.K. Born, and R.L. O'Brien, Limited receptor repertoire in a mycobacteria-reactive subset of γδ T lymphocytes, *Nature 342*: 696 (1989).
31. W. Born, L. Hall, A. Dallas, J. Boymel, T. Shinnick, D. Young, P. Brennan, and R. O'Brien, Recognition of a peptide antigen by heat shock reactive γδ T lymphocytes, *Science 249*:67 (1990).
32. Y.-X. Fu, R. Cranfill, M. Vollmer, R. van der Zee, R.L. O'Brien, and W. Born, In vivo response of murine γδ T cells to a heat shock protein–derived peptide, *Proc. Natl. Acad. Sci. USA 90*:322 (1993).
33. R.L. O'Brien, Y.-X. Fu, R. Cranfill, A. Dallas, C. Reardon, J. Lang, S. R. Carding, R. Kubo, and W. Born, 1992. Heat shock protein Hsp-60 reactive γδ cells: a large, diversified T lymphocyte subset with highly focused specificity, *Proc. Natl. Acad. Sci. USA 89*:4348 (1992).
34. C.E. Roark, M.K. Vollmer, R.L. Cranfill, S.R. Carding, W.K. Born, and R.L. O'Brien, Liver γδ T cells: TCR junctions reveal differences in HSP-60 reactive cells in liver and spleen, *J. Immunol. 150*:4867 (1993).
35. K.D. Heyborne, R.L. Cranfill, S.R. Carding, W.K. Born, and R.L. O'Brien, Characterization of γδ T lymphocytes at the maternal-fetal interface, *J. Immunol. 149*: 2872 (1992).
36. C. Nagler-Anderson, L.A. McNair, and A. Cradock, Self-reactive, T cell receptor-γδ$^+$, lymphocytes from the intestinal epithelium of weanling mice, *J. Immunol. 149*:2315 (1992).
37. C.L. Reardon, M. Vollmer, R. Cranfill, R.L. O'Brien, and W.K. Born, Response

of a murine epidermal Vγ1/Vδ6-TCR$^+$ hybridoma to heat shock protein, HSP-60, *J. Infect. Dis.* (in press).
38. R. Rajasekar, G. Sim, and A. Augustin, Self heat shock and γδ T-cell reactivity, *Proc. Natl. Acad. Sci. USA 87*:1767 (1990).
39. J. Holoshitz, F. Koning, J.E. Coligan, J. De Bruyn, and S. Strober, Isolation of CD4$^-$ CD8$^-$ mycobacteria-reactive T lymphocyte clones from rheumatoid arthritis synovial fluid, *Nature 339*:226 (1989).
40. A. Haregewoin, G. Soman, R.C. Hom, and R.W. Finberg, Human γδ T cells respond to mycobacterial heat-shock protein, *Nature 340*:309 (1989).
41. A. Haregewoin, B. Singh, R.S. Gupta, and R.W. Finberg, A mycobacterial heat shock protein responsive γδ T cell clone also responds to the homologous human heat shock protein: a possible link between infection and autoimmunity, *J. Infect. Dis. 163*:156 (1990).
42. P. Fisch, M. Malkovsky, B.S. Klein, L.W. Morrissey, S.W. Carper, W.J. Welch, and P.M. Sondel, Human Vγ9/Vδ2 T cells recognize a groEL homolog on Daudi Burkitt's lymphoma cells, *Science 250*:1269 (1990).
43. I. Kaur, S.D. Voss, R.S. Gupta, K. Schell, P. Fisch, and P.M. Sondel, Human peripheral γδ T cells recognize hsp60 molecules on Daudi Burkitt's lymphoma cells, *J. Immunol. 150*:2046 (1993).
44. S. Rajagopalan, T. Zordan, G.C. Tsokos, and S.K. Datta, Pathogenic anti-DNA autoantibody-inducing T helper cell lines from patients with active lupus nephritis: Isolation of CD4$^-$ CD8$^-$ T helper cell lines that express the γδ T-cell antigen receptor, *Proc. Natl. Acad. Sci. USA 87*:7020 (1990).
45. D. Kabelitz, A. Bender, S. Schondelmaier, B. Schoel, and S.H.E. Kaufmann, A large fraction of human peripheral blood γδ+ T cells is activated by Mycobacterium tuberculosis but not by its 65kD heat shock protein, *J. Exp. Med. 171*:667 (1990).
46. K. Pfeffer, B. Schoel, H. Gulle, S.H.E. Kaufmann, and H. Wagner, Primary responses to human T cells to mycobacteria: a frequent set of γ/δ T cells are stimulated by protease-resistant ligands, *Eur. J. Immunol. 20*:1175 (1990).
47. K. Pfeffer, B. Schoel, N. Plesnila, G.B. Lipford, S. Kromer, K. Deusch, and H. Wagner, A lectin-binding, protease-resistant mycobacterial ligand specifically activates Vγ9$^+$ human γδ T cells, *J. Immunol. 148*:575 (1992).
48. K. Söderström, E. Halapi, E. Nilsson, A. Grönberg, J. Van Embden, L. Klareskog, and R. Kiessling, Synovial cells responding to a 65-kDa mycobacterial heat shock protein have a high proportion of a TcRγδ subtype uncommon in peripheral blood, *Scand. J. Immunol. 32*:503 (1990).
49. W. Born, Y.-X. Fu, H. Kalataradi, T. Ellis, C. Reardon, K. Heyborne, and R. O'Brien, Germ-line-encoded recognition of certain short peptide antigens?, *Chem. Immunol. 55*:185 (1992).
50. K.W. Beagley, K. Fujihashi, C.A. Black, A.S. Lagoo, M. Yamamoto, J.R. McGhee, and H. Kiyono, The *Mycobacterium tuberculosis* 71-kDa heat-shock protein induces proliferation and cytokine secretion by murine gut intraepithelial lymphocytes, *Eur. J. Immunol. 23*:2049 (1993).
51. J.-P. Rosat, M. Schreyer, T. Ohteki, G.A. Waanders, H.R. MacDonald, and J.A.

Louis, Selective expansion of activated Vδ4$^+$ cells during experimental infection of mice with *Leishmania major*, *Eur. J. Immunol. 24*:496 (1994).
52. Y.-X. Fu, G. Kersh, M. Vollmer, H. Kalataradi, K. Heyborne, C. Reardon, C. Miles, R. O'Brien, and W. Born, Structural requirements for peptides that stimulate a subset of γδ T cells, *J. Immunol. 152*:1578 (1994).
53. R.L. O'Brien, M.P. Happ, A. Dallas, R. Cranfill, L. Hall, J. Lang, Y.-X. Fu, R. Kubo, and W. Born, Recognition of a single HSP-60 epitope by an entire subset of γδ T lymphocytes, *Immunol. Rev. 121*:155 (1991).
54. M. Bigby, J.S. Markowitz, P.A. Bleicher, M.J. Grusby, S. Simha, M. Siebrecht, M. Wagner, C. Nagler-Anderson, and L.H. Glimcher, Most γδ T cells develop normally in the absence of MHC class II molecules, *J. Immunol. 151*:4465 (1993).
55. M. Zijlstra, M. Bix, N.E. Simister, J.M. Loring, D.H. Raulet, and R. Jaenisch, β2-microglobulin deficient mice lack CD4-8+ cytolytic T cells, *Nature 344*:742 (1990).
56. P. Fisch, K. Oettel, N. Fudim, J.E. Surfus, M. Malkovsky, and P.M. Sondel, MHC-unrestricted cytotoxic and proliferative responses of two distinct human γ/δ T cell subsets to Daudi cells, *J. Immunol. 148*:2315 (1992).

Index

About the Editors

WILLEM VAN EDEN is a Full Professor of Immunology and Head of the Immunology Section of the Faculty of Veterinary Medicine at Utrecht University, The Netherlands. The author, coauthor, editor, or coeditor of over 100 professional publications, Dr. van Eden is a member of the Dutch Society of Immunology, the Society for Infectious Diseases, the American Society for Microbiology, and the Immunocompromised Host Society, among others. He received the M.D. degree (1978) from the University of Groningen, The Netherlands, and the Ph.D. degree (1983) in immunogenetics from Leiden University, The Netherlands.

DOUGLAS B. YOUNG is the Fleming Professor of Medical Microbiology at St. Mary's Hospital Medical School, Imperial College of Science, Technology and Medicine, London, England. The author or coauthor of over 100 journal articles and book chapters, Professor Young is currently chairman of the World Health Organization's Steering Committee responsible for the coordination of molecular and immunological research on tuberculosis and leprosy. He received the B.Sc. degree (1975) in biochemistry from the University of Edinburgh, Scotland, and the D.Phil. degree (1978) in biochemistry from the University of Oxford, England.